FORAGES AND FODDER
INDIAN PERSPECTIVE

FORAGES AND FODDER
INDIAN PERSPECTIVE

Anil Kumar Singh
M.A. Khan
Natraja Subash
Krishna Murari Singh

2024
Daya Publishing House®
A Division of
Astral International Pvt. Ltd.
New Delhi – 110 002

First Published, 2011

Reprinted, 2024

ISBN: 978-81-70356-67-7 (HB)

Published by : **Daya Publishing House®**
A Division of
Astral International Pvt. Ltd.
– ISO 9001:2015 Certified Company –
4736/23, Ansari Road, Darya Ganj,
New Delhi-110 002
Ph. 011-43549197, 23278134
E-mail: info@astralint.com
Website: www.astralint.com

Printed at : **Replika Press Pvt. Ltd.**

Foreword

Although forage and *fodder* have similar connotation for all practical purposes, they differ in finer details. *Forage crops* can be defined as those plants which are directly or indirectly consumed by animals while *fodder* consist of species of plants which are generally used for feeding livestock either raw as a green and cut fodder directly or after minor processing like hay and silage. As per data available, the area under forage production is about 5.0 per cent of the total cultivated area of the country while the experts suggest that it should be a minimum 10 percent. The principal crops which are commonly grown in India include forage sorghum, pearl millet, lablab, cowpeas, soybeans Egyptian clover, grain sorghum and maize. India is an agro-based country and livestock and dairy are the life-line of our rural population largely comprising small, marginal and landless farmers. Their livelihood depends a great deal on livestock production. Considering the important role played by forage and fodder in India for quality livestock production, these crops have surprisingly been neglected. This is a cause for serious concern and results in low productivity and production of milch cattle and other livestock. Despite having one of the largest population of livestock (cattle, poultry, pigs, sheep, goats etc.), India still lags behind many countries in terms of availability of livestock products. Milk availability per person is only around 100g/day. Similarly, in the case of egg, the availability is only 12 eggs/ capita /year, much below the minimum per capita requirements prescribed by Indian Council of Medical Research (ICMR) for milk and eggs at 210 gm and 1 egg respectively. This shortfall may be largely attributed to the low availability of quality forage and fodder.

Realizing the need and importance of forage and fodder in livestock production, improvement in agriculture in general and livelihood of rural poor and landless population in particular, the present book, "*Forages and Fodder: Indian Perspective*" covers aspects related to production, protection, PGR management, consumption, preservation and socio-economic aspects of these crops. The book provides the latest and updated information on forage and fodder crops in the current situation. I am glad that the chapters by various contributors are quite informative and would be beneficial to the persons directly or indirectly interested in learning more on the subject. The present volume would also act as a good reference material for researchers and all those who are interested in the development of forage and fodder crops or are concerned with animal husbandry and dairy production sectors. The editors have done a distinctive service to the agrarian economy of the country, especially to the rural poor having dairy and animal husbandry as a major source of livelihood. I firmly believe that this book would be useful for all concerned directly or indirectly with this vital sector and I applaud the efforts of the the contributors and editors for bringing out this volume.

Ramadhar

IAS (Retired)

Former Chairman, State Farmers Commission, Bihar

Former Member, Commission for Agricultural Cost and Prices, Govt. of India

Former Service Chief, FAO of the UN

Former Chairman, Bihar State Agricultural Marketing Board

Preface

This book is intended as a professional textbook for students of agriculture/ veterinary degree programme. It will also serve as an updated valuable reference to agricultural scientists/researches working in different aspects of forages and fodder. In addition, the book will also be valuable reference for the candidates appearing in competitive examinations for the Agricultural Research Services/Assistant Professors in various State Agricultural Universities (SAUs), Agricultural Institutions, Colleges, Krishi Vigayan Kendras, Bank, State Department and other Line Department, Development Originations and Institutes. *Forages and Fodder: Indian Perspective*, covers all the aspects of forages and fodder production viz., production, protection, PGR Management, consumption, preservation and socioeconomic aspects. This is one of the latest and updated manuscripts related to forage and fodder in Indian context. The book covers latest information on forages and fodder genetic resources in India, management and varietal aspects of forages and fodder grown under different climatic zones of India, organic forage production, nutrient management, pest and disease management, weed control in forage crops, economic aspects related with forage production as well as socio-economic aspect of fodder production were also dealt in this presentation. Indian economy is still an agricultural based, contributes handsomely to the national GDP to the tune of 25 percent and its share to the export is around 15 percent. About 67 per cent of their people lives in villages hence their livelihood is directly depend on agriculture and animal husbandry. Though India has a huge livestock population of over 343 millions, besides poultry, yet the production of milk and other livestock products is about the lowest in the world. Due to efforts various levels enables the country to increase the production of milk six times and eggs 27 times since 1950-51, thus making a visible impact on the national food and nutritional security. India is house of about one-fourth of the total cattle population of the world but we are highly deficient in various livestock products. The main reasons for the low productivity of our livestock are malnutrition, under-nutrition or both, besides the low genetic potential of the animals. Availability and requirement of green-fodder crops, crop residues and concentrates, that there is a huge gap between demand and supply of all kinds of feeds and fodders. Cultivated area devoted to fodder production is estimated to only 4.4 per cent of the total cropped area. Likewise, permanent pastures and cultivable wastelands area is approximately 13 and 15 million hectares respectively.

On-farm water-use efficient techniques if coupled with improved irrigation management options, better crop selection and appropriate cultural practices, genetic make-up, and timely socio-economic interventions would help achieving good production in water scarce condition. According to an estimate the shortages in dry fodder, green fodder and concentrates are 40.4 per cent, 24.7 per cent and 47.1 per cent against the requirements of 650.7, 761.5and 79.4 million tons for the current livestock population, respectively. In certain areas in India, women play key roles as seed selectors and in seedling production. Their knowledge on seeds and seed storage contribute to viability of agricultural diversity and production. Women prepare and apply green and farmyard manure. In case of livestock more than 90% of the work related to animal care is done by women. In Livestock management their role vary widely ranging from care of animals, grazing, fodder collection, cleaning of animal sheds, processing milk and livestock products. Weed growth should not be overlooked as it hampers yield, lower forage quality, increase the incidence of disease and insect problems, cause premature stand loss, and create harvesting problems. Some weeds are unpalatable and may be poisonous for livestock. Insect pest problem is poorly documented. Insect pest cause yield loss, both quantitative and qualitative, its effect depend on many factors like the level of infestation, crop variety, agronomic practices and other related factors. Insect pests induced losses are not only in term of green or dry fodder yield but also quality factors which affect the re-growth period and canopy structure. Updated information on insect pest problem is well documented. Insect pest cause yield loss, both quantitative and qualitative, its effect depend on many factors like the level of infestation, crop variety, agronomic practices and other related factors. While every effort has been made to acknowledge the sources of information, any omission therein is inadvertent.

Contents

Introduction

M.A. Khan

Director,

ICAR Research Complex for Eastern Region, Patna – 800 014, Bihar

Term forage and fodder carries similar meaning in all practical purposes, there is no clear cut demarcation because they are primarily use for livestock feeding. Forage includes plant species directly or indirectly consumed by animals. They belong to cultivated ecosystem of rainfed and irrigated and range grasses and legumes growing in stress ecosystem and specific habitats like wetlands, bunds and terrace risers as well as wild and weedy form. Foraging is done on the wild barren wastelands, rangelands and pastures. Cultivated forage crops occupied about 5.0 per cent of the total cultivated area of India. The animal fodder consists of several species of plants which are used to feed the cattle either raw (green and cut fodder) or after minor processing (such as silaging). In other words Forage are bulky food like grass or hay for browsing or grazing of horses or cattle and fodder are coarse food (especially for livestock) composed of entire plants or the leaves and stalks of a cereal crop. Forage and fodder crops include forage sorghum, pennisetum, millet, lablab, cowpeas, soybeans, grain sorghum and maize. Forage and fodder crops have always been neglected to such an extent that some scientists called them as 'abandoned crops' or 'orphan crops'.

Indian economy is still an agricultural based, contributes handsomely to the national GDP to the tune of 25 per cent and its share to the export is around 15 percent. About 67 per cent of their people lives in villages hence their livelihood is directly depend on agriculture and animal husbandry. Though India has a huge livestock population of over 343 millions, besides poultry, yet the production of milk and other livestock products is about the lowest in the world. Due to efforts at various levels enables the country to increase the production of milk six times and eggs 27 times since 1950-51, thus making a visible impact on the national food and nutritional security. Despite of commendable progress and achievements there is still a huge gap between the availability per head and the minimum nutritional requirement set by the nutritionists in respect to livestock product in India. The Per-head availability of milk (100g/head/day), meat (1 million tonnes annually) and eggs (12 eggs/head/ year) where as the minimum requirements of milk 201g/head/day), meat (7,122 million tonnes annually) and eggs (1 egg/head/day). India is house of about one-fourth of the total cattle population of the world but we are highly deficient in various livestock products. The main reasons for the low productivity of our livestock are malnutrition, under-nutrition or both, besides the low genetic potential of the animals. Availability and requirement of green-fodder crops, crop residues and concentrates, that there is a huge gap between demand and supply of all kinds of feeds and fodders. Cultivated area devoted to fodder production is estimated to only 4.4 per cent of the total area. Likewise, permanent pastures and cultivable wastelands area is approximately 13 and 15 million hectares respectively. The area under open grazing during monsoon season is about 21 million hectares; major share comes under forest department land. The animals have to be maintained on the crop residues or straws of *jowar*, *bajra*, *ragi*, wheat, barley, etc. either in the form of whole straw or a *bhusa*, supplemented with some green fodder, or as sole feed in remaining part of year. The crop residues are available mainly from wheat, paddy, *bajra*, *jowar*, *ragi*, sugarcane trash, etc. The present graminaceous (Poaceae) plant wealth although accounts for about 15,000 species distributed in different phyto-geographic regions of the world, only 40-44 species account for 99 per cent of the earth sown grass pastures in the tropics and subtropics. The priority forage crops for India could be accounted cereals sorghum, maize, oats guinea grass and other grasses.

The Indian sub-continent poses, Gramineae (Poaceae) is about 245 genera and 1256 species. One-third of the grasses of India are considered of some value as forage plants. Wild relatives of crop plants represent a part of crops gene pool particularly resistant to biotic and abiotic stresses and have been the donors of many other useful traits, new techniques of molecular biology can be used to transfer these genes to develop resistant crops. Poor health of wild animals and low productivity of livestock due to lack of proper nourishment of livestock is a nagging problem for biodiversity conservators and livestock managers respectively. One of the important reasons behind poor nourishment of livestock is poor quality production of fodder. Turner, Ignace's (2000) review showed that traditional ecological knowledge of indigenous people has fundamental importance in the management of local resources, in the husbandry of the world's biodiversity, and in providing locally valid models for sustainable life. The entire worker should know what kind of genetic stock (germplasm) already available in India in respect to forage/fodder crops, tagged with some prominent features expressed in particular agro-climatic conditions. The flow of plant genetic resources from everywhere to India is slowing down gradually in the era of strong Intellectual Property Right (IPR).National Bureau of Plant Genetic Resources (NBPGR), New Delhi is the nodal institution in India, which has the mandate of germplasm exchange for research including introduction. NBPGR has linkages with over 115 countries. However, the scenario has changed from free flow to limited access of genetic resources in view of national and international agreements; Intellectual Property Rights (IPR) and expanded scope of these agreements to plant varieties.

With increase in urbanization, the demand of coarse grain is replaced by meat, dairy product and processed food. Globalization puts pressure on the export market to increase the quality of fodder products. This enhanced demand of animal products should be met by improving the productivity of animals rather than rearing more animals. India has a large number of cattle and milch animals, but the quality of products is still unsatisfactory, malnutrition being the primary cause. Areas for fodder production are able to meet the forage requirements of the grazing animals only during the monsoon season. But for the remaining periods of the year, the animals have to be maintained on the crop residues or straws of jowar, bajra, ragi, wheat, barley, etc. either in the form of whole straw or a bhusa, supplemented with some green fodder, or as sole feed. On-farm water-use efficient techniques if coupled with improved irrigation management options, better crop selection and appropriate cultural practices, genetic make-up, and timely socio-economic interventions would help achieving good production in water scarce condition. According to an estimate the shortages in dry fodder, green fodder and concentrates are 40.4 per cent, 24.7 per cent and 47.1 per cent against the requirements of 650.7, 761.5 and 79.4 million tons for the current livestock population, respectively.

Pasture which is grassland, usually with improved species, carefully managed with inputs such as irrigation and manures for grazing livestock. Good pasture management strives to keep overall productivity of pastures without sacrificing quality. The management practices play an important role in determining the productivity of cultivated fodders/forages and grasslands, yet this has been the last priority of the farmers in the country. Presence of inferior and unproductive grass species, lack of balance fertilization, and absence of legume component, improper cutting and indiscriminate grazing are some of the factors responsible for poor productivity of the grasslands. If soil where you are producing fodder lacks important nutrients, then the fodder/forage will also lack in nutrients which are deficient in the soil. The productivity of crops depends on the type and concentration of salts in the soil and in the irrigation water at germination and other active phases.

The National Commission on Agriculture (1976) recommended that a minimum 10 per cent of the arable area in the country (about 16.5 million ha) should be under improved forage crops to meet the green forage needs of the livestock population. Principal forages are grasses in the family [oaceae and legumes in the family leguminaceae. In dairy sector, cost of feed alone constitutes about 60-65 percent, of the total cost of milk production. This cost must be brought down by 30-40 per cent, if quality roughages are available. Nutritionally legumes are 2-3 times richer in protein than cereal grains and many also contain oil. Leguminous mulches have always been used as a source of nutrient-rich organic matter and nitrogen for crops. The nutritive value of legumes is measured in terms of the potential intake of digestible dry herbage and, in general, legumes have both higher digestibility and higher intake than grasses and their nutritive value tends to remain higher as plants mature. Cow pea forms excellent forage. On dry weight basis cowpea grains contains 23.4 per cent protein, 1.8 per cent fat and 60.3 per cent carbohydrates. It is also a rich source of calcium and iron. The crop gives such a heavy vegetative growth and covers the ground so well that it checks the soil erosion in problem area and can later be ploughed as a green manure.

Weed growth should not be overlooked as it hampers yield, lower forage quality, increase the incidence of disease and insect problems, cause premature stand loss, and create harvesting problems. Some weeds are unpalatable and may be poisonous for livestock. Insect pest problem is poorly documented. Insect pest cause yield loss, both quantitative and qualitative, its effect depend on many factors like the level of infestation, crop variety, agronomic practices and other related factors. Insect pests induced losses are not only in term of green or dry fodder yield but also quality factors which affect the re-growth period and canopy structure. Among the various constraints, the diseases have always been the major limiting factor for fodder cultivation. In India, diseases alone can cause losses up to 72 per cent in Lucerne (Ahmad *et al.*, 1977); 74 per cent in cowpea (Chester, 1950); 50 per cent sorghum (Sunderam, 1970); 30 per cent in bajra (Ahmad, 1969; Sunderam, 1970); 75 per cent in clusterbean (Chester, 1950) and 55 per cent in oats (Ahmad, 1969). Various diseases of different non leguminous fodder crops like sorghum are smuts(*Sphacelotheca sorghi, Sphacelotheca cruenta, Sphacelotheca reiliana, Tolyposporium ehrenbergii*), rusts (*Puccinia pupurea*), downy mildew (*Peronosclerospora sorghi*), Leaf spot or leaf blight (*Exerohilum tercicum*), Anthracnose and red rot(*Colletotrichum graminicola*), Zonate leaf spot (*Gloecercospora sorghi*), Sooty stripe (*Ramulisporia sorghi*); pearl millet are Green ear or downy mildew (*Sclerospora graminicola*), Rust (*Puccinia penniseti*), Smut (*Tolyposporium penicillariae*), Ergot (*Claviceps microcephala*); maize are Downy mildew (*Peronosclerospora sorghi*), Leaf blight (*Helminthosporium maydis*), Stalk rot (*Erwinia chrysanthemi* var. *zeae*); oat are Loose smut (*Ustilago avenae*), Covered smut (*Ustilago kolleri*), Stem rust (*Puccinia graminis* f. sp. *avenae*), Crown rust (*Puccinia coronata* var. *avenae*), Leaf blotch (*Drechslera* (*Helminthosporium*) *avenae*). The diseases of leguminous fodder crops like berseem are Root rot complex (*Rhizoctonia solani, Fusarium semitectum, Tylenchrhynchus vulgaris*), Stem rot (*Sclerotinia trifoliorum*); lucerne are Downy mildew (*Peronospora trifolii*), Rust (*Uromyces striatus*), Common leaf spot (*Pseudopeziza medicagensis*), Bacterial wilt (*Aplanobacter insidiosum*); cluster bean are Bacterial blight (*Xanthomonas campestris* pv. *cyamopsidis*), Alternaria leaf spot (*Alternaria cyamopsidis*); and of cow pea are Dry root rot (*Macrophomina phaseolina*), Anthracnose (*Colletotrichum lindemuthianum*), Cowpea mosaic (Cowpea mosaic virus)etc. Pasture ecosystem in arid tracts provide ideal and safe abode to rodents, where these species can easily fulfill their food requirement. They cause damage to mature trees by debarking them. By consuming seeds and small sapling, they inflict severe damage to nurseries. Being fussorial habit, rodents cause severe damage to fibrous root system of the grasses by tunneling activities. While digging the rodents gnaw at the roots and kill them by exposing roots to dry air. As the most of the rodents are seedivorous, they consume seeds of the grasses trees and shrubs, hampering the further regeneration of the vegetation. They also cause severe damage to barseem, cowpea, Lucerne and great napier. Grasses, if fertilized properly and harvested in the vegetative stage of maturity, can have more than 20 per cent CP. An exception to this general relationship is corn silage, which is low in CP but is a high-quality forage because of its energy content. Agronomic conditions that the producer can control include pest control and fertilization. Forage fields infested with weeds or insects generally yield less per acre, but quality may or may not be affected. Maintaining proper soil fertility (based on soil test) increases yield per acre. Nitrogen (N) is the plant nutrient that has a large influence on forage quality. Legumes do not need to be fertilized with N; however, grasses should receive N fertilizer to improve both quality and yield. Nitrogen fertilization increases CP content of grasses markedly, but slightly reduces the concentration of available energy. Recommendations concerning fertilization regimens and pest control can be found in the Ohio Agronomy Guide. The production, productivity and availability of quality seed are the most important factors for forage crops. It is more so important because the crops have been bred substantially for vegetative purpose and the forage crops are not allowed mature and cut at the vegetative stage and as such the opportunity of producing seed is limited. The lower milk production per cattle in India is mainly attributed to production and feeding of poor quality forage and that too inadequate amount (46.6 per cent) of requirement. Majority of our animals depends largely on agricultural wastes *i.e..*, dry leaves, straw etc. Green fodder is the essential component of feeding high yielding milch animals to obtain desired level of milk production. To increase the milk production in the country, emphasis has to be given on bridging the gap between supply and demand of fodder. Therefore, efforts is being required in order to increase the area, production, productivity and creating awareness among the farmers regarding importance of forage to their milch as well as draft animal. Thus, to produce targeted quantity of green fodder, the best option is to maximize the fodder production per unit area per unit time.

In eastern region of India, only 4 per cent cropped area is under fodder crop while it is more than 10 per cent in Punjab and Haryana. Lack of precise information on area under fodder, improved varieties used, nutritive value of different cultivars are lacking in different agro-climatic zones which are hampering the effective feeding policies for livestock sector. It is recognized in developing countries that preparation of high quality silage can be a valuable component for the development of a high-performing and low-cost system of animal production, using a relatively low level of purchased concentrates. The quality or nutritional value of forage can be defined as its ability to support a certain level of animal performance. Dairy cows fed high-quality forage produce more milk with less supplemental concentrate than cows fed lower-quality forage. The nutrient or chemical composition of forage largely determines its quality. Chemical composition of forage depends on plant characteristics and harvesting and storage methods. Generally, CP content is positively correlated with quality. In other words, high-protein forages generally are high-quality forages. Alfalfa, if harvested in the late bud stage of maturity, can contain 20 to 25 per cent CP (DM basis).

When forage yield exceeds fodder requirement of herd, then it is effective way to conserve fodder as silage and hay. Silage is the material produced by controlled fermentation of forages or crop residues with high moisture content. Hay refers to forage that are harvested and dried and stored as 85-90 per cent dry matter. Hay is a leafy dry fodder, green in colour, and free from moulds. Hay is made to reduce the moisture level of the green crop to a level low enough so that it can be safely be stored in mass without undergoing fermentation or becoming mouldy.

India alone has more than 25 per cent of the world's bovine population which has resulted not only democratic ownership of cattle but also an inseparable cultural and symbiotic relationship between rural families and their farm animals. Although, livestock rearing is an important occupation of farmers since immemorial but the forage cultivation has remained almost neglected. Grazing in forest areas and pastures is the mainstay for livestock but fodder trees and shrubs also contribute significantly. Now-a-days, uncontrolled cutting of trees, indiscriminate use of grazing areas and absence of rehabilitation programmes has lead to denudation of hill slopes, which has resulted in critically low biomass availability and adverse effects on livestock rearing. Livestock production has been the primary source of energy for agriculture operation and major source of animal protein for masses. Animal husbandry practices is affected by agricultural method (like cropping system, water resources, diversification of crops, intensification of agriculture), increasing use of mechanical power, transformation from sustenance farming to market oriented farming, changing food habits etc. In India, there is no practice of fodder production in rural areas and animals generally consume naturally grown grasses and shrubs which are of low quality in terms of protein and available energy, they are thus heavily dependent on seasonal variations and this results in fluctuation in fodder supply round the year affecting supply of milk round the year.

In certain areas in India, women play key roles as seed selectors and in seedling production. Their knowledge on seeds and seed storage contribute to viability of agricultural diversity and production. Women prepare and apply green and farmyard manure. In case of livestock more than 90 per cent of the work related to animal care is done by women. In Livestock management their role vary widely ranging from care of animals, grazing, fodder collection, cleaning of animal sheds, processing milk and livestock products. Despite their considerable involvement and contribution, women's role in livestock production has been underestimated, undervalued and widely ignored. This is due to paternal bias by society and culture.

Chapter 1

Forages and Fodder Genetic Resources in India

Anil Kumar Singh

ICAR Research Complex for Eastern Region,
ICAR Parisar (P.O. - B.V.College), Patna – 800 014, Bihar
E-mail: aksingh_14k@yahoo.co.in

Forage includes plant species directly or indirectly consumed by animals, they belonging to cultivated ecosystem of rainfed and irrigated and range grasses and legumes growing in stress ecosystem and specific habitats like wetlands, bunds and terrace risers as well as wild and weedy form. They serve both the economic and environmental use and sustain agro-ecosystems. Cultivated forage crops occupied about 5.0 per cent of the total cultivated area of India. These together with fodder trees and grasses meet about 60 per cent of the green fodder requirement and, hence, there is need to bridge the gap between a demand of about 900 MT and supply of about 550 MT. One of the constraints to achieve this is the poor genetic resource of forage crops. The area under cultivated forage crops (about 18 in number) is about 9 m ha, and the area under grasslands is about 12 m ha covering about 28 species of range grasses and legumes which need to be improved upon. In addition about 49 m ha of wastelands and degraded soils could be used for increasing forage resources provided that proper tree, grass and legume species are identified and improved upon to thrive in such areas. Major thrust, therefore, needs to be given to re orient the 'research related to germplasm collection, evaluation and utilization. The priority forage crops for India could be accounted cereals sorghum, maize, oats guinea grass and other grasses as well.

Trait Specific Forage/Fodder Germplasm Available in India

Qualty seed is mother of successful crop production. Germplasm has potential to produce the same. In the era of modern crop improvement and agriculture classical as well as new and innovative tools of breeding has got immense power to realize and break the yield barrier. India has got diverse agro climatic condition and poor and marginal soil as well, for efficient forage crop production technologies. India having great potential to change the agricultural scenario by ever green revolution with effective inclusion of forage crops to boost dairy sector in particular and agricultural sector as whole. Promising forage/fodder germplasm for specific traits were selected and most promising accession was categorized and presented under the suitable sub head *viz.* germplasm for agronomic trait, germplasm resistance to biotic and abiotic stress and germplasm for quality traits etc. The most significant accessions of different crops were presented. All the exotic (introduced) germplasm were accessed as exotic collection no (EC No.) and all the indigenous collection were assigned as indigenous collection no. IC/NIC and the ICRISAT materials were assigned as ICC numbers.

Promising Germplasm for Agronomic Traits

Every crop plant has some unique agronomic traits, which definitely influence the economy produce of the crop. The trait may vary from crop to crop to grate extent and upto some extent by the agroclimatic situation and micro climate. In this presentation we have manually emphasis, plot status, duration of crop, tilling/ branching behaviors, flowering and maturity, crop wise details are as under.

Eleusine indica

Accession nos. EC 516241(Zimbabwe), EC 516242 (Belgium), EC 516243 (China) and EC 516244 EC 516245(South Africa) evaluated at GKVK Campus, Bangalore were found promising for profused tillering and early flowering good for fodder proposed and are shy seed produces.

Eleusine multiflora

EC 516251 received from Kenya and evaluated at GKVK Campus was found early flowering and profuse tillering good for fodder proposed.

Hordeum vulgare

EC 497614 received from Canada and evaluated at DWR Karnal is six rowed husk less variety with good grain type, medium plant height and medium maturity. IBON (LRA-M)-26, IBON (LRA-M)-27 evaluated at CCSHAU Hissar were hulled, medium dwarf and promising for yield and quality.

Sorghum

Sorghum accessions evaluated at Marathwada Agriculture University, Prabhani were IC 188369, 40768, 40769, 40855, 40864 were Bold seeded, Rabi type; IC 23448 Kharif type.

Zea mays

Accession nos. EC 303525, EC 310335, EC 303527, EC 306692, EC 310337 evaluated and found to be good for specific combinations (SCA).

Cicer arietinum

ICC 17123 (ICRISAT) evaluated at IARI, New Delhi high branching with spreading habit and high yielding.

Cyamopsis tetragonoloba

EC 470310 (Senegal) evaluated at CCSHAU Hisar, more leaf, EC470311 (South Africa), early maturity, EC470312 Late maturing, EC470313, 14 Vegetable type, EC470322 Unbranched, EC470323, 24 Branched, Fodder type, EC470326 late maturing.

Macrotyloma uniflorum

Promising genotypes identified for different characters in horse gram after evaluation at HPKV Palampur for different traits are Biological yield/plant: Guglada –3, VL-Gehet-1, Rajain –1, Rajain –4,Plant height: Rajain-4, VLG-1, Rajain-5, IC 120808, Leaf area: Rajain-4, Rajain-5, Rajain-3, Chalwada-4 Early flowering: Chalwada-4, Chalwada-3, Papand-2, Guglada-5 Early maturity: Chalwada-3, Chalwada-4, Guglada-5, Papand-2.

Promising Germplasm for Biotic Stresses

Realization of potential of economic yielding capacity is depending on numerous factors, being a polygenic character is also influence by biotic stresses are one of important one. Biotic stress has potential to vanish total crops as it was seen in past to name few like potato famine and Bengal rice blast etc. is only because of biotic stresses causing agent. In Modern agriculture, the genetic base is named and narrowing for their, which is potential threat under adverse condition for crops and favorable conditions for biotic agent. Insect, pest, disease causing elements and needs are commonly known as biotic stress causing agent. There is urgent need to broader the base of genetic makeup of the varieties of crops to counter affect on this trait. Hence in this citation we are mentioning the some peculiar accession which was tolerant/resistant to pest is given below, which can be utilized by crop improvement man in their programme against the particular pest.

Hordeum vulgare

Accession nos. EC 493169, EC 493921 (USA) evaluated at DWR, Karnal was resistant to yellow rust; IBLSGP-1 evaluated at CCSHAU, Hisar tolerant to smut under partially reclaimed soils.

Oryza sativa

Accession nos. EC 364-11730, EC366-12507 (Sri Lanka), EC 367-33964 (Bangladesh) evaluated at IGKV Raipur were resistant to rice brown plant hopper, IC 206220 resistant to leaf folder, IC 210057 resistant to leaf folder IC 210078, resistant to leaf folder, erect panicle were evaluated at NBPGR RS Thrissur; EC 453841, EC 453843-47, EC 453860 (Philippines) highly resistant to panicle blast; EC 453842, EC 453855, EC 453861 (Philippines) highly resistant to brown spot, EC 453848, EC 453852(Philippines) highly resistant to Sheath blast were evaluated at IGKV Raipur.

Sorghum

Accessions evaluated at NRC for Sorghum, Hyderabad were IS 1054, 1055, 1082, 1122, 1151, 2123, 2146, 2195, 2312, 3962, 4283, 4522, 4553, 4567, 4646, 4664, 4777, 5030, 5469, 5470, 5476, 5480, 5483, 5490, 5566, 5604, 5613, 5615, 5622, 5642, 5801, 5826, 8315, 18551; Varieties CSV 5, CSV 6, CSV 7R, CSV8R, CSV14R; Hybrids CSH 7R, CSH 8R; Parental lines 104A/B, M 148-1 38 Germplasm resistant to shoot fly (*Atherigona soccata Rond.*) among the germplasm testes were found against shoot fly.

IS 1044, 1054, 1096, 1151, 2122, 2123, 2146, 2195, 2205, 3962, 4213, 4337, 4405, 4660, 4881, 5480, 5613, 8315, 10364,, 10370, 10711, 12447, 18323, 18326, 18427, 18479, 18517, 18527, 18551, 188676; Varieties CSV 3, CSV 6, CSV 8R; Hybrids CSH 13R; Parental lines VZM 2A/B were found Germplasm resistant to Stem borers (*Chilo partellus* Swin., and *Sesamia inferens*) among the germplasm screened.

IS 61, 703, 1121C, 1151, 1309C, 2579C, 2664, 2761, 3073, 3272, 3472, 3696, 4076, 4114, 4307, 4308, 4411, 4757C, 4808, 4832, 4870, 4876, 4955, 5230, 5940, 5977, 6170, 6174, 6179, 6392, 6394, 6446, 6810, 7005, 8134C, 8231C, 8232C, 8237, 8262C, 8263C, 8284, 8313, 8571, 8711, 8713, 8721, 8724, 9327, 9333, 9530, 9807, 10712, 11117, 12213, 12573C, 12593, 12612C, 12664C, 12666C, 12683C, 14864, 14871, 14876, 14889, 15107, 18733, 18836, 119474, 19512, 20506, 21873; Varieties DJ 6514, DSV 3 (ICSV 745); Hybrids CSH 6 were found resistant to Midge (*Stenodiplosis sorghicola* Coq.), IS 2741, 17645, 17610, 21444 were found resistant to Head bug (*Calocoris angustatus* Leth.)

IS 18567, 19349, 18657 were resistant to Corn plant hopper (*Peregrinus maidis* Ashm.)

IS 44, 84, 1133C, 33843, 718, 1063, 1117, 1840, 2312, 4657, 5490, 14048 were found resistant to Sugarcane aphid (*Melanaphis sacchari* Zehnt.),

IS nos: 625, 3612, 3691, 6365, 8283, 8763, 9470, 9487, 10892, 12622, 14332, 14375, 14384, 18758, EC 35-1; Varieties (SPV nos): 386, 462, 475, 615, 881, 938, 946, 1010, 1041, 1 and 333; Hybrids (SPH nos): 821, 822, 842, 832; Hybrids (CSH nos): 9, 10, 13, 14, 16, 17, 18, 19; Parental lines SRT 18B, SRT 26 B, SR 155-12, SR 330-20, SR 367, TNS 30, RS 29, CS 3521, SB 1085, NSV 13, GMRP NOS.: 4, 5, 8, 9, 27, 28, 30, 35, 44, C43, C85 were resistant to Grain molds (A complex of several fungi)

IS nos: 84, 2195, 2328, 2217, 2415, 3443, 6265, 8283, 12622, 18348, 18551, 18758; Varieties (SPV nos): 462, 475, 881, 938, 946, (CSV 15), 1018, 1025, 1231; Hybrids (SPH nos): 660, 815,; Hybrids (CSH nos): 6, 11, 14; Parental lines: SRT 26B, SB 101B, RS 29, M 12B, TNS 30, MR 750, C 43 were found resistant to Downy Mildew *Perenosclerospora sorghi.*

IS nos: 3433, 3286, 6365, 4332, 8758, 22129; Varieties (SPV nos): 462, 475, 678, 881, 1010, 1231; Hybrids (SPH nos): 822 Hybrids (CSH nos): 5; Parental lines: SRT 18B, SRT 26B, MR 750, AKMS 14, NSV 13, RS 29, C 43 were found resistant to Sugary Disease caused by *Sphacelia sorghi.*

IS nos: 84, 2217, 3443, 3691, 8283, 14332; Varieties (SPV nos): 462, 475, (CSV 13), 692, 932, 946 (CSV 15); Hybrids (SPH nos): 815, 821, 843; Parental lines: CS 3541, SB 101, SB 401B, SRT 18B, MR 750, RS 29 were screened resistant to rust caused by *Puccinia purpurea.*

IS nos: 84, 1054, 1085, 3443, 4332, 18432, E-36-1, BP53; Varieties (SPV nos): 462, 738, 932, 946; Hybrids (SPH nos): 504 Hybrids (CSH nos): 13R, 6; Parental lines: RS 29, CS 3541, SB 101B, SB 401B, SB 1085 were screened resistant to charcol rot *Macrophomina phaseolina*.IS nos: 1085, 3443, 5501, 6365, 4332, 18758; Varieties (SPV nos): 678, 736, 881; Hybrids (CSH nos): 13k; Parental lines: RS 29, C 43, CS 3541, SB 1085, SB 101B, SB 401B were resistant to leaf spot *Gleoscercospora sorghi*

IS nos: 3443, 1085, 14332, 18758; Varieties (SPV nos): 678, 736, 881; Hybrids (CSH nos): 13; Parental lines: RS 29, C43, CS3541, SB 101B, SB 401B resistant to Anthracnose *Colletotrichum* spp. IC 23454 evaluated at Marathwada Agriculture University, Prabhani was Shoot fly tolerant

Cicer arietinum

EC 469512 - 13 (Australia) evaluated at AAU Jorhat and found to be tolerant to *Callosobruchus chinensis* (stored gain pests).

Glycine max

EC 450628, 30, 32, 34-42, 45-47 (USA) tollebt to CYMP.

Promising Germplasm for Abiotic Stress

Abiotic stresses are one of the major causes of hindrances for crop production in everywhere. These stresses are highly influenced by adaphic, topographic, climatic and geographical conditions. Drought, flood/submergences, heat, cold, salinity and alkalinity, acidic and physiological stress are common and frequent in nature. Information on this aspect is presented here.

Sorghum

Accessions evaluated at NRC for Sorghum, Hyderabad were IS lines: IS 301, Naga white, D 71463, D 71464; Vatieties: IS 2877, IS 1045, D 38061, D 38093, D 38060, ICSV 88050, 88065, SPV 354; Parental lines: VZM1-B, 2077B drought tolerance in rabi season.

IS nos:824, 1037, 3477, 6928, 8370, 10596, 10701, 12611, EC 36-1, DJ 1195; Varieties: ICSV88056, ICSV88057, ICSV 88059, ICSV 88063, IS 24025, SAR 35, DKV 3, DKV4, DKV17, DKV18; Parental lines: ICSB nos. 3, 6, 11, 37, 54, 88001, 2219B were found drought tolerance in rabi season during early stage

IS nos:1347, 13441, DJ 1195; Varieties: ICSV 213, ICSV 221, ICSV 210, ICSV272, ICSV273, ICSV295, ICSV378, ICSV572, D71463, 71464, DKV1, DKV 3, DKV 7; Parental lines: ICSB nos. 58, 196B, 2077B were found drought tolerance in rabi season during mid season stage

IS nos: DJ 1195, M 35-1, IS 22314, IS 22380, EC 185-2, IS 12611, is 6928; Varieties: D 38001, D 71238, D 71464, IS 13441, DKV 3, DKV 4, DKV 17, DKV 18; Parental lines:

ICSSB nos. 17, 296B were found drought tolerance in *rabi* season during terminal stage. VRBT-77 leaves were found drought tolerant.

Promising Accession on Quality Traits

Quality of produce is one of the important parameter which is going to attract more prices. The data recorded by researcher in this aspect is presented here. The most promising accessions in respect to quality traits were categorized into the different sets groups are given below.

Pennisatum typhoides

EC 469937 Mature in 74 days, 2-3 effective tillers, 124cm. in height, thin stem with good grain and fodder yield; EC 469941 matures in 70 days, 3-4 effective tillers, 114cm. in height, thin stem, compact ear heads with good grain and fodder yield - BEST LINE were received from Nigeria and evaluated at RAU Bikaner.

Zea mays

EC 474843 have modified opaque, bold, luster, yellow kernels; EC 474844 Plants dark green with good foliage, plant and ear height medium to tall, maturity late, ear size medium to long, yielding ability 45-50q/ha were received from Mexico and evaluated at GBPUA&T Pantnagar; IC 3400979, IC 355841, have maximum numbers of ears per plant; IC 337349 have maximum ear length, IC 355857 have maximum ear length, IC 355857 have maximum ear width, IC 355848 maximum 100 seed weight were evaluated at NBPGR Reg. Stat., Bhowali; In IC 265350 maximum seed yield was 31.2q/ha and it was evaluated at NBPGR Reg. Stat., Shillong. EC 474844 (Mexico) evaluated GBPUA&T Pantnagar Plants dark green with good foliage, plant and ear height medium to tall, ear size medium to long, yielding ability 45-50q/ha; EC 497834 (Mexico) evaluated at Sher-E-Kashmir Univ. of Agri. Sci. and Tech., J&K produced highest grain yield (62.14q/h).

Methi

IC 143793 Mt., semi erect, m. vigour, broad, sparcy and light green leaf. Two cuttings observed; IC 143794 Mt., semi erect, poor vigour, sparcy, green leaf. Medium bolter and three cuttings observed; IC 143798 Tall., erect, m. vigour, green leaf with purple shade. Medium bolter and three cuttings observed; IC 143799 Mt., erect, m. vigour, green tinny leaf. Two cuttings observed; IC 143807 Tall, semi erect, poor vigour and green broad leaf with purple shade. Late bolter and two cuttings observed; IC 143835; Mt, erect, medium vigour, dark green broad lea and tender shoot. Medium late bolter and three cuttings observed; IC 144302 Mt, erect, medium vigour, tinny dark green leaf with purple shade and tender shoot. Early bolter and two cuttings observed; IC 31644 Tall, erect, v. poor vigour, tinny dark green leaf and tender shoot. Late bolter and four cuttings observed

References

Arora, R K, K L Mehra and M W Hardas. 1975. The Indian gene centre : prospects for exploration and collection of herbage grasses. Forage Res. 1 : 11-12.

Bhagmal. 1985. Breeding approaches for varietal improvement in tropical grasses. Proc. Summer Institute on Recent Advances in Forage Breeding for Farming Systems, IGFRI, Jhansi.

Chakravarty, A K, Ram Rattan and K Murari. 1970. Selection go grasses and legumes for pastures of the arid and semi-arid zones. III. Valuation of morphological and physiological characters in different starins of Cenchrus ciliaris Linn. And selection of high yielding and nutritious strains for forage production. Indian J. Agric. 40 : 192-196.

Chattopadhyay, P. 2001. Evaluation and breeding of Coix sp. As forage crop in saline and non-saline lowlands of West Bengal, Summary Report, BCKV, Kalyani.

Chopra, D P, T A Thomas and K L Mehra. 1980. Catalogue on moth bean germplasm (Series 1). NBPGR. Publ Regional Station. Jodhpur. pp. 20 p.

Dujardin, M. and W. W. Hanna. 1986. An apomictic polyhaploid obtained from a pearl millet x *Pennisetum squamulatum* apomictic interspecific hybrid. *Theor. Appl. Genet*. 72:33-36.

Gupta, J N, S R Gupta and U P Singh. 1986. Collection, evaluation and maintenance of genetic resources of forage crops-range legumes. Annual Report, IGFRI, Jhansi, p. 16.

Gupta, J Nand S R Gupta. 1988. Genetic resources of forage plants in dry and humid tropics of India. In : P. Singh (Ed.) Pasture and Forage Crops Research - A State of Knowledge Report, Range Management Society of India, IGFRI, Jhansi, pp. 121-128.

Gupta, V P. 1975. Fodder improvement in Pennisetums. Forage Res. 1: 54-60.

Harlan, J. R. 1976. *Genetic resources in wild relatives of crops.* Crop Sci. 16:329-333.

IGFRI 1985. Proc. of Summer Institute on Recent Advances in Forage Breeding for Farming Systems. Expand, Jhansi.

Katewa, S S. 2001. Identification, evaluation and germplasm collection of grasses of Rajasthan, Report, Mohan Lal Sukharia University, Udaipur (Rajasthan).

Katiyar, P K, U P Singh, D K Agarwal, A S Negi and P Saxena. 1999. IGFRI. Germplasm Catalogue of Siratro (*Macroptilium atropurpureum*) : Evaluation and Analysis, IGFRI, Jhansi, 35 p

Khoshoo, T N. 1996. Concern mounts over rapid erosion of Himalayan biodiversity. Diversity 12(3) 24-25.

Ladizinsky, G. 1988. Biological species and wild genetic resource in oats. *In* B. Mattsson and R. Lyagen (eds.) Proc. 3rd Intl. Oat Conf., Svalof AB, Sweeden.

Mehra, K L and M L Magoon. 1973. Cytogenetics in the evolution and improvement of tropial forage grasses. Adv. Forntier of Cytogenetics. Hindustan Publ. Group, Delhi. Pp. 86-94.

Mehra, K L and M L Magoon. 1974. Gene centers of tropical and sup-tropical pasture legumes and their significance in plant introduction. Proc. XII. Internal Grassland Cong. Moscow. 25 : 1-256.

Mehra, K L and R S Paroda. 1986. Genetic resources activities in forage plants. In : R S Paroda, R K Arora and K P S Chandel (Eds.) Plant Genetic Resources : Indian Perspective, NBPGR, New Delhi, pp. 274-179.

Mehra, K L, Bhagmal and D S Katiyar. 1970. Metroglyph analysis of fodder attributes in oats. Indian J. Heree. 2 : 81.

Paroda, R S and Bhagmal. 1988. Forage crop improvement through genetic resources. In : P. Singh (Ed.) Pasture and Forage Crops Research – A State of Knowledge Report, Range Management Society of India, IGFRI, Jhansi, pp. 111-120.

Paroda, R S. 1991. Genetic improvement: Achievement in Crop sciences. In : R S Paroda and R K Arora (Eds.) Plant Genetic Resources : Conservation and Management, NBPGR, New Delhi, pp. 183-210.

Singh, C B, K L Mehra and K S Kohli. 1971. Evaluation of a world collection cowpea, Vigna sinensis (L.)Savi, Pod characters, SABRAO Newsletter. 3 (1): 11-16.

Singh, C B, K L Mehra, A Singh, K S Kohli and L L Magoon, 1981. Genetic variability and correction between quantitative characters in luceme. Jap. J. Genet. (Nov.)

Singh, I P and P L Gautam. 1999. Chhara Fasalo ke sudhar hetu prajanan sansadhan. In : P S Tomer, B K Trivedi, S N Tripathi and J N Gupta (Eds.) Chhara Anusandhan Avm Pasudhan vikas Ke Naye Aayam (In Hindi), IGFRI, Jhansi, pp. 15-26.

Singh, K A, B P S Yadav ans S N Gowwami. 1996. Forage Resource Management for Sustaind Livestock Productivity in Himalayan Agro-ecosystems. Res. Bull. No. 39. ICAR Research Complex for NEH Region, Barapani, 94 p.

Singh, U P, A Singh and J N Gupta. 1983. Range in fodder attributes in guar germplasm at IGFRI, Jhansi, Guar Newsletter 3: 21-23.

Yadav, M S and P R Sreenath. 1975. Factor analysis of fodder yield components in Pennisetum pedicellatum Trin. Andhra Agric. J. 2 (1 and 2) 5-9.

Yadav, M S, K L Mehra and M L Magoon. 1974. Variability and heritability of fodder yield components in berseem. Trifolium alexandrianum L. Indian J. Hered. 6 : 57-61.

Chapter 2

Indian Needs of Forage Genetic Resource

Anil Kumar Singh[1] and Aditya Kumar Singh[2]

[1]Senior Scientist, Agronomy, Division of Land and Water Management, ICAR Research Complex for Eastern Region, ICAR Parisar (P.O.- B.V.College), Patna – 800 014, Bihar

E-mail: aksingh_14k@yahoo.co.in

[2]College of Post Graduate Studies, Central Agricultural University, Barapani, Umiyam, Shillong

Acquisition of more exotic germplasm as early as possible has become the priority due to latest development under convention on Biological Diversity and other International treaties and laws, which makes flow of genetic resources bit more difficult than it was before enforcement of new laws. The germplasm/ genetic material contain promising traits related to yield and yield attributing characters, quality characters and also resistance to various biotic and abiotic stresses. Germplasm of diverse crops were required for our National needs which will further distributed to various potential users (breeders/crop improvement workers) in the country for its evaluation and further utilization in their ongoing/ ensuing crop improvement works for our food and nutritional security.

Introduction

The main objective of this chapter is to generate awareness among the crop improvement worker by its vide circulation. Let all the worker should know what kind of genetic stock (germplasm) already available in India in respect to forage/

fodder crops, tagged with some prominent features expressed in particular agro climatic conditions. Now stage is set to reorient pre breeding/breeding/advanced breeding works by utilizing it in right direction. It will going to be transformed Indian agricultural scenario with in short span of time, which can boost our dairy based economy and will have synergistic effect on the mega national goal of second green revolution. The flow of plant genetic resources from everywhere to India is slowing down gradually in the era of strong intellectual property right (IPR). Secondly we are not getting the trait specific materials as it was previously seen. It may be due to several reasons but reservation by most of donor country is one of them. Under this present circumstance we have to judiciously evaluate our existing germplasm/genetic resources to plan and conduct crop improvement programme for our current requirement and future use as well. Achievement of Indian agriculture science 1960 and 70s with respect to the staple crop production is one of the most remarkable success in the history of technological development at global level. Now we are facing many faced challenged posed to our agricultural production system. In order to feed the burgeoning population there is urgent need to double the cereal crop production by the year 2050. There is utmost need to breed potentially high yielding forage varieties to match up the requirement along with corrective measure to bridge the huge gap between attainable yield (yield obtained in demonstration trails) and the average yield harvested by farmers. India has diverse agro-climatic regions comprising tropical, sub-tropical and temperate regions, where we can grow various kinds of crops including fruit plants.

We have achieved a lot particularly in the improvement of field crops and almost reached at peak. However, breeders are trying their level best to increase the income of the farmers by applying various latest techniques for improving the productivity. Hence, there is a need to diversify the base of gene pool of Indian agriculture by various means and sources, introduction of germplasm from abroad is one of the major source for broadening genetic base. To sustain our National current requirement and future needs we have to widen our genetic resources for stronger India, as we have to march ahead words feed secure Nation to proud prevailed nutritional secure India, for that there is need to plan and execute systematic introduction and exchange programme for required plant genetic resources with the desired country through bilateral or multilateral programme.Following trait specific forage/fodder genetic material is needed to sustain our long term needs.

Maize

Characteristics	*Countries of Import*
Fodder types, dual purpose (fodder and grain types) and early maturing types	Indonesia, Vietnam, Egypt, Korea, Mexico, USA
Resistant to stalk rot, pink stem borer, shoot flies, Cereal thrip, army worm and multiple disease resistant type	USA
Salinity, water lodging, drought tolerant	USA,UK
Frost/cold tolerant lines	Korea, USA
Leaf blight resistant lines	Korea

Barley

Characteristics	*Countries of Import*
Six rowed, two rowed, high yielding forage type, hulless varieties, dual purpose type (fodder and grain), and land races, early maturing, male sterile lines	China, Japan, USA (Minnesota), Canada
Lodging resistant, drought, heat and cold tolerant lines	USA
Resistant to powdery mildew, downey mildew, stripe rust, yellow, brown rust, bacterial leaf spot and aphid resistant lines	Mongolia, Iraq, Syria, Sweden, USA

Sorghum

Characteristics	*Countries of Import*
Dual type (fodder and grain types), CMS lines, sweet sorghum having high sugar content, superior grain quality	China, Australia, Sudan, USA, Canada
Different land races	Namibia
Cold tolerant for rabi sowing, acidic soil tolerant	China, Sudan, USA, Canada
Resistant to anthracnose, rust, grey mould, shoot fly, downy mildew	Sudan, USA

Oat

Characteristics	*Countries of Import*
Fodder types and dDual type (fodder and grain types),	China, Iran, USA,

Pearl Millet

Characteristics	*Countries of Import*
Dual type (fodder and grain types), CMS lines, Early maturing (below 35 days), large grain type, dual type, yellow grain types rich in Vitamin A.	Togo, Ghana, Mali, Niger, Nigeria, Chad, South Africa, Senegal, USA
Long bristles and long ear types	West Africa

Foxtail Millet

Characteristics	*Countries of Import*
Dual type (fodder and grain types), CMS(Cytoplasmic/genetic male sterility) lines	China

Proso Millet

Characteristics	*Countries of Import*
Dual type (fodder and grain types), long compact panicle types. Lines having different coloured *viz.* orange, red, yellow/cream and grey types elite lines	Russia

Finger Millet

Characteristics	*Countries of Import*
Dual type (fodder and grain types)	Ethiopia, Uganda, Zambia, Zimbabwe

Rice

Characteristics	*Countries of Import*
Dual purpose (fodder and grain types), New plant types (Tropical *japonica* crossed with *indica*), Profuse tillering type	Vietnam, Thailand
Lowland land races	SriLanka, Nepal, Myannmar, Laos, Thhailand, Malaysia, Philippines
Land races and elite lines	Korea, Japan
Early maturing types	China, Korea, Japan
Deep water types/submergence tolerant	Bangladesh, Thailand, Laos, Vietnam
Salinity/alkalinity and acid tolerant lines	Laos, Thailand, Vietnam
Cold tolerant lines	Korea
Resistant to bacterial blight, blast, leaf smut, narrow brown leaf spot, sheath rot resistant	USA
Resistant to leaf and neck blight, sheath rot	Korea

Wheat

Characteristics	*Countries of Import*
Short duration, dwarf types, dual purpose (fodder and grain types)	Russia, Canada
Drought tolerant, heat tolerant, lodging resistant, salinity/alkalinity and acid tolerant lines	Australia, UK, USA, Canada
Resistant to powdery mildew, stripe rust, yellow, black, brown rust, spot blotch, karnal bunt, bacterial rot, *Helminthosporium* blight, leaf bight, collar blight	UK, USA, Canada
Cold tolerant varieties	Russia, France, UK, Canada

Cowpea

Characteristics	*Countries of Import*
Photo and thermo insensitive, bushy type, dual purpose (fodder and grain), resistant to blight, yellow vine mosaic virus, cucumber mosaic virus, dry root rot, cowpea mottle virus, cowpea aphid borne mosaic virus, cowpea black eye mosaic virus, cowpea black eye mosaic.	Nepal, Sri-Lanka, China, Myanmar, Korea, Thailand, Indonesia

Mothbean

Characteristics	*Countries of Import*
High yielding, early maturing dual type (grain and fodder), drought tolerant	Bangladesh, Thailand, Australia

Soybean

Characteristics	*Countries of Import*
Grain and fodder type CMS,GMS lines for hybrid development	Taiwan, China, Japan, USA
Tolerant to acidity and alkaline soils, drought tolerant	China, Japan, USA
Resistant to rust, stem borer, soybean mosaic virus, soybean cyst nematode, root knot nematode, downy mildew, multiple foliar disease, phytopthora rot, white mould, stem canker, frog eye leaf spot resistant lines	Taiwan, Nigeria, USA

French Bean

Characteristics	*Countries of Import*
High yielding, fodder type dual purpose (grain and fodder)	Columbia, USA, Mexico, Costa Rica, El Salvador, Brazil, Argentina
Heat tolerant	Columbia, USA
Resistant to bean golden mosaic virus, bean common mosaic virus, anthracnose, rust, angular leaf spot	Columbia, USA
High leaf crude protein content type	USA

Pea

Characteristics	*Countries of Import*
Fodder type, high dual purpose type (grain and fodder),	Egypt, Russia, Syria, Turkey, Sweden
Cold, drought, heat and moisture stress tolerant types	Egypt, Russia, Syria, Turkey, Sweden

Lentil

Characteristics	*Countries of Import*
Fodder type dual purpose (grain and fodder) drought tolerant, cold tolerant, acidity and alkalinity tolerant lines	Bangladesh, Ethiopia, Mexico
Resistant to root rot, wilt	Israel

Lathyrus

Characteristics	Countries of Import
Drought tolerant dual purpose (grain and fodder) and fodder type	China, Iran, Iraq, Uzbekistan

Groundnut

Characteristics	Countries of Import
Drought resistant high foliage type	Australia
Tolerant to heat, salt and acidic soils	Egypt,Sudan,
Cold tolerant	Russia, Ukraine, Canada

***Brassica* sp. (Mustard, Rapeseed Mustard)**

Characteristics	Countries of Import
Multicut, dual purpose (grain and fodder) drought, frost, cold and salt tolerant germplasm	Sweden, Canada, USA
Resistant to alternaria blight,aphid, white rust, downey mildew	Spain, Sweden, USA, Canada
Wilt resistant	China

Conserving the Gene(s) for Future Uses

Genetic resource/germplasm is easiest, cheapest and natural source for creating variation, transferring desired gene from donor to receiver, is an essential pre-requisite for the future forage improvement programmes. Considerably less is known about the preservation of forage germplasm than the field crops and much more research in needed to raise the level of knowledge about forages. Forages are multiplies by sexual and a sexual means due to its shy seed producing nature. Vegetative propagated forages germplasm can be maintained through vegetative clones and the technique of meristem tissue culture can be applied to maintain homogenous populations free from diseases and insect-pests. In case of seed producing forages poor quality of seed is likely to lose viability more quickly than the good quality seed even in good storage conditions hence the quality of seed being preserved in a germplasm bank is of great importance. By nature in many grasses are predominantly immature, even when the stalk is ripe and ready for harvest so every precaution should be taken while harvesting such crops, it is practically true for tropical forages. It is always better to delay the harvesting in seed multiplication plots for germplasm materials, since there is evidence that mature seeds remain viable longer than immature seeds. The major responsibility of maintaining the active collections of different forage species lies with the Indian Grassland and Fodder Research Institute, Jhansi which is National Active Germplasm Site (NAGS), Coordinated Centres of the AICRP on Forage Crops. Germplasm for long-and short-term under controlled conditions of temperature and humidity, are to be maintained at NBPGR New Delhi.

References

Bhagmal, 1985. Breeding approaches for varietal improvement in tropical grasses. Proc. Summer Institute on Recent Advances in Forage Breeding for Farming Systems, IGFRI, Jhansi.

Chakravarty, A K, Ram Rattan and K Murari. 1970. Selection go grasses and legumes for pastures of the arid and semi-arid zones. III. Valuation of morphological and physiological characters in different strains of *Cenchrus ciliaris* Linn. And selection of high yielding and nutritious strains for forage production. Indian J. Agric. 40 : 192-196.

Chattopadhyay, P. 2001. Evaluation and breeding of *Coix* sp. As forage crop in saline and non-saline lowlands of West Bengal, Summary Report, BCKV, Kalyani.

Dujardin, M. and W. W. Hanna. 1986. An apomictic polyhaploid obtained from a pearl millet x *Pennisetum squamulatum* apomictic interspecific hybrid. *Theor. Appl. Genet*. 72:33-36.

Gupta, J Nand S R Gupta. 1988. Genetic resources of forage plants in dry and humid tropics of India. In : P. Singh (Ed.) Pasture and Forage Crop Research - A State of Knowledge Report, Range Management Society of India, IGFRI, Jhansi, pp. 121-128.

Harlan, J. R. 1976. *Genetic resources in wild relatives of crops*. Crop Sci. 16:329-333.

IGFRI 1985. Proc. of Summer Institute on Recent Advances in Forage Breeding for Farming Systems. Expand, Jhansi.

Katiyar, P K, U P Singh, D K Agarwal, A S Negi and P Saxena. 1999. IGFRI. Germplasm Catalogue of Siratro (*Macroptilium atropurpureum*) : Evaluation and Analysis, IGFRI, Jhansi, 35 p

Khoshoo, T N. 1996. Concern mounts over rapid erosion of Himalayan biodiversity. Diversity 12(3) 24-25.

Ladizinsky, G. 1988. Biological species and wild genetic resource in oats. *In* B. Mattsson and R. Lyagen (eds.) Proc. 3rd Intl. Oat Conf., Svalof AB, Sweeden..

Mehra, K L and M L Magoon. 1974. Gene centers of tropical and sup-tropical pasture legumes and their significance in plant introduction. Proc. XII. Internal grassland Cong. Moscow. 25 : 1-256.

Mehra, K L and R S Paroda. 1986. Genetic resources activities in forage plants. In : R S Paroda, R K Arora and K P S Chandel (Eds.) Plant Genetic Resources : Indian Perspective, NBPGR, New Delhi, pp. 274-179.

Mehra, K L, Bhagmal and D S Katiyar. 1970. Metroglyph analysis of fodder attributes in oats. Indian J. Heree. 2 : 81.

Paroda, R S and Bhagmal. 1988. Forage crop improvement through genetic resources. In : P. Singh (Ed.) Pasture and Forage Crops Research – A State of Knowledge Report, Range Management Society of India, IGFRI, Jhansi, pp. 111-120.

Paroda, R S. 1991. Genetic improvement: Achievement in Crop Sciences. In : R S Paroda and R K Arora (Eds.) Plant Genetic Resources : Conservation and Management, NBPGR, New Delhi, pp. 183-210.

Singh, C B, K L Mehra, A Singh, K S Kohli and L L Magoon, 1981. Genetic variability and correction between quantitative characters in luceme. Jap. J. Genet. (Nov.)

Singh, I P and P L Gautam. 1999. Chhara Fasalo ke sudhar hetu prajanan sansadhan. In : P S Tomer, B K Trivedi, S N Tripathi and J N Gupta (Eds.) Chhara Anusandhan Avam Pasudhan Vikas Ke Naye Aayam (In Hindi), IGFRI, Jhansi, pp. 15-26.

Singh, K A, B P S Yadav ans S N Gowwami. 1996. Forage Resource Management for Sustaind Livestock Productivity in Himalayan Agro-ecosystems. Res. Bull. No. 39. ICAR Research Complex for NEH Region, Barapani, 94 p.

Singh, U P, A Singh and J N Gupta. 1983. Range in fodder attributes in guar germplasm at IGFRI, Jhansi, Guar Newsletter 3: 21-23.

Yadav, M S, K L Mehra and M L Magoon. 1974. Variability and heritability of fodder yield components in berseem. *Trifolium alexandrianum* L. Indian J. Hered. 6 : 57-61.

Chapter 3
Biodiversity in Indian Forage Species

***Anil Kumar Singh*[1] *and Birendra Prasad*[2]**

[1]*Senior Scientist, Agronomy, Division of Land And Water Management, ICAR Research Complex for Eastern Region, ICAR Parisar (P.O.- B.V.College), Patna – 800 014, Bihar*
E-mail: aksingh_14k@yahoo.co.in

[2]*Assistant Professor, Department of Seed Science and Technology, GBPUA&T, Hill Campus, Ranichauri, Tehri Garhwal, Uttarakhand*
E-mail: bprasadsst@yahoo.co.in

The Indian agro-biodiversity is distributed in 15 agro-climatic regions or zones, these distinguish areas having different agro-ecosystem, posse's unique gene pools, comprising of landraces, primitive forms and wild relatives of crop plants, ornamental and medicinal plants, and forage species as well. India is treasurer of forage biodiversity, its proper management and utilization in sustainable manner, conservation with futuristic approach, equitable sharing are some basic rule to follow for its environmental friendly and sustainable evolution of plant genetic resources. Forages are generally treated as orphan crops but for sustainable development of Indian dairy and allied sectors. Every possible effort should be taken to explore, collect, evaluate, conserve, manage and restoration for sustainable use of indigenous forage genetic resources in India. Equal efforts should also be taken for enrichment of forage gene pool by introducing them. Priorities have been worked out for forage/fodder crops for introduction along with some potential wild and weedy relatives of these crops for their utilization in National crop improvement programme. Emphasis of this article is to generate awareness among the workers engaged in the breeders/crop improvement works. By utilizing information available here they can plan/reorient their respective crop improvement works to meet out the national security on feeds and fodder and vis-à-vis quality of fodder.

Introduction

The Indian sub-continent is one of the world's mega centres of crop origins and named as "Hindustani center" of diversity. The Eastern Himalaya is recognized as one of the 18 "hotspots" of biodiversity in the world. The present graminaceous (Poaceae) plant wealth although accounts for about 15,000 species distributed in different phyto-geographic regions of the world, only 40-44 species account for 99 per cent of the earth sown grass pastures in the tropics and subtropics. As regards legumes, about 30 species comprise most of the managed pastures and forage legumes. The priority forage crops for India could be accounted cereals sorghum, maize, oats guinea grass and other grasses. As per conservative estimate the north-eastern region of India holds about 50 per cent of India's total floral species diversity. The Indian sub-continent poses, Gramineae (Poaceae) is about 245 genera and 1256 species. One-third of the grasses of India are considered of some value as forage plants. Genetic resources are a crucial component of agro-biodiversity and total biological diversity as well playing pivotal role in the National food security. Genetic resources deals with that part of biodiversity which economic value has been known are having potential for it.

Forage includes plant species directly or indirectly consumed by animals, they belonging to cultivated ecosystem of rainfed and irrigated and range grasses and legumes growing in stress ecosystem and specific habitats like wetlands, bunds and terrace risers as well as wild and weedy form. Cultivated forage crops occupied about 5.0 per cent of the total cultivated area of India. The major cultivated forages crops is limited to only 20 in number, the area under cultivated forage crops is about 9.0 million hectare, however the area under grasslands is about 12.0 million hectare covering about 28 species of range grasses and legumes which need to be improved upon. Apart form this about 48 million hectare of wastelands and degraded soils could be used for increasing forage resources provided that proper tree, grass and legume species are identified and improved upon to thrive in such areas. These together with fodder trees and grasses meet about 60 per cent of the green fodder requirement and hence, there is need to bridge the gap between a demand of about 1000 million ton and supply of about 600 million ton to sustain 17 per cent of total world livestock population. There are several factors influencing the productivity and carrying capacity of forages and ranges land and makes it's very difficult to brig the gap between demand and supply of quality forages in India, one such major constraints is the poor genetic resource of forage crops, therefore, needs to be given to reorient the 'research related to germplasm collection, evaluation and utilization.

Perspectives and Strategies for Forage Genetic Resources

Biodiversity in forage species include species belonging to rainfed and irrigated ecosystem forage crops to range grasses, legumes, agro-forestry trees and rhizobium thriving in stress conditions such as cold and hot arid ecosystems, grazing and degraded Land and specific habitats like wetlands, field bunds, watershed and terrace risen. This vast diversity provides forages for animal production and serves to the socio-economic, soil conservation and environmental needs. Forage genetic resources management and conservation programme has not been paid due attention in the past. Research efforts have also been defused in the sense that numerous crops/species are involved to work with and by the limited scientific manpower. Collection and introduction efforts were confined largely to improve upon fodders/feeds meant for intensive dairy development programme undermining the pasture based animal raising and other alternative feed resources. Concurrently, improvement and cultivar developments, in other than forage crops made in past two decades, have grossly neglected the importance of straw/stover quality and quantity. Consequently, the country has gone down to the deficit of about 53 per cent of roughages and 68 per cent of green forage. The issue of joint products analysis *i.e.* trade off between straw/stove quality, quantity and grain yield of various crops has now become forthright. The huge deficit can be minimized largely through augmentation of forage crops germplasm more systematically based on crops and/or regional priorities and tapping the non-conventional or potential under-utilized/agro-forestry trees species adaptable in marginal, sub-marginal and poor agro-climatic conditions. Augmentation, evaluation and utilization of all these forage genetic resources in fact, fraught with complexities and also beset with many-prolinged activities, such as soil conservation/reclamation, over grazing of pasture lands and multi purpose utilization of fodder/straw/stover/grain consequent upon which the germplasm holdings held with different organization/institutions could not be hitherto properly accounted and documented. Therefore, the primary task before us is to inventories the forage germplasm available with different organizations/institutions cutting the across the discipline boundaries or mandated works. The wholesome task is certainly voluminous and should, therefore, be viewed wisely based on crops and/or regional priorities, severity of erosion and nutritional/commercial value of joint product analysis. We have too many species to study and too few specialists to study them. We must increase our efficiency by developing a body of theory and general principles to improve our chances of success. We must screen enough material of enough species for biological patterns to become evident. We must learn what to look for and where to look for it. A massive effort in this regard is urgently needed.

1. Western Himalayan Region

This agro-climatic region/zone consists of three sub- zones of Jammu and Kashmir, Himachal Pradesh and U.P. hills composed of skeletal soils of cold region, podslic soils, mountain meadow soils and hilly brown soils mostly silty loam in nature. Erosion slides and slips are quite common problems. The highest cropping intensity has been observed in H.P.. whereas lowest is in Jammu and Kashmir. Prominent forage and range genetic resources of the region are Red and White clover, *Medicago* spp, *Arundila napelensis*, Chrysopogon, *Lotus corniculatus, Dactylis glomerata, Festuca*, Rye grass and *Pennisetum, Festuca, Pennisetum, Eleusine*, maize, *Echinochloa*, soybean range grasses and legumes *and* Kikui grass.

2. Eastern Himalayan Region

This agro-climatic region/zone consists of Sikkim and Darjeeling Hills, Arunachal Pradesh, Meghalaya, Nagaland, Manipur, Tripura, Mizoram, Assam and Jalpaiguri and Coochbehar district of West Bengal characterised with high rainfall and high forest covers. Widespread shifting cultivation (Jhum) in around 30 per cent of the area is of greatest concern as it causes denudation and degradation of soils with heavy runoff, massive soil erosion associated with floods in the lower reaches and basins. This region is one of best known for genetic resources of range and wetland grasses including important leguminous forages. Presently available prominent forage and range genetic resources of the region are rice bean, maize range grasses, legumes, *Brachiaria* broom grass, coix, minor millets sub-tropical grasses and beans, guinea, Lablab bean, maize, Indigenous wet land grasses, *Ficus, b*rowsing species, preferred by Mithun.

3. Lower Gangetic Plains

This agro-climatic region/zone consist of four sub-regions *viz.* Basin plains, central alluvial plains, alluvial coastal plains and Rarh plains. This zone is famous for rice cultivation and accounts for 12 per cent of the country's total rice production. Floods and inundation of fields in basin and central plains often destroy standing crops. Mustard, winter maize and potato are the relatively newly introduced crops of this zone. Forages and range genetic resources of the region are rice bean, guinea grass, coix, range grasses and legumes.

4. Middle Gangetic Plains

This zone consists of 12 districts of eastern U.P. and 27 districts of Bihar plains. Depending upon heterogeneity in soil, land use, topography and climatic factors it has been sub-divided into several smaller sub zones. The area is characterized with high rainfall and frequent floods. About 39 per cent area is irrigated having a cropping intensity of 142 per cent. Bihar plains consist of 17 lakh hectares as flood prone and 10 lakh hectares as chaur, tal and diara. Main crop of the zone is rice although the average productivity is very low. Forage genetic resources available in this agro-climatic region/zone are maize, cowpea, ricebean, berseem, *Panicetum pedicellatum,* coix.

5. Upper Gangetic Plains

This zone consists of 32 districts of U.P. divided into three sub-zones *viz.* Central, south-western and northern-western U.P. The soil is mostly canal or tube-well irrigated and has about 144 per cent cropping intensity. Rice and wheat are major crops. About 9 lakh hectares area is saline - alkali or problem soils. Available forage genetic resources of in this agro-climatic region/zone are maize, sorghum, cowpea, berseem, Senji, *Dicanthium, Sehima,* and *Heteropogon.*

6. Trans-Genetic Plains

This zone consists of Punjab, Haryana, Union Territories of Delhi and Chandigarh and Sriganganagar district of Rajasthan which is divided into three sub-zones *viz.* foothills or Shiwalik and Himalayas; Semi-arid and arid plains bordering Thar desert. This zone is characterised with highest net sown area and highest irrigated area, least poverty level, high cropping intensity and high ground water utilization. The area has got highest productivity in the country. Available forage genetic resources under scare water supply conditions in this agro-climatic region/zone are as under. Prominent forage and range genetic resources of the region are guar, maize, bajra, berseem, Lucerne, guinea grass, sorghum, carrot, cowpea, methi moth, kulthi, range grasses and legumes.

7. Eastern Plateau and Hills

This region consists of following sub-regions : (*i*) Sub-region of Wainganga, M.P. eastern hills and Orissa inland; (*ii*) Orissa northern and M.P. eastern hills and plateau; (*iii*) Chhotanagpur north and eastern hills and plateau; (*iv*) Chhotanagpur south and West Bengal hills and plateau, and (*v*) Chhattisgarh and south-western Orissa hills. The soils are undulating with shallow to medium depths. Tank irrigation is very common in sub-zone (*ii*) and (*v*) while tube well irrigation is significant in sub-zone (*i*) and parts of sub-zone (*ii*) and (*v*) Kharif season consists of 82 per cent rice, 6 per cent oilseeds and 6 per cent pulses whereas Rabi has 28 per cent cereals, 53 per cent pulses and 12 per cent oilseeds. Prominent forage and range genetic resources of the region are cowpea, rice bean, berseem, *Panictum pedicellatum*, guinea grass, coix soybean, maize sorghum, dicanthium *spp. Atylosia,* bajra n, range grasses and legumes.

8. Central Plateau and Hills

It comprises 46 districts of Uttar Pradesh, Madhya Pradesh and Rajasthan having soils of variable topography with predominance of ravines and hills. Hardly 30 per cent of the total land is under cultivation with very poor irrigation and very low cropping intensity, literacy is very low and poverty ratio is very high. Low value crops are to be replaced by high value crops having advanced technological backup and crop diversification. Forage genetic resources available under this agro climatic region with low moisture condition are maize, cowpea, rice bean, berseem, *P. pedicellatum* coix, *Atylosia*, sorghum, bajra, guar, *cenchrus*, range grasses and legumes.

9. Western Plateau and Hills

It comprises the major part of Maharashtra, parts of M.P. and one district of Rajasthan forming a major part of peninsular India and receives 904 mm annual rainfall. About 65 per cent of the area is under crops and 11 per cent under forests. 12.4 per cent, area is irrigated by canals. Cotton and sorghum are grown in more than 50 per cent of the area of the zone. About 50 per cent of the country's sorghum production and 20 per cent of the country's cotton production are obtained from this zone. This zone is famous for best quality oranges, grapes and bananas though the area under these crops is hardly one lakh hectares, which needs to be increased. Available important forage genetic resources under scare water ecosystem are soybean, maize sorghum, *dicanthium spp.* pearl millet, *Dichanthium carzacosum, vicia*, cowpea, rice bean, berseem, *P. pedicellatum, Coix* a*tylosia,* guar, cenchrus range grasses and legumes.

10. Southern Plateau and Hills

This zone comprises 35 districts of A.P., Karnataka and Tamil Nadu which are semi-arid in nature. Nearly 81 per cent area is rainfed having 111 per cent cropping intensity consisting of mainly low value cereals and minor millets. Prominent forage and range genetic resources of the region are small millet, sorghum, *Heteropogon, Dichanthium*, Sehima, pearl millet, *Stylosanthes sp*, range grasses and legumes.

11. East Coast Plains and Hills

This consists of six zones: (*i*) Orissa Coast, (*ii*) North Coastal and Gujrat (*iii*) South Coastal Andhra, (*iv*) North Coastal Tamil Nadu, (*v*) Thanjavur and (*vi*) South Coastal Tamil Nadu. This is main rice and groundnut producing area which accounts for about 21 per cent of country's rice and 18 per cent of country's groundnut productions. Saline and alkaline soils are found in coastal areas to the extent of 4.9 lakh hectares. About 70 per cent area is rainfed and needs better watershed management. Forage genetic resources existing in this agro climatic condition are maize, cowpea, rice bean, berseem, *P. pedicellatum*, guinea grass, coix, small millet, sorghum, *Heteropogon, Dichanthium*, sehima, pearl millet, Stylosanthes *sp* range grasses and legumes.

12. West Coast Plains and Ghats

This includes western coast of Tamil Nadu, Kerala, Karnataka, Maharashtra and Goa which is famous for plantation crops and spices. The strategies include: rain water management, minor irrigation development, crop diversification, fisheries development especially prawn culture, reclamation of Pokhali levels and promoting spices production. Prominent forage and range genetic resources of the region are congo, signal grass, *Paspalum, Panicum, Digitaria, Brachiaria, P. mailma* gumea gran, soybean, maize sorghum, *Dicanthium* spp., pearl millet, range grasses *Dichanthium carzacosum, Lsilema vicia* and other range grasses and legumes.

13. Gujarat Plains and Hills

This zone consists of 19 districts of Gujarat having arid climate. Only 22.5 per cent area is irrigated and 50 per cent area is used for production of crops although it is an important oilseed zone. The cropping intensity ranges to around 114 per cent and 60 per cent of the cropped area is drought prone. Forage species existing in this agro-climatic condition are lucerne, sorghum, small millet, pearl millet, chioori, range grasses and legumes.

14. Western Dry Region

It consists of nine districts of Rajasthan and is characterized by hot sandy desert, erratic rainfall, high evaporation, no perennial river and scanty vegetation. Ground water is very deep and brackish. Frequent famine and drought force people and animal to migrate to other places in search of water, feed and feeder. Land/man ratio is high (1.73 ha/person). Average annual rainfall is 395 mm with high fluctuation from year to year. The forest area is only 1.2 and that of pasture is 4.3(70. Cultivable waste and fallow lands are nearly 42 per cent of geographical area and net irrigated area is only 6.3 per cent of net sown area (44.4 per cent). The cropping intensity is hardly 105 per cent. bajra, guar and moth are major crops in kharif and wheat and gram in rabi season, though the yields are very poor. Forage species existing in this agro climatic condition are lucerne, moth, maize, guar, carrot, cowpea, sorghum, sorghum, small millet, pearl millet, chioori, range grasses and legumes.

15. Island Region

It covers the Island territories of the Andaman and Nicobar Islands and Lakshdweep having an annual rainfall of 3000 mm spread over 8 to 9 months. It is smallest zone and largely a forest zone having highest literacy and least poverty. Productivity of rice and other crops has to be boosted by developing suitable varieties, arranging inputs and adopting package of new practices. This is one zone having good amount of biodiversity and forage species prominent forage and range genetic resources of the region are rice bean, cowpea, guinea grass, small millet sehima and other grasses.

Search for Potential Forage/Fodder Crop Across the Globe

The new species having potential value, have resulted in to the identification of several promising types which can adapt themselves to the harsh environmental and degraded soil conditions, give economic yield in different agro-climatic regions and, thereby, ensure feed and nutritional quality, and provides additional income to the resource poor farmers of this country particularly farmers of remote, backward, tribal, hilly and other difficult areas of the country. Consequently, there has been an emphasis on their germplasm collection, introduction, evaluation and utilization. Some of the domesticated wild and weedy relatives of earlier domesticated forages/ fodder plant are listed which were introduced in to India and performing well in specific agro-climatic conditions are mentioned in Table 3.1.

Needs of Systematic Exploration and Collection of Forage and Range Plants Available in India

Although an extensive ecological survey of Indian grassland has been completed and contradictory compiled and the geographic sources of collections of several native species of grasses and legumes have been worked out, the follow-up action on the re-survey of grasslands and collection of germplasm has been rather slow. Species priorities and gaps in collection from specific areas may be identified to collect and conserve the genetic diversity of forage plants. Required attention in collection, evaluation, documentation and conservation of forage species from stress environs, such as ravine, riverine, alkaline, saline-sodic, acidic, hot and cold and environ is wanted. The systematic work on the collection, valuation, documentation and conservation of germplasm of forage species, including wild and weedy taxa were paid serious attention since last few decades. Initially the activities related to germplasm resources in forage plants was started with the collection and evaluation of the local ecotypes of selected species by State Departments of Agriculture/ Agricultural Colleges of the State Agricultural Universities in the states of various states. Collecting must be done and massive screening will be needed. We need to adapt modern technology to give us efficient evaluation techniques. Rapid and automated chemical analyses; improved *in vitro* digestibility techniques and more rapid and accurate assessment of nutritive values are needed, there is need for serious research on the ecology of establishment and maintenance of pastures as well.

Table 3.1: Wild Species Domesticated for Use as Potential Commercial Crop

Species	*EC No.*	*Special Traits*	*Country*
Winged bean (*Psophocarpus tetragonoloba*)	EC114273	A multipurpose plant used as vegetable (green pod) yields edible tubers, green parts provide useful forage	Indonesia
Casuarina equisetifolia	EC168821	Suitable for growing in coastal sandy wastelands and alkaline soils	Australia
Acacia senegal	EC177144	Best grade edible gum and forage	Australia
Cassia sturtii	EC171975	Forage/fodder type in semi arid habitats	Australia
Atirplex nummularia	EC129766	The plant remains green through out year, grows well in deep soils and resist temperature as low as –10°C	Australia
Atriplex halimus	EC129767	Salt tolerant evergreen potential forage	Tunisia
A. canescence	EC129768	Perennials, suitable for arid zones potential forage	Tunisia
Cucurbita foetidissima	EC176511-15	Adapted to arid lands, yields edible tubers and oil from seeds useful for food industry also contain 30-35 per cent protein	USA
Atylosia acutifolia	EC198406	Well adapted on infertile soil, potential forage	Australia
Brachychiton populneum		Drought tolerant provide fodder and also useful as wind breaker when planted in multiple rows, potential forage	Australia

Needs of Introduction of Forage and Range Plants

National Bureau of Plant Genetic Resources (NBPGR) New Delhi is nodal organization engaged in the various activities related to enrichment of plant genetic resources in the country. Introduction of cultivated species and improved varieties of forages and range is required to strengthened the forage and range improvement programme. Wild and weedy relatives of these plants play vital role in improvement because they represent a part of crops gene pool particularly resistant to biotic and abiotic stresses and have been the donors of many other useful traits. Therefore, National Bureau of Plant Genetic Resources (NBPGR) continued its efforts to introduce important wild relatives of crop plants for their utilization in crop improvement programmes in the country. Acquisition of more exotic germplasm as early as possible has become the priority due to latest development under convention on Biological Diversity and other International treaties and laws. The germplasm/genetic material contain promising traits related to yield and yield attributing characters, quality characters and also resistance to various biotic and abiotic stresses. The flow of plant genetic resources from everywhere to India is slowing down gradually in the era of strong intellectual property right (IPR). Secondly we are not getting the trait specific materials as it was previously seen. It may be due to several reasons but reservation by most of donor country is one of them. Under this present circumstance we have to judiciously evaluate our existing germplasm/genetic resources to plan and conduct crop improvement programme for our current requirement and future use as well, we needs specific germplasm of forage from different country for our national needs mentioned in Table 3.2.

Table 3.2: Forage and Range Species Needs to be Introduced Form Abroad

Genus	*Species including Wild and Weedy Relatives*
Acacia sp.	*A. difficilis, A. mangiu, A. torulosa* and *A. tumida var. tumida*
Agropyron sp.	*Agropyron cristateum, A. cimmericum*
Astragalus	*Astragalus alpinus, Astragalus adsurgens*
Avena sp.	*A. abyssinica kochisi, A. barleota, A. byzantina, A. canariensis, A. claida, A. faba, A. hirtula, A. insalavis, A. longiglumis, A. lucovicana, A. macrostachya, A. pilosa, A. saviloviana, A. strigosa and A. weistii*
Brachiaria sp.	*B. ruziziensis*
Brassica sp.	*Brassica abyssinica B. orientalis*
Centrosema	*Centrosema pubescens*
Casuarina sp.	*Casuarina cunninghamiana ssp. cunninghamiana*
Chenopodium sp.	*C. capitatum, C. gigantenum, C. pallidicaule* and *C. quinoa*
Clitoria sp.	*Clitoria ternetea*
Centrosema	*Chloris gayana, Chloris dactylis*
Cyamopsis sp.	*C. senegalensis* and *C. serrata*
Dactylis	*Dactylis glomerata,*
Desmodium sp.	*Desmodium barbatum, D. intortum* and *D. uncinotum*
Eleusine sp.	*Eleusine indica, E. africana E. fleceifolia, E. jaegeri, E. multiflora and E. tristachya*
Eragrostis	*Eragrostis curvul*
Elymus sp.	*Elymus haffmannii*
Eucalyptus sp.	*E. argophloia, E. camaldulens var. obtusa E. tereticornis ssp. tereticornis*
Festuca	*Festuca arundinacea*
Glycine sp.	*G. argyrea, G. canescens, G. centennial, G. clandestina, G. cyrtoloba, G. falcata, G. javanica, G. latifolia, G. latrobeana, G. microphylla, G. soja, G. tabacina* and *G. tomentella*
Hevea sp.	*H. camarguana, H. colina, H. fousiflora, H. guyanensis, H. nitida* and *H. pauciflora*
Lablab	*Lablab purpureeus*
Lathyurs sp.	*Lathyrus sativus, L. parenne, L. inconspicuus, L. annuus, L. aphaca, L. blepharicarpus, L. ciliolatus, L. clymenum, L. gorgoni, L. hierosalymitanus, L. hierosolymitanus, L. hirsutum, L. inconspicuus, L. latifolius var. ensifolius, L. odoratus, L. orchus and L. tuberosus, L. pratensis, L. pseudo, L. sylvestris, L. tingitanus, L. vernus*
Lolium	*Lolium multiflorum*
Lotus sp.	*Lotus corniculatus, L. conimbricensis, L. coribricensis, L. ornithopodioid, L. pendunoulatus, L. purshianus* and *L. uliginosus*

Contd...

Table 3.2–Contd...

Genus	Species including Wild and Weedy Relatives
Macroptilium sp.	*M. atropurpureum* and *M. axillare*
Medicago sp.	*Medicago sativa, Medicago faclcata, M. arabica, M. ciliaris, M. coronata, M. intertexta, laciniata, M. littoralis, M. lupulina, M. minima, M. orbicularis, M. polymorpha, M. radiata, M. rigidula, M. rugosa, M. scutellata, M. tornata, M. truncatula, M. turbinate, M. liciruata* and *M.littoralis*
Melilotus sp.	*Melilotus alba, M. indicus, M. messanensis, M. officinalis, M. segetalis, M. sulcatus*
Panicum sp.	*P. antidotale, P. coloratum, P. coloratum var. makarikariensis, P. lanipe, P. milioides, P. queenslandicum, P. schinzii and P. virgatum, P. stapfianum*
Pennisetum. sp.	*Pennisetum flassidum, Pennisetum purpureum, P. alopecuroides, P. basedowii, P. flaccidum, P. latifolium, P. macrourum, P. orientale, P. pedicellatum, P. polystachyon, P. setaceum, P. squamulatum, P. unisetum and P. violaceum*
Phleum	*Phleum pratense*
Poa	*Poa pratansis, P. Annua, P. pratensis*
Pueraria	*Pueraria phaseoloides, P. javanica*
Setaria sp.	*S. asphacelata var. anceps, S. australiensis, S. incrassata, S. italica ssp. viridis, S. lachnea, S. neglecta, S. sphacelata and S. verticillata*
Sorghum sp.	*Sorghum sudanenses*
Stylosanthes sp.	*Stylosanthes capitata, Stylosanthes pilosa, S. scabra* and *S. viscosa*
Trifolium sp.	*Trifolium pratense, T. repens, T. resupinatum, T. subterraneum, T. alpestre, T. angustifolium, T. arvense, T. aureum, T. baccarinii, T. balansaa, T. billardiari, T. campestre, T. clypeatum, T. dasyurum, T. diffusum, T. fomentosum, T. fragiferum, T. grandiflorum, T. incarnatum, T. lappaceum, T. leucanthum, T. ligusticum, T. michelianum, T. nigrescens, T. pilulare T. scabrum, T. squamosum, T. striatum, T. striatum var. spinescens, T. subterraneum* and *T. tomentosum*
Trigonella sp.	*Trigonella caerulea*
Vicia sp.	*V. hirsuta, V. hyaeniscyamus, V. hyracanica, V. johannis var. johannis, V. johannis var. procumbens, V. lutea ssp. lutea, V. melanops, V. michauxii, V. monttana, V. narbonensis, V. narbonensis var. aegyptiaca, V. narbonensis var. affinis, V. narbonensis var. jordanica, V. narbonensis var. narbonensis*
Vigna sp.	*V. filicaulis var. filicaulis, V. filicaulis var. pseudovenulosa, V. racemosa var. racemosa, V. radicans, V. reflexo-pilosa var. glabra, V. reflexo-pilosa var. reflexo-pilosa, V. reticulata, V. riukiuensis, V. schimperi, V. stipulacea, V. subramaniana, V. trilobata, V. trinervia var. trinervia, V. triphylla, V. umbellata, V. umbellata var. gracilis, V. umbellata var. umbellata, V. venulosa, V. vexillata*
Zea sp.	*Z. diploperennis, Z. luxurians, Z. mays ssp. huehuetangenesis, Z. mays ssp. mexicana* and *Z. mays ssp. parviglumis*

Sustainable Uses of Forage Genetic Resources

The National Bureau of Plant Genetic Resources (NBPGR) with the help of other ICAR research organization, state and central agricultural universities, state department of agriculture, other autonomous body and non governmental organization (NGO) etc. is dedicated to save and conserve the bio diversity of forages which is very difficult rather impossible without active cooperation and all short of support by all the stakeholder, for the purpose questionnaire on forage genetic resources is prepared for collecting information relevant to forages systematically. Such information would enable to consolidate the work and strengthen the National Network System Approach with decentralizing the crops/species responsibilities to various lead Institutes/centres most likely the 18 coordinated centres functional under All India Coordinated Research Project on Forage Crops. Our prime responsibility is to hand over the all the natural resources be it land water or biodiversity to our offspring in ever possible best condition otherwise in condition if not better to the level of what we have received from our ancestors, to make future happiest and glorious. To make sound footing on that conservation of biodiversity is utmost needed in general and forage crop in particular. A National action plan should be prepared in a mission mode approach by the core group of forage workers to seek information (passport and ITKs) and germplasm materials from various institutions/organizations for conservation in the National Gene Bank, to reveal the national current status, sensitize the forage workers all over and set forth future course of action in achieving this national endeavor. The promising genetic stocks should be registered for their uniqueness or value addition traits, so as to make out the earnest gains and identify the contributions made by the forage germ psalm curators/breeders. Endemic and endangered species of forage grasses from Himlalaya and the Western Ghats may be conserved in situ. It sis desirable to grow the accessions at a location as nearer as possible to the place of origin/original habitat so that the agronomic characters are fully expressed. Species-wise inventory of genetic resources of forage plants maintained by different organizations may be prepared and a duplicate set may be deposited in the National Repository at NBPGR, New Delhi. Perhaps, rare availability of the taxonomists in regard to forage corps especially, should also be major concern in germplasm – related scientific programmes.

The calls for a shift in crop improvement programme from grain type alone to grain-cum-fodder type cultivars. In context to Himalayan mountainous region, for an improved agriculture that balance optimum productivity of mixed farming systems and conservation of genetic diversity on sustainable basis, the strategic priorities need to be focused on the following:

- ✰ Linking agrobiodiversity with biodiversity of natural ecosystems like rangelands/grasslands/wetlands and forests.
- ✰ Recognition to gender issues and traditional knowledge and technologies and local farmers, partnership in explorations and collection of indigenous germplasm.
- ✰ Improvement in the scientific knowledge of the ecological and socio-economic functions of agrobiodiversity, especially range grasses and legumes and fodder trees.
- ✰ Improvement in the traditional agronomic practices and technologies for soil tertility management.
- ✰ Recognition to villages and the farm community, especially women, who still maintain traditional agrobiodiversity and/or eroded the least.

- ✰ Establishment of "crop parks" and "gardens" and "(agro) biodiversity villages" analogous to national parks, botanic gardens of biosphere reserves.
- ✰ Exploration of avenues for value addition to under utilized dual purpose crops and of potential market within the country and abroad.

Unfortunately, forage crops, especially true fodder type being indirect use of humankind, does not receive priority in most of the organization. This mindset limitation need to be changed through motivation and concentrated efforts of popularizing growing significance of forage crops for animal productivity on one hand and soil plant animal human relationship on the other.

Summary

India is a country having large amount of biodiversity in almost all the crops, thanks to its geographical position and its agro climatic conditions obviously India enjoys diverse climatic condition which play a pivotal role in evolution of various flora and fauna over centuries. Its cultural diversity also play significant role in enriching its diversity by introducing new crops in India by visitors. Forage is some time treated as orphan crops, now time has come to take all out efforts to for sustainable management for that one should know the current status of forage biodiversity. For sustainable management proper planning is needed for its exploration and survey, collection, evaluation, conservation and equitable utilization of priceless forage genetic resources to match with National demand of forages for Dairy and other allied sector. To achieve our ultimate objective *i.e.* to achieve self sufficiency in quality forage production to feed our live stock and consequently to realize its full potential in dairy and other allied sector, their is urgent need for explore, collect, evaluate and utilized these genetic resource judiciously for current requirement and conserved the all for future needs by application of pre-breeding and germplasm enhancement approaches that increase the use efficiency in the crop improvement/production/conservation programme.

References

Ahmed, S T, M L Magoon and K L Mehra, 1974. Evaluation of certain forage crop introductions for resistance to plant pathogens. Proc. XII Internat. Grassland Cong. Moscow. Pp. 9-14.

Arora, R K and E R Nayar. 1983. Wild relatives and related species of crop plants in India-their diversity and distribution. Bull. Bot. Surv. India 25 (1-4): 35-45.

Arora, R K and K P S Chandel. 1972. Botanical source areas of wild herbage legumes in India. Trop. Grasslands 6 : 213-21.

Bhagmal, M S Yadav, B D Patil and U S Mishra. 1985. Forage yield attributes in *Dichanthium annulatum* (Forsk.) Stapf – Biometrical analysis of variability. Trop. Agric.

Chandel, K P S. 1996. Hindustani center provides India and the world a trove of genetic diversity Diversity 12 (3): 21.

Chandel, K P S. 1996. Hindustani center provides India and the world a trove of genetic diversity Diversity 12 (3): 21.

Harinarayana, G, S Appa R.ao, Mengesha and H. Melak.1987. Prospects of utilizing genetic diversity in pearl millet. Nat. Symp. on Plant Genetic Resources, March 3-6, NBPGR, New Delhi.

Harlan, J R. 1971. Agricultural origins: Centers and non-centers. Science 174: 468-474.

Khoshoo, T. N. 1988. Conservation of Biological Diversity. Plant genetic Resources Indian Perspective Ed. Paroda R.S., R.K. Arora and K.P.S. Chandel pp76-87 NBPGRpub.

Kochhar, S, R K Arora and E R Nayar. 1992. Bamboo in North Eastern Region, ICAR Bulletin, ICAR Research Complex for NEH Region, Barapani, Meghalaya.

Kohli, K S, C B Singh and D K Agarwal. 1999. IGFRI Cowpea Germplasm Catalogue, IGFRI, Jhansi, 53 p.

Mehra, K L, Bhagmal, P R Sreenath, M L Magoon and D S Katiyar. 1969. Metroglyph analysis of complexes in fodder sorghums. Indian J. Sci. and Indust. 3 (4) : 211-214.

Mehra, K L, Bhagmal, P R Sreenath, M L Magoon and D S Katiyar. 1971. Factor analysis of fodder yield components in oats. Euphytica 20 : 297-301.

Mehra, K L, C B Singh, K S Kohli and M L Magoon. 1975. Divergence and distribution of pigmentation of plant parts in a world collection of cowpea *Vigna genesis*. Genetica Lberica 26-27 : 79-97.

Paroda, R S and G P Lodhi. 1980. Genetic improvement in forage sorghum. Forage Res. 7A : 17-56.

Patil, B D and Ghosh. 1962. Research on Grassland Development. Studies on Pasture grass, *Pennisetum pedicellatum*, Agro-botanical survey, Classification and Type description. (Mimeograph). IARI, New Delhi. 23. p.

Patil, B D and S Singh. 1960. Studies on Anjan grass, *Cenchrus ciliaris*, M.Sc. Thesis, IARI, New Delhi.

Rai, M. 1997. Benefit sharing, compensation modes and financial mechanisms for agro-biodiversity management in India. Keynote Adress, NAAS – NBPGR (ICAR) Workshop on "National Concern for Management, Conservation and use of Biodiversity, Oct. 15-16, 1997, Shimla, India, 8p.

Reeves, R. G., 1950. The use of Teosinete in the improvement of corn inbreds. Agron. J. 42:248-251

Singh RV, Deep Chand, Vandana Tyagi, AK Singh, SP Singh, Surender Singh and PC Binda (2005) Priorities for Introduction of Fruit Crops in India.Indian J.Plant Genet. Resour 18 (1): 67-68.

Singh RV, IP Singh, Deep Chand, Vandana Tyagi, AK Singh, SP Singh and Surender Singh 2008. Important Crop Germplasm Introduced in India during 2006. Indian J. Plant Genet. Resour. 20 (1): 64-69.

Singh, A.K. JP Moss, and J. Smartt. 1990. Ploidy Manipulations for Interspecific Gene Transfer. Advances in Agronomy 43:191-240

Singh, C B, A Singh, K S Kohli, K L Mehta and M L Magoon. 1971. Correlation and factor analysis of fodder yield components in cowpea. Acta Agronomica Scientiarum Hungaricae. (Hungary). 26 ; 378-387.

Singh, V. 1998. Organizing mountain farmers to carryout *in-situ* conservation of their agricultural resources' diversity. In: Tej Pratap and B. Sthapit (Eds.). Managing Agrobiodiversity-Farmers' Changing Perspectives and Institutional responses in the HKH region. ICIMOD, Kathmandu, Nepal, pp. 341-349.

Sri Ram, C B Singh, K L Mehra and M L Magoon. 1973. Tolerance to leaf hopper, flea beetle and semi-looper ingestation in a world collection of cowpea : inter-regional comparisons. Genet. Lber. 28 : 115-130.

Stalker, H. T. 1980. Utilization of wild species for crop improvement. Adv. Agron. 33:111-147.

Vavilov, N I. 1950. The origin, variation, Immunity and breeding of cultivated plants. Chron. Bot. 13 : 364

Whyte, R O. 1964. The Grassland and Fodder Resources of India. ICAR. Sci. Monogr. 33. ICAR, New Delhi.

Yadav, M S and P R Sreenath. 1975. Factor analysis of fodder yield components in *Pennisetum pedicellatum* Trin. Andhra Agric. J. 2 (1 and 2) 5-9.

Yadav, M S, K L Mehra and M L Magoon. 1974. Heritability and correlation among fodder yield components in pasture grass, *Dichanthium annulatum*. Indian Forester. 102 91) : 64-68.

Chapter 4

Utilization of Wild and Weedy Relatives in Forage Improvement

Anil Kumar Singh[1] and Birendra Prasad[2]

[1]*Senior Scientist, Agronomy, Division of Land And Water Management, ICAR Research Complex for Eastern Region, ICAR Parisar (P.O.- B.V.College), Patna – 800 014, Bihar*
E-mail: aksingh_14k@yahoo.co.in

[2]*Assistant Professor, Department of Seed Science and Technology, GBPUA&T, Hill Campus, Ranichauri, Tehri Garhwal, Uttarakhand*
E-mail: bprasadsst@yahoo.co.in

Wild and Weedy relatives of crop plants represent a part of crops genepool particularly resistant to biotic and abiotic stresses and have been the donors of many other useful traits. Therefore, National Bureau of Plant Genetic Resources continued its efforts to introduce important wild relatives of crop plants for their utilization in crop improvement programmes in the country. An account of introduction of wild species alongwith their special features such as resistance to different biotic and abiotic stresses are presented in this article. During recent past, more than 1000 wild species of 85 genera in different group of crops have been introduced from 24 countries. The cultivated species often lack some important characters generally not available in cultivated forms, but are present in wild relatives such as resistance to biotic, abiotic, quality etc. Although interspecific hybridization has beer undertaken for the transfer of genes from wild species into commercial varieties of some important crops. The crops in which great efforts have been attempted to utilize wild species germplasm are cotton, tobacco, sugarcane, potato, wheat, corn and more recently in oat, rice, rapeseed, mustard and tomato. There is need for proper screening of the wild and weedy forms for identifications of various unique traits is essential for effective utilization in the crop improvement programmes. Since the potentials of wild germplasm are unknown until collections are screened and tested e.g. only one accession of *Oryza nivera* was found resistant to grassy stunt virus after thousands of cultivated and thousands of wild accessions were screened for this purpose. The information given in this paper may be useful for the breeders/researchers in their crop improvement programmes by selecting the desired germplasm of these introduced wild species.

Introduction

With the advent of new techniques in the field of molecular biology, the transfer of genes from wild to cultivated forms are now possible at will. Increasing cost of agricultural production due to highly susceptible nature of prevailing high yielding varieties, has led the scientists to look for varieties, which could be grown under low input conditions and could be resistant to various biotic and abiotic stresses. Wild relatives of crop plants represent a part of crops genepool particularly resistant to biotic and abiotic stresses and have been the donors of many other useful traits. The accidental discovery of *Zea diploperennis* in 1977 by a student of botany in Mexico has been regarded as the 'botanical finding of the century'. This plant is a wild relative of the cultivated maize, which was thought to be extinct. The species is perennial and diploid and is immune to at least seven major diseases, which pose a serious threat to maize all over the world, especially in the USA, South America and Africa.

Objectives

The main objective of this exercise is to analyse the present status of wild relatives introduced in various crop plants, their extent of utilization, so that the needs may be assessed for future crop improvement programmes as cultivated species often lack some important characters generally not available in related cultivars but are present in wild relatives such as resistance to diseases, insect pests, drought tolerance, quality and new characters such as transfer of genes conferring resistance to salinity, cold, diseases to bread wheat from related genera like *Aegilops, Agropyron* and *Secale* etc.

Introduction of Wild Genetic Resources in India

India has introduced a number of wild relatives of different crop plants to enrich the gene pool and their further exploitation in the breeding programmes in the country. The account of such wild species is given in the Table 4.1.

Trait Specific Wild Species Introduced from Different Countries

About 80,000 species of plants have been used to meet the routine needs by the human being. Of the 30,000 species so for have been identified as edible and about 7,000 species have been cultivated and/or collected for food at one time or the other. Presently only 20 to 30 crops such as cereals (wheat, rice, maize, millets, sorghum), root/tuber crops (potato, sweet potato, cassava), legumes (pea, bean, peanut, soybean), sugarcane, sugar beat, coconut and banana are mainly used to feed the world.

Table 4.1: Wild and Weedy Species Introduced in Different Forage/Fodder Crops

Genus	*Species including Wild and Weedy Relatives*
Acrocomia	*Acrocomia sclerocarpa*
Aegilops sp.	*A. biuncialis, A. columnaris, A. comosa, A. crassa, A. cylindrica, A. geniculata, A. juvenalis, A. kotschyi, A. longissima, A. markgrofii, A. neglecta, A. peregrina, A. peregrina var. brachyathera, A. searsii, A. sharonensis, A. speltoides, A. speltoides var. speltoides, A. tauschii, A. triuncialis, A. triunicialis var. persica, A. umbellulata, A. uniaristata* and *A. ventricosa*
Agropyron sp.	*A. cimmericum*
Arachis sp.	*A.apressipila, A. archeri, A. batizocoi, A. burkartii, A. cardensaii, A. chiguitana, A. cruziana, A. cryptoptamica, A. dardani, A. diogoi, A. duranensis, A. fenthamii, A. glabrata, A. glandulifera, A. hermanii, A. herzogii, A. ipaensis, A. kempfmoicadoi, A. kretschmeri, A. lutescens, A. major, A. matiensis, A. monticola, A. otavioi, A. palustris, A. paraguariensis, A. pinto, A. praecox, A. pseudovillosa, A. rigonii, A. simpsonii, A. stenophylla, A. stenosperma, A. sylvestris, A. tristeminata, A. villosa* and *A. williamsii*
Avena sp.	*A. abyssinica kochisi, A. barleota, A. byzantina, A. canariensis, A. claida, A. faba, A. hirtula, A. insalavis, A. longiglumis, A. lucovicana, A. macrostachya, A. pilosa, A. saviloviana, A. strigosa* and *A. weistii*
Brachiaria sp.	*B. ruziziensis*
Brassica sp.	*Brassica abyssinica* and *B. orientalis*
Chenopodium sp.	*C. capitatum, C. gigantenum, C. pallidicaule* and *C. quinoa*
Clitoria sp.	*Clitoria ternetea*
Cyamopsis sp.	*C. senegalensis* and *C. serrata*
Desmodium sp.	*D. intortum* and *D. uncinotum*
Eleusine sp.	*E. fleceifolia, E. jaegeri, E. multiflora* and *E. tristachya*
Elymus sp.	*Elymus haffmannii*
Glycine sp.	*G. argyrea, G. canescens, G. centennial, G. clandestina, G. cyrtoloba, G. falcata, G. javanica, G. latifolia, G. latrobeana, G. microphylla, G. soja, G. tabacina* and *G. tomentella*
Lathyurs sp.	*L. inconspicuus, L. annuus, L. aphaca, L. blepharicarpus, L. ciliolatus, L. clymenum, L. gorgoni, L. hierosalymitanus, L. hierosolymitanus, L. hirsutum, L. inconspicuus, L. latifolius var. ensifolius, L. odoratus, L. orchus* and *L. tuberosus, L. pratensis, L. pseudo, L. sylvestris, L. tingitanus, L. vernus*
Lens sp.	*L. ervoides, L. lamottei, L. nigricans, L. odemensis, L. orientalis, L. slamottei, L. sorientalis* and *L. tomentosus*
Linum sp.	*L. altaicum, L. angustifolium, L. austracum, L. bienne, L. decumbens, L. flavum L. grandiflorum, L. lewisii* and *L. pereune*
Lotus sp.	*L. conimbricensis, L. coribricensis, L. ornithopodioid, L. pendunoulatus, L. purshianus* and *L. uliginosus*
Macroptilium sp.	*M. atropurpureum* and *M. axillare*
Medicago sp.	*M. arabica, M. ciliaris, M. coronata, M. intertexta, laciniata, M. littoralis, M. lupulina, M. minima, M. orbicularis, M. polymorpha, M. radiata, M. rigidula, M. rugosa, M. scutellata, M. tornata, M. truncatula, M. turbinate, M.liciruata* and *M.littoralis*

Contd...

Table 4.1–Contd...

Genus	*Species including Wild and Weedy Relatives*
Melilotus sp.	*M. indicus, M. messanensis, M. officinalis, M. segetalis* and *M. sulcatus*
Oryza sp.	*O. alta, O. australiensis, O. barthii, O. brachyantha, O. eichingeri, O. glumaepatula, O. grandiglumis, O. granulata, O. latifolia, O. lingiglumis, O. longistaminata, O. lumaepatula, O. meridionalis, O. minuta, O. officinalis, O. punctata, O. rhizomatis, O. ridleyi* and *O. rufipogon*
Panicum sp.	*P. antidotale, P. coloratum, P. coloratum var. makarikariensis, P. lanipe, P. milioides, P. queenslandicum, P. schinzii and P. virgatum, P. stapfianum*
Pennisetum. sp.	*P. alopecuroides, P. basedowii, P. flaccidum, P. latifolium, P. macrourum, P. orientale, P. pedicellatum, P. polystachyon, P. setaceum, P. squamulatum, P. unisetum* and *P. violaceum*
Rhynchosia sp.	*R. minima, R. phaseoloid, R. phramidalis, R. rapatabunda* and *R. retculata*
Sesbania sp.	*S. aculeata, S. afraspera, S. bispinosa, S. cannabia, S. cinerascens, S. emerus, S. exasperata, S. formosa, S. goetzea, S. grandiflora, S. greenwayi, S. hirtistyla, S. javanica, S. keniensis, S. leptocarpa, S. macrantha, S. microphylla, S. nilotica, S. pachycarpa, S. quadrata, S. rostrata, S. sericea, S. speciosa, S. tetraptera, S.varadera* and *S.virgata*
Setaria sp.	*S. asphacelata var. anceps, S. australiensis, S. incrassata, S. italica ssp. viridis, S. lachnea, S. neglecta, S. sphacelata* and *S. verticillata*
Sinapsis sp.	*Sinapsis arvensis.*
Sorghum sp.	*Sorghum sudanenses*
Stylosanthes sp.	*S. scabra* and *S. viscose*
Trifolium sp.	*T. alpestre, T. angustifolium, T. arvense, T. aureum, T. baccarinii, T. balansaa, T. billardiari, T. campestre, T. clypeatum, T. dasyurum, T. diffusum, T. fomentosum, T. fragiferum, T. grandiflorum, T. incarnatum, T. lappaceum, T. leucanthum, T. ligusticum, T. michelianum, T. nigrescens, T. pilulare T. scabrum, T. squamosum, T. striatum, T. striatum var. spinescens, T. subterraneum* and *T. tomentosum*
Trigonella sp.	*Trigonella caerulea*
Triticum sp.	*T. ispahanicum, T. monococcum ssp. aegilopoides, T. monoeoccum ssp. monococcum, T. tauschii, T. timopheevii ssp. armeniacum, T. timopheevii ssp. timopheevii, T. turgidum ssp. carthlicum, T. turgidum ssp.. dicoccoides, T. turgidum ssp. durum, T. turgidum ssp. polonicum, T. turgidum ssp. turgidum, T. urartu, T. vavilovii,* and *T. zhukovskyi*
Vicia sp.	*V. angustifolia, V. arbonensis var. salmonea, V. articulata, V. bengha, V. ervilia, V cortala, V.disperma, V. dumetorum, V. flugens, V. hirsuta, V. hyaeniscyamus, V. hyracanica, V. johannis var. johannis, V. johannis var. procumbens, V. lutea ssp. lutea, V. melanops, V. michauxii, V. monttana, V. narbonensis, V. narbonensis var. aegyptiaca, V. narbonensis var. affinis, V. narbonensis var. jordanica, V. narbonensis var. narbonensis, V. narbonensis var. salmonea, V. pannonica, V. peregrina, V. venosa* and *V. villosa,*
Vigna sp.	*V. ambacensis var. ambacensis, V. ambacensis var. pubigera, V. angivensis, V. angularis var. nipponensis, V. benuensis, V. comosa ssp. comosa var. comosa, V. filicaulis var. filicaulis, V. filicaulis var. pseudovenulosa, V. friesorum var. friesorum, V. frutescens ssp. frutescens var. frutescens, V. frutescens ssp. incana, V. glabrescens, V. gracilis var. gracilis, V. grandiflora, V. heterophylla, V. hirtella, V. hosei var. hosei, V. kirkii, V. lanceolata,*

Contd...

Table 4.1–Contd...

Genus	*Species including Wild and Weedy Relatives*
	V. laurentii, V. luteola, V. macrorhyncha, V. marina, V. membranacea, V. minima, V. monophylla, V. multinervis, V. mungo var. silvestris, V. nakashimae, V. nepa, V. nigritia, V. nyangensis, V. oblongifolia var. oblongifolia, V. oblongifolia var. parviflora, V. parkeri ssp. maranguensis, V. racemosa, V. racemosa var. racemosa, V. radicans, V. reflexo-pilosa var. glabra, V. reflexo-pilosa var. reflexo-pilosa, V. reticulata, V. riukiuensis, V. schimperi, V. stipulacea, V. subramaniana, V. trilobata, V. trinervia var. trinervia, V. triphylla, V. umbellata, V. umbellata var. gracilis, V. umbellata var. umbellata, V. venulosa, V. vexillata, V. vexillata var. angustifolia, V. vexillata var. lobatifolia, V. vexillata var. macrosperma, V. vexillata var. ovata and *V. vexillata var. vexillata*
Zea sp.	*Z. diploperennis, Z. luxurians, Z. mays ssp. huehuetangenesis, Z. mays ssp. mexicana* and *Z. mays ssp. Parviglumis*

The special features of some wild species are presented in Table 4.2 for the general awareness of the breeders in the country, so that the same may be utilized in their breeding programmes by transferring the desirable genes.

Table 4.2: Special Features of Important Wild Species Introduced from Different Countries

Species	*EC No.*	*Special Features*	*Country*
Cereals			
Aegilops sp.	EC 380605-08	Resistant to cereal cyst nematode, root knot and eye spot disease	Australia
	EC, 169756	Crossable with cultivated wheat for possible transfer of salt tolerant genes	Germany
Triticum boeticum	EC 168521	Salt tolerant species	Germany
Zea mays ssp. *mexicana*	EC 180027	Photo insensitive	USA
Vigna narbonensssis	EC 389358	Resistant to drought combined with responsiveness to irrigated conditions	Syria
Vicia villosa ssp. *dasycarpa*	EC 389362	Resistant to drought	Syria
Acrocomia sclerocarpa	170454	Kernel oil is used for soap industry, where as refined oil is suitable for cooking	Brazil

Utilization of Wild Species in Crop Improvement

Man uses 3000 or more plant species for food and cultivates about 150 species. The crops that produce at least 20 million metric tons include; wheat, rice, corn, potato, barley, sweet potato, cassava, grape, soybean, oat, sorghum, sugarcane, millet, banana, tomato, sugar beet, rye, orange, coconut, cotton, apple, yam, peanut and watermelon. Although interspecific hybridization has been attempted in all these crops, the transfer of genes from wild species in to commercial varieties has varying success. The crops in which great efforts have been executed to utilize wild species germplasm are cotton, tobacco, sugarcane, potato, wheat and corn and more recently in rice, rapeseed, mustard, tomato, oats etc. The brief review of these crops is as follows

Trait Oriented Utilization of Wild Species in Crop Improvement

The most common reason to attempt of utilization of wild species is to transfer disease resistance. However, the genes for quantitative and qualitative traits were also transferred from wild species into cultivated forms (Stalker, 1980).

Disease and Insect Resistance

Utilization of the rootstock of wild species has eliminated many diseases and insect pests which are now commonly grafted such as grapes, citrus and rubber. Genetic resistance to many insect pests has been found in wild species. The successful gene transfer in red mite and aphid resistances from *Fragaria chiloensis*; resistance to leaf- chewing insect in *Arachis hypogaea* derived from *A. monticola*.

Yield Traits

Hybrids between wild and cultivated species of wheat, maize, sorghum, pearl millet, rice, tobacco and cereal forage plants have resulted in increased production of leaves and stems. Cotton fibres may be lengthened as a result of interspecific hybridization. Increased tuber yields were achieved developing interspecific hybridization between wild and cultivated potato species. Yield increases have also been reported for species of *Vigna, Zea, Ribes, Vanilla, Arachis* and *Avena*. *Avena sativa* x *A. sterilis* resulted in 25-30 per cent yield increase over the recurrent parent. This was twice the yield obtained from 1905 to1960 utilizing the germplasm of Mid Western United States. Best example is in the genus *Saccharum*. All modern varieties are progenies of *S. officinarum* x wild sp. of this genus. With the resulting resistance to sugarcane pests, the high polyploid derived varieties greatly out yielded the noble canes.

Quality Traits

Wild species germplasm has had limited use in this area because of the genetic complexity of most quality characters, However, potential exists for making alternations in chemical composition and morphological traits by using wild species germplasm.

In *Poaceae*, protein quality and quantity has been altered in several species, however the protein quantity and seed size are negatively correlated and advantage sought is often lost by the time, large seeded types are selected from hybrid derivatives. Seeds with increased protein percentage have been selected from crosses with wild species of rye, rice and oats. Oil quality has been improved by utilization of wild palm species. In tomatoes, soluble solids content of commercial varieties has been increased substantially by hybridizing cultivated varieties with a wild green-fruited species. Leaf quality of tobacco is improved by using *N. debneyi*. Fiber qualities in cotton and starch content in potato were improved by utilizing wild species germplasm. Quality of plant products has also been improved by utilization of species hybrids.

Hybrids between widely separated species may be better adapted in forages than other groups because 50 per cent of the grasses are polyploids and some sterility can be accepted for example the *Cynodon interspecific* hybrid derivative coastal Bermuda grass is widely used as forage grass in the southeastern region of the United State. Interspecific hybrids are more palatable than parental species *e.g.* quality of tall fescue giant fescue hybrid derivates is superior to the parental species and has promise as new forage. Other examples include -Maize used as silage with teosinte in its pedigree, sorghum-sudan grass hybrids; pearl millet-elephant grass hybrids are used as fodder.

Earliness and Adaptation

The land is becoming limited for agriculture and the crops are moved into less suitable environments. Variability found in wild species offer a valuable resource for genes conferring wider adaptations. Because of severe climatic conditions in Russia, their breeders have probably utilized wild species to improve winter hardiness of their crops in Wheat. Cold tolerance transferred from wild species of peppermint, tomato, grape, strawberry, wheat, rye, onion, and potato to their cultivated relatives. Breeding for earliness in Brassica and *Glycine* species have been used to introgress genes for earliness into cultivated varieties. Breeding for drought and heat tolerance has been accomplished by utilizing wild species of peas and wheat. Other environmental characters for which wild species have been utilized include increased salt tolerance in tomato, tolerance to calcareous soils in grapes and lack of photosensitivity in *Pennisetum*.

Development of CMS Lines for Exploitation of Hetrosis and Seed Production

Most common alternation in reproduction is sterility, which is resulted from interspecific hybridization. CMS is important derivative of this, used for hybrid seed production and exploitation of hybrid vigour in Pearl millet, sorghum, maize tomato, brinjal etc. The examples include wheat, cotton, barley, tobacco, ryegrass, and sunflower. Other modes of reproduction have been altered in crops by introducing wild germplasm. *e.g.* Maize-*Tripascum* hybrid derivatives. Cleistogamy and self-fertility traits of wild *Secale* species may also be transferred to cultivated rice.

Miscellaneous Uses

Genes transfer for hard seededness to cultivated bramble fruits; to introgress dark green color and excellent leaf texture into lettuce; for carotenoid synthesis to cultivated tomato. Interspecific hybrids of oil palm resulted in shorter plants; Selection among *Triticum* x *Agropyron* hybrid derivatives has resulted in semi dwarf wheat.

Utilization of Wild Relatives in Improvement of Different Crop

Rice

Rice genetic resources of Indian subcontinent have contributed greatly to the improvement, production and productivity worldwide. Indian rice germplasm has been used extensively for various characters such as resistance to biotic and abiotic stresses to consumer quality parameters. A recent analysis shows that IRRI's 35 breeding lines adopted as varieties in rice growing countries include in their ancestry one or more varieties or wild species of India. Wild *O. nivara* and *O. rufipogon* are important sources of disease resistance. An Indian accession of *O. nivara* which is the only source of resistance to grassy stunt virus: occurs in as many as seven varieties.

Triticum

The cultivated hexaploid specie *T. aestivum* contains the genome of three different species and is also a classical example of utilizing wild species for crop improvement. Resistance to several diseases has been incorporated in to cultivated wheat from closely related species of *Triticum, Aegilops* and *Agropyron*. Wild species have been a source for improved winter hardiness, lodging resistance and cytoplasmic male sterility. Some of the earliest successful attempts to transfer genes from wild to cultivated relate to crosses between tetraploid emmer wheat (*T. tauschii*) to hexaploid bread wheat (*T. aestivum*) that led to the development of 'Hope', the most popular rust free variety in the USA (Mc Fadden, 1930) and cross between maize (*Zea mays*) and teosinte (*Z. mexicana*) (Reeves, 1950).

Barley

A cooperative programme between Swedish researchers achieved one of the most successful public-private research partnerships for introgression of *Hordeum vulgare* ssp. *spontaneum* into barley breeding. Private breeders carried out the disease screening against powdry mildew (*Erysiphe polygoni*) while the public researchers did the crosses, manipulation and documentation. A backcross strategy was applied using cultivated barley lines as the maternal recurrent parent. The second disease screening was carried out in the BC1 F1 to select the parents of the BC2, and plants of the BC1 F2 were screened to select the resistant material for the BC3, which was then selfed for one generation (BC3F2) before providing this advance generation to the private plant breeders. As a result of this cooperative project, several hundred BC3F2 plants were developed from 89 accessions of *H. vulgare* ssp. *spontaneum* with many unknown resistance genes to powdry mildew. This resulted in broadening the genetic base for host plant resistance to a disease in which the recessive *mlo* gene accounts for 30 per cent of the barley grown area in Europe. Progress was also made in the transfer of resistance to spot blotch, net blotch, leaf rust and scald from H. *vulgare* ssp. *spontaneum*. However, sterility barriers and chromosome pairing in hybrids have restricted the access to genes from other wild *Hordeum* species.

Zea

Hybrid derivatives of tetraploid *Tripsacum* x diploid maize are now being utilized in commercial breeding programmes.

Sorghum

Diseases such as downy mildew, anthracnose and grain mould affect this crop. Immune host response to these appears to be available in *Sorghum sudanense* and species of other section. *S. sudanense* belongs to the primary genepool and easily crosses with sorghum but not the species from the other sections that are in the tertiary genepool. It seems that pre-fertilization crossing-barriers exist, owing to either lack of pollen germination or very slow and irregular pollen tube growth. *Sorghum timorense* shows immunity to downy mildew. However, crossing barriers are still hindring an easy gene transfer to the breeding pool

Pearl Millet

The four major diseases of this crop are downy mildew, smut, ergot and rust. Though sources of resistance to these diseases are available in the cultivated species, the resistances break at high inoculums or at early inoculation. The inbred line 'Tift 85' was developed by transforming dominant resistance genes to rust and leaf spot from *P. glaucum* ssp. *monodii.* This inbred line also led to a new source of cytoplasmic male sterility in pearl millet hybrid breeding. Napier grass (*P. purpureum*), which belongs to the secondary genepool and whose progeny crosses with pearl millet is often sterile hybrid, possesses resistance to many pests plus an outstanding forage yield potential. Genes for earliness, long inflorescens, leaf size and male fertility restoration were transferred from this wild species to pearlmillet. Apomixis has been restored in the tertiary genepool of pearlmillet, but crosses with species of this genepool are difficult. Besides apomixis, other interesting characteristics available in this genepool are perennial growth habit, drought and cold resistance, pest resistance and new cytoplasm sources.

Some researchers (Dujardin and Hanna, 1986) have developed interspecific hybrids of this gene pool with pearlmillet through ploidy manipulations. Pollen from species such as *P. ramosum* or *P. mezianum* may produce shrivelled and immature seeds, and thus they do not germinate. Embryo rescue may be an alternative path to obtain interspecific hybrids in crosses between the tertiary genepool and pearl millet.

Asian *Vigna* species

The Asian *Vigna* in the subgenus *Ceratotropis*, with 16 to 17 recognized species distributed across Asia, constitute an economically important group of cultivated and wild species for which a rich diversity occurs in India. The cultivated species with conspecific wild forms are urdbean, V. *mungo* (wild types- V. *mungo* var. *silvestris*); mungbean. V. *radiata* (more diversity in wild types- V. *radiata* var. *sublobata* and V. *radiata* var. *setulosa*); rice bean, V. *umbellata* (wild type occurring, conspecific with cultigen typesV. *umbellata* var. *gracilis*); mothbean, V. *aconitifolia* (wild occurring variability conspecific with cultigen types) and the semi-domesticated V. *trilobata* (wild occurring variability conspecific with cultigen type).

Mungbean yellow mosaic virus (MYMV) has been a major problem in mungbean. The wild species *V. radiata* var. *sublobata* has been an important source to incorporate resistance into cultivated varieties (Singh, 1990). Some accessions of wild *V. radiata* var. *sublobata* have also been reported to be highly resistant to bruchid, *Callosobruchus chinensis* (L.) at the AVRDC.

Wild Species Domesticated as Potential Commercial Crop

The ancestral species had undergone selection through domestication for alternative purposes, resulting in different types of species. For example in *Brassica* species many crops developed as a result of selection for different edible parts, that is, cabbage (*Brassica oleracea* var. *capitata*), kale (*Brassica oleracea* var. *alboglabra*) for leaves, kohlarabi (*Brassica oleracea* var. *gongylodes*) for stem, broccoli (*Brassica oleracea* var. *italica*), cauliflower (*Brassica oleracea* var. *botrytis*) for shoots and brussel sprouts (*Brassica oleracea* var. *gemmifera*) for buds.

The new species and those having potential value have resulted in the identification of several promising types which can adapt themselves to the harsh environmental and degraded soil conditions, give economic yield in different agro-climatic regions and, thereby, ensure food and nutritional security, and provides additional income to the resource poor farmers in remote, backward, tribal, hilly and other difficult areas of the country. Consequently, there has been an emphasis on their germplasm collection, introduction, evaluation and utilization.

Some Important Concerns

There is need to study the reproductive biology of various wild relatives of different crop plants for their judicious utilization for the transfer of desirable genes. Because hybrids obtained between cultivated and wild species. The first generation hybrids are generally partially sterile and after a few cycles of selection they continued to be sterile, low yielder or poor quality characteristics of hybrids derivatives. In the past many programmes were abandoned due to these problems. Proper screening of the wild and weedy forms for identifications of various traits is essential for effective utilization in the crop improvement programmes. Because the potentials of wild germplasm are unknown until collections are screened and tested *e.g.* only one accession of *Oryza nivera* was found resistant to grassy stunt virus after thousands of cultivated and thousands of wild accessions were screened. The emphasis should also be given to ethnobotanical studies as well as traditional uses of the wild species occur in different eco-geographical regions, to exploits their potential particularly for medicinal purposes so that these can be domesticated for cultivation after few selection cycles. The proper understanding of gene centers of diversity and species relationship is also needed for collection and to enhancement of the germplasm. Suitable measures should also be taken for the maintenance/conservation of the wild relatives of crop plants to get maximum benefit.

References

Arora, R K, K L Mehra and M W Hardas. 1975. The Indian gene centre: prospects for exploration and collection of herbage grasses. Forage Res. 1: 11-12.

Bhagmal, K L Mehra and D S Katiyar. 1970. Heritable and non-heritable variability in fodder oats. SABRAO Newsletter 2 (1) : 31-36.

Bhagmal, K L Mehra, M L Magoon and D S Katiyar. 1975. Heritability, Correlations and factor analysis of fodder yield components in Sorghum, Indian J. Hered. 7 (1): 59-67.

Bor, N L 1960. The Grasses of Burma, Celylon, Indain an Pakistan, Pergamon Press, Londan.

Borthakur, D N. 1992. Agriculture of the North Eastern Region with Special Reference to Hill Agriculture, Beecee Prakashan, Guwahati.

Chatterjee, D. 1939a. Studies on the endemic flora of India and Burma. J. Royal Asiatic Soc. Bengal Sci. 5: 16-67.

Chatterjee, D. 1939b. Studies on the endemic flora of India and Burma. J. Royal Asiat. Soc. Beng. Sci. 5 : 19-67.

Chopra, D P, T A Thomas, K L Mehra and R Singh. 1981. Catalogue on moth bean germplasm (Series 2). NBPGR. Publ. Regional Station. Jodhpur, 27 p.

Chopra, D P, T A Thomas, K L Mehra and R Singh. 1983. Catalogue on guar germplasm, NBPGR Publ. Regional Station, Jodhpur. No. 3 : 55 p.

Coubey, R N, P K Katiyar, P Saxena, S Singh and N K Shah. 1999. IGFRI Teosinte Germplasm Catalogue, IGFRI, Jhansi, 41 p.

Dabadghao. P M and K A Shankarnarayan. 1972. The Grasscover of India, ICAR, New Delhi, 713 p.

Dabas, B S, T A Thomas and K L Mehra. 1981. Catalogue on guar [Cyamopsis tetragonoloba (L.) Taub.] germplasm. NBPGR Publ., New Delhi, 148 p.

Dujardin, M. and W. W. Hanna. 1986. An apomictic polyhaploid obtained from a pearl millet x *Pennisetum squamulatum* apomictic interspecific hybrid. *Theor. Appl. Genet*. 72:33-36.

Dujardin, M. and W. W. Hanna. 1986. An apomictic polyhaploid obtained from a pearl millet x *Pennisetum squamulatum* apomictic interspecific hybrid. Theor. Appl. Genet. 72:33-36.

Harlan, J. R. 1976. Genetic resources in wild relatives of crops. Crop Sci. 16:329-333.

Hawkes, J. G. 1977. The importance of wild germplasm in plant breeding. *Euphytica* 26:615-621.

IARI, 1970. Collection, maintenance and assessment of indigenous, cultivated and wild genetic stocks of grasses and leumes for food, forage and conservation purposes. Final Technical Report. PL. 480 Scheme, Division of Plant Introduction, IARI, New Delhi.

Khoshoo, T N. 1996. Concern mounts over repid erosion of Himalayan biodiversity. Diversity 12(3) 24-25.

Khoshoo, T. N. 1988. Conservation of Biological Diversity. Plant genetic Resources Indian Perspective Ed Paroda R.S., R.K. Arora and K.P.S. Chandel pp76-87 NBPGRpub.

Khoshoo, T. N. 1988. Conservation of Biological Diversity. Plant genetic Resources Indian Perspective Ed Paroda R.S., R.K. Arora and K.P.S. Chandel pp76-87 NBPGRpub.

Kohli, K S, C B Singh, A Singh, K L Mehra and M L Magoon. 1971. Variability of quantitative characters in a world collection of cowpea: inter-regional comparison. Genetica Agraria 25 : 231-242.

Ladizinsky, G. 1988. Biological species and wild genetic resource in oats. *In* B. Mattsson and R. Lyagen (eds.) Proc. 3rd Intl. Oat Conf., Svalof AB, Sweeden..

Ladizinsky, G. 1988. Biological species and wild genetic resource in oats. In B. Mattsson and R. Lyagen (eds.) Proc. 3rd Intl. Oat Conf., Svalof AB, Sweeden..

Mc Fadden, E. S. 1930. A successful transfer of emmer characters to vulgare Wheat. *J. Am. Soc. Agron.* 22, 1020-1034.

Mc Fadden, E. S. 1930. A successful transfer of emmer characters to vulgare Wheat. J. Am. Soc. Agron. 22, 1020-1034.

Mc Fadden, E. S. 1930. A successful transfer of emmer characters to vulgare Wheat. *J. Am. Soc. Agron.* 22, 1020-1034.

Mehra, K L and R S Paroda. 1986. Genetic resources activities in forage plants. In : R S Paroda, R K Arora and K P S Chandel (Eds.) Plant Genetic Resources : Indian Perspective, NBPGR,

Mehra, K L. 1964. The *Dichanthium annulatum* complex I., Origin and artificial synthesis of

Mehra, K L. 1978. Oats. ICAR Publ. New Delhi. 152 p.

Mehra, K L.1962. The *Dichanthium annulatum* complex I., Morphology. Phyton. 18 (1): 87-94.

Mehra, K Land M L Magoon. 1973a. Collection, conservation and exchange of gene pools of forage crops. SABRAO Second Cong. Indian. J. Genet. 34 A : 26-32.

Melkania, N P. 1988. Status and potentials of sub-alpine and alpine meadows – A Central Himalayan study. In : P. Singh (Ed.) Pasture and Forage Crops Research – A State of Knowledge Report, Range Management Society of India, IGFRI, Jhansi, pp. 97-110.

MOFF, 1994. Ethnobiology of India. A Status Report. Ministry of Environment and Forests, Govt. of India, New Delhi.

Murthy, B R 1962. Cataloguing and Classifying Genetic Stocks of Sorghum. Final Technial Report. PL. 480 Scheme, IARI, New Delhi.

Murthy, B R, M K Upadhya and P L Manchanda. 1967. World collection of genetic stocks of Pennisetum. Indian J. Genet. 27A : 313-394.

NBPGR, 1984. Annual Report. expand, New Delhi.

Pant, K C, K P S Chandel and B S Joshi. 1982. Analysis of diversity in Indian cowpea genetic resources. SABRAO Journal 14(2): 103-111.

Paroda, R S, R K Arora and K P S Chandel. 1986. Plant Genetic Resources : Indian Perspective, NBPGR, New Delhi, India.

Patil, B D and A Singh. 1963. Studies on Anjan grass, Cenchrus setigerus Vahl. Interselection variation for forage and seed production. Indian J. agric. Sci. 33 (1): 44-51.

Reeves, R. G., 1950. The use of Teosinete in the improvement of corn inbreds. Agron. J. 42:248-251

Seetaram, A. 1987. Genetic resources of small millets in India. Nat. Symp. on Plant Genetic Resources, March 3-6, 1987, NBPGR, New Delhi.

Sharma, H. C., G. Pampathy and L. J. Reddy. 2003. Wild relatives of pigeonpea as a sources of resistance to the pod fly (*Melanagromyza obtuse* Malloch) and pod wasp (*Taraostignodes cajanianae* La Salle). *Genetic Resources and Crop Evolution* 50:817-824.

Sharma, H. C., G. Pampathy and L. J. Reddy. 2003. Wild relatives of pigeonpea as a sources of resistance to the pod fly (Melanagromyza obtuse Malloch) and pod wasp Taraostignodes cajanianae La Salle). Genetic Resources and Crop Evolution 50:817-824.

Singh RV, Deep Chand, Vandana Tyagi, AK Singh, SP Singh and Surender Singh (2004) Important Crop Germplasm Introduced into India during 2004 Indian J.Plant Genet. Resour. 17 (3): 196-205.

Singh RV, Deep Chand, Vandana Tyagi, AK Singh, SP Singh and Surender Singh 2003 Important Crop Germplasm Introduced into India during 2003 Indian J.Plant Genet. Resour. 16 (3): 214-221.

Singh, A.K. JP Moss, and J. Smartt. 1990. Ploidy Manipulations for Interspecific Gene Transfer. *Advances in Agronomy* 43:191-240

Singh, I P and P L Gautam. 1999. Chhara Fasalo ke sudhar hetu prajanan sansadhan. In : P S Tomer, B K Trivedi, S N Tripathi and J N Gupta (Eds.) Chhara Anusandhan Avm Pasudhan vikas Ke Naye Aayam (In Hindi), IGFRI, Jhansi, pp. 15-26.

Singh, I P and P L Gautam. 1999. Chhara Fasalo ke sudhar hetu prajanan sansadhan. In : P S Tomer, B K Trivedi, S N Tripathi and J N Gupta (Eds.) Chhara Anusandhan Avm Pasudhan vikas Ke Naye Aayam (In Hindi), IGFRI, Jhansi, pp. 15-26.

Singh, J and J V Williams. 1984. Maintenance and multiplication of plant genetic resources. In : YJHW Holden and J T Williams (Eds.) Crop Genetic Resources : Conservation and Evaluation, IPGR, A. Johns Allen and Unvin, Ltd.

Stalker, H. T. 1980. Utilization of wild species for crop improvement. *Adv. Agron.* 33:111-147.

Stalker, H. T. 1980. Utilization of wild species for crop improvement. *Adv. Agron.* 33:111-147.

Sundararaj, D. 1967, FAI Technical Conference on Exploration. Utilization and Conservation of Plant Gene Resources : Exploration and collection of Tropical Plants. FAO, Rome, Italy, 19 p.

Swaminathan, M S and S Jana. 1992. Biodiversity : Implications for Global Food Security, MacMillon, Madras, India.

Tyagi Vandana, AK Singh, Deep Chand, RV Singh and BS Dhillon. 2008. Plant Introduction in India during Pre-and Post-CBD periods- An analysis Indian J. Plant Genet. Resour. 19 (3): 436-441.

Vavilov, N I. 1950. The origin, variation, Immunity and breeding of cultivated plants. Chron. Bot. 13 : 364

Whyte, R O. 1964. The Grassland and Fodder Resources of India. ICAR. Sci. Monogr. 33. ICAR, New Delhi.

Yadav, M S. 1974. Genetic variability and correlations among fodder yield components in pasture grass, Pennisetum pedicellatue. Indian Forester. 100 (2): 108-117.

Yadav, M S. 1981. Gene environment response for fodder yield and quality components in Cenchrus complex. Ph. D. Thesis, Jiwaji Univ., Gwalior, M.P.

Yadav, M S. 1985. Genetic variability and productivity of pasture grasses. Proc. Summer Instiute on Recent Advances in Forage Breeding for Farming Systems, IGFRI, Jhansi.

Chapter 5

New Potential Forage/Fodder Crops for India

Aparajita Mohanty[1], Nidhi Verma[2] and Babeeta Chrungu Kaula[3]

[1]*Department of Botany, Gargi College, Siri Fort Road, New Delhi – 110 049*

[2]*National Bureau of Plant Genetic Resources, IARI Campus, New Delhi – 110 012*

[3]*Zakir Husain College, Jawaharlal Nehru Marg, New Delhi – 110 002*

Introduction

Importance of forage and fodder crops for livestock as well as wild herbivorous animals is well established in India. Livestock form a crucial part of Indian agriculture and wild animals are important component of biodiversity. However, poor health of wild animals and low productivity of livestock due to lack of proper nourishment of livestock is a nagging problem for biodiversity conservators and livestock managers respectively. One of the important reasons behind poor nourishment of livestock is production of poor quality fodder. There is also an acute shortage of fodder for livestock. The huge gap between demand and supply of fodder is evident from Table 5.1.

Table 5.1: Demand, Supply and Deficits of Fodder in the Country (million tonnes): Past, Present and Future

Year	Supply		Demand		Deficit as % Demand	
	Green	Dry	Green	Dry	Green	Dry
2003	387.7	437.3	1006	560.1	61.51	21.81
2005	389.8	441.6	1021	568.0	61.83	22.12
2010	395.2	452.7	1057	588.2	62.63	22.91
2020	406.6	475.7	1134	630.9	64.26	24.57

Source: Handbook of Agriculture, 2006.

Therefore, there is an urgent need for increment in supply and for improved quality production of forage crops. For this some of the measures that may be considered are:

1. Increase in acreage under forage production.
2. Increase in yield by developing new varieties.
3. Improvement in nutritional quality and other traits *e.g.*, resistance for biotic and abiotic stresses.
4. Search for new potential forage crops which can supplement the existing list of forage crops.

The new potential forage crops maybe those crops or plants already in cultivation or they can be wild and weedy allies of existing crops. There are multiple advantages of introducing wild and weedy relatives of existing forage crops, the chief amongst them being:

1. Wild species are comparatively more resistant to biotic and abiotic stresses than their cultivated counterparts; hence they will be more successful ecologically.
2. Such species grow in wilderness, so agricultural inputs like water supply, fertilizers, pesticides etc and manual labour required will be much less.
3. Wild and weedy relatives of existing forage crops can undergo natural hybridization or can be crossed artificially to broaden the genetic base of forage plants.

Based on the ecological requirement of the wild species as potential forage crops, they can be grown/introduced in the following areas:

Grasslands

Several grasses like guinea grass and its wild relatives like *P. antidotale*, *P. coloratum*, *P. lanipe*, *Poa pratansis*, *Poa annua* etc can be tried as potential forage grasses.

Sanctuaries and Biodiversity Parks

It has been observed that elephants preferred certain forage plants to others (Santra *et al.*, 2008). This selective preference could be correlated to the nutritional quality of the forage plants. Such studies can suggest the choice of potential forage plants to be grown in biodiversity parks and national parks.

Wastelands

The best forage plants that can help in reclamation of wastelands would be legume forages. *Cytisus scoparius*, a wild shrub is a potential legume which can be used as forage crop in India. Other weedy and wild legumes that can be grown are: *Lathyrus parenne, L. annuus, L. hirsutum, Medicago falcate, Vigna radicans, Vicia hirsuta* etc.

Marshy Lands

Saltgrass or *Distichlis spicata* is a salt resistant potential forage crop (Bustan *et al.*, 2005) which may be tried on Indian soil.

Agricultural Land and Surrounding Areas

It is important to assess the extent of competition for the pollinators between the forage crops and the actual cultivated crop, when any new potential wild forage species are tried out near agricultural fields.

Brief Description of Some Potential Wild and Weedy Forage Species

The perspectives and strategies of genetic resources of Indian forage crops have been described by Singh *et al.*, 2009. The genera which include forage species and their wild and weedy relative that can be further assessed for utilization at a larger scale has been listed in Table 5.2. Some of the important potential forage species are briefly described here.

Table 5.2: List of Genera and Some of their Wild and Weedy Relatives that have been/can be Used as Potential Forage

Genus	*Species including Wild and Weedy Relatives*
Acacia sp.	*A. difficilis, A. mangiu, A. torulosa* and *A. tumida var. tumida*
Agropyron sp.	*Agropyron cristateum, A. cimmericum*
Astragalus sp.	*Astragalus alpinus, Astragalus adsurgens*
Avena sp.	*A. abyssinica kochisi, A. barleota, A. byzantina, A. canariensis, A. claida, A. faba, A. hirtula, A. insalavis, A. longiglumis, A. lucovicana, A. macrostachya, A. pilosa, A. saviloviana, A. strigosa* and *A. weistii*
Brachiaria sp.	*B. ruziziensis*
Brassica sp.	*Brassica abyssinica B. orientalis*
Centrosema	*Centrosema pubescens*
Casuarina sp.	*Casuarina cunninghamiana* ssp. *cunninghamiana*
Chenopodium sp.	*C. capitatum, C. gigantenum, C. pallidicaule* and *C. quinoa*
Clitoria sp.	*Clitoria ternetea*
Chloris sp.	*Chloris gayana, Chloris dactylis*

Contd...

Table 5.2–Contd...

Genus	Species including Wild and Weedy Relatives
Cyamopsis sp.	*C. senegalensis* and *C. serrata*
Dactylis sp.	*Dactylis glomerata,*
Desmodium sp.	*Desmodium barbatum, D. intortum* and *D. uncinotum*
Eleusine sp.	*Eleusine indica, E. africana E. fleceifolia, E. jaegeri, E. multiflora* and *E. tristachya*
Eragrostis sp.	*Eragrostis curvul*
Elymus sp.	*Elymus haffmannii*
Eucalyptus sp.	*E. argophloia, E. camaldulens var. obtusa, E. tereticornis* ssp. *tereticornis*
Festuca sp.	*Festuca arundinacea*
Glycine sp.	*G. argyrea, G. canescens, G. centennial, G. clandestina, G. cyrtoloba, G. falcata, G. javanica, G. latifolia, G. latrobeana, G. microphylla, G. soja, G. tabacina* and *G. tomentella*
Hevea sp.	*H. camarguana, H. colina, H. fousiflora, H. guyanensis, H. nitida* and *H. pauciflora*

- ☆ *Avena abyssinica*: It is an annual plant. The plant requires well-drained soil and can grow in heavy clay soil. They can grow on acid, neutral and basic soils. It cannot grow in the shade. It requires dry or moist soil and can tolerate drought.
- ☆ *Brachiaria ruziziensis*: It is also known as congosignal grass and can be grown solo or as an intercrop. It is a creeping perennial with dense foliage. It prefers tropical climate and can tolerate shade. However, it cannot tolerate water logging.
- ☆ *Brassica orientalis*: It is an annual growing up to 0.45m. It requires moist soil.The plant can tolerate strong winds but not maritime exposure. It can grow on very clayey soils.
- ☆ *Chenopodium giganteum*: It is an annual growing up to 2.4m. The flowers are hermaphrodite and are pollinated by wind. It cannot grow in shade and requires moist soil.
- ☆ *Desmodium intortum*: It is an annual/perennial plant and may grow as a herb or a vine. Nitrogen fixation capacity is very high and has a long lifespan. Its palatability to grazing animals are high and can grow on all types of textured soil.
- ☆ *Eleusine indica*: It is a monocot weed in Poaceae family and is commonly known as goosegrass. In Brazil these weed first evolved resistance to Group A/1 herbicides in 2003 and infests soybean.
- ☆ *Festuca arundinaceae*: The wild varieties are seen growing native in Baltic states. It is grown for creation of rough lawns and can be used for fodder.
- ☆ *Glycine canescens*: This wild relative of the legume *Glycine max* is resistant to powdery mildew caused by *Microsphaera diffusa*. It is seed oil; protein and

fatty acid content are similar to the other species of the subgenus (Newell and Hymowitz, 1975)

- *Melilotus indicus*: This less exploited legume species can be utilized as potential forage crop where environmental stresses restrict the use of more conventional forage crops. This species is commonly known as yellow sweet clover and it occurs as a weed in different habitats in Egypt. It grows in moderately saline areas, where traditional forage legumes cannot be cultivated.
- *Poa pratensis*: It is known as blue grass. These grow wild all over Russia.
- *Stylosanthes scabra*: This plant is commonly known as shrubby. It is a perennial herb with deep root system. This enables the plant to remain green in dry seasons as well.
- *Setaria anceps*: It is also known as Golden Timothy. This grass can grow in tropics and subtropics with medium rainfall. It is a tufted perennial grass with erect stems and grows up to 2m. It is cold tolerant and at the same time survive long hot and dry seasons.
- *Distichlis spicata*: This is a halophytic grass which is salt resistant. It may be initially introduced at the primary stages of land reclamation during which it can provide fodder to livestock. There are several ecotypes of this salt grass which may be tried for their suitability in different environments (Bustan *et al.*, 2005).
- *Cichorium intybus*: It is commonly known as chicory and is a dominant weed in fields where *Trifolium alexandrianum* grows. Chicory is a potential forage crop with its higher organic matter digestibility (Rao S, 2008). It can grow on low fertile soils and can tolerate a wide range of pH (4.8-6.5).
- *Commelina diffusa*: This weed has been evaluated for its potential as a ruminant feed (Lanyasunya *et al.*, 2006). It is eaten by cows in rural Mauritius. In moist climate, this species regenerates fast and ensure a sustainable source of nutrients to cattle.

References

Bustan A, Pasternak D, Pirogova I, Durikov M, Devries T, El-Meccawi S, Degen AA (2005). Evaluation of saltgrass as a fodder crop for livestock. *Journal of Science of Food and Agriculture*. 85: 2077-2084.

Lanyasunya TP, Wang HR, Abdulrazak SA, Mukisira EA and Zhang jie (2006). The potential of the weed, *Commelina diffusa* L. as a fodder crop for ruminants. *South African Journal of Animal Sciences* 36: 28-32.

Newell CA and Hymowitz T (1975). *Glycine canescens* F.J. Herm., a Wild Relative of the Soybean. *Crop Science* 15: 879-881.

Rao DS (2008). Chicory-a case of weed into forage crop transition. *Current Science* 95: 1532-1533.

Santra AK, Pan S, Samanta AK, Das S and Halder S (2008). Nutritional status of forage plants and their use by wild elephants in South West Bengal, India. *Tropical Ecology*. 49 (2): 251-257.

Singh AK, Verma N, Yadav SK, Mohanty A, Singh SP and Singh S (2009). Indian Forage Genetic Resources: Perspective and Strategies. *Prog. Agri.* 9 (2): 250-256.

Chapter 6
Ethnobotany Related to Forages and Livestock

Anjali Kak, Meenakshi Bhardwaj and Vandana Tyagi

National Bureau of Plant Genetic Resources, New Delhi – 110 012

Ethnobotany is the study of the direct relationship between plants and people: Ethnobotany studies the complex relationships between (uses of) plants and cultures. The focus of ethnobotany is on how plants have been or are used, managed and perceived in human societies and includes plants used for food, forage, medicine, divination, cosmetics, dyeing, textiles, for building, tools, currency, clothing, rituals and social life. Ethnobotanical studies are important for the conservation of biological and cultural diversities as well as sustainable utilization of resources. This conservation and sustainable utilization will not be successful without the full participation of indigenous people and the application of their ethnobotanical and ecological knowledge.

Traditional botanical knowledge of Indigenous communities relating to the uses and management of wild plant resources is extensive (Cotton 1997).The need to conserve the biological diversity and its associated indigenous knowledge has been emphasized in the contemporary studies of ethnobotany (Rossato, 1999). Turner, Ignace's (2000) review showed that traditional ecological knowledge of indigenous people has fundamental importance in the management of local resources, in the husbandry of the world's biodiversity, and in providing locally valid models for sustainable life. Other studies (Jin *et al.*, 1999; Luoga, Witkowski 2000) also emphasized documenting indigenous knowledge. This ethnobotany and related traditional knowledge has been in vogue for centuries and extensively practiced. However, much traditional knowledge is being lost before it has been touched by modern science (Huai and Xu, 2000). Many times, the reason being the absence of comprehensive documentation of their traditional practices. The present communication is a comprehensive account of ethnobotany related to forage and livestock. The plants have been classified into two categories.

1. Ethnobotany of forage plants
2. Ethnoveterinary uses of plants

Ethnobotany Related to Forage Plants

Ajuga bracteosa Wall. Ex. Benth. (Neelkanthi), Lamiaceae

Root bark is anti helmintic, antisyphilitic and is used in fever and skin diseases, stomach ache, earache, bronchitis. (Srivastava and Kumar, 2003)

Amaranthus virdis L. (Jungli chulai), Amaranthaceae

Plant is emollient, stomachic, laxative, cooling, cures burning sensation, diarrhea, leuchorrea and gonorrhea. (Paria, 2005)

Boerhavia diffusa L. (Punarnava), Nyctaginaceae

Roots are diuretic and expectorant. Dried root powder applied on the effected organs to treat leprosy, while paste is given to women to hasten delivery (Paria, 2005)

Cyperus rotundus L. (Nut grass), Cyperacceae

Roots are used ti treat leprosy, fever, blood diseases, epilepsy, dyspepsis and urinary diseases. Decoction is given 4-5 times to cure indigestion and dysentery. Juice boiled in ghee and applied as ointment in ulcers (Paria, 2005)

Desmostachya bipinnata (L.) Stapf. (Dubha)

Whole plant juice is used in dysentery, diarrhea, colic worm, bowel complains, cold and cough of children (Bhattacharjee and De, 2006).

Galium aparine L. (Goose grass), Rubiacceae

Plant is used in fever and Vit.C deficiency. (Bhattacharjee and De, 2006).

Medicago lupulina L. (Black Medic), Fabaceae

Plant is used in arthritis, boils, cancer, dysuria, fever, heart ailments and to cure tumours (Duke, 1992).

Oxalis corniculata L. (wood sorrel), Oxalidaceae

Paste of the top shoots with black pepper is appliled to abscess boils, weeping eczema and wounds (Sharma, 2003)

Plantago major R.Br. var plebium, (Tarakmana), Polygonaceae

Fresh shoots are cooked and eaten in stomach disorders, bowel complaints and pneumonia (Sinha, 1996, Singh *et al.*, 2003)

***Euphorbia hirta* L. (Mitnailei), Euphorbiaceae**

The 2-3 drops leaf juice is applied to the eyes to cure redness and burning. (Halam *et al.*, 2008)

***Brassica rapa* L. spp. *campestris* (L) Clapham, (Tilgogul), Brassicaceae**

Oil extracted from the seeds is mixed with salt and sodium bicarbonate and then applied into the tooth cavity to relieve pain. (Ganai and Nawchoo, 2003).

***Physalis minima* L. (Tulatipati), Solanaceae**

The fruits are dried, powdered and rolled in a dry leaf and smoked like a cigar to relieve toothache. (Sharma and Rana, 1999)

***Bidens pillosa* L. (Kuri), Asteraceae**

Dried fruits are boiled to give a typical kind of flavor and colour to the tea. (Bhattachacharya and Chandrakala, 2008)

***Fragaria vesca* L. (wild strawberry), Rosaceae**

Ripe fruits are eaten. They are also used to prepare local wine after fermentation with the help of marcha (yeast) (Bhattacharya and Rai, 2008)

***Nasturtium officinale* R.Br. (Chuch), Brassicaceae**

The leafy stem is eaten as vegetable. The plant is effective against jaundice, cardiac diseases, low blood pressure, gout and tuberculosis

***Setaria italic* (L.) P. Beauv. (Mondia), Poaceae**

Seeds of the mondia are made into powder and a drink is prepared with water and sugar, which is prescribed to the patients of jaundice for 7 days for its cure.

***Cyperus iria* L., Cyperaceae**

Plant is considered stimulant, tonic, astringent and useful in stomachache (Singh *et al.*)

***Polygonum plebeium*, Polygonaceae**

Root extract is applied on head to avoid baldness; plant is also used as manure.

Ethnoveterinary Uses of Plants

Ethnoveterinary medicines are used extensively, and quite effectively for primary health care treatment and maintaining animals productive.

***Terminalia bellerica* (Harad), Combretaceae**

Harad seeds are mixed with mustard oil and given orally to the animal to which heat is to be induced.

***Dendrocalamus strictus* (Bamboo), Poaceae**

The rhizomes of the plant are very rich in minerals, vitamins and are given to induce heat in the cows and buffaloes.

Urtica dioca L. Urticaceae (Cajourina)

The plant is fed to the lactating animal to enhance the milk production in Uttranchal.

Micromeria biflora Benth. Lamiaceae

The plant is given to cure prolapsed of vagina and uterus in cows and buffaloes.

Helinus lanceolatus Brand Rhamnaceae

The plant is given to cure prolapsed of vagina and uterus in cows and buffaloes.

Tricholepis indica (Willd.) S.M.Almeida Asteraceae

The plant is given to cure prolapsed of vagina and uterus in cows and buffaloes.

Acanthus ilicifolius L. Acanthaceae

The plant is used as an effective paster. The paste is prepared by crushing the plant and a thick layer is applied over the broken bone of the livestock as a plaster.

Phaseolus radiatus L. (Syaru) Fabaceae

The plant paste is used as a plastering material to cure the broken bones in the livestocks. To give support to the plaster, 4-5 branches of the plant are tied by a cloth or rope and hence the bone get cured very early.

Ulmus wallichiana Planch (Jaamar, Chamarmeva) Ulmaceae

The plant is used as an effective paster. The paste is prepared by crushing the plant and a thick layer is applied over the broken bone of the livestock as a plaster.

Crataeva nurvala L. Capparaceae

Castor oil is applied on the fractures part of the livestock body and then this part is bandaged with *Crataeva nurvala* plant paste. Fresh paste is applied every third day to enhance the effectiveness of the herb to cure fractured bone.

Bombax ceiba L. (Semul), Bombacaceae

Semul bark is used to cure the haematuria (blood start coming along with urine) problem of cow and buffalo. The semul bark is given along with isabgoal, sugar and water are mixed thoroughly and given orally.

Dalbergia sissoo Roxb. (Shisham), Leguminosae

The shisham leaves are given along with barley flour twice or thrice a day to animal to cure rinder pest disease.

Tephrosia purpurea (L.) Pers. Fabaceae

The leaves of the plant are crushed to extract juice, which is applied with tobacco dust to the skin of the animal against the sucking ticks.

Adhatoda vasica Nees (Basuti), Acanthaceae

The basuti leaves are fed to the cattle to break the habit of drinking their own urine.

Holostemma rheedei **Wall. Asclepiadaceae**

The plant roots are fed along with *Embilia ribes* roots to cure swelling in livestock.

Emblica officinalis **Gaertn Euphorbiaceae**

The dried fruit powder is given along with jiggery twice a day for a month to cure joint pain in animals.

Ocimum gratissimum **L. Lamiaceae**

The tulsi leaves are crushed and extracted juice is applied over the skin allergy of the animal.

Ricinus communis **L.(Castor) Euphorbiaceae**

Leaf and root extract of the plant is used to increase the milk production in milking animals.

Strychnos potatorum **L.f Loganiaceae**

Leaves are crushed and soaked in water and then the mixture is fed to the animal to cure diarrhea.

Ficus hispida **L.f. Moraceae**

Roots of the plant are crushed and soaked in water. The extract is filtered through a cloth and given orally to the animal to cure diarrhea.

Leucas lanata **Benth. Lamiaceae**

The leaves are fed to the animal to cure diarrhea.

Schleichera oleosa **Merr(Kusum) Sapindaceae**

The powder of the seed is applied on the ulcers of hooves for its cure.

Clerodendron phlomides **Burm (Arni) Verbenaceae**

The leaf paste of the plant is applied along with tobacco leaves to cure secondary bacterial infection during foot and mouth disease.

References

Bhattacharjee, S.K and L.L.De. 2005. Medicinal herbs and flowers. Aavishkar Publlisher and Distributors, Jaipur.

Bhattacharya, R.P, Chandrakala, Rai, 2008. Ethnobotanical studies of srikhola Lepchajagat Tonglu and Pankhabari area of Darjeeling district, West Bengal. Journal of Economic and Taxonomic Botany 32(Supplement): 65-71.

Cotton, C.M. 1997.Ethnobotany. Principles and applications. John Wiley & Sons, Chichester,UK.

Duke, A. James, 1992. Handbook of Edible Weeds.

Ganai, K.A and Nawchoo, I. A. 2003. Traditional treatment of toothache by the Gujjar and Bakerwal tribes of Kashmir in India. Journal of Economic and Taxonomic Botany. 27(1): 105-107.

Halam, R., Saha R. and Datta, B.K. 2008. An ethnobotanical study of Halam tribes of Dhalai district, Tripura, India. Journal of Economic and Taxonomic Botany. 32(Supplement):8-12.

Huai, H.Y.and Xu, J.C., Indigenous knowledge: An inexhaustible information bank to taxin research, *Toxicon*, 2000, 38, 745-746.

Jin, C.,S. Yin-Chun, C. Gui-Qin, and W. Wen-Dun 1999.Ethnobotanical studies on wild edible fruits in southern Yunnan: Folk names, nutritional value and uses. Economic Botany 53(1):2-14.

Luoga, E.J.,T.F.Witkowski, and K. Balkwil. 2000. Differential utilization and ethnobotany of trees in Kitulanghalo forest reserve and surrounding communal lands, eastern Tanzania. Economic Botany 54(3):328-343.

Paria, N.D. 2005. Medicinal plants Resources of south west Bengal. Directorate of Forest Govt. of west Bengal, Kolkatta.

Rosatto, S. C., H. de F.Leitao-Filho, and A.Begossi. 1999. Ethnobotany of Caicaras of the Atlantic forest coast (Brazil). Economic Botany 53(4):387-395.

Sharma, B.D and Rana, J.C. 1999. Traditional Medicinal used of plants of Himachal Pradesh. Journal of Economic and Taxonomic Botany 23(1): 173-176.

Sharma, Ravindra, 2003. Medicinal Plants of India: An encyclopedia. Daya Publishing House, delhi.

Singh, H.B.,R.S Singh and J.S. Sandhu, 2003. Herbal medicinal plants of Manipur. Daya Publishing House, Delhi.

Singh, Lokesh, Sharma, Neelam, Joshi, S.P, Manhas, R.K and Joshi, Venita. @008. Ethnobotanical uses of some weeds in some agro-ecosystems of Doon valley. Journal of Economic and Taxonomic Botany 32(Suppl.) 97-103.

Srivastava, N and Kumar, S. 2003. Drug plant resources of Doon valley. Annals for. 11(1): 63-84.

Turner, N.J., M.B. Ignace, and R. Ignace. 2000. Traditional ecological knowledge and wisdom of aboriginal people in British Columbia. Ecological Applications 10:1275-1287.

Chapter 7

Flow of Forages/Fodders Genetic Resources in India: Impact Due to Various International Treaties/ Laws/Rules

Vandana Tyagi, Anjali Kak and Vandana Joshi

National Bureau of Plant Genetic Resources, New Delhi – 110 012

Introduction

Flow/exchange of genetic resources offer enormous opportunity for better economic growth, potential stability, human health and for a sustainable environment. There is a continuous need of genetic resources for developing varieties resistant to various pests and diseases and to improve quality and quantity. Genetic resources constitute undeniable interdependence between countries and continents. Even biodiversity rich regions depend for more than 30 per cent of their food production on crops originating from other countries (Cooper, 1994) and this interdependence plays a very important role in international collection and exchange of germplasm.

Forage is any vegetative material eaten by livestock. It includes live grasses and legumes grazed directly by pasture animals as well as cut-and-carry biomass and fodder such as hay, leaves, shredded sugarcane, chopped maize cobs, and dried cassava chips. As global demand for meat, milk, and other animal products grows dramatically in the coming decades, so too will the need for improved forages. Small-scale livestock producers in developing countries, faced with stiff competition from highly efficient industrial operations, both domestic and foreign, will have to look at new technical options, such as combination of superior grasses and legumes, to replace native pasture forages, which tend to be low in nutritional value. Hence, it becomes imperative that adequate efforts are made to develop high yielding varieties and transfer the available technology to farmer's field in order to boost the forage productivity and production in the country. Acquisition of diverse and superior germplasm in forage germplasm and their conservation is thus an important concern.

The forage species are distributed widely in different agro-ecological regions of India. The concentration of species lies in the (i) tropical humid belt of south- western India, particularly for tropical types (ii) north-west arid/semi arid belt and north-eastern moist belt for subtropical species and (iii) western Himalayan tract for the temperate species. Thus, these regions offer great opportunity for the collection of indigenous grasses and legume genetic resources for forages and pasture lands. The four major tropical and subtropical types reported to occur in India are species of *Dicanthium, Cenchrus, Lasiurus, Sehima, Phragmites, Saccharum, Imperata, Themada* and *Arudinella.* Fodder cereals include oats *Avena sativa* and *A. strigosa,* maize *Zea mays,* sorghum *Sorghum bicolor,* pearl millet *Pennisetum americanum,* barley *Hordeum sativum,* rye *Secale cereale,* proso millet *Panicum miliaceum* and finger millet *Eleusine coracana. Saccharum officinarum* is also used as forage.

National Bureau of Plant Genetic Resources (NBPGR) is the nodal institution in India, which has the mandate of germplasm exchange for research including introduction. The Bureau introduces plant genetic resources of forages on the basis of specific requests from the scientists across the country and also at the initiative of its own scientists, from foreign countries, IARCs and other agencies with which the country has joint protocols/memoranda of understanding (MoUs) on reciprocal basis. NBPGR has linkages with over 115 countries. However, the scenario has changed from free flow to limited access of genetic resources in view of national and international agreements; intellectual property rights (IPR) and expanded scope of these agreements to plant varieties. Thus, it is very important to understand these issues in the changing scenario. For regulated flow of genetic resources developing countries need to understand the implications of IPR relevant to genetic resources.

Historical Perspectives of Exchange and Introduction of Exotic Germplasm in Forage Crops

In the past, agriculture has been tremendously benefited through unrestricted exchange, utilization and marketing of germplasm. Farmers were the breeders and conservers of the seed material. They had full rights to do anything with the seeds without restriction. Thus every country gained more than what it contributed. It promised self sufficiency in food, food security and improved economy. Green revolution was made possible under such international germplasm exchange. Earlier biodiversity including PGR, was regarded a common heritage of mankind, (everybody should have access to the resources to meet the basic needs). The free availability of germplasm especially from *ex situ* collection of CGIAR centres helped most countries to strengthen crop improvement programs. India has benefited tremendously by importing forage/fodder germplasm, varieties and new crop species from other countries, which have improved productivity and lead to diversification of crop species and had tremendous impact on Indian agriculture (Brahmi 2005).

The NBPGR continues to work towards identifying promising trait specific germplasm through literature search and personal contacts and introduce the same for utilization by Indian plant breeders. Since 1976 a total of 13,181 accessions were imported/introduced mainly from countries *viz.*, Australia, Brazil, Italy, Japan, Kenya, New Zealand, Russia, UK and USA. Major genera introduced are namely- *Agropyron, Amblyopyrum, Austialopyrum, Avena, Brachiaria, Clitoria, Cnidoscylus, Corymbia, Dasypyrum, Desmanthus, Desmodium, Elymus, Elytrigia, Eragrostis, Eremspyrum, Eucalyptus, Festuca, Henradia Hordelymus, Hordeum, Lespedeza, Leymus, Lolium, Lotus, Macroptilum Medicago, Melilotus, Pascopyrum, Poa, Psathyrostachys, Pseudorengeria, Sorghum, Taeniatherum, Thinopyron, Trifolium, Trigonella* and *Triticale.* A large number of varieties of fodder crops were developed in India using the *exotic germplasm.* These include Giant Bajra (evolved through a cross between Australian bajra and local bajra) of *Pennisetum americanum;* Pusa Deenanath Grass (a selection from African material) of *Pennisetum pedicellatum;* Punjab Guinea Grass-1 (an introduction from Australia under the name CPI-59985) of *Panicum maximum,* Nandi (selection from African germplasm) of *Setaria anceps;* Marwar Dhaman (a clonal selection from exotic material EC 017655) of *Cenchrus setigerus;* Marwar Anjan (a clonal selection from exotic material EC 014369) of *Cenchrus ciliaris;* Kent (an introduction from USA), UPO-90 (a single plant selection from American material) of *Avena sativa;* Mescavi and Fahl (introductions from Egypt), BL-1 (a selection from mescavi) of *Trifolium alexandrium;* Kohinoor (a selection from material obtained from Iran) and UPC-5286 (a single plant selection from germplasm line 5286) of *Vigna unguiculata,* etc.

Wild species are generally more variable than the corresponding crop. Plant breeders utilize wild species mainly for sources of resistance to biotic and abiotic stresses as well as are breeding for genetic enhancement. 986 wild species in forage crops were also introduced are detailed in Table 7.1.

Table 7.1: Wild Species Introduced in Forage Crops (1976-2008)

Genus	Species
Agropyron	*A. cimmericum*
Brachiaria	*B. ruziziensis*
Clitoria	*Clitoria ternetea*
Desmodium	*D. intortum* and *D. uncinotum*
Elymus	*Elymus haffmannii*
Lotus	*L. conimbricensis, L. coribricensis, L. ornithopodioid, L. pendunoulatus, L. purshianus* and *L. uliginosus*
Macroptilium	*M. atropurpureum* and *M. axillare*
Medicago	*M. arabica, M. ciliaris, M. coronata, M. intertexta, laciniata, M. littoralis, M. lupulina, M. minima, M. orbicularis, M. polymorpha, M. radiata, M. rigidula, M. rugosa, M. scutellata, M. tornata, M. truncatula, M. turbinate, M. liciruata* and *M.littoralis*

Contd...

Table 7.1–Contd..

Genus	Species
Melilotus	*M. indicus, M. messanensis, M. officinalis, M. segetalis, M. sulcatus*
Stylosanthes	*S. scabra* and *S. viscosa*
Trifolium	*T. alpestre, T. angustifolium, T. arvense, T. aureum, T. baccarinii, T. balansaa, T. billardiari, T. campestre, T. clypeatum, T. dasyurum, T. diffusum, T. fomentosum, T. fragiferum, T. grandiflorum, T. incarnatum, T. lappaceum, T. leucanthum, T. ligusticum, T. michelianum, T. nigrescens, T. pilulare, T. scabrum, T. squamosum, T. striatum, T. striatum var. spinescens, T. subterraneum* and *T. tomentosum*

However when CBD came into force, things changed fast and the paradigm shift from free flow of genetic resources to a restricted exchange was introduced. The concept behind was, that profits accrued from developing commercial products such as new varieties or new compounds should be shared with the provider of the genetic material/resources used to develop the product and thus various issues related to access of germplasm came into forefront.

Enforcement of CBD, 1992 and provision of TRIPS in 1990 led to apprehension that exchange of germplasm would get restricted and thus an analysis was done to interpret the impact of these on the flow of genetic resources (Table 7.2). The study was done under two parts pre CBD period (1988-1992) and post-CBD period (1997-2001). For this study, period in between was not considered to avoid the effects of transit phase. Pre CBD, a total of 3,303 accessions was introduced however post-CBD the number dropped to 1645. Thus depicting that of the total forage germplasm introduced, 67 per cent was received pre-CBD and 33 per cent post (CBD) since forage crops are not being maintained by CG centres, which are the major source of accessions introduced or imported for various crop improvement programmes. For procurement of forage crops, one has to search for germplasm sources in the National Genebank of other countries. Amongst them, USDA is the main supplier of germplasm to India and continues to be the major contributor (Tyagi, 2006).

Table 7.2: Introduction of Germplasm in Forage Crops

Pre- CBD (1988-1992)		*Post- CBD (1997-2001)*	
Origin Different from the Supplier Country	*Origin Same as the Supplier Country*	*Origin Different from the Supplier Country*	*Origin Same as the Supplier Country*
2452	851	1352	293
74.25	25.77	82.81	17.82

Total Germplasm Introduced–Accession (numbers)	
Pre- CBD	*Post- CBD*
3302	1645
67 per cent	33 per cent

Contd...

Table 7.2–Contd..

Origin Different from the Supplier Country	Origin Same as the Supplier Country
3804	1144
76.87	23.13

An interested finding is that 76.87 per cent of total germplasm introduced was imported from countries which were not the centre of origin of that particular crop, *i.e.* these were introduced from other sources and conserved in their genebanks and only 23.13 per cent was received directly from the countries or the centres of origin of that particular crop (Table 7.2). Regular flow of material is from nations which have conserved accessions in their gene banks. The trend may however change with the facilitated access to plant genetic resources under legally binding Treaty on PGRFA.

In the present scenario, where the access to germplasm is restricted due to many national and international treaties/Acts, it is very important to search for and introduce on priority trait specific germplasm for use in various crop breeding programmes. The details of trait specific germplasm introduced since 2000 are presented Table 7.3.

Table 7.3: Important Trait Specific Introductions in Forage Crops

Crop/EC No. and Source Country	Traits	Distribution
Brachiaria hybrid variety Mulato EC 549024-25 USA	Stoloniferous growth, excellent forage production, vigorous re-growth, excellent palatability, drought tolerant, and produces forage round the year	IGFRI Regional Station, Dharwad, Karanataka
Deschampsia antarctica Antartic hair grass EC631954 Chile	New crop, native to Antarctica	Avesthagen Limited, Bangalore
Elytrigia repens EC 586940 USA	Var. Eversett, advanced generation synthetic cultivar for high rhizome production ability to spread by rhizomes and used for land stabilization and reclamation	IGFRI, Jhansi
Medicago sativa EC 499771-72 USA	Highly resistant to Aphanomyces root rot and northern root knot nematode, resistant to *Phytopthora*, pea aphid, spotted aphid and moderately resistant to Verticillium wilt, anthracnose race 1 and stem nematode, High forage yield under dry land condition	IGFRI, Jhansi
Medicago sativa EC 596671 USA	Var. OK 190, unique combination of broad genetic base, resistance to blue alfaalfa aphid, spotted alfaalfa aphid and *Phytopthora* rot	IGFRI, Jhansi
Medicago sativa EC 596673 USA	Var. OK 207, broad genetic base population provide resistance to blue alfa alfa aphid biotype BAOK 90 and spotted alfa alfa aphid	IGFRI, Jhansi

Contd...

Table 7.3–Contd..

Crop/EC No. and Source Country	Traits	Distribution
Medicago spp. EC 271425-29 USA	Produce high forage, cold tolerant and resistant to aphids	IGFRI, Jhansi
Pennisetum typhoides ssp *montana* EC 473259 USA	Potential for increase growth rate, resistance to rust, smut and leaf spot	NBPGR, Regional Station, Jodhpur
Trifolium pratense EC560447-448 USA	Short internodes, plants usually present a rosette appearance but flower sparingly under long day conditions	IGFRI, Jhansi
Trifolium pratense EC 578957 USA	Var. Freedom, free from pubescence, good for hay making as it permits faster drying and reduces dustiness	IGFRI, Jhansi
Trifolium pratense EC 560447-448 USA	Short internodes, plants usually present a rosette appearance but flower sparingly under long day	IGFRI, Jhansi NBPGR RS, Bhowali
Triticale EC 537921 Mexico	Hexaploid winter triticale, good forage quality	IGFRI, Jhansi
Triticale EC 537922 Mexico	Hexaploid intermediate triticale, good forage quality	GCD, NBPGR
Triticale EC 534274 USA	High yielding, superior forage quality	IGFRI, Jhansi
Triticale EC 467937 Canada	Variety Bobcat, sprouting tolerant and resistant to stem rust and leaf rust	NBPGR, Regional Station, Bhowali

The germplasm introduced/imported into India is being multiplied, characterized, evaluated and conserved at different NBPGR regional stations located in different agro-climatic zones and also at Indian Grassland and Fodder Research Institute (IGFRI), Jhansi. IGFRI networks with its three Regional Centres and All India Coordinated Research Projects (AICRP) on forages with 18 coordinated centres under ICAR system and seven Regional Stations on Forage Production and Demonstration and one Fodder Seed Production Farm under Department of Animal Husbandry is the National Active Germplasm Site for forages. It was established by Government of India in 1962. Mandate of IGFRI is collection, evaluation, documentation and conservation of forage genetic resources and to carry basic and strategic research on improvement, production and utilization of fodder crops and grasslands. The function of the AICRP is to coordinate multi –location testing programmes at the national level for identification of appropriate varieties and production technologies for different agro-ecological conditions. IGFRI has a total germplasm holding of 3,390 in grasses and 2,200 in forage legumes. Significant headway has also been made in varietal development. One hundred and eighty eight varieties of 30 fodder crops, range legumes and grasses have been released and notified through IGFRI.

Regulatory Aspects of Germplasm Exchange (Inflow/Outflow of Forage Genetic Resources)

Transboundary movement or flow of germplasm of plant material on a world wide basis had been mostly carried out without regard to well defined procedures, but several countries have established plant introduction services / organization for regulating, import and export (exchange) of plant materials. The National Plant Introduction Services of most countries have some features in common which facilitate their efficient operation.

- ✰ Plant Introduction Services usually forms part of the Department of Agriculture or a major research organization and hence is able to draw upon extensive resources in carrying out its work.
- ✰ Plant Introduction Services either has direct responsibility for plant quarantine or works in close collaboration with the quarantine authority.
- ✰ Plant Introduction Services has at its disposal good testing facilities in all the major climatic zones of the country concerned.
- ✰ Plant Introduction Services maintains records of introduced plants, usually including plants introduced by other agencies and is thus able to ensure, that national requirements are met without unnecessary duplication.
- ✰ Plant Introduction Services provides a contact point for the International Collaboration through FAO and other agencies.

By maintaining extensive collection, and by establishing close collaboration with its counter parts abroad, the plant introduction service is able to provide plant material needed by specialists, across the country.

In India, there is a single window system for the exchange of small samples of plant germplasm (including transgenics) meant for research, and NBPGR is the nodal institution for that. It regulates the import of seeds / planting material for research under the provisions of Plant Quarantine (Regulation of Import into India) Order, 2003 of the Destructive Insects and Pests (DIP) Act of 1914. The plant introductions include germplasm, elite strains, improved varieties, genetic stocks and related species in from various parts of the world. In order to prevent the introduction and spread of pests associated with plant genetic resources and to ensure that only pest-free material is supplied to the indentors, strict plant quarantine protocol is followed including post-entry quarantine (Ram Nath, 1996; Parakh and Gautam, 1999). NBPGR is empowered by the Ministry of Agriculture to undertake quarantine processing of all germplasm and research materials including transgenics under exchange for both public and private sectors.

Procedure for Import/Inflow of Forage/Fodder Germplasm for Research Purposes

As per New Policy on Seed Development (1989) and PQ Order 2003, the Government of India has made it obligatory for all plant breeders and researchers intending to receive seed/planting materials, from other countries to fulfill the following two mandatory requirements:

1. Import permit before import of any material.
2. Phytosanitary certificate from the country of origin.

These two documents must accompany every seed/planting material consignment imported from abroad.

Issuance of Import Permit for Import of Germplasm into India

Director, NBPGR has been authorized to issue import permit and receive imported materials from custom authorities for its quarantine inspection and clearance. The recipient desirous of importing seed/planting material has to apply to the Director, NBPGR on a prescribed application form (*PQ Form 08*). For obtaining the Import permit the recipient is required to duly fill in the form and submit to Director, NBPGR, New Delhi. The IP is issued in form PQ 09 in triplicate. IP is valid for six months from the date of issue and valid for successive shipment provided the exporter and importer, bill of entry, country of origin and phytosanitary certificate are the same for the entire consignment. Validity may be extended up to one year on request, if adequate reasons in writing are justified. Import permit is non-transferable.

After obtaining import permit the recipient should send it to the concerned official that has agreed to supply the required germplasm for use in research with the request that the import permit in duplicate must be enclosed alongwith the seed/planting material.

Director, NBPGR is authorised to issue IP to import the seeds/plant meant for research purpose only as per clause 6 (2) of PQ Order 2003.

The application form (PQ 08) can be downloaded from NBPGR website www.nbpgr.ernet.in.

Alongwith the application form PQ 08, a demand draft of Rs 150/- (for govt./ public organizations) or Rs 300/- (for private organizations/seed companies), as the case may be plus service charges, in favour of Director, NBPGR, New Delhi as processing fee for the issuance of IP should be sent. The fee is non-refundable. It should also be ensured that the consignment must be addressed to the Director, NBPGR. Also, the seed or planting material should not be treated with any chemical until and unless asked to do so in the import permit.

Phytosanitary Certificate (PS)

The second mandatory requirement is that of Phytosanitary certificate which is to be issued by the official agency of the donor country. Every consignment shall be accompanied by PS issued by authorized officer at country of origin/supplier country with additional declarations for freedom from specific pests and diseases as specified or that the pests specified do not occur in the country or state of origin as supported by documentary evidence thereof. PS is a document regarding the health status of consignment.

It is also issued by the NBPGR for all germplasm material meant for export to foreign countries. Also every import should be accompanied with phytosanitary certificate (original copy) issued by Govt. official from country of origin in the prescribed format of Food and Agriculture Organization (FAO).

Issuance of Import Permit for Transgenic Crops

Gazette of India extra ordinary Part II Section-3 sub section (1) published by authority no. 621, New Delhi has defined -Rules for the Manufacture, Use, Import, Export and Storage of Hazardous Microorganism/Genetically engineered organisms or cells made under sections 6,8 and 25 of the Environment (Protection) Act, EPA 1986 (29 of 1986). EPA plays a major role in minimizing the risk from pollutants and contaminants affecting flora and fauna, human and animal health and preserving the environment. In accordance with this Act, all transgenic plants are regulated items. The provisions of Plant Quarantine (Regulation of Import into India) Order 2003 are applicable to import of transgenic seeds as well.

Department of Biotechnology (DBT), Ministry of Science and Technology and the Ministry of Environment and Forests (MoEF) have separate set of prescribed procedures for providing permission for import of transgenics. An indentor who wishes to import transgenics has to submit the proposal through the Institutional Biosafety Committee (IBSC) to Review Committee on Genetic Manipulation (RCGM). RCGM is an authorized agency of the Government of India, functions under DBT, assesses the import indents of transgenic material for research purposes and issues Seed Transfer Clearance Letter (valid for one year). RCGM examines the desirability of import of transgenic line, from the bio-safety point of view. It includes representatives from Department of Biotechnology (DBT), Indian Council of Medical Research (ICMR), Indian Council of Agricultural Research (ICAR), Council of Scientific and Industrial Research (CSIR) and other experts in their individual capacity. Genetic Engineering Approval Committee (GEAC) functioning under the MoEF, regulates the commercial introduction (large-scale use/experiment) of transgenic material (Singh, 2001)

RCGM recommends to GEAC for consideration of release of the transgenic material, which is environmentally safe. After getting the technical clearance for import, from RCGM, the indentor has to apply to Director, NBPGR, and New Delhi for the issuance of Import Permit.

Alongwith the application form, the indentor is also required to furnish the information and provide undertaking as per para 4, 5 and 6 of the DBT permission. Undertaking with the contents of para 4, 5 and 6 should be typed on letter head and signed by the applicant.

Para No. 4

- ☆ No transgenic material is permitted for experimentation in open environment without prior authorization from the Government of India.

- Full account of transgenic plants raised from the imported seeds is to be kept in a bound book, which should be available for inspection by the authority in case such a need arises.
- All transgenic materials prescribed by the indenters may be available for inspection, whenever required.
- All the unwanted transgenic materials may be destroyed by burning after the experiments are conducted.
- For any use of the transgenic material for propagation in the open environment, the applicant will make a separate application to RCGM through Institutional Biosafety Committee (IBSC).
- All precautions would be taken to prevent the escape of the genetic material into the open environment and shall follow the Recombinant DNA Safety Guidelines of the Government of India (http://dbtindia.nic.in/uniquepage.asp).
- The applicant shall certify to the NBPGR that to his/her best of knowledge and belief the material being imported confirm to the description given in the import clearance.
- NBPGR shall retain up to 5 per cent of the transgenic seeds in its facility and keep them under safe custody. NBPGR shall further issue a receipt to that effect to the applicant.
- The seeds retained by the NBPGR shall be used by the Government for future reference.

Para No. 5

- The supplier of the transgenic material shall certify that transgenic materials has the genes as has been described in the permission.
- The supplier shall also certify that these transgenic materials do not contain any embryogenesis deactivator gene sequence.

Para No. 6

- The transgenic material shall be handled, packaged and transported as specified in "r-DNA safety guidelines and revised guidelines for research in transgenic plants" and guidelines for toxicity and allerginicity evaluation.

Release of the Consignment of Forage Crops to the Indentor

After obtaining import permit the recipient should send it (both the copies) to the concerned official/supplier (abroad) that has agreed to supply the required germplasm for use in research. The supplier should be instructed to send two copies of IP (one pasted outside the seed parcel and other inside the seed packet) and other relevant documents with the consignment. The import permit in duplicate must be enclosed alongwith the seed/planting material for custom clearance. It should be clearly instructed to the sender/supplier that consignment should be addressed only tò the Director, NBPGR, New Delhi. The port of entry of germplasm is New Delhi Airport only.

The material so introduced shall after quarantine clearance is assigned an Exotic Collection (EC) number which remains unchanged and then material is forwarded to the recipient after necessary quarantine clearance.

Registration, National Accessioning and Import Quarantine of Germplasm Imported

All indents for import of germplasm are registered for assigning the case number and then forwarded to the Plant Quarantine (PQ) Division without opening the parcel alongwith duly filled Import Quarantine (IQ) form for detailed quarantine inspection and clearance. After clearance from PQ Division, the samples are first arranged taxonomically indicating their genus, species, common name and cultivar name etc. for national accessioning in the national record. Each introduction/ accession is assigned an EC (Exotic Collection) number which remains unchanged with information like name and address of donors, characteristics of the germplasm, relevant references, date of arrival, condition and distribution of the materials. All assembled healthy plant material is regularly transmitted to the various researchers to make use of these valuable genetic resources.

Procedure for Outflow/Export of Germplasm from India

Under the provisions of the Convention on Biological Diversity (CBD), Government of India enacted legislation called Biological Diversity Act (BDA), 2002 and also notified the Biological Diversity Rules, 2004. As per section 3 of the Act, no person from outside India or a body corporate, association, organization incorporated or registered in India having non -Indian participation in its share capital or management, can access any biological resources or knowledge associated, for research, commercial utilization, bio-prospecting or bio-utilization, without prior approval of National Biodiversity Authority (NBA).As per section 5 of BDA, 2002, exchange of germplasm for research under the bilateral agreements/Collaborative projects are however exempted which confirm to the policy guidelines issued by the Central Government or approved by the Central Government.

The person who shall be required to take the approval of the National Biodiversity Authority are the following, namely: (a) a person who is not a citizen of India;(b) a citizen of India, who is a non-resident as defined in clause (30) of section 2 of the Income-tax Act, 1961; (c) a body corporate, association or organization- (i) not incorporated or registered in India; or (ii) incorporated or registered in India under any law for the time being in force which has any non-Indian participation in its share capital or management All such persons or organizations seeking approval of the Authority (NBA, India) for access to biological resources and associated knowledge for research or for commercial utilization shall apply in the prescribed form.

However, the provisions of Section 3 and 4 shall not apply to collaborative research projects involving transfer or exchange of biological resources or information relating thereto between institutions including Governments sponsored institutions of India, and such institutions in other countries, if such collaborative projects conform to the policy guidelines issued by the Central Government in this behalf or approved by the Central Government.

Export of PGRs in the light of notification of the Biological Diversity Act 2002 is now decided by the PGR Export Facilitation Committee (PGR-EFC) which submits its recommendations to DARE for approval.

Documentation of Information in the National Database

All indents for import of germplasm are registered for assigning the case number and then forwarded to the Plant Quarantine (PQ) Division without opening the parcel alongwith duly filled Import Quarantine (IQ) form for detailed quarantine inspection and clearance. A flow chart of the different steps involved in quarantine processing of imported germplasm at NBPGR is given in Figure 2. After clearance from PQ Division, the samples are first arranged taxonomically indicating their genus, species, common name and cultivar name etc. for national accessioning in the national record. Accessioning of exotic accessions are being done online (http://www.nbpgr.ernet.in/geq) and each introduction/accession is assigned an EC (Exotic Collection) number which remains unchanged and documented alongwith all the detailed information such as:

- ☆ Import Quarantine (IQ) number and date
- ☆ Source country/country of origin
- ☆ Complete postal address of the source country
- ☆ Import Permit number and date of issue
- ☆ Phytosanitary Certificate number and date
- ☆ Biological status/variety name/type/category of the material
- ☆ Distribution details
- ☆ Indenter of the material
- ☆ Whether the material was introduced under MoU/restored/transgenic ?
- ☆ Additional information/specific traits/remarks if any.

Issues Related to Access and Flow of Germplasm

In 1970's Plant Breeder Rights (PBR) were implemented in many developed economies of the world which encouraged an expansion of private sector breeding. Inclusion of IP rights in the Trade –Related Aspects of Intellectual Property Rights (TRIPS) section of General Agreement on Tariffs and Trade now World Trade Organization (WTO), were widely discussed in the context of international trade, agriculture and development, The WTO TRIPS agreement required all signatories to implement IPR protection for a range of biologically based material that they had not previously been obliged to protect. Many developing countries were required to address Intellectual Property issue for the first time. With stronger patent rights incorporated into TRIPS and biodiversity being regarded as a treasure under national sovereignty apprehensions on free exchange of germplasm were raised. Apprehensions related to limited access to PGR raised several questions such as ownership issues and availability. Questions such as who owns biological resources/ living things (Hamilton 2005) and can them be shared or sold. Can we use it for breeding and research without any obligations for using them? The dilemma of doing only what we are allowed to do and also to ensure that others do what they are allowed. These issues were addressed to the policy makers and concerns were far beyond scientific communities. Thus understanding the issues and complying with them is therefore, a priority to meet the ever-changing needs of crop improvement programs and sustainable agriculture.

Issues were raised on the ownership of PGR and benefit sharing, as the vast collections of germplasm conserved in various genebanks in CG system were collected from gene rich/economically developing nations and stored away from the place of collection. Continued free flow of germplasm was the main concern but more importantly the most critical issue is how the farmers can benefit from these resources who are the true discoverers, conservers, producers and breeders of these invaluable resources.

For addressing these issues and concerns several international agreements and conventions entered into force which directly or indirectly had implications on the conservation and access of genetic resources. These international and national agreements and conventions which address the issues of regulated flow to biological resources are discussed hereunder briefly.

Trade Related Aspects of Intellectual Property Rights (TRIPS)

Main objective is to recognize and protect monopolistic and private IPR's. Enjoins to grant patents for any inventions in all field of technology which covers biotechnology, Hence, biodiversity falls firmly under the legal regime. But article 27(2) and (3) provide important exceptions in favour of protecting environment and thereby the biodiversity. There are two preconditions to exclude inventions from patentability- (i) commercial exploitation of the invention should be disallowed(such prevention is necessary for the purpose of avoiding serious prejudice to the environment) (ii) exclusion should not merely because the exploitation is prohibited by domestic law.,

International Undertaking on Plant Genetic Resources (IUPGR)

Based on the universally made principle that PGR is heritage of mankind and consequently should be available without restriction, FAO in 1983 adopted a non binding International Undertaking on Plant Genetic Resources (IUPGR) with the objective to ensure that PGR are of economic and/or social interest particularly for agriculture, will be explored, preserved, evaluated and made available for plant breeding and research purposes. FAO Commission on Genetic Resources for Food and Agriculture (CGRFA) monitored the implementation of IUPGR. However, the principle gradually narrowed and 1989 amendment made it consistent with Plant Breeder's right favouring technology rich countries.

Convention on Biological Diversity (CBD)

Most widely adopted UN agreement ever aimed at conservation and sustainable use of the components of biodiversity, with fair and equitable sharing of benefits arising from the utilization of genetic resources was adopted in the background of increased threat to genetic resources by the developments in biotechnology. Legally binding Convention on Biological Diversity (CBD) entered into force in 1993. Access is to be determined by national governments subject to Prior Informed Consent (PIC) and Mutually Agreed Terms (MAT). Accordingly, patents on genetic material can only be consistent with the CBD if the resources are acquired legally (with national approval). Thus country of origin and proof of PIC together known as disclosure issues be indicated in patent application. Though it was meant to encourage exchange of germplasm, it also raised the required level of negotiations.

International Treaty on Plant Genetic Resources for Food and Agriculture (ITPGRFA)

Legally binding ITPGRFA was thus negotiated as a direct response to CBD in 2001, came into force in 2004 to facilitate access to PGRFA in harmony with CBD, through an efficient mutually agreed multilateral system (MS) of access and benefit sharing. Access is only for research, breeding and training and not for chemical, pharmaceutical nor non-food/feed industrial use. No Intellectual Property rights can be claimed on the form received from MS that limit the facilitated access to PGRFA/ genetic parts or components. To ensure that the germplasm is used only legally the Material Transfer Agreement (MTA) binds terms and conditions on the recipient to use it. MTA is signed for both inflow and outflow of germplasm. The treaty works through inter-governmental agreement as required by CBD. Two key components of treaty being:

- ✰ Each party should facilitate germplasm exchange for a list of crops covered in Annexure I (includes 35 food crops and 29 forages) which are important for food security and for which countries are interdependent.
- ✰ Provides a system to enforce equitable share of benefits, through a central fund for obligatory payments to country of origin.

Despite the position taken by CBD that IP issues must not conflict with the conservation and sustainable use of biodiversity and must be supportive, conflicts do arise. India enacted two laws to implement TRIPs and CBD- Protection of Plant Varieties and Farmers Right Act (PPV& FR Act); Biological Diversity Act (BDA)

Plant Varieties and Farmers Right Act (PPV and FRA)

The Indian PPV and FR Act provides effective system for protection of plant varieties, and protects rights of farmers and breeders. The Act has recognized farmer as a conserver, provider of genetic resources, breeder and as a producer and consumer of seed and is effective from January 2005 (www.plantauthority.in)

Biological Diversity Act (BDA)

Under the provisions of the Convention on Biological Diversity (CBD), Government of India enacted legislation called Biological Diversity Act (BDA), 2002 and also notified the Biological Diversity Rules, 2004. The objective is to provide for access to biological resources of the country and equitable share in benefits arising out of the use of biological resource together with sustainable use and conservation of biological diversity. As per the Act (Section 3), no person from outside India or a body corporate, association, organization incorporated or registered in India having non -Indian participation in its share capital or management, can access any biological resources or knowledge associated, for research, commercial utilization, bio-prospecting or bio-utilization, without proper approval of National Biodiversity Authority (NBA). No person can apply for any IPR in or outside India for any invention based on any research or information on a biological resource obtained from India without obtaining approval from NBA.

Collaborative research projects involving transfer or exchange of biological resources of information between Governments sponsored institutions of India, and such institutions in other countries, if they conform to the policy guidelines or approved by the Central Government are however exempted.

The person who shall be required to take the approval of the National Biodiversity Authority [Section 3 (2)] are the following, namely (a) a person who is not a citizen of India; (b) a citizen of India, who is a non-resident as defined in clause (30) of section 2 of the Income tax Act, 1961; (c) a body corporate, association or organization- (i) not incorporated or registered in India; or (ii) incorporated or registered in India under any law for the time being in force which has any non-Indian participation in its share capital or management.

Conclusion

Flow of genetic resources (import and export) of germplasm is now viewed as subject nationally and legally controlled, keeping in view the international agreements and national legislations. The responsibility for import and export of plant genetic resources for research purposes has been delegated by the Ministry of Agriculture to the Indian Council of Agricultural Research (ICAR)/Department of Agricultural Research and Education (DARE) and NBPGR is the nodal agency recognized for PGR exchange within the country as well as outside the country. After enactment of the Biological Diversity Act, Protection of Plant Varieties and Farmers Rights Act (PPVFRA) and ratification of International Treaty on Plant Genetic Resources for Food and Agriculture by India, NBPGR has to follow new rules for flow of plant genetic resources for food and agriculture. The import of forage germplasm for research purposes are regulated under the provision of PQ Order, 2003.

Regulation of access to forage germplasm and information under the new regime in India has to take into account the established institutional mechanism and various acts in force relating to agro- biodiversity. The requests of indenters are to be dealt with depending on the status of requesting party and the conditions for access under different categories. Conditions of MTA and SMTA also need to be strictly followed and monitored. What one is supposed to do now is to learn more about the changing scenario of IP regime which is guiding the management of PGR and exchange of these resources following all if's and but's of the policy. To regulate access to the germplasm and to ensure/conform that it is used legally the exchange is done under Material Transfer Agreement (MTA) in and out of country through single window system. The MTA defines the terms and conditions which are binding on the parties who are signatory to it. The terms and agreements specified speaks what can be done and what cannot be done with the acquired germplasm and thus ensures its compliance with all relevant national and international agreements and treaties. The MTA is signed for all germplasm designated in different groups *viz.* germplasm governed either by ITPGRFA, or by CBD and for germplasm not covered by either of them.

References

Brahmi Pratibha, R K Khetarpal and BS Dhillon (2005) Access to Plant Genetic Resources : The changing Scenario In *Biodiversity: Status and Prospects* (eds) Pramod Tandon, Manju Sharma and Renu Swaroop, Narosa Publication 124-133 pp

Cooper D, J Engels and E.Frison (1994) A Multilateral System for plant genetic resources: imperatives, achievements and challenges. Rep.2. Int. Plant Genetic Resources Institute, Rome, Italy

Dhillon BS and Anuradha Agarwal (2004) Plant Genetic Resources: Ownership, Access and Intellectual Property Rights In : *Plant Genetic Resources: Oilseed and Cash Crops* (eds) Dhillon BS, R K Tyagi, S Saxena and Anuradha Agarwal, Narosa Publication pp 1-20

Evenson R E (1999) Intellectual Property Rights, Access to Plant Germplasm, and Crop Production Scenarios in 2020. *Crop Sci* 39: 1630-1635

Hamilton S. Ruraidh and Edwin Javier (2005) Intellectual Property Rights and Germplasm Exchange: the new rules In Planning Rice breeding programs for IMPACT Plant Quarantine (Regulation of Import into India) Order (2003), The Gazette of India, Extraordinary, Part II-Section 3-subsection (ii) (www.plantquarantineindia.org)

Tyagi Vandana, AK Singh, Nidhi Verma, Deep Chand, RV Singh and B S Dhillon (2006) Plant Introduction in India during Pre- and Post-CBD Periods-An Analysis Indian *Journal of Plant Genetic Resources* 19 (3): 436-441

Chapter 8
Botany and Taxonomy of Forage and Fodder Crops

Vikas and Babeeta Chrungu Kaula*

Zakir Husain College, University of Delhi, Delhi

**E-mail: vikasbot@gmail.com*

Introduction

The economy of India is largely dependent upon agriculture sector. A large chunk of population dwells in villages. The rural people have been traditionally fulfilling their daily demands directly or indirectly from the forests, agricultural fields, pastures and natural rangelands. The current increase in human population by 80 per cent since 1960 (Chrispeels and Sadava, 2003) has put a tremendous pressure on agriculture sector directly for food and indirectly for other products. Thus, the sustainable growth of this sector has become the forefront area of interest for plant scientists, geographers, sociologists, economists, and policy makers. Most of the investment in agricultural research and development (R&D) is done by government only. In agriculture, the investments are huge and the profit and returns come after a long time (nearly ten years). This has always kept the private companies away from investing. Most of the previous research was based upon the cumulative traditional knowledge of farmers over the centuries. Therefore, the fodder industry, which is a small sub-sector, is still in its infancy in India.

With the increase in urban population, the lifestyles have undergone major change. The coarse grains, millets, and roots are being replaced by dairy products, meat, and processed food. The demand of milk and meat has increased considerably.

More feed is required for the livestock and cattle to be used for meat and milk. The increase in income of a person also stimulates a shift from vegetarian to non vegetarian diets. The requirement of maize is predicted to rise not only for human consumption but mainly to feed the poultry and animals to be consumed by man.

The animal fodder consists of several species of plants which are used to feed the cattle either raw (green and cut fodder) or after minor processing (such as silaging). Foraging is done on the wild barren wastelands, rangelands and pastures. Green revolution made the country self sufficient in food production and provided temporary solution to the giant problem of food scarcity. The focus of attention was mainly on wheat, rice, and other staple food crop cereals. No effort was made to increase the fodder production; even the need was not realized. Forage and fodder crops have always been neglected to such an extent that some scientists called them as 'abandoned crops' or 'orphan crops'. Globalization puts pressure on the export market to increase the quality of fodder products. Therefore, the quality and quantity of fodder in India has to be greatly enhanced.

With the intensive growth in livestock production, fodder industry has shown similar growth trends. The demand for nutritious, palatable, and overall high quality cattle feed has increased tremendously. Good feed is necessary to maintain the quality of the final products obtained from these animals. Human health is also affected by the quality of feed given to the animals used for food. The consumption of meat in India is likely to increase by 2.8 per cent and consumption of milk and milk products by 3.3 per cent in near future. This enhanced demand of animal products should be met by improving the productivity of animals rather than rearing more animals. India has a large number of cattle and milch animals, but the quality of products is still unsatisfactory, malnutrition being the primary cause. According to Chakravarti (1987), the density of cattle in India may be as high as 7000 per hundred hectares (ha) in some parts. The share of land in pastures is only 4 per cent of the total cultivated area while only 2 per cent of this is dedicated to fodder crops. The green fodder products are deficit by a huge 61.2 per cent, and the dry residues by 21.9 per cent at present (Kaushal *et el*, 2008). This wide gap in the demand and supply has opened various vistas for the fodder industry to establish itself as a sustainable, profitable industry. This also requires specialized producers to play important role in domestic and international markets to ensure the availability of assured good quality animal feed. However, the scenario is not encouraging because, so far the pace of R&D has been slow and the production has not reached even near to its maximum capacity. There is a need to stimulate the scientific research and other developmental activities in Indian fodder and forage sector. The area under fodder and forage crops cultivation may not increase significantly because the cereals put a lot of pressure on land. India has wide range of soil, and climatic conditions which make one variety successful at one place and failure at other. New high yielding and resistant varieties have to be produced. The fodder and forage crops will not only help in improving quality of animal products but can also restore the deserts in India. So far, fodder is taken from the natural grasslands and pastures. This puts enormous pressure on the desert ecosystem such as the 2.34 million km^2 Thar desert. Overgrazing has been a problem in recent years and has put the desert ecosystem in danger (Chauhan, 2003). Therefore, fodder and forage crops are now occupying an important place.

This chapter attempts to review the botany and taxonomy of major forage and fodder crops of India. The list of species having potential as good fodder and forage crops is comprehensive. There is a wide range of species being used for fodder in different parts of the country but most important species belong to Fabaceae and Poaceae. These families are amongst the largest families of flowering plants supplying man with food and other needs. Recently, some weeds such as chicory are being considered as feed. Some tree species such as *Prosopis juliflora, Aegle marmelos, Ficus* sp. are used as feed for goat and other animals. The major forage crop are covered in this chapter for a quick reference.

Fabaceae

The members of Fabaceae are the keystone species in most of the grasslands because of their ability to fix atmospheric nitrogen. This is not only advantageous to them but also adds nitrogenous compounds to the soil used by neighboring plants. Legumes are fast growing, high yielding, and nutritious forage crops. Such crops are economically beneficial for the farmers as compared with other forages. They are variously adapted to the local soil conditions and therefore, preferred by the farmers. The family Fabaceae has 700 genera and 18,000 species, such an enormous family can not be handled by taxonomists, and therefore, it is divided into three subfamilies: the Faboideae, Caesalpiniodeae, and Mimosoideae. In Mimosoideae, the inflorescence is a prominent head and the anthers are generally protruding out and most visible. In Caesalpiniodeae, the flowers are zygomorphic and have various irregular shapes. The third subfamily Faboideae is the largest with nearly 500 genera and 12,000 species (Langer and Hill, 1991). The leaves have pulvinus and are compound but the first pair of leaves in the seedling is generally simple.

The flowers in Faboideae are mostly hermaphrodite, zygomorphic, the five sepals are jointed; the five free petals are distinct and showy. The posterior petal called the standard or vexillum is the largest and most prominent. The two lateral petals called the wings gradually narrow down from tip towards their base. The other two smaller anterior petals form a boat shaped structure called the keel which is enclosed within the wings. The ten stamens have their filaments fused to present either a monadelphous or diadelphous condition where the filaments of nine stamens are fused and the tenth is free. The pistil has a long, protruding out, curved and pubescent style, simple stigma, and the ovary has numerous marginal ovules. The abortion of fertilized ovules is a common phenomenon in this family. The major pollinators are honey bees, attracted by the nectar and pollen. The ability to fix nitrogen makes this family distinct. The crops do not require nitrogenous fertilizer and also add usable forms of nitrogen to the soil, making the soil fertile. This family is important source of proteins. The major forage and fodder crops belonging to Fabaceae are discussed below.

Medicago (Alfalfa, Medicks, Lucerne)

The genus *Medicago* is large, having more than 55 species of which several are commercially utilized as fodder crops across the globe. The literal meaning of the word alfalfa is "the best fodder" which suits the crop, since it is considered the best fodder today. There are two main species, *Medicago sativa* and *Medicago falcata* (tribe Trifolieae) which are dominant over the other species of the genus. *M. sativa* has blue flowers and *M. falcata* has yellow flowers; the plants can grow 70-120 cm tall, with much branched stem and up to 20 foot deep tap root. The leaves are pinnately trifoliate, leaflets tapering towards base. These two species are inter-fertile and have given rise to several hybrids which pose difficulty in classification. *M. sativa* is native to the southern area of Black Sea and Caspian Sea and *M. falcata* to northern Asia. From these parts, the species have spread rapidly across the globe. Alfalfa is adapted to well drained, dry and deep, fertile soils rich in phosphorus, potassium, boron and lime. It is a highly productive cold season perennial crop used mainly as nutritious, protein rich hay, but also suitable for green manure and forage. The leaves are variegated or simple, inflorescence is a compact, axillary pedunculate raceme with 20-40 flowers. Some varieties are hardy while the others are relatively fragile and can not be grown in frost susceptible areas. The seeds are cheap and easily available. It is grown alone or in combination with other crops. In full bloom, it gives 2-3 cuttings per year. The crop has got considerable attention in recent years but the systematic work still needs to be done to overcome the weaknesses of this crop and to produce resistant varieties. Some other species of alfalfa with limited utility are *M. tornata, M. truncatula, M. laciniata,* and *M. lupulina.*

The problem of classification has been addressed by Small *et al.* (1999) by using isozyme analysis. Lesins (1952) has studied the cytogenetics of alfalfa for taxonomic purpose. A comprehensive account of classification of the entire genus using floral characters is given by Gunn (1978) in his book. Valizadeh (1996) studied nine species of Medicago and found their relatedness to each other using DNA polymorphism studies (RFLP) and grouped them together according to their phylogeny.

Lotus corniculatus (Birdfoot Trefoil)

It is a very successful, perennial, very leafy herbaceous plant with thinner stems than alfalfa. The plant grows upright 10-50 cm, therefore there is a tendency for lodging, to avoid this any perennial grass can be grown together for support. The root system is deep and branching. There are five leaflets and the flowers are yellow. The crop establishes slowly and the seed germination is also slow. The genus *Lotus* has nearly 100 species in the world. India has three species among which *Lotus corniculatus* is the most variable. The variations are continuous and observed among the populations. This has led to confusion and many species have been named wrongly (Chaudhary, 1996). In India, three varieties in the species *Lotus corniculatus* are recognized: var. *corniculatus*, var *glabrous*, and var. *minor* which differ mainly in their hairiness.

Trifolium (Clovers)

The genus *Trifolium* (tribe Trifolieae) has nearly 240 species all over the world. Nearly 25 species are grown as fodder and forage crops, commonly called true clovers with characteristic trifoliate leaves. It is native to Mediterranean region, later spread to America, Asia and Africa. The stipules vary in shape, size, and colour and are taxonomically important. However, some species within the genus are closely related. They may be environmental variants of the same species rather than separate species. Clovers are annual or perennials. For good growth, cool, moist climate and calcium, phosphorus, and potassium rich soil is suitable. Under unfavorable conditions, perennials may become biennial or annual. Generally they are long day species. The genus has a wide adaptability and the species show variability in the number of flowers in the inflorescence (head) from 5 to 200, the seeds in a pod may be 1-8. The genus shows self and cross pollination both (by honeybees and bumble bees). The maximum diversity of clovers is observed in the Mediterranean region. Some work has been done to reveal the taxonomic relationships between the different species of this genus (Lange and Schifino-Wittmann, 2000). They studied isozyme variation in a few species of this genus. However, the correct identification of species in this genus remains difficult and an elaborate task (Zohary and Heller, 1984). Some important species of the genus *Trifolium* are:

Trifolium pretense (Red Clover, Tripatra, Tipatiya)

It is widely cultivated, short lived herbaceous annual or perennial plant growing to a height of 60-90 cm. It is native to Europe and Asia. The flowers are pink, up to 12 mm long. The literal meaning of "pratense" is the one found in meadows. It is an important fodder crop and a green manure.

Trifolium hybridum (Alsike Clover)

It is not a hybrid as the name might suggest but a distinct species. It is generally a perennial, sometimes biennial crop, generally 30-70 cm tall but under favourable conditions, the plant may grow up to 5 foot tall but in stressed conditions may not grow beyond 25 cm. the stem is erect or ascending, without adventitious roots at the base. Pollinated by honey bees and bumble bees, it is self as well as cross pollinated. The flower heads are pink and white borne acropetally. The mature pod remains enclosed within the calyx and has 1-4 seeds. There is no terminal inflorescence; therefore the stem continues to grow. These clovers perform better in cold climate and wet soils than other clovers. The secondary roots on the main tap root bear the nodules.

Trifolium repens (Ladino Clover, White Clover)

This is a perennial herb, suitable for forage as well as seed production, and to all soil conditions and all types of cattle. Within the species *T. repens* some variants are called 'ladino' clover and the others are called 'white' clover. The species is self-incompatible and pollinated by bees. The white clover grows best in combination with grasses such as ryegrass, and tall fescue. The many adventitious roots and branched stolons make it suitable for grazing and help in quick regeneration. The stem is glabrous, rooted at the nodes. Flowers are borne in compact heads with 30-90 flowers. The fruit is linear, terete, generally with 4-5 seeds.

Trifolium incarnatum (Crimson Clover, Italian Clover, Tripatra, Tipatra)

It is a winter annual, growing up to 60 cm tall. The stem is erect, hairy and branched near the base. Stipules are broad, ending in a tip. Leaves are trifoliate, with hairy leaflets and a long petiole. It is a good pasture crop, fixes nitrogen and protects the soil by binding the large particles together. This plant is multipurpose, yields hay and seeds both in large quantity. The plant bears showy crimson colored heads, nearly 5 cm across. Cool and humid weather is suitable for a good harvest. There is some tolerance for cold but extreme or too severe cold is highly damaging. The seeds have no dormancy after they are shed, which is undesirable. The stiff hairs on calyx may be a problem for grazing animals. Therefore; the crop should be utilized before the appearance of flowers. Neither it is very hardy nor has the capacity to recover heavy grazing, it is gradually being replaced by other species. It grows well on light sandy soil. It is used for soil improvement because after one cutting, the roots remain firm in the soil but the shoots regenerate very slowly.

Trifolium subterraneum (Subclover)

This is herbaceous winter annual, well suited to the cold weather. The ideal climate for this crop is dry summer to moist winters but without heavy rains. Once established, it produces seeds every year and can self survive for years if the pasture is properly used. It grows easily in the non irrigated, dry pastures and meadows. Extreme winters are detrimental. The stem is prostrate, up to one foot long. Flowers are white, but sometimes pink, borne in close clusters. During seed maturation, the inflorescences bend downwards and try to burry themselves in the ground. The seeds mature below the soil surface; hence it is called subterranean clover. This forms a protective mat on the soil. It should be grown on well drained soil along with a grass. It is highly variable crop with three varieties namely, the early, mid and the late season. There is a lot of variation in this genus and several varieties, subspecies are distinguished which are given different positions by taxonomists.

Trifolium resupinatus (Persian Clover)

It is a winter annual herb, up to 70 cm tall, native to Southern Asia Minor and Mediterranean countries. The stem is generally hollow and branched. It yields fodder in late winter and early spring when grasses are scarce. Fodder has high nutrition values and relished by all animals. It is primarily a pasture and hay crop, it is also used as green manure. Once sown, this plant maintains itself if not over exploited. However, Persian clover recovers quickly from grazing, and gives at least three cuts. The seed set takes place by early summer when the plant dies. The calyx forms a balloon like structure around the mature fruit which bursts open helping in a wide dispersal of seeds. It performs better in low-lying, heavy, moist soils. It is self pollinated and self compatible plant, pollinated by honeybees, produces prolific olive green or black seeds. Seeds are viable for three years, although viability declines after each year of storage.

Trifolium alexandrinum (Egyptian Clover, Berseem)

This plant is native to Asia Minor. It is slightly hairy, winter annual non-reseeding crop, grows up to 60-70 cm tall. Berseem is a good fodder crop having very high yields under good growth conditions. The stem is erect, branching; leaves are alternate. It is self pollinated plant pollinated by honeybees. It is an ancient Egyptian crop introduced in northern India from where it has rapidly spread. Berseem is multicut (up to eight cuts), palatable crop, preferred over alfalfa by the cattle. The available varieties of berseem show high variability in growth rate, fodder yield, and nutritional values, therefore there is ample scope of producing high yielding varieties in India (Naeem *et al.*, 2006). Although it is a winter crop, it can not tolerate extremes, which drastically slow down the growth. It grows well in sandy, loamy and drained soils with pH 6 and above. It is an important crop for north, north east, north-west and central India. Among all the *Trifolium* species, berseem has been the best and got maximum attention. The flowers are yellow, borne in heads which attract honeybees for pollination in large groups. The high moisture content of the plant makes it unsuitable as a hay plant but good as a green manure.

Trifolium hirtum (Rose Clover)

Rose clover is a winter annual, grows a few inches to 70 cm tall, native to Mediterranean region. The stem is semi erect and branched, less hairy than the branches. Stipules are narrow, needle tipped. Leaves are alternate with very long petioles. Inflorescence heads are rose colored, spherical, up to one inch across, covered with hairs, bear 25-45 sessile flowers in its life span. This crop is best suited when grown as a mixture with other clovers and grasses in pasture. Seeds are yellowish, smooth, very light (3 gram has nearly 1000 seeds) with a scar at the tip, up to 2 mm long. Seeds germinate asynchronously. This clover is valuable in poor soil fertility and low rainfall condition. Animals relish it green or even when completely dry.

Glycine max (Soybean, Soyabean)

It is one of the most ancient plants grown by man at the beginning of agricultural practices. Ancient Chinese records (4700 years back) have descriptions of soybean (Wheeler and Hill, 1957). There are numerous popular verities available to suit the variable Indian climate and soil types. It is a fast growing, summer annual, grown for fodder and seeds both; hence it is a multipurpose crop. Seeds are source of soybean oil. The disease and insect resistance of soybean is better than other popular legumes. The plant is erect, variable in height (30-120 cm), branched, and very hairy. The root system is short and much branched. The seeds are generally yellow. Sometimes, green, brown or multicolored seeds are also seen. The healthy seeds show nearly 90 per cent germination. They grow well on fertile loam, although grow on mild acidic and limy or poor soil also. The soil should be warm when the seeds are sown to avoid pathogen attack. Soybean is grown with Sudan-grass when hay is required and with sorghum or corn when silage is needed. For seeds, it is used in monoculture. In India, it is very useful fodder since it remains green in November and December. The taxonomy of the genus *Glycine* is stormy (Hymowitz and Newell, 1981) and needs revision. The cultivated soybean was put in the genera *Phaseolus max* and *Dolichos soja* by Linnaeus. Since then, the varieties, subspecies and even several species have been renamed. A comprehensive review of the classification, domestication and ethnobotany of soybean has been done by Hymowitz and Newell (1981).

Melilotus (Sour Clover and Sweet Clover)

Generally, melilots are common in grasslands and agricultural fields as weeds. The Genus *Melilotus* has nearly 20 species. It is a straight, branched winter annual, used as a green manure and cover crop. The stem is harder and becomes much vigorous when the plant is present singly. *M. indica* is specifically called sour clover. The seeds are cheap and it is a wonderful pasture and soil improving crop suitable for almost all climatic and soil condition. It is used alone or in combination with other legumes and grasses. There are several intermediates, between the species which may not have arisen due to environmental effects alone. So far, there is no comprehensive taxonomic study on this important plant (Aboel-Atta, 2009).

Vicia (Vetches)

Several varieties of vetches are grown of which hairy vetch is most useful as a systematic crop. It is a leafy winter annual, with winy stem and fibrous root system. It is a good source of all types of forage, grown alone for soil improvement and with ryegrass for forage. The genus *Vicia* has nearly 130 species but on a world level, systematic studies are not completely done. All the species are grouped into six sections and 15 series (Yamamoto and Plitmann, 1980), but the classification remains unsatisfactory.

Pisum (Field Peas)

Field peas are common crop of winter. The plants are weak and lodge easily; therefore generally oats are sown together which provide support. But, the crop is susceptible to rots and the seeds are costly which makes it less preferable as a fodder crop in India. It is a palatable, green fodder for all types of cattle. The nutrition value of green pea is high.

Other Legumes

There are numerous other legumes which may be grown as crops but have been used locally. *Acaia* is a large genus with variable habits from shrubs to big trees. *Acacia auriculiformis* and *Acacia leucophloea*, *Acacia senegal* are suited for the great India desert where herbs and shrubs are scarce. These two trees are a good source of fodder for the goats. An edible gum is also extracted from the natural stands of *Acacia*. The other species which are useful are *A. indica*, *A. nilotica*, and *A. tortilis*. These trees are used from the natural patches, there is no systematic cultivation. Few species of *Lespedeza* are also utilized in the arid zones for hay, green fodder, and seeds. The summer annual *Lespedeza* self-establishes by reseeding annually, therefore it has a special importance for permanent pastures. They are very hardy and more heat and drought tolerant than alfalfa, clovers, and soybean.

The cowpea (*Vigna*) is a fast growing summer annual crop adapted to arid conditions and poor soil but under extreme drought seed production is drastically reduced. The stem is weak, but leaves are large. The seedlings are hardy and competent over weeds. The yield of seeds is low and harvesting is costly, which make this crop less preferable by farmers. The hay is of excellent quality although it takes long time to cure due to thick, hard stem and big pods. Cowpeas are generally used as mixed crops with corn, sorghum, or millets.

The lupines (*Lupinus*) are grown for fodder and for soil improvement. The different varieties differ in their adaptability to soil and weather; in general they require cool weather. *Crotolaria* is a tall summer annual, grown on very poor and sandy soils as a green manure and fodder. The *Sesbania sesban*, *S. bispinosa* are commonly grown as green manure and a source of cheap and excellent fuel wood. The green plant is used as a high quality, palatable fodder in northern India. These plants grow very fast on poor soil in arid region. The Subabool (*Leucaena leucocephala*) is known for it extremely fast growth and green cover. Reproduction is vigorous; in a short time the plants cover the entire barren area and fertilize the soil.

The wild winter pea, also called rough pea (*Lathyrus*) is a winter herbaceous annual. The stem is weak and not erect, grows up to three foot long. It is used for soil improvement and fodder all over the Indian plains.

Poaceae (Grasses)

Grasses have a great economic and ecologic significance. The major staple food of man comes from wheat, rice, maize, and sugarcane. Soil binding, eco-remediation, and restoration of heavily degraded ecosystems are possible only by grasses. Poaceae (previously Gramineae) is a huge family with 610 genera and nearly 10,000 species adapted to every climate and soil (Langer and Hill, 1991). The highly modified and variable flowers make this family difficult for taxonomists; therefore a completely satisfactory classification is not in place. Grasslands cover 24 per cent of the earth's vegetation cover and comprise one of the four major terrestrial biomes. Grasslands are predominantly covered by Poaceae members. The flowers are arranged in spikelets, having one or more florets. There are two bracts at the base of the spikelet, called the glumes. Further, the usually bisexual flower is surrounded by two bracts, the outer lemma and the inner palea. The glumes and lemma may taper down and form a bristle like structure called the awn. Ovary has only one basal ovule. The seed like structure is a fruit (caryopsis) where the fruit wall and seed wall are not distinguishable due to fusion. There are twelve subfamilies in Poaceae. Festucoideae (temperate grasses), Panicoideae and Chloridoideae (tropical species), Oryzoideae, Bambusoideae (bamboos) are the five most important subfamilies of grasses (Langer and Hill, 1991).

There are two important growth forms of grasses which should be considered for making suitable combinations when grown with non grass species. The bunchgrass habit is characterized by formation of bunches or clumps which spread only by seeds. This habit is more productive but demands more management. The sod formation is another habit which is less productive. Such grasses propagate by underground rhizomes which help in the binding of soil, and hold the soil together by filling the gaps. The nodes and internodes are very distinct in the stem which may be hollow, solid or pithy. The most important species from Poaceae are described below.

Zea (Maize)

The field corn remains the most important and widely used fodder for cattle. It is a source of grain but equally valued as a silage crop. It is a tall (height greatly variable), annual, monoecious grass. There are separate pistillate (axillary spadix or cob) and staminate (terminal panicle) inflorescences. Corn is a wind pollinated plant where self and cross, both fertilizations are possible. The above ground mass produced by maize is higher than many other crops which make it an excellent fodder. The digestibility and palatability of processed fodder is better than dried fodder. The green plant parts as well as the grains are fed to livestock, thus, it is a dual crop. When maize is grown with a legume, the productivity is considerably enhanced. During droughts, when grasses are dried up, maize gives hope since it remains green. Weed control and use of fertilizers and other agrochemicals greatly enhances the yield of corn.

The entire genus *Zea* is divided into two sections: the *Luxuriantes* section and the *Zea* section. Further, the *Zea* section has several subspecies. There are numerous varieties within the genus. The presently cultivated maize has derived from teosinte (Z. *mexicana*).

Sorghum (Sorghum, Jowar)

The sorghum is tropical fast growing summer annual crops, grown for fodder and grain. The grain colour is variable from red, yellow, and brown to black. The fodder is soft, palatable, commercially used all over India. The fodder can be used green or the plants can be dried, stored for a long time (up to 2 years). Sorghum can substitute corn for grains used as poultry and animal feed. They are generally grown alone but can be mixed with soybean or other legumes as minor crops. Sorghum is extremely drought resistant adapted to arid and semi-arid regions in India. In a year two crops of sorghum can be taken (both rabi and kharif).

Avena (Oats, Javi, Jaii)

It is a winter season crop of rain fed and irrigated fields, adapted to cool moist climate but susceptible to heat. Oats grow as weeds in wheat and barley fields mainly in the Himalayan region where they reduce the yield by competing with the crop. Plant height, breadth of leaf, and number of tillers are important parameters for selecting a suitable cultivar. Oat is the sixth most important cereal crop in the world, adapted to various soil types. It is the best choice on acidic soil among all the fodder grasses. Systematic oat cultivation started only recently, although it is an ancient plant (nearly 4000 years old oat grains discovered). Oat has got ample attention as a grain crop but has never been considered seriously for fodder, therefore, high fodder yielding varieties are not there. The taxonomic studies on *Avena* genus are summarized by Baum (1977). Vavilov Institute of Plant Industry has maintained a magnificent collection of different accessions of *Avena* (10,000 accessions of cultivated, and 2000 accessions of wild oats). The inflorescence is a loose, spreading, terminal panicle, up to 40 cm long. Spikelets are large, have 2-3 flowers which shed the seeds singly when mature. Seeds have poor dormancy.

Pennisetum and Other Millets (Order Poales)

Millets are a variable group of annual or perennial grasses mainly from the Tribe Paniceae. Finger millet (*Eleusine coracana*) belongs to Chlorideae and teff millet (*Eragrostis tef*) belongs to Festuceae. Most common millets are *Paspalum, Setaria, Pennisetum, Panicum,* and *Echinochloa.* The plants have variable height, small and round grains. Japanese millet (*Echinochloa esculenta*) is grown generally as a summer green manure crop. It is palatable and has hard stem. The Japanese millet performs very well on wet soils. The foxtail millet (*setaria*) is used a fodder when other crops are not available. It matures fast and grows well in warm conditions. Bajra or pearl millet (*P. glaucum*) is a fast growing, tall, erect, annual grass, very popular feed grown for leaves and grain in arid and semi-arid regions of India (Haryana, Rajasthan) under low rainfall, sandy conditions. The inflorescence is dense, spike-like panicle, up to 30 cm long, green when young and gradually turns brown. The spikelets have two flowers, one being staminate. Finger millet (*Eleusine coracana*) is grown in areas where rice grows under moist conditions. Millets have good balance of amino acids.

Other Grasses

There are a number of grasses used as fodder, collected from the pasture lands. They are winter and summer annuals and perennials. Grasses are called the backbone of any pasture. These grasses are not cultivated but form important constituent of the pasture fodder in India. The timothy grass (*Phleum*) is one of the most important and easiest grass to handle. It survives well in a mixture but can not stand very heavy grazing. Timothy is cold resistant but susceptible to high heat. Under favourable conditions, timothy grass attains a height of 3-4 foot and produces enormous amount of seeds. The hay is palatable, nutritious, and multipurpose. Timothy is generally grown in combination with alfalfa and clovers (legumes).

Saccharum (sugarcane) has some importance as a fodder crop in India. The sugarcane tops (few internodes of cane and upper tuft of green leaves) are the left over product of sugarcane harvesting. The canes are used for juice extraction while the cane tops are a good green fodder and also dried for hay. However, they are not too rich in nutrition and have to be supplemented with cereals. When pesticides such as organochlorines are applied to sugar cane, the leaves become toxic for cattle. Therefore, care must be taken to ensure that the fodder is free from pesticides.

The species of *Dactylis* (orchard grass) are tough, leafy, tall and hardy perennials generally grow in summer. They perform best when sown with clovers and alfalfa. The seeds remain viable for three years. It is grown alone or with clovers and alfalfa. It is a palatable grass in vegetative stage but after flowering, the palatability decreases, which is undesirable characteristic.

Several species of *Bromus* (bromegrass) are utilized as fodder. It is a tall, perennial, drought resistant and more palatable grass than timothy, slowly getting popularity among farmers. The grass propagates by underground stem and seeds both. Bromegrass has considerable hardness for cold and drought.

The tall fescue (*Festuca arundinacea*, synonym: *Lolium arundinaceum*) is a deep rooted, vigorous, perennial, and leafy grass, well suited for meadows. It grows as a bunchgrass and partial sod grass with rough stem and leaves. The root system is very expanded which gives high soil binding quality to this grass. This also provides a wide adaptability in variable climate and soil conditions (acidic and alkaline). The tall fescue has a good resistance to heat and cold, dry and wet weather.

The Kentucky bluegrass (*Poa pratensis*) also known as smooth meadow or common meadow grass is grown on permanent meadows, pastures, and lawns. It is a leafy grass attaining height up to 60 cm. The underground rhizome of the grass forms extensive sod, which is good for soil binding. Kentucky grass is suitable for close and continuous grazing by cattle. This grass grows well in cool and moist conditions and becomes dormant in hot summer and dry seasons. The Canada bluegrass (*Poa compressa*) is closely related to Kentucky grass and has similar characteristics.

Redtop grasses (*Agrostis* sp) are cool season grasses; grow to a height of 2-3 foot. The mode of propagation is through rhizomes. *Agrostis* is commonly seen in unmanaged lands, it is not grown alone. Although, it has limited utility as fodder grass but it can grow in extremely poor soil where all other plants would not survive.

Bermuda grass (*Cyanodon dactylon*, Synonym: *Panicum dactylon*, *Capriola dactylon*) is a warm season perennial grass, widely distributed throughout the country. It reproduces by seeds, rhizomes, and stolons, and may pose difficulties even as weed. The ryegrasses (*Lolium*) are cool season grasses. The most important species are *L. multiflorum* (annual ryegrass) and *L. perenne* (perennial ryegrass) which are vigorous and fast spreading. Some members of Cyperaceae (called sedges, not grasses) such as *Cyperus rotundus*, *Scirpus maritimus* grow only during rainy season and therefore have only limited value as a constituent of animal fodder.

References

Aboel-Atta A-MI, 2009. Isozyme, RAPD and ISSR variation in Melilotus indica (L.) All. and M. siculus (Turra) BG Jacks. (Leguminosae). *Academic Journal of Plant Sciences* 2(2): 113-118.

Baum BR, 1977. *Oats: wild and cultivated.* Canada Department of Agriculture. Ottawa.

Chakravarti AK, 1987. Availability of Cattle Fodder in India, Geographical Review 77(2): 209-217.

Chauhan SS, 2003. Desertification control and management of land in the Thar desert of India. *The environmentalist* 23(3): 219-227.

Chaudhary LB, 1996. A taxonomic evaluation of *Lotus corniculatus* Linn. (Leguminosae-Papilionoideae) in India. *Taiwania* 41(2): 168-173.

Chrispeels MJ, Sadava DE, 2003. *Plants, Genes, and Crop Biotechnology*. Jones and Bartlett Publishers, Canada.

Gunn CR, 1978. *Classification of Medicago sativa L. using legume characters and flower colours*. Department of Agriculture, Agricultural Research Service, Washington.

Hymowitz T, Newell CA, 1981. Taxonomy of the genus *Glycine*, domestication and uses of soybean. *Economic Botany* 35(3): 272-288.

Kaushal P, Malaviya DR, Roy AK, 2008 (Eds). *Biotechnological approaches for forage crops Improvement: Prospects, achievements and road map*. IGFRI, Jhansi (India).

Lange O, Schifino-Wittmann, 2000. Isozyme variation in wild and cultivated species of the genus *Trifolium* L. (Leguminosae). Annals of Botany 86: 339-345.

Langer RHM, Hill GD, 1991. *Agricultural Plants.* Cambridge University Press, Cambridge.

Lesins K, 1952. Some data on the cytogenetics of Alfalfa. Journal of Heredity 43: 287-292.

Naeem M, Kainth RA, Muhd Chohan SM, Khan AH, 2006. Performance of berseem, *Trifolium alexandrinum* varieties for green fodder yield potential *Journal of Agricultural Research* 44(4): 285.

Small E, Warwick SI, Brookes B, 1999. Allozyme variation in relation to morphology and taxonomy in *Medicago* sect. *Spirocarpos* subsect. *Intertextae* (Fabaceae). *Plant Systematics and Evolution* 214(1-4): 29-47.

Valizadeh M, Kang KK, Kanno A, Kameya T, 1996. Analysis of genetic distance among nine *Medicago* species by using DNA polymorphism. *Breeding Science* 46: 7-10.

Wheeler WA, Hill DD, 1957. Grassland Seeds. D. Van Nostrand Company Inc, Princeton.

Yamamoto K, Plitmann U, 1980. Isozyme polymorphism in species of the genus *Vicia* (Leguminosae). *Japanese Journal of Genetics* 55: 151–164.

Zohary M, Heller D 1984. The genus *Trifolium*. Israel Sci. Humanitiel 71: 2-145.

Chapter 9

Forages/Fodders of Different Agro Climatic Situation

Natraja Subash
Division of Crop Research
ICAR Research Complex for Eastern Region
ICAR Parisar, Bihar Veterinary College P.O., Patna – 800 014, Bihar
E-mail: n_suby@rediffmail.com

Introduction

Climate Change refers to statistically significant variation in either the mean state of the climate or in its variability, persisting for an extended period. The change may be due to natural or anthropogenic origin. Intergovernmental Panel of Climate Change (IPCC) (2007) in its fourth assessment indicated with very high confidence (90 per cent probability of being correct) that human activities, since industrialization have caused the planet to warm by about 1°C. Future projections of climate change using global and regional climate models, run by Indian Institute of Tropical Meteorology (IITM) with different IPCC emission scenarios, indicate temperature changes of about 3 – 5°C and increase of about 5-10 per cent in summer monsoon rainfall (NATCOM, 2004). It is also projected that number of rainy days may decrease by 20 to 30 per cent, which would mean that the intensity of rainfall is expected to increase. Extremes in temperature and rainfall also show increase in their frequency and intensity by the end of the year 2100. A more recent study using daily rainfall data for over 50 years shows significant increasing trend in extreme rainfall events over central India (Goswami *et al.*, 2006). India is mainly agriculture dominant country.

About 70 per cent of the population lives in villages and their main occupation is agriculture and animal husbandry. The animal products depend on the fodder and forage given to them.

Today the area for fodder production is 4.4 per cent of the total area. The area under permanent pastures and cultivable wastelands is approximately 13 to 15 million hectares, respectively. Similarly out of 2.51 crore hectare land, only 2.1 crore hectare is open to grazing. All these resources are able to meet the forage requirements of the grazing animals only during the monsoon season. But for the remaining periods of the year, the animals have to be maintained on the crop residues or straws of jowar, bajra, ragi, wheat, barley, etc. either in the form of whole straw or a bhusa, supplemented with some green fodder, or as sole feed. The semi-arid climate in many parts of India and the pressure on land use have made tree and shrub fodders a more important component of feeds for ruminants compared to grasses or grass-legume pastures. Dry, deciduous vegetation is mostly found in semi-arid regions and is confined to the north-west area of the subcontinent. Many of the fodder trees are not cultivated and the landless population which owns small herds of sheep and goats depends on shrubs and tree feed resources growing near the villages, roadsides and community lands (Raghavan, 1990). When the sources in the vicinity of villages are depleted the rural women frequently reserve forest areas in the hills to obtain the daily requirements of their livestock. It is also a common practice throughout India to lop and dry tree leaves when they are abundant and store them for feeding during periods when feed is scarce. Although most of the trees and shrubs used for animal feed are not cultivated, there are traditional farming systems in India where they are deliberately planted by farmers in an agrosilvicultural system.The natural grasslands and the cultivable waste and fallow lands provide some grazing during the favourable growth periods in the monsoon season. The system of fodder production vary from region to region, place to place and farmer to farmer, depending upon the availability of inputs, namely fertilizers, irrigation, insecticides, pesticides, etc. and the topography. An ideal fodder system is that which gives the maximum outturn of digestible nutrients per hectare, or maximum livestock products from a unit area. It should also ensure the availability of succulent, palatable and nutritive fodder throughout the year.

Two types of fodder crops are grown, one as temporary crops and the other as permanent crops. Permanent fodder crops are crops especially grown for fodder. Permanent fodder crop may be cultivated or growing wild. Temporary fodder crop include three major group of fodder, which are grasses including cereals, legume crops and root crops. These are given to animals either as green or dried as hay or as silage. Nutritive values of the three fodder groups are different. Grasses are rich in crude fiber and crude protein. Legumes are rich in proteins and minerals and roots are rich in starch and sugars and poor in fibers.

Following are important climatic requirement of some of the fodder crops:

Berseem (*Trifolium alexandrinum*)

Berseem is winter legume crop of central, north and north-west part of India which include Punjab, Himachal Pradesh, Western UP, Maharshtra, Karnataka, and Tamil Nadu. Berseem is an important fodder for milk industry, as it is said to be milk multiplier. It has 70 per cent crude protein and 70 per cent dry matter digestibility. Being a leguminous crop, it increases nitrogen content of soil. Also organic carbon and phosphorous content of soil is increased. Diploid and tetraploid varieties are found. Diploid varieties include BL-1, Miscavi and Khadaravi, Chhindwara, IGFRI-99-1, whereas tetraploid varieties are Pussa, Giant, and T-378. Berseem grows very well in heavy soil. It is tolerant to alkaline and soil with high salt concentration but sensitive to acidic soil. It also grows very well in medium loam soil rich in phosphorous and calcium. Berseem requires dry and cool climate for its normal growth. It needs mild temperature for germination and establishment. The crop is not sown unless average daily temperature reaches 13-15.5 degree centigrade. Temperature above this will hamper the germination. The crop growth is very fast at 18-21 degree centigrade. Its growth is checked during very cold or frosty periods. It is also not suitable for high rainfall areas. It is grown as rabi crop in north India and best time for sowing is first fortnight of October. The final cutting of the crop is done at the end February if left for seed purposes. The crude protein content is maximum (27 per cent) at 60 day cutting, but it drops to 17 per cent by 105 day. Its green fodder yield ranges from 100 to 120 t/ha in 6 to 7 cuts.

Lucerne (Alfalfa or *Medicago sativa*)

Lucerne is leguminous fodder crop and it is considered worldwide as Queen of fodder crop. It is generally grown in temperate zones of the world and in tropics at higher altitude. It is grown as perennial crops in drier regions and as annual crop in hot and humid regions. In India it is grown extensively in rabi season. In North India it is sown from early October to November end but best time for sowing is middle of October. It is grown from October to December in southern India. Harvesting is done before full bloom because nutritive value and digestibility decreases after full bloom. The crop is ready after 50 to 60 days of sowing. It can supply fodder continuously for 3 years. It generally grows in areas were condition is unfavorable for growing berseem. Its root grows vigorously deeper in the soil, thus can draw water from deep in the soil. It is a long day plant but minimum photoperiod required to initiate flowering varies among different cultivars. It is harvested for hay, soilage, silage, dehydrated meal and medicinal purposes. Its nutrient content is about 15 to 20 per cent crude protein with 72 per cent digestibility, 1.5 per cent calcium and 1.2 per cent phosphorous (on dry weight basis) and high amount of vitamin A, B and D. Being rich in nutrients; it is primarily given to dairy cattle. Sirsa No-8k, Sirsa No-9, Co-1, Moopa, Ramber, Anand-2, NDRI Selection No-1, IGFRI-S-244 (chetak) and IGFRI-S-54 are few good varieties of alfalfa plant. It grows well on well drained, deep and loam soil with high soil fertility. Its favourable pH range is 6.0 to 6.5. Lime application is necessary in acidic soil for optimum yield. It does not grow well on heavy water logged soil. It grows well in various climatic condition but best on warm, dry and sunny season. First crop is harvested after 2 to 3 months and subsequent harvesting at an interval of 20 to 30 days.

Oat (*Avena sativa*)

Oat is 6th most cultivated cereal in the world. It is used as food by humans and also given to animals as green fodder, silage and hay. Oat grains are given to horses, sheep and other livestock whereas hay or straw is given to cattle. It is a temperate region crop but also grown in tropics at higher altitude. In India it is grown in Rabi season. In the southern India it is grown at higher altitude under rainfed condition. In the northern and central India it is also grown in rotation with Kharif crops like sorghum, maize. It is widely grown in Uttar Pradesh, Madhya Pradesh, Haryana, Punjab, Himachal Pradesh, Rajasthan, Bihar, Gujarat, and Andhra Pradesh. Its nutrient contents is 7.6 per cent crude protein at 50 per cent flowering stage and about 14.6 per cent at early stage of plant growth. Palampur1 and Kent are the two most grown variety of oat; some other varieties are OS-6, OS-7, OL-9, OL-125, UPO-212 and JHO-851. Their annual green fodder yield is about 500 quintals/hectare and 360 quintals/hectare respectively. The crop is harvested at 50 per cent flowering stage for one cut management system, whereas in multi cut management system first cut is done at 65 days from date of sowing. It grows well in all types of soils except water logged, however it grows well in well drained friable loams. Optimum temperature for its growth is between 5°C to 30°C, however optimum temperature required is 25°C. The crop matures in about 130 to 150 days. From December to April two to four cuts are done. If no cut is done for fodder, its seed production is 15 quintals per hectare.

Cowpea (*Vigna unguiculata*)

Cowpea is leguminous fodder crop grown for both food and fodder. It is a very fast growing crop and thus gives large amount of fodder in very short duration. As fodder it is used for green feeding, hay making and ensiling mixture with sorghum or maize. It is best grown in moderately humid areas of tropics and sub tropics. It grows well in areas with height 1500m above sea level and areas receiving annual rainfall of 750mm. It can be grown as both annual and perennial herb. Its protein content is 18 per cent. Varieties suitable for fodder are Russian giant, Karnataka local, RS-9, UPC-1956, UPC-5827, and UPC-9805. It is grown in both rainy and summer season. It is best grown in 21° to 35°C temperature range. It is a drought tolerant crop and can be grown in wide variety of climatic condition but it cannot withstand frost. Being leguminous crop, it is also grown as intercrop with maize, sorghum, pearl millet, guinea grass etc. Cowpea is best grown in loam soil, but it can also grow from sandy loam to heavy clay. It also requires well drained soil, as it cannot withstand water logging and flooding. It prefers growing in slightly acidic soil having pH range 5 to 6.5. It is not suitable for waterlogged, alkaline and saline soil. A single cut crop gives green matter yield of 25 to 30 ton/hectare while multicut crop yield 40 ton/hectare.

Hybrid Napier (*Pennisetum perpureum*)

Napier is also known as elephant grass. It provides green fodder all around the year. Napier grass is coarse texture with leaf blade hairy and less juicy. In 1951 it was crossed with bajra, the resultant hybrid napier is fine textured, juicy, palatable, fast growing and drought resistant. Hybrid napier is a sterile hybrid and it is grown by rooted slips or stem cutting. Its nutrient content is 8.2 per cent protein, 34 per cent crude fiber, 10.5 per cent ash with both calcium and phosphorus in proper balance. It is good for green fodder, silage and hay. Pusa Giant Napier, Gajraj, NB-5, NB-6, NB-21, NB-35, Co-1, Co-2, Co-3 etc are most grown variety. Among all these Co-3 is high yielder. The grass grows throughout the year in the tropics. The optimum temperature is about 31°C. Light showers alternated with bright sunshine are very congenial to the crop, but the best time for sowing is by the onset of south west monsoon. Total water requirement is about 800-1000 mm. It grows on variety of soils, however grows best on light loams and sandy soils with good drainage system. It also grows best in soil rich in organic matter and nutrients. Its ideal pH range is 5 to 8. Water logged and flood prone areas are not suitable for napier grass. The crop is ready for first harvest after 75 days from date of sowing, and subsequent cut is done at an interval of 30-45 days. About 7-10 cuts can be done in one year. Green fodder yield ranges from 200 to 250 ton per hectare from 6 to 8 cutting.

Maize (*Zea mays*)

Maize is grown for both cereals and fodder. It gives high yield and good quality green fodder in very short duration. It is grown as mixed crop or as intercrop with compatible grasses and legumes. The crop is utilized a grazing soilage and silage but it is best crop for silage making. In India maize is grown over 5.7 million hectare area. Varieties cultivated for fodder is Deccan, African tall, Jawahar, Ganga-5, Ganga safed-2, and Ganga-3 and composite variety Vijay. It is warm season crop but can grow in diverse climatic condition ranging from tropical to temperate region. Maize is cultivated during kharif season in north India. In the south the best time is between April and October. It grows well in regions with warm climate and annual rainfall 60 to 100cm, but cannot tolerate heavy rains and dry hot winds. It gives best yield in areas receiving 1200-1500mm annual rainfall. It is drought tolerant up to five weeks and thereafter it is susceptible. Maize is a warm weather plant. It grows from sea level to 3000 meter altitudes. It can be grown throughout the year and under diverse conditions. Maize requires considerable moisture and warmth from germination to flowering. The most suitable temperature for germination is 21 degree centigrade and for growth 32 degree centigrade but can tolerate minimum temperature of 10°C and maximum temperature of 45°C. 50-75 centimeter of well distributed rain is conducive to proper growth.. It grows well in drained sandy loam and silty loam soils. Its ideal pH range is 5.5 to 7.5. It can give fodder yield of 600quintals green fodder per hectare. Harvesting is done after 60 days from date of sowing. Second cut can also be done in good moisture condition. The crude protein content is maximum at tasseling stage and fall sharply after milk stage. For silage making it can be cut when grain is full and glazed. Its nutrient content is 7-10 per cent crude protein and 25-35 per cent crude fibre. Its yield range is 35-55 ton/hectare.

Sorghum (*Sorghum* sp)

Sorghum is tropical forage crop. It is also grown as cereal crop. It is sometime called 'crop camel' as it is highly drought tolerant and grows well in semi-arid areas and areas with less rainfall. It is used as both food and fodder. After harvest of grain, sorghum straw is feed to cattle. 'Sargos' is group of sorghums grown specifically for fodder or syrup production. Modern cultivars have been produced by crossing male sterile grains with sargos or sudan grass. Important varieties grown for fodder are M.P.Chari, SSG-59-3, MPK V-I, JS-20, S-1O49 and JS-3. It is grown as Rabi crop in southern and western India, and in northern India is kharif crop. It grows well in warm condition (26-30C°) with annual rainfall 60 to 100 cm. Minimum temperatures for germination of seed is 6 to 10°C. It is generally grown in areas too dry. It grows very well clayey loam soil rich in humus with good water retention capacity. Its ideal pH range required is 5 to 8.5. Harvesting is done at 50 percent flowering stage, as after this stage crude protein content and digestibility decreases sharply. Less amount of HCN is present at flowering stage. Sorghum is very palatable at young and flowering stage. In multicut varieties first harvesting is done after two month of sowing and subsequent cut at an interval of 35 to 40 days. Its fodder yield on dry land is 20-30 ton/hectare whereas on irrigated land 50-60 ton/hectare. Its nutrient content is 7.8 percent crude protein and 32.3 percent crude fibre.

Subabul (*Leucaena leucocephala*)

Subabul is perennial evergreen shrub. It is multipurpose plant whose green leaves are used as fodder, woods are used for firewood and timber, for afforestation of hill slopes, for soil erosion control through wind breaks and fire breaksand as good source of manure. It is highly nutritious fodder with protein content 27 to 34 per cent. It is also rich in vitamin A and carotene. Provitamin A is highest among all plant species. Its leaf contains amino acid mimosine which is toxic to non ruminants at level 10 per cent in diet. It improves quality of silage when mixed with hybrids napier or guinea grass in proportion of 1:1 or 1:2. There are four type of subabul depending upon its morphological character, which are Hawaiian type, Salvador type, Peru and Cunningham. It grows in warm regions were temperature range is 22 to 30°C. It is found in areas with altitude below 500m and rainfall above 750mm. It can withstand variation in rainfall, temperature, wind storm, drought, salinity and land terrain. It cannot tolerate water logging. It grows well in all type of soil but grows best on deep drained neutral soil. It can also be grown on saline and acidic soil. It also grows in gravely areas and sandy loams. Its favourable pH range is 5.0 to 8.0. Branches are harvested after five to six months of planting, subsequent cutting are done after 50-60 days depending upon regrowth. Its annual yield in non irrigated condition is 25 to 30 ton/year whereas in irrigated condition it can give 100 ton/hectare green fodder in seven to eight cutting.

Guinea Grass (*Panicum maximum*)

Guinea grass is a popular fodder grass of the tropics. It can be profitably grown as a component of agro-forestry systems and comes up well under coconut and other trees. As an excellent fodder it is much valued for its high productivity, palatability, good persistence, herbage yield and good response to fertilizer. The important varieties are Makueni, Riversdale, Hamil, PGG-4, FR-600, Haritha, and Marathakom. Makueni is a drought resistant cultivar. Guinea grass thrives well in warm moist climate. In India it is best suited to agro-climatic condition of Kerala and places similar to Kerala climate. It can grow from sea level to 1800 m altitude. It is frost sensitive. It thrives well between temperature ranges of 15 to 38 °C. It does not tolerate drought and requires annual rainfall of 1000 mm and more. The grass tolerates shade and grows under trees and bushes and is best suitable as an intercrop in coconut gardens, however it does not tolerate continuous shade. The grass is adapted to a wide range of soils. It usually grows on well-drained light textured soil, preferably sandy loams or loams, but is better suited to medium to highly fertile loams. It cannot tolerate heavy clays or prolonged water logging. Normally it does not tolerate acidic and saline conditions too but it can tolerate acidic soil in proper drainage condition. Under Kerala conditions, the best season of planting is with the onset of southwest monsoon during May-July. As an irrigated crop planting can be done at any time of the year. At planting two irrigations are required within seven to ten days for quick establishment. The crop should be subsequently irrigated depending upon the rainfall and soil type. Usually irrigation once in 7-10 days is required. Irrigation with cowshed washing or sewage water within 3-4 days after cutting gives better growth. The crop becomes ready for harvest in 75 days from date of sowing; subsequent cuts are made in every 45-60 days. About 7 to 9 harvests are possible in one year. Its average yield is 80-100 ton per hectare, certain cultivars can give yield of 140ton per hectare.

Fodder Bajra (*Pennisetum americanum*)

Bajra is cultivated for both food and fodder purpose, mainly in arid and semi arid regions of the world. It hybridizes freely with napier grass, which is known as hybrid napier. Fodder bajra is grown in the same way as bajra grain. It tolerates droughts and thus grown in semi arid region of world without irrigation. It can grow in areas receiving annual rainfall of 250mm but grows better with annual rainfall of 400-750mm. Some of its root can penetrate deep in the ground (up to 360cm); although 80 percent of root weight is at top 10 cm. It can grow in wide variety of soil but prefers growing in well drained fertile sandy loam and loams. It cannot tolerate water logging. It is grown as rainfed kharif crop or irrigated crop during summer season. In India it is grown as Kharif crop and sown during July. Depending upon monsoon the crop can be grown during May-June, September-October in southern India. Commonly grown fodder bajra cultivars are Giant Bajra, L-72, L-74, TNSC-1 and Rajko. The crop is ready to harvest by 60-75 days from date of sowing. Average green fodder yield is 25-30 ton per hectare. Its nutrient content is 6.8-12.8 per cent crude protein and 29-34 per cent crude fiber.

Cluster Bean or Guar (*Cyamopsis tetragonoloba*)

Cluster bean is a multipurpose legume crop, used for vegetable, fodder, green manure, and in industries. It is drought tolerant crop and thus grown in semi arid tropics of the world. It grows in all types of soil, but grows well in alluvial and sandy loams. It is sensitive to water logging and excess moisture. Commonly grown cultivars are FS-277, HFS-119, HFS-156, HGS-156 and Guara-80. Harvesting is done from flowering to pod initiation stages. The plant becomes ready to harvest within 65-75 days from date of sowing. Delayed harvesting may result in woody and fibrous plant. Green godder yield is in range of 25-30 ton per hectare. It is used as green fodder, hay and silage. Its nutrient content is 14-90 per cent crude protein and 22-32 per cent crude fibre.

Para Grass (*Brachiaria mutica*)

Para grass is perennial grass widely grown in tropics and subtropics. The grass is easy to propagate from cutting, gives high yield, good quality and palatability, and is tolerant to water logged and flooded conditions. it has less tolerance to droughts and in wet areas it may become problem for blocking drainage channel. It responds well to sewage irrigation and is generally found in areas near sewage disposal. It has vigorous and competitive nature and competes with coexistence of other crops, thus it is not suitable as mixed and intercrop. It grows well in hot and humid areas with rainfall above 1000mm. It is sensitive to cold environment and its growth ceases in this condition. It may also become dormant in dry season, if water is not provided. It prefers alluvial and hydromorphic soil. It also grows well in wide range of moist soils excluding highly wet soil. It also grows better in fertile clay and clay loam textured soil with high moisture retention capacity. It can also grow in sandy soil with irrigation facility. It is highly tolerant to saline or sodic soil condition but less tolerant to slightly acidic to alkaline soil. All these features make this grass excellent for soil reclamation. It does not tolerate shade and thus cannot be grown as intercrop with coconut or other trees. The crop is ready to harvest in 3 months from date of sowing. Subsequent cuts are made at 30-40 days time interval. It attains height of 60-75cm during harvesting period. About 10 cuts are done in southern India, whereas in northern India only 4-5 cuts are done because of winter. Thus yield in south India is 120 ton per hectare and in north India it is 70 ton per hectare. The grass does not have any toxic effect but nutritive value is less. In pre bloom stage, it has 5.9-15.4 per cent crude protein and 27.2-41.5 per cent crude fibre.the grass is used as green fodder and soilage.

Congosignal Grass (Prostrate Signal Grass or *Brachiara ruziziensis*)

It is creeping perennial grass with dense foliage. It forms dense mat in the field and thus used for soil erosion control. It is commonly found in cleared rain forest of Africa. It can be grown as solitary crop or as intercrop in coconut garden. It is more shade tolerant than signal or para grass. It can also be grown with legumes and other grasses. It prefers growing in warm moist tropical climate receiving rainfall above 1000mm. It has good growth in 28-30°C day/night temperature and minimum night temperature of 19°C. It is drought tolerant too. It grows in all types of soils but requires good soil fertility. It can be grown in acidic soil but it is sensitive to water logging and needs proper drainage. Its preferred soil pH range is 5-6.8. The crop is ready to graze after 50 days from date of plantation, and subsequent grazing at interval of 30-40 days interval. Sometimes silage and hay are aloes made. It can give 35-45ton/hectare green fodder under rainfed condition and 50-100 ton/hectare under irrigated condition. Its nutrient content is 13 per cent crude protein and 27 per cent crude fibre. It is very palatable nad no toxic principles have been reported.

Setaria Grass (*Setaria sphacelata*)

Setaria is nutritious and palatable grass, brought to India in the year 1950 through Indo-Swiss project. It can be grown alone or with perennial grasses and legumes. It grows well in places where annual rainfall ranges 900-1825mm. It can survive in long, hot and dry season. It grows well in sea level to 3300 altitude. It is drought tolerant plant and can withstand water logging, but it is not shade tolerant. Some of the cultivars are Kazungala, Narok, Bua River, Du Toits Kraal. Optimum temperature for its growth is 19-22°C. First harvesting is done after 70-80 days from date of planting, subsequent cuts after every 40-50 days. About 7-8 cuts are done in a year. The grass is lightly grazed during early stage of planting and then heavily grazed because in early stage heavy grazing can make plants to get out with their roots and in later stages plant may become stemmy and coarse under light grazing. Its annual fodder yild is 75-150ton per hectare. The fodder can be used as silage, hay and green fodder. Its nutrient content is 4.8-18.4 per cent crude protein and 24-34 per cent crude fibre. Fresh fibre is acidic with pH 4.8 due to high oxalic acid content.

Deenanath Grass (Kyasuwa grass or *Pennisetum pedicellatum*)

It is a common grass found along roadside. It is capable of growing in poor management condition and thus it is called Dheenanath *i.e.* friend of poor. Annual types are cultivated while perennial types are grown as pasture grass. It can be intercropped with legumes, rice, soybean etc. It requires warm climate and is capable of growing in areas with rainfall 500-1500mm. Optimum temperature for its growth is 30-35°C. It is tolerant to acidic, alkaline and saline soil. Important cultivars grown in India are PS-3, IGFRI-4-2-1, IGFRI-43-1, IGFRI-3808 and Pusa Deenanath. As rainfed crop it is sown after onset of south west monsoon and as irrigated crop during early march. First cut is done after 70-100 days of plantation, subsequent cut at an interval of 35-45 days. Its total green fodder yield is 120 ton per hectare. It is genrally used a sgreen fodder but sometime hay and silage are also made.

Mission Grass (*Pennisetum polystachyon* (L.)

Like dinanth grass, mission grass is also major grass in tropics. It is common along road side and waste places. It grows from sea level to 1500m altitude. It is abundant in both heavy rainfall areas and semi arid areas. It is found in light to medium textured soil. It is also tolerant to both acidic and alkaline soil. It is sown by onset of monsoon. The crop is ready to harvest after 70 days from date of sowing, subsequent cuts at an interval of 45 days. Usually three cuts are done in one season. It is used as green fodder, hay and silage. Its nutritive value lessens after seeding. Its nutritive value is 14.5 per cent crude protein and 30 per cent crude fibre.

Conch Flower Creeper (Butterfly Pea or *Clitoria ternatea*)

Butterfly is widely grown as ornamental creeper in tropical regions. It suppresses growth of other weeds when grown for fodder. It is widely grown on grasses on which it can climb. The plant produce large amount of seed, thus it is self grown. It is also faster to establish and takes shorter time to first grazing and production. The plant is very palatable and persists well, when lightly grazed. It grows well in areas with 1500mm rainfall. It is drought tolerant but grows well under irrigation. It cannot tolerate water logging. It grows well from sandy to deep loam and heavy clays soil. It grows on wide range of soil pH 5.5-8.9. It produces 30-40ton per hectare green fodder under cut and carry system. It has 10.5 to 25.5 per cent crude protein.

Gliricidia [Madre Tree or *Gliricidia sepium* (Jacq.) Walph]

Gliricidia is tropical tree grown as a shade tree, green manure, fodder, conservation hedge, live fence, and fire wood. It is preferable fodder tree over subabul, as it is tolerant to moderate shade, very poor and acidic soil, and leaf psyllid

(*Heteropsylla cubana*) attack, which has proved fatal to many areas. It is generally grown in areas with 900 to 1500mm annual rainfall and five month dry period. It can also grow in very less rainfall region with 600-3500mm annual rainfall and 20-30°C temperature range.

Calliandra (*Calliandra calothyrsus*)

Calliandra is leguminous tree, generally cultivated for fodder, reforestation, soil stabilization and improvement, and as shade tree. It grows on areas with 1000mm or more annual rainfall. It can grow on varying altitudes from sea level to 1800m but prefers growing in 250 to 800m altitude with 2000 to 4000mm annual rainfall. The tree is evergreen in humid climate, whereas semi deciduous in areas with long dry season. It can grow on deep loams to acidic sandy clay soils, but prefers light textured, slightly acidic, humid tropical soils. The plant cannot tolerate waterlogged situation. The plant attains 3-5m height after 1 year. This is the stage for first harvesting. The cutting is made 20-50cm above the ground, so that plant has fast re-growth. Its annual fresh fodder yield is 46.2 ton per hectare. Its nutrient content for dry leaves is 22 per cent crude protein, 30-7- per cent crude fibre. It can act as carotene and protein source, which maintain yolk colour in commercially produced eggs.

Teosinte (Buffalo grass or *Zea mexicana*)

Teonsite is wild relative of maize, and crosses freely with maize. It has dark green and narrow leaves than maize. It is hot and humid region crop and requires annual rainfall of about 1000mm. It is sown either in May-June or September-October. The plant is harvested before appearance of tassel. It can give two cutting if sown early. Its fodder yield is 700 to 900 quintal per hectare. Its nutrient content depends upon stages of growth and growth condition, which is in range of 4.5-12 per cent crude protein and 19.6-32.2 per cent crude fibre. It is used as green fodder, hay and silage. For silage making harvesting is done when tassels are fully formed and seeds are in watery to milk stages.

Velvet Beans [*Mucuna pruriens* (L)]

Velvet beans are leguminous plant which can be grown as annual or perennial plant. It is of herbaceous and climbing nature and sometime bushy forms also exists. It can be grown with other crops (like sugarcane, maize) or as cover crop or for green manure and also as anti-erosion crop. It is best grown in hot and humid climate receiving annual rainfall of 650 to 2500mm. It can be grown from sea level to 2100m altitude. Velvet bean are generally cultivated in tropical and sub tropical regions. Its favorable temperature range is 20-30°C and 21°C night temperature stimulate flowering. It is susceptible to frost. It can grow on sandy to clay soils. It can also grow in well drained heavy clay soil. Its favorable pH range is 5-6.5. It can be grown on soil unsuitable to grow cowpeas but has disadvantage of longer growth period and difficult to thresh. The crop is ready to harvest after 90-120 days from date of sowing, if grown for fodder. Its leaves and vines are palatable and given to cattle. Hay and silage are also made from this plant. It can give 300 quintals per hectare green fodder. It is rich in calcium, phosphorous and iodine, thus very good fodder for all farm animals. Its nature seed contain alkaloid, thus not recommended for livestock feed.

Centro (*Centrosema pubescens* Benth.)

Centro is climbing perennial herb, widely found in tropics and sub tropics. It was first grown as cover crop with trees but now widely grown as fodder crop because of its palatability, vigorous and productive growth, and good quality herbage. It is widely grown as mixed crop with grasses and legumes, as it conserves soil and water. It becomes shade tolerant when mature. It is generally grown in areas with 1750mm rainfall but can also grow in areas with 750-1000mm rainfall. Being deep rooted, it is fairly drought tolerant. It can grow on wide range of soils, from sandy loams to clays. Its favorable pH range is 4.9-5.5. it takes one year to become well established, and then it can withstand heavy grazing. It is grown for grazing, soilage and hay. It is cut before flowering, if planted for hay. Its nutrient content is 25 per cent crude protein and 30 per cent crude fibre.

Fodder Production in Arable Farming

There is ample scope for fitting in the short-duration fodder crops, either single or in mixture, with the other crops during the gap period between two main crops. Two distinct fallow periods are available for raising short-duration fodder crops, provided adequate resources are available. In the case of the wheat-jowar rotation, gap periods between April and June and between October and November are available for each crop as fodders. M.P. chari + cowpea, maize + cowpea, *bajra* + cowpea is successfully grown and an additional green-fodder yield to the tune of 300-350 q per ha is obtained. Similarly, in the second gap period (October-November), which is rather short, the growing of fodder turnips and short-duration mustard varieties helps to get 250-300 q per ha of fodder without disturbing the normal cropping systems.

Fodder Production Under Dryland Farming

A large proportion of the area of our country is located in the dryland regions. In these areas, the farmers usually grow at least one crop in the *rabi* season after conserving the soil moisture. Thus there is a great scope for raising two crops under such situations. First, the growing of a fodder crop which gets ready in 45-50 days after sowing (cowpea, *jowar*, *guar*, *sanwa*, *moth*, etc.), yield 150-250 q per ha of green fodder. After harvesting the fodder crops, crops such as gram, linseed, barley, wheat and safflower are raised on the conserved moisture

Intensive Fodder Production Under Relay Cropping

There is ample scope for increasing fodder production from the high-input areas, either by growing high-yielding fodder crops singly or in mixture. The growing of three or four successive fodder crops, helps to boost fodder production per unit area. Some of the important intensive fodder-crops rotations and the expected yields from each are summarized in below:

Maize + cowpea - maize + cowpea + teosinte - *berseem* + mustard (300 q/ha) - (450 q/ha) - (1,000 q/ha)

Sweet sudan + cowpea - *berseem* + oats (1,000 q/ha) - (1,000 q/ha)

Hybrid Napier + lucerne (1,250 q/ha) - (850 q/ha)

Maize + cowpea - *jowar* + cowpea - *berseem* + mustard (300 q/ha) - (400 q/ha) - (1,000q/ha)

Teosinte + *bajra* + cowpea - *berseem* + oats (1,000 q/ha) - (1,000 q/ha)

Sweet sudan + cowpea - mustard - oats + peas (1,000 q/ha) - (250 q/ha) - (500 q/ha)

Jowar - turnips - oats - 1800 q/ha

Adapted from www.krishiworld.com.

Limitations and Future Strategy for Cultivation of Fodder and Forages

Despite urgent need, cultivation of fodder crops, grasses, etc. not being given due attention on account of stress on more remunerative agricultural crops like grains, oilseeds, pulses, Sugarcane etc. Even crop residues are being diverted for other by products. In certain states crop residues are being burnt or left in field after harvesting through combines leading to loss of precious fodder. Pressures on grazing areas are increasing on account of increase in population, encroachments of pasture areas; tendency of farmer to keep large herds due to low productivity; and Non-remunerative prices for fodder and fodder like produces. There is a deficit of approx. 62 per cent of green fodder and 22 per cent of dry fodder and this in absolute terms, shortage is 635 million tonnes of green fodder and 126 million tonnes of dry fodder. The deficit situation becomes worse in the event of extreme climatic conditions drought, floods, cyclones etc.

Need for a paradigm shift from only drought management to adopting long term strategy to develop fodder as an integral part of agricultural practice and crop diversification. Efforts should be made to increase the area under fodder cultivation, through cultivation of Ussar and Wastelands, propagation of high yielding varieties particularly those enabling multicuts. Planned cultivation of fodder trees, including fodder shrubs as an integral part of agroforestry programmes. Better agriculture residue management should be adopted. Burning of crop residues to be prohibited. Residue to be collected, compressed and nutritionally enhanced, scientifically stored, distributed and marketed. Establish Fodder banks to store surplus forages and this can help in reducing the regional imbalances in fodder availability. Involvement of NGOs, Cooperatives, existing Gowshalas and Gowsadans in increasing fodder cultivation, their processing and storing. There is an urgent need to enhance the production of feed grains especially maize, which is extensively used in poultry. Presently only 3 per cent of the grains is utilized for livestock and poultry. The proportion of feed grains for organized compounded livestock and poultry feed must be increased. Fodder development should be taken as a part of the integrated agriculture practice, land use planning and management.

References

Goswami, B.N., Venugopal, V., Sengupta, D., Madhusoodanan, M.S., Prince K. Xavier., 2006. Increasing Trend of Extreme Rain Events over India in a warming Environment. Sci. 314, 1442-1444.

International governmental panel on climate change, 2007. Summary for policymakers. Climate change 2007: the physical science basis. Contribution of Working Group I to the Fourth Assessment Report of the Intergovernmental Panel on Climate Change. Cambridge University Press, Cambridge, United Kingdom and New York, NY, USA.

NATCOM, 2004. India's Initial National Communication to the United Nations Framework Convention on Climate Change. National Communication Project, Ministry of Environment and Forests, Govt. of India.

Raghavan, G.V. 1990. Availability and use of shrubs and tree fodder in India. In: Devendra, C. (ed.), *Shrubs and tree fodders for farm animals*. Proceedings of a workshop in Denpasar, Indonesia, 24–29 July 1989. IDRC-276e, Ottawa, Ontario, pp. 196–210.

Thomas, C.G. 2003. Forage Crop Production in the Tropics. Kalyani Publications, Ludhiana, India. pp:259.

Chapter 10

Crop Selection: An Efficient Tool for Sustainable Forage Production

Anil Kumar Singh[1], Lal Singh[2] and Sushil Dimree[3]

[1]Senior Scientist, Agronomy, Division of Land And Water Management, ICAR Research Complex for Eastern Region, ICAR Parisar (P.O.- B.V.College), Patna – 800 014, Bihar

E-mail: aksingh_14k@yahoo.co.in

[2]Assistant Professor, Division of Agronomy, SKUAST-K, Shalimar, Srinagar – 191 121, J&K

E-mail : drlalsingh@rediffmail.com, drlalsingh72@rediffmail.com

[3]Assistant Professor, Department of Soil Science and Agriculture Chemistry, CSAUA&T, Kanpur – 208 002, U.P.

E-mail : dimri.astro.gmail.com

Indian crop production system excels in front of several biophysical adversities commonly known as biotic and abiotic stress. Water is the principal factor limiting crop yield and about 2/3 of total cultivated land in India comprise of dryland agriculture primarily under rainfed conditions with no or limited water supply system. The Indian production systems are dominated by cereals, primarily rice maize, sorghum, pearl millet other coerces and fine cereals leguminous (urd bean, mung bean, peanut) and oil yielding crops during kharif and wheat barley chickpea, lentil and forage legumes in rabi season. The production systems are generally characterized by cereal/legume mixed-cropping dominated by maize, millet, sorghum, and wheat. The major constraints in both regions to crop production are low soil fertility, insecure rainfall, low-productive genotypes, low adoption of improved soil and crop management practices, and lack of appropriate institutional support. Choice of crop is important but selection of right varieties is utmost important because varieties which have proven excellent in irrigated or high rainfall areas are generally not suited for limited irrigated conditions. An attempt has been taken to screen out crops and their suitable varieties which could perform better with limited irrigation in different agro climatic condition. In this manuscript a brief account on criteria for selection of particular crops in general and their varieties were discussed in the light of Agro climatic conditions and requirement of crops.

Background

India has diverse agro-climatic regions comprising tropical, sub tropical and temperate regions, where we can grow various kinds of crops including fruit plants. We have achieved a lot particularly in the improvement of field crops and almost reached at peak. However, breeders are trying their level best to increase the income of the farmers by applying various latest techniques for improving the productivity. The increasing scarcity of water is now a well-recognized problem. High rate of population growth and development, require continuous diversion of agricultural water to higher priority sectors. The need to produce more food with less water poses enormous challenges to transfer existing supplies, encourage more efficient use and promote natural resources conservation.

On-farm water-use efficient techniques if coupled with improved irrigation management options, better crop selection and appropriate cultural practices, genetic make-up, and timely socio-economic interventions would help achieving this objective. There are two options for the management of crops in water limiting environments, the agronomic and the genetic management.

Criteria for Selection of Crop and their Variety for Stress Environment

Choice of crops and varieties is important. Varieties which have proven excellent in irrigated or high rainfall areas are generally unsuited for dry land conditions. Many attempts at dry land farming have failed, largely due to lack of recognition of the requirements for the variety selection. Variety requirements for limited irrigation conditions Short-stemmed varieties with limited leaf surface minimize transpiration. Deep, prolific root systems enhance moisture utilization. Quick-maturing varieties are important in order that the crop may develop prior to the hottest and driest part of the year and mature before moisture supplies are completely exhausted.

Land Resources Indian Scenario

Indian location in world map is unique in several contexts; it is located in Asian Region particularly in South Eastern Region, which is again much diversified in all respect having almost all kind of climate right from tropical to temperate high land to below see level cultivated area in Kerala. India having 7.0 per cent of total geographical area of world with 16 percentage of whole universe population. India having 329 M ha total geographical area out of which 142 M ha area under cultivation with less than 150 per cent cropping intensity due to lack of assured irrigation system. Apart from having more area in rain fed cultivation total area under non suitable category, relatively less or unsuitable for general cultivation is listed in Table 10.1 which is ultimate produce of agro ecological evolution process, which needs suitable intervention to brought back gradually under partial or full cultivation.

Table 10.1: Type and Extent of Land Degradation in India

Type of Degradation	*Area (Million ha)*
Water erosion	57.16
Wind erosion	10.46
Salt Affected and water logging	9.52
Shifting cultivation	2.38
Ravines	2.68
Degraded forest	24.90
Others	1.34
Total	107.43

Crops and Cropping Pattern

Moisture is basic necessity for existence of any farm of life including agriculture. Crops and cropping pattern is interdependence and interchangeable largely influenced by growing period of particular region which is again an outcome of soil and climatological considerations. An attempt has been made to quantify the agricultural/ cropping activities based on length of growing period basically depends on availability of water/moisture to support successful crop production (Table 10.2).

Table 10.2: Choice of Cropping Pattern for Different Growing Periods

Length of Growing Period	*Efficient Cropping System*
<75 days	Perennial vegetation
	Monocropping of short duration pulses
75-140 days	Monocropping
140-180 days	Intercropping
> 180 days	Double cropping

Criteria for Selecting Crop for Water Scanty Situation

Selection of crop/species based on their tolerance to dry spell/condition, temperature salinity etc. Fallowing crops mentioned in Tables 10.1 and 10.2 may be selected as per their tolerance power (degree of tolerance) to environmental abiotic stress and requirement. Crop succeeds under series of event, interaction with surrounding environment particularly with soil, water and weather conditions upon which biotic stress is buildup. Based on variety of environmental conditions a list of crops is presented in Table 10.3. One should select crops for their agroclimatic situation based on consideration mentioned elsewhere in this article.

Table 10.3: Crops with Tolerance to Various Stresses

Characteristics	*Crop Names*
High temperature tolerance	Cotton, Ground Nut, Chilies, (favor Jute and Yams only in humid tropics)
Drought tolerance	Common Millet, Barley, Chickpeas, Safflower (lower temperatures) Sorghum, Bullrush Millet, Phaseolus crops Radiatus (gram mung bean), Cassava, Castor Bean, Sesame, Ground Nut (Spanish variety), Pigeon peas, Sunflower Lower temperatures favor: Wheat, Potato, Sugar, Tomato, Safflower
Wide climatic tolerance	Maize, Soybean, Ground Nut (Valencia and Virginia type), Phaseolus lunatis, Kenaf, Hemp, Sweet Potato, Sugar cane, Tobacco
Very high rainfall tolerance	Rice, Cassava, Yam

Survival of fittest and adoption to the extreme condition is two widely accepted theories in this modern biological system but in real situation both are seen in combination because nature is great leveler in one or other respects for coexistence of above said theories partially. In agricultural production system some crops requires more water where as some need comparatively less than others. Economic plants which are more concern in agricultural systems are annual, biennial or some perennial herbs, based on agro-ecological conditions and availability of water crops are simply divided in to two category water loving plants (hydrophytes, cryophytes etc) and some of them having less affinity to water are (sandophytes). Based on experience gained due to experimentations and evolves from centuries lists of crop are given in the Table 10.4 which needs less water than others. Crops are also categorized based on their relative degree of tolerance to the limited water. Generally grasses and beans are hardy in nature and considered more tolerant than others.

Table 10.4: Crops for Limited Water Supply Condition in India

Scientific Name	Common Name	Degree of Tolerance*
Cereal		
Zea mays	Maize	1.0
Sorghum bicolor	Sorghum	1.5
Pennesitum americanum	Pearl Millet	2.5
Grasses		
Digitaria decumbens	Pangola Grass	1.0
Sorghum sudanense	Sudan Grass	1.0
Legumes		
Gliricidia sepium	Mother of Cacao	1.0
Ceratonia siliqua	St. John's Bread	1.5
Prosopsis sp.	Mesquite	2.0
Leucaena leucacephala	Leucaena	2.0
Phaseolus vulgaris	Common Bean	1
Vigna unguiculata	Cowpea	1.5
Cajanus cajan	Pigeon Pea	2.0
Dolichos lablab	Lablab Bean	2.5
Vigna radiata	Mung Bean	2.0
Phaseolus acutifolius	Tepary Bean	2.5
Vigna aconitifolius	Mat Bean	2.5
Tylosema esculentum	Marama Bean	3.0

Rated from 0 (no tolerance) to 3 (high tolerance).

Temperature Requirements of Crops

Successful crop production can only be done if all the required input and ambient environment are conducive to produce economic produce. Crops can be selected based on their agroclimatic requirement. One of the major limiting factors of successful crop production is temperature requirement of different crops during different phenological stages. This may vary from crop to crop, season to season and even varietal requirement can also differ in this respect. Temperature required by major crops of the world used for various food and feed purposes is listed in Table 10.5.

Table 10.5: Temperature Requirements for Selected Annual, Drought Resistant Crops

Crop Name	*Mean Monthly Temperature (Degrees C)*			
	For Germination	*For Growth*	*Minimum for Growth*	*Maximum for Growth*
		Cereal		
Maize	10	15	Frostless	47
Sorghum	15	18	Frostless	45
Common Millet	5	–	–	55
Bulrush Millet	10	15	Frostless	55
Barley	5	23	30	50
		Legumes		
Groundnut	18	20	Frostless	50
Phaseolus radiatus	15	18	Frostless	40
Chickpea	5 –	12-22	15	–
Pigeon Pea	10	15-26	Frostless	
		Root Crops		
Cassava	–	20-29	–	–

Growing Season and Rainfall Requirements

Under limited irrigation condition, duration of crops and their total water requirement and breakup of requirement is very much essential. According to availability of scanty water one should select crops as per in Table 10.6. List may be endless but some of most widely grown and used only are listed with respect to days taken to complete vegetative phase and post vegetative(reproductive) phase for long duration (late maturing) and as well as short duration (early maturing) genotypes. Requirement for rainfall under sandy soils and clay soils from sowing to full vegetative growth, at flowering and at grain formation and harvesting time.

Soil Consideration

Before selecting particular crops and their specific variety we must considered the type of soil, depth of soil, fertility status, salt tolerant capacity, pH of the soil and minimum depth of ground water along with the capacity of tolerance to short periods of water logging and moisture stress. According to above consideration, requirements for different principle crops are laid down in Table 10.7.

Table 10.6: Growing Season and Total Water Requirements for Selected Drought Tolerant Crops

Crops Name	Number of Days				Rainfall after Sowing[1]				Rainfall in Month of Flowering		Rainfall at Harvest
	From Sowing to 50 per cent Flowering		From Sowing to Harvest		Early Maturing		Late Maturing		(Sand) No Moisture Storage	(Sand) No Moisture Storage	
	Early Maturing	Late Maturing	Early Maturing	Late Maturing	(Sand) No Moisture Storage	(Clay) Maximum Moisture Storage	(Sand) No Moisture Storage	(Clay) Maximum Moisture Storage			
Maize	50-65	65-90	90-110	110-140	400 mm/ 3 months	200 mm/ 3 months	500 mm/ 4 months	300 mm/ 4 months	40 mm/ 10 days	25 mm/ 10 days	70 mm/ 10 days
Sorghum	50-65	65-90	90-110	110-140	200 mm/ 2 months	100 mm/ 2 months	400 mm/ 3 months	200 mm/ 3 months	30 mm/ 10 days	15 mm/ 10 days	50 mm/ 10 days
Common Millet	35	60	60	90	125 mm/ 1 month	50 mm/ 1 month	200 mm/ 2 months	200 mm/ 2 months	30 mm/ 10 days	15 mm/ 10 days	50 mm/ 10 days
Barley	70-80	80-90	95-110	110-130	225 mm/ 2 months	60 mm/ 2 months	350 mm/ 3 months	150 mm/ 3 months	30 mm/ 10 days	15 mm/ 10 days	50 mm/ 10 days
Ground Nut	60	90	95-110	110-140	300 mm/ 3 months	125 mm/ 3 months	500 mm/ 4 months	300 mm/ 4 months	60 mm/ 10 days	60 mm/ 10 days	30 mm/ 10 days
Chick Pea	–	–	120	180	200-300 mm/ 3 months	0 mm/ 3 months	300-400 mm/ month	100 mm/ 4 months	25 mm/ 10 days	25 mm/ 10 days	25 mm/ month
Pigeon Pea	100	210-300	150-180	270-360	500-1000 mm/ 6 months	300 mm/ 6 months	700-1300 mm/ 12 months	700-1300 mm/ 12 months	40 mm/ 10 days	40 mm/ 10 days	50 mm/ month
Castor Bean	–	–	100-150	150-180	450-750 mm/month	250 mm/ 4 months	500-900 mm/ 6 months	300 mm/ 6 months	–	–	20 mm/ 10 days
Cassava	–	–	180-450	700	900 mm/ 12 months	100 mm/ 6 driest months	–	1500 mm/ 12 months	70 mm/ 2 driest months	–	40 mm/ 10 days

Table 10.7: Soil Requirements for Selected Drought Tolerant Crops

Soil Characteristics		*Cereal Crops*			*Legumes*				*Root Crops*
		Maize	*Sorghum*	*Millet*	*Ground nut*	*Pea*	*Gram*	*Cowpea*	*Cassava*
Texture	Heavy		✓				✓		✓
	Medium	✓	✓	✓	✓	✓	✓	✓	✓
	Light		✓	✓	✓	✓		✓	
Minimum	Deep (90+)								✓
rooting	Med. (60-90)	✓	✓		✓	✓	✓	✓	
Depth (cm)	Shallow (30-60)			✓					
Fertility									
	High	✓							
	Medium		✓	✓	✓	✓	✓	✓	✓
Salt	Good								
Tolerance	Moderate								
	Poor	✓	✓	✓	✓		✓	✓	✓
pH Range		5.5-7.5	4.5-8.5	5.0-6.0	6.0-8.0	5.5-7.5	5.5-7.5	5.5-7.5	5,5-6.5
Tolerance to short periods of water logging		Low	Medium to High	Medium	Low	Medium to Low	Medium to Low	Medium to Low	Low
Minimum depth of ground water (cm)		75	50	60	60	30-50	30-50	40	60

Crops Suitable for Salt Streets Conditions

Based on response of crop to the soil salinity crops are categorized broadly in to two category sensitive (High and medium) and tolerant (High and medium). One should select crops and their varieties based on salt concentration in the field to produce optimum. Crops like lentil, chickpea, pea, carrot and onion should simply avoided if the soil is affected by salt in such condition it would be always be better to take crops like barley rice (irrigated) cotton, tobacco sugar beet etc for profitable crop husbandry (Table 10.8).

Crops of Acidic Soils

Some of the soils in the high rainfall area or containing acid farming parent material are in due course of time turned in to acidic nature. Under this circumstance one should prefer to produce tea, potato, paddy, oat etc, it is always better to avoid wheat, barley tobacco for economically sustainable productivity (Table 10.9).

Crops of Water Logging

Water logging or water stagnant at stage when crops are still standing the field may cause severe damage to crop to the extent of total failure due to several physiological adversity including lake of aeration. Based on capacity of crops to survive under standing water condition they are enlisted in Table 10.10. Farmers are advised to select crops and their appropriate varieties which can fight better under water logging conditions.

Table 10.8: Crop Group Based on Response to Salt Stress

Sensitive		*Tolerant*	
High Sensitive	*Medium Sensitive*	*High Tolerant*	*Medium Tolerant*
Lentil	Cowpea	Spinach	Barley
Beans	Broadbean	Sugarcane	Rice(transplanted)
Peas	Vetch	Indian Mustard	Sugar beat
Carrot	Cabage	Rice (Direct seeded)	Turnip
	Cauliflower	Pearl Millet	Safflower
	Sorghum	Oats	Taramira
	Minor Millets	Alfaalfa	Karnal Grass
	Maize	Para grass	Casurina
	Clover	Rhodes grass	Salvadora
	Berseem	Sudan grass	
		Accacia	

Table 10.9: Crop Group Based on Response to Acid Stress

Sensitive	*Tolerant*
Wheat	Paddy
Barley	Oat
Sorghum	Rye
Lucerne	Maize
Cabbage	Cow Pea
Cauliflower	Ground nut
Carrot	Soy bean
Berseem	Sugar beat
	Sugarcane

Table 10.10: Relative Tolerance of Crops to the Water Logging

Threshold (t) (Hr)	*Crops*
0<t<12 (1day)	Wheat, groundnut, maize(*) and mustard
12<t<36 (1-2 days)	Cowpea and barley
36<60 (2-3 days)	Wheat, Beets, forage grass, sunflower

Threshold value indicates that if water stagnation does not exceed this value, their would be no decrease in the yield.

Summary

This Nation is simply enjoy the twenty one agro -ecological zones, eight phyto-geographical regions and fifteen ago- climatic zone, all these classification numerated is just to know that Indian environment is result of rich and complex ecological, environmental diversity which enable to evolve several crops and its various species, due diversity at greater length and same time one can find similarity too leads to accommodate all most all type of plant in its environment successfully. Only a limited number of crops are adapted to the climatic conditions and the farmer must sow the crop best suited to the moisture conditions encountered at that time. Moisture is so dominantly limiting, that "soil improving" crops are much less effective than in more humid areas. Success with rigid or complex sequences is difficult in the face of widely varying rainfall. To achieve the optimum production potential under different abiotic and biotic stress condition, selecting crops and varieties based on criteria described in this presentation would certainly boost crop production efficiently in India.

References

Abraham Blum. 2004. The Mitigation of drought stress, plant stress.com;-Henry T. Nguyen, Abraham Blum (Ed.), Physiology and Biotechnology Integration for Plant Breeding.

Allen, R.G., Pereira, L.S., Raes, D., Smith, M. 1998. Crop Evapotranspiration guidelines for computing crop water requirements. FAO Irrig. Drain. Pap. 56, FAO, Rome, 300 p.

Ashraf, M. 1994. Breeding for salinity tolerance in plants. Critical review of Plant Science 13: 17-42.

Ayars, J.E., Phene, C.J., Hutmacher, R.B., Davis, K.R., Schoneman, R.A., Vail, S.S., Mead, R.M., 1999. Subsurface drip irrigation of row crops: a review of 15 years research at the Water Management Research Laboratory. Agric. Water Manag. 42: 1-27.

Batchelor, C., Rama, M.R.; and Rao, M. 2003. 'Watershed development: A solution to water shortages in semi-arid India or part of the problem?. LUWRR 3: 1-10.

Bray, E.A. 1993. Molecular responses to water deficit. Plant Physiology, 3 : 61-66.

Ceccarelli S., S. Grando and A. Impiglia.1998. Choice of selection strategy in breeding barley for stress environments. Euphytica 103: 307-318, 1998.

Cooper, P.J.M. 1991. Fertilizer use, crop growth, water use and WUE in Mediterranean rainfed farming systems. pp 135-152 in: H.C. Harris, P.J.M. Cooper and M. Pala (Eds.), Soil and crop management for improved water use efficiency in rainfed areas. Proceedings of an International Workshop, Ankara, Turkey, May 1989. Aleppo, Syria: ICARDA.

Erskine, W. and Malhotra, R.S. 1997. Progress in breeding, selecting and delivering production packages for winter sowing chickpea and lentil. Pages 43-50 in Problems and prospects of Winter sowing of grain legumes in Europe, AEP Workshop Dec 1996 Dijon. AEP, France.

Fan, S.; Hazell, P., Haque, T. 2000. Targeting public investments by agro-ecological zone to achieve growth and poverty alleviation goals in rural India', Food Policy (25): 411–428.

FAO. 1984, Land, food and people. FAO Economic and social development series 30. Food and Agriculture Organization of the United Nations, Rome.

Government of India..2000. Warsa Jan Sahbhagita: Guidelines for national watershed development project for rainfed areas (NWDPRA)' Ministry of Agriculture, Deptt. of Agriculture and Co-operation. Rainfed Farming Systems, New Delhi.

Haloph; A. 1989. Database of salt tolerant plants of the world, by James A. Aronson, Office of Arid Lands Studies (The University of Arizona, 845 North Park Ave., Tucson, AZ 85719).

Haramata. 2001. Drylands programme newsletter no. 39. IIED, London.

ICRISAT. 1998. 'Sustainable rainfed agriculture research and development: database development, typology construction and economic policy analysis', Mimeo prepared for the Indian Council of Agricultural Research (ICAR),ICRISAT: Patancheru.

Indira P. 1989. Drought tolerant traits I sweet potato genotypes. Journal of Root Crops, 15: 139-146.

IPCC 2001. Climate change 2001. The science of climate change. Contribution of working group I to the second assessment report of the Inter-governmental Panel on Climate Change. Adaptation to climate change in the context of sustainable development and equity. Cambridge University Press, Cambridge, U.K.

IPCC, 2007. Climate Change 2007: Mitigation. Contribution of Working Group III to the Fourth Assessment Report of the Inter-governmental Panel on Climate Change. B. Metz, O.R. Davidson, P.R. Bosch, R. Dave, L.A. Meyer (eds), Cambridge University Press, Cambridge, United Kingdom and New York, NY, USA.

Kasuga, M., Q. Liu, S. Miura, K. Yamaguchi-Shinozaki and K. Shinozaki. 1999. Improving plant drought salt and freezing tolerance by gene transfer of a single stress inducible transcription factor. Nature Biotechnology, 17: 287-291.

Keller, A.; Sakthivadivel, R.; Seckler, D. 2000. 'Water scarcity and the role of storage in development' IWMI research report 39, IWMI: Colombo.

Kerr, J.; Pangare, G.; and Lokur Pangare, V. 2000. An evaluation of dryland watershed development projects in India',Research report 127 IFPRI: Washington DC.

Lee, J.H., A. Hebel and F. Schoffl. 1995. Derepression of the activity of genetically engineered heat shock factor causes constitutive synthesis of heat shock protein and increase thermo tolerance in transgenic Arabidopsis. Plant Journal, 8 : 603-612.

Matthew, B. (2008). Planning for Change - Guidelines for national programmes on sustainable consumption and production. United Nations [UN] Environment Program. Waterside Press. pp 106.

Nilelas, Holmberg and Leif, Bulow. 1998. Improving stress tolerance in plants by gene transfer - review. Trends in Plant Sciences, 3 : 61-66.

Pala M. and C. Studer. 1999. Cropping systems management for improved water use efficiency in dryland agriculture. Paper presented at the International Conference on: Water Resources Conservation and Management in Dry Areas. 3-6, December 1999. Amman, Jordan.

Pala M. and Mazid, A. 1992. On-farm assessment of improved crop production practices in Northwest Syria: Chickpea. Experimental Agriculture 28: 175-184.

Pala, M., A. Matar and A. Mazid. 1996. Assessment of the effects of environmental factors on the response of wheat to fertilizer in on farm trials in a Mediterranean type environment. Experimental Agriculture 32: 339?349.

Puskur, R.; Bouma,J.; and Scott, C. 2004. 'Sustainable Livestock Production in semi-arid Watersheds', Economic andPolitical Weekly: 3477-3483.

Putter, de, J. 2003. Land and Water for Livestock – Biophysical Characterization and Water Balance of semi-arid watersheds in India' LEAD project report (unpublished), IWMI: Hyderabad.

Rockström, J.; and Falkenmark, M. 2000. Semi-arid crop production from a hydrological perspective – Gap between potential and actual yields', Critical Reviews in Plant Sciences 19 (4): 319 – 346.

Rosegrant, M.; Cai, X.; Cline, S.; and Nakagawa, N. 2000. 'The role of rainfed agriculture in the future of global food production', EPTD discussion paper 90, IFPRI: Washington DC.

Ryan, J.; and Spencer,D. 2001. 'Future challenges and opportunities for agricultural land in the semi-arid tropics,ICRISAT: Patancheru.

Schoffl, F., R. Pand and A. Reindl. 1998. Regulation of heat shock response. Plant Physiology, 117 : 1135-1145.

Studer, C. and W. Erskine. 1999. Integrating germplasm improvement and agricultural management to achieve more efficient water use in dry area crop production. Paper presented at the International Conference on Water Resources Conservation and Management in Dry Areas. 3-6, December, 1999. Amman, Jordan.

Thomashow, M.F. 1998. Role of cold responsive genes in plant freezing tolerance. Plant Physiology, 118 : 1-7.

Unger, P.W. (2006) Soil and Water Conservation Handbook – policies, practices, conditions and terms. Haworth Food and Agricultural Productes Press. London. p 248.

Wani, S., Pathak, P., Sreedevi, T., Singh, H., and Singh, P. 2003. Efficient management of rainwater for increased crop productivity and groundwater recharge in Asia', in Water productivity in agriculture: Limits and opportunities for improvement, Kijne, J.W.; Barker, R.; and Molden, D. (Eds). Wallingford: CAB International: 199-215.

Water Organization Trust (WOTR). 2003. Collection and Analysis of Baseline Data for Vaiju Babulgaon Cluster Watershed villages Ahmednagar, Maharashtra, unpublished project report, IWMI: Hyderabad.

World Resources Institute. 1999. "World Resource: A Guide to the Global Environment. Special focus on Climate change, Data on 146 Countries". Oxford University Press, New York.

Yadav, R. L., Panjab Singh, Rajendra Prasad and I.P.S. Ahlawat (1998) *Fifty years of Agronomic Research in India.* Indian Society of Agronomy, New Delhi 270p.

Chapter 11

Forages/Fodder Crops in Cropping System

***Lal Singh*[1] *and Anil Kumar Singh*[2]**

[1]*Division of Agronomy, Sher-e-Kashmir University of Agricultural Sciences and Technology of Kashmir, Shalimar, Srinagar – 191 121, J&K*

[2]*Senior Scientist, Agronomy, Division of Land and Water Management, ICAR Research Complex for Eastern Region, ICAR Parisar (P.O.- B.V.College), Patna – 800 014, Bihar*

E-mail: aksingh_14k@yahoo.co.in

The present fodder crop area of 8.3 million hectare could not be increased due to increasing pressure on cultivated land for food and commercial crops. The sustainability of dairy industry in India largely depend on the quality of herbage based animal feed and fodder. To produce the targeted quantity of green fodder, the best option is to maximize the fodder production per unit area and per unit time. To ensure the availability of quality fodder as per requirement throughout the year there is need to incorporate suitable fodder crops in the cropping systems. An ideal fodder system is that which gives the maximum outturn of digestible nutrients per hectare, or maximum livestock products from a unit area. It should also ensure the availability of succulent, palatable and nutritive fodder throughout the year. Some of the important fodder crop production systems are described in this chapter like intensive cropping system *viz.* the overlapping cropping and the relay-cropping, fodder production in gap of two crops in a rotation (opportunity cropping system), normal cropping system, intercropping system, under dry land farming system, multi-story cropping system, lay farming system, allay farming system, under agro-forestry system, horti-postoral system, and silvi-postoral system etc.

India is basically an agricultural country and about 70 per cent of its people live in villages. Their livelihood is dependent mainly on agriculture and animal husbandry. Though India has a huge livestock population of over 343 millions, besides poultry, yet the production of milk and other livestock products is about the lowest in the world. The figures regarding the availability of milk 100 g/head/day, meat 1 million tonnes annually and eggs 12 no./head/year, however, the minimum nutritional requirement set by the nutritionists are milk 201 g/head/day, meat 7,122 million tonnes annually and eggs 1 no./head/year is necessary. It is evident that we are highly deficient in various livestock products, though we have about one-fourth of the total cattle population of the world. The analysis of this situation reveals that one of the main reasons for the low productivity of our livestock is malnutrition, under-nutrition or both, besides the low genetic potential of the animals. This fact is adequately supported by the figures given in Table 11.1.

Table 11.1: Balance Sheet of Animal Feeds and Fodders[1]

Feeds and Fodders	*Availability*	*Requirement*	*Deficit*
Green fodder	224.08 m tonnes	611.99 m tonnes	387.91 m tonnes
Crop residues	231.05 m tonnes	869.79 m tonnes	638.74 m tonnes
Concentrates	31.6 m tonnes	65.4 m tonnes	81.8 m tonnes

It is seen from the figures of availability, *vis-a-vis* the requirement of green-fodder crops, crop residues and concentrates, that there is a huge gap between demand and supply of all kinds of feeds and fodders.

On the other hand, if we examine the land resources available for growing fodder and forage crops, it is estimated that the average cultivated area devoted to fodder production is only 4.4 per cent of the total area. Similarly, the area under permanent pastures and cultivable wastelands is approximately 13 and 15 million hectares respectively.

Presently the chronic shortage of feed coupled with the poor quality of fodder is widely regarded as the major constraint in animal production. It has been estimated that with the present feed and fodder resources we are able to meet only 46.6 per cent of animal requirements, which in turn resulted in 50 per cent of the desired products. It is imperative to say that in the present system of intensive livestock production, increasing concentrate feeding has increased the milk production cost and substantially decreased the profits of farmers. Green fodder is the essential components of feeding high yielding milch animals to obtain desired level of milk production. The present fodder crop area of 8.3 million hectare could not be increased due to increasing pressure on cultivated land for food and commercial crops. The Sustainability of dairy industry in India largely depend on the quality of herbage based animal feed and fodder. To produce the targeted quantity of green fodder, the best option is to maximize the fodder production per unit area and per unit time. To ensure the availability of quality fodder as per requirement throughout the year there is need to incorporate suitable fodder crops in the cropping systems.

Systems of Fodder Production

The system of fodder production vary from region to region, place to place and farmer to farmer, depending upon the availability of inputs, namely fertilizers, irrigation, insecticides, pesticides, etc. and the topography. An ideal fodder system is that which gives the maximum outturn of digestible nutrients per hectare, or maximum livestock products from a unit area. It should also ensure the availability of succulent, palatable and nutritive fodder throughout the year. Some of the important intensive fodder crop production systems are discussed here.

Fodder Production in Intensive Cropping System

The requisites for intensive cropping system for fodder/forage production are that (*i*) fodder is required in uniform quantity throughout the year, (*ii*) the fodder crops in the rotation should be high-yielding, (*iii*) the area for production of fodder should be fully irrigated, and (*iv*) other inputs, such as fertilizers and pesticides, should be available in optimum quantity. The different intensive systems of fodder production fall into two categories, *viz.* the overlapping cropping and the relay-cropping. In the overlapping system, a fodder crop is introduced in the field before the other crop completes its life-cycle. In relay-cropping, the fodder crops are grown in successions, *i.e.* one after another, the gap between the two crops being very small.

Intensive Fodder Production Under Overlapping Cropping System

The overlapping cropping system evolved by taking advantage of the growth periods of different species ensures a uniform supply of green fodder throughout the year. One such system continues for three years. The best rotation in this system is *berseem* + Japan *sarson* - Hybrid Napier + cowpea - Hybrid Napier; (October-April) - (April-June) - (June-October).

How to Adopt Overlapping Cropping System

1. In this cropping system, *berseem* + Japan *sarson* seed mixed in the ratio of 25 : 2, are sown in the first week of October, using a basal fertilizer dose of 20 kg of N and 80 kg of P_2O_5 per ha. The sowing is done by broadcasting the mixed seed in the seedbeds, flooded with water. Care should be taken to inoculate the *berseem* seed with *Rhizobium* culture before sowing, especially when the crop is being sown for the first time. However, if the culture is not available, soil from the top 5 to 7 cm layer is collected from the field in which *berseem* was grown in the previous year and broadcast along with the seed. Irrigation may be given at intervals of 7-8 days, depending upon the soil and climatic conditions.
2. The first cut from the mixture is taken in 50-55 days after sowing. Japan *sarson* being quicker in growth boosts the yields in the first cut, whereas in the subsequent cuts *berseem* takes over.

3. Hybrid Napier is introduced in the standing crop of *berseem* after taking the third or fourth cut from *berseem*. Rooted slips are planted in February (central India) and in March (northern and north-western parts) in lines by keeping a distance of one metre between the rows and 30-40 cm between the plants.

   The planting of a hectare would need about 33,000 rooted sets of Hybrid Napier. Hybrid Napier starts growing actively after March and should be cut 8-10 weeks after transplanting and the subsequent cuts are taken at intervals of 40-45 days. After the *berseem* crop is over, a basal dose of 100 kg of P_2O_5 and 50 kg of N per ha is applied.
4. *Berseem*, being an annual crop, completes its lifecycle in April and then the inter-row spaces of Hybrid Napier are prepared with a *desi* plough and cowpea is sown in lines, 25 cm apart. In this way, in each set of two rows of Hybrid Napier, there will be two rows of cowpeas. Cowpea is cut 60 days after sowing and thereafter Hybrid Napier does not allow any other legume to grow along with it.
5. Hybrid Napier continues to supply green fodder during the monsoon season. At the time of the last cutting in October, the inter-row spaces are again ploughed up and the land is prepared for sowing *berseem* and Japan *sarson* to start the second cycle of the rotation.
6. This system of intensive fodder production is economically viable only for 3 years. After three years. Hybrid Napier is uprooted and fresh planting is taken up. The stumps of Hybrid Napier become old and the tillering capacity diminishes considerably.

Advantages Overlapping Cropping System

1. This system ensures green fodder throughout the year.
2. It takes care of the dormancy period of Hybrid Napier during winter.
3. The inter-row spaces of Hybrid Napier are efficiently utilized for raising *berseem* in winter and cowpea in summer.
4. The growing of legumes enriches the soil.
5. Hybrid Napier gets established without much care and cost.
6. Green fodder in the first cut is increased up to 50 per cent by Japan *sarson*.

Intensive Fodder Production Under Relay Cropping System

There is ample scope for increasing fodder production from the high-input areas, either by growing high-yielding fodder crops singly or in mixture. The growing of three or four successive fodder crops helps to boost fodder production per unit area. Some of the important intensive fodder-crops, crop- rotations and the expected yields from each are given below.

1. Maize + cowpea - maize + cowpea + teosinte - *berseem* + mustard (300 q/ha) - (450 q/ha) - (1,000 q/ha)
2. Sweet sudan + cowpea - *berseem* + oats (1,000 q/ha) - (1,000 q/ha)

3. Hybrid Napier + lucerne (1,250 q/ha) - (850 q/ha)
4. Maize + cowpea - *jowar* + cowpea - *berseem* + mustard (300 q/ha) - (400 q/ha) - (1,000q/ha)
5. Teosinte + *bajra* + cowpea - *berseem* + oats (1,000 q/ha) - (1,000 q/ha)
6. Sweet sudan + cowpea - mustard - oats + peas (1,000 q/ha) - (250 q/ha) - (500 q/ha)
7. *Jowar* - turnips - oats - 1800 q/ha.

Fodder Production in Gap of Two Crops in a Rotation (Opportunity Cropping System)

There is ample scope for fitting in the short-duration fodder crops, either single or in mixture, with the other crops during the gap period between two main cash crops. Two distinct fallow periods are available for raising short-duration fodder crops, provided adequate resources are available. In the case of the jowar-wheat rotation, gap periods between October and November and between April and June are available for each crop as fodders. Thus in the first gap period (October-November), which is rather short, the growing of fodder turnips and short-duration mustard varieties helps to get 250-300 q per ha is obtained. Similarly, in the second gap period (April to June) M.P. chari + cowpea, maize + cowpea, *bajra* + cowpea is successfully grown and an additional green-fodder yield to the tune of 300-350 q per ha of fodder without disturbing the normal cropping systems.

Fodder Production in Normal Cropping System

Green fodder is a essential input in livestock feeding thus it is necessary to grow on the farmer fields along with other crops simultaneously to fulfill the requirement. It should be included in general cropping system which was adopted by the farmers in such a way green fodder is getting throughout the year or it preserved as hay or silage for the lean period. Some systems is given below :

Paddy-berseem-moong, Maize-berseem-multicut sorghum, sorghum-potato-wheat, paddy-wheat- sudan/M.P. chari, bajra-toria-wheat-cowpea

Fodder Production in Inter-cropping System

In this system growing of two or more dissimilar crops simultaneously on the same piece of land. The base crop necessarily in distinct row arrangement and its recommended optimum plant population is suitably combined with the additional plant density of the associated crop. The objective is the intensification of cropping both in time and space dimensions and to raise productivity per unit area by increasing the pressure of plant population. It has better utilization of growth resources than sole cropping. The option of intercrop may select as per requirement of fodder crops. There is two option of intercrop which can be adopted by farmers as per requirement of fodder crops.

1. Fodder crop intercropped with fodder crops *e.g.* sorghum + cowpea/ soybean/moong, maize + cowpea/soybean/moong, Bajra + cowpea/

soybean/moong, sudan/M.P. chari + cowpea/soybean/moong, berseem + japani sarson, oat + hairy vetch etc.

2. Other crops intercropped with fodder crops *e.g.* sugarcane + cowpea/soybean/moong, sorghum + arhar, maize + arhar/castor, bajra + arhar/castor, cotton + soybean/cowpea/moong, potato + radish/turnip, wheat + mustard etc.

Fodder Production in Multi-Story Cropping System

Multi-store or multi-level cropping in a system of growing crops of different heights together at the same time on the same piece of land and thus using land, water and space most efficiently and economically. It is aimed at maximum production per unit area per unit time where in economic yields of compatible crop species are harvested from different height *e.g.* sugarcane + sunflower + cowpea/moong, sorghum + arhar + cowpea/moong, maize + arhar + soybean/cowpea, castor + arhar + moong/cowpea, bajra + arhar + cowpea/soybean, coconut + pepper + grasses etc.

Fodder Production in Lay Farming System

Lay farming is a system in which crops and pastures are alternated on the same fields (Hall *et al.*, 1979). A rotation of arable crops requiring annual cultivation and artificial pasture occupying field for two years or longer (Ruthenberg, 1976). In this system farmers can grow perennial fodder crops to maintain an artificial pasture for two three years. Crops like hybrid napier, deenanath grass, marvel grass, orchardgrass, tallfesque, setaria, paragrass, fulwagrass, white clover, red clover, perennial vetch, sainfoin, stilo, lucern etc. can be grown alone or in combination of grasses and legumes.

Fodder Production in Allay Farming System

Alley cropping system is the system of growing any arable crops in the alleys (passage in between two crops of leguminous shrubs).Competition for light remains a key factor in the suitability of different species in mixed shrubs (Chatterjee and Maiti, 1984). In this system farmers should be select leguminous fodder shrubs to make the alleyes for production of arable crops. Some species like subabul (*Leucaena leucocephala*), khejri (*Prosopis cineraria*) etc.

Fodder Production Under Dryland Farming System

A large proportion of the area of our country is located in the dryland regions. In these areas, the farmers usually grow at least one crop in the *rabi* season after conserving the soil moisture. Thus there is a great scope for raising two crops under such situations. First, the growing of a fodder crop which gets ready in 45-50 days after sowing (cowpea, *jowar*, *guar*, *sanwa*, *moth*, etc.), yield 150-250 q per ha of green fodder. After harvesting the fodder crops, crops such as gram, linseed, barley, wheat and safflower are raised on the conserved moisture.

Fodder Production Under Agroforestry System

King and Chandler (1978) defined agro-forestry as a sustainable land management systems which increases the over all yield of land, combines the production of crops and forest plants/or animals simultaneously or sequentially on the same unit of land and applies management practices that are compatible with the cultural practices of the local population. Thus both horizontal and vertical interspaces can be used to maximise not only food grains production but also to supply farmers the fuel, fodder, fiber and timber to augment his meager income. Majority of farmers possess small and scattered holdings, agro-forestry is the only way to ameliorate his socio-economic condition. The objective of any ideal agro-forestry system is to maintain or improve the productivity of a field. Subabul (*Leucaena leucocephala*) is an ideal plantation crop that can be used by farmers for agro-forestry. Subabul is used as fodder because of high protein content in leaves, fuel, timber wood and is a rich source of organic manure. For fodder purpose varieties cunninghum, peru and K-8 are supposed to be better while the Hawaiian type are generally low yielding and are suitable mostly to arid and semi arid conditions (Relwani and Rangenkar, 1981). A good number of grasses like Guatemala grass (*Tripsacum laxum*), paragrass (*Brachiaria mutica*), dhoral grass (*Chrysopogon montanus*) anjan grass (*Cenchrus ciliaris*), senior grass (*Sehima nervosum*), marvel grass (*Dichanthium annulatum*), hybrid napiar (*Pennisatum purpurium*), deenanath grass, orcardgrass (*Dactylis glomerata*), Tallfesque (*Festuca arundinacy*), and other may also be included in the agro-forestry with success in accordance with the soil and climate where they do well. Cultivation in agro-forestry system would be particularly appropriate for marginal farmers who do not have access to large land holdings for fodder grass cultivation [5].

Fodder Production in Horti-Postoral System

It is a land management system for the concurrent production of fruits and fodder crops. In this system fodder crops are grown in inter spaces of horti cultural crops without any adverse effect on the fruit production. Gupta *et al.* (1982) have reported lamon (*Citrus lemon*) and pine apple were found to grow very well with fodder cowpea. Several anual and perinial fodder garsses and legumes can be grown successfully in between rows of fruit trees.

Fodder Production in Silvi-postoral System

The tree species are planted in pasture lands to provide fuel and fodder requirements of the local population. However, agriculture crops and leguminous grasses (stall feeding) can be introduced which will form agri-silvi-pastoral system of management. Sesbania graniflora and sesbania sesban have increased forage production of *Senchrus ciliaris, Setaria anceps, Desmanthus* and *Siratro* under arid condition of rajsthan (Patil *et al.*, 1979). Paroda and Muthana (1979) reported the best tree-grass combination of *Azadirachta indica + Senchrus* and *Acacia tortlis + Cenchrus* species. They also obtained higher yield of *Vigna radiatus* and *Cyamopsis tetragonoloba* when intercroped with these trees. Gupta (1983) reported a higher yield of *Crysopogon fulvus* by growing it with Eucalyptus hybrid.

References

http://www.krishiworld.com/html/for_crop_grass1.html

http://www.hatsun.com/shows/dairy_framing

http://www.ikisan.com/links/ap_fc_sorghum.shtml

http://www.igfri.ernet.in/crop_profile_oat.htm

http://www.indiaenvironmentportal.org.in/node/23059

http://igfri.ernet.in/crop_profile_berseem.htm

www.Our Agriculture Editor/New Delhi September 27, 2005

Chatterjee, B.N. and Maiti, S. 1984. Cropping system (theory and practice). Oxford and IBH Publication, Culcutta-16.

Hall, A.E. ;Carnell, G.H. and Hawton, H.W. 1979. Agriculture in semi-arid environment. Ecological studies 34, Springer Verlag, Berlin Heidilberg, New York, pp.200-223.

King, K.F.S. and Chandler, M.T. 1978. The wasted lands. Nairboi ICRAF.

Relwani, L.L. and Rangenkar, D.V. 1981. Social forestry, the second green revolution. Science Today (October) : 48-50.

Ruthenberg, H.1976. Farming system in the tropics. Clarendon Press Oxford. Second edition, 1976.

Chapter 12

Year-Round Forage/Fodder Production in Different Agroclimatic Zones of India

Sant Kumar Singh

ICAR Resesarch Complex for Eastern Rregion, Patna – 800 014

E-mail: santpatna@yahoo.co.in

Only during monsoon season green fodder is available for the animals in sufficient quantity. Whereas, during other seasons animals are fed on crop residues or dry fodders. There is need to identify the crops that can be grown in between the two crop seasons. The overlapping system of year round green fodder production system can lead to obtain economically viable maximum forage yield. Only 8.3 million hectare area are cultivated under fodder crops in our country and further there is a little scope for its expansions Thus, to produce targeted quantity of green fodder, the best option is to maximize the fodder production per unit area per unit time. Bunds, waterways, pond, embankment, backyards are some of the areas where small and marginal farmers can be encouraged to cultivate the fodder crops. Agroforestry technologies such as silvopasture, integrate pasture and/or animals with trees.

India is house to 15 per cent world cattle population and 16 per cent of human population to be sustained and progressed on 2 per cent of total geographical areas. Due to ever increasing population pressure of human, arable land is mainly used for food and cash crops, thus there is little chance of having good quality arable land available for fodder production, until milk production is remunerative to the farmer as compared to other crops. According to an estimate the shortages in dry fodder, green fodder and concentrates are 40.4 per cent, 24.7 per cent and 47.1 per cent against the requirements of 650.7, 761.5 and 79.4 million tones for the current livestock population, respectively. The pattern of deficit varies in different parts of the country. For instance, the green fodder availability in Western Himalayan, Upper Gangetic Plains and Eastern Plateau and Hilly Zones is more than 60 per cent of the actual requirement. In Trans Gangetic Plains the feed availability is between 40-60 per cent of the requirement and in the remaining zones the figure is below 40 per cent. In case of dry fodder, availability is over 60 per cent in the Eastern Himalayan, Middle Gangetic Plains, Upper Gangetic Plains, East Coast Plains and Hills Zones. In Trans Gangetic Plains, Eastern Plateau and Hills and Central Plateau and Hills the availability is in the range of 40-60 per cent, while in the remaining zones of the country the availability is below 40 per cent.

Only during monsoon season green fodder is available for the animals in sufficient quantity. Whereas, during other seasons animals are fed on crop residues or dry fodders. Only 8.3 million hectare area are cultivated under fodder crops in our country and further there is a little scope for its expansions Thus, to produce targeted quantity of green fodder, the best option is to maximize the fodder production per unit area per unit time. Bunds, waterways, pond, embankment, backyards are some of the areas where small and marginal farmers can be encouraged to cultivate the fodder crops.

Integrated approach however has several advantages as mentioned below:

1. Security against complete failure of the system
2. Minimization of dependence for external inputs
3. Optimum utilization of farm resource
4. Efficient use of natural resources sunlight, water and land etc.

ICAR has delineated the country into 20 agro-eco-regions (AER) and 60 agro-eco-subregions (AERS) using criteria of soils, physiography, bio-climate (climate, crops, vegetation) and length of growing period. ICAR under NATP has identified five major ecosystems, within each of the major ecosystem 2-4 different production systems which require support in an interdisciplinary mode responsive to farmer's specific needs were identified which is as given in Table 12.1

Table 12.1: Different Production Systems within the Five Major Agro-ecosystems

Sl.No.	*Major Ecosystems*	*Production Systems*
1	Arid Agro-ecosystem	1. Agri-silvi-horti pastoral production system
		2. Livestock and fish production system
2	Coastal Agro-ecosystem	1. Fish and Livestock production system
		2. Agri-Horti production system
3.	Hill and Mountain	1. Agri-horti production system
		2. Livestock and fish production system

Contd...

Table 12.1–Contd...

Sl.No.	Major Ecosystems	Production Systems
4.	Irrigated Agro-ecosystem	1. Rice-wheat production system
		2. Cotton based production system
		3. Sugarcane based production system
		4. Dairying and fish production system
5.	Rainfed Agro-ecosystem	1. Arable farming system
		2. Agroforestry production system
		3. Livestock based farming system

Management for Maintaining Lean Period Fodder Supplies

In fodder supplies, the paradoxical situation is that in many parts there is surplus fodder during monsoon and a deficit occurring during lean season. This is especially so in the far flung areas. Thus there is need to identify the crops that can be grown in between the two crop seasons. This would help to meet the forage requirements of the lean periods. Utilizing the leaf meal of the leguminous species (both woody perennials and herbaceous) such as *lucerne, stylo, Leucaena* hold promise to overcome the lean period fodder deficits. The post harvest technologies such as biomass processing, enrichment, densification etc. hold the key for better animal husbandry in the deficit zones. India is a mosaic of agro-environments ranging from extremely humid to extremely arid tropical and temperate zone. Thus a different approach is required to feed the animals during lean period. The present practice depends largely on grains that is costly and also competes with the human requirement. The Main concern in developing dairy industry in India is to feed the high yielding animals economically on green fodder. The IGFRI has developed the overlapping system of year round green fodder production system to obtain economically viable maximum forage yield. The recommended highest yielding crop rotation is:

OUTPUTS		
Production/ha	Forage	220 t/ha green fodder
PLANT MATERIAL REQUIRED		
Hybrid napier		30000 slips/ha
Cowpea		40 kg seed/ha
Maize		30 kg seed/ha
Berseem		25 kg seed/ha
Sorghum		20 kg seed/ha
Mustard		1.5 kg seed/ha
FERTILIZER/FYM		
FYM		10 tonnes/ha

FERTILIZER	
Nitrogen	150 kg
Phosphate	150 kg
Potash	60 kg
MACHINERY	
Weeder - cum - mulcher	
ECONOMICS	
(Net profit Rs./ha) Rs. 21,000=00/ha/year	

Benefits

The Indian Grassland and Fodder Research Institute, Jhansi has evolved high yielding fodder crop sequences to harness year round green fodder production. In central region berseem along with mustard in winter followed by hybrid napier intercropped with cowpea in summer and rainy season has produced high yield (240 - 280 t/ha green fodder). The Institute recommends to grow berseem with recommended inputs in October and planting of high yielding perennial grasses like hybrid napier or guinea grass in the month of February/March in rows of 100 cm apart. The berseem crop is over in April/May. The row space now available is sown and cultivated with cowpea. The added advantage of the system is that the cowpea is raised on residual soil fertility left by berseem. In the second year the berseem is sown in the beginning of November after thinning of grass tussocks and the same operational cycle is repeated for three years. The space between the grass rows can be adjusted to facilitate the use of appropriate farm machinery.

The technology of year round fodder production has been tested in All India Coordinated Research Project on Forage Crops and different crop sequences have been identified on regional basis.

- In northern region the green fodder yield varies from 123-201 t/ha/year.
- Similarly, the green fodder yield varies as 168-253, 85-131 and 75-225 t/ha/year in western, eastern and southern regions respectively.
- The year round fodder production system developed by this Institute promises for continuous flow of green forages throughout the year from the same piece of land and thus provides opportunities for dairy development at medium farm level.
- The technology of growing year round fodder production has helped the farmers/dairy owners to sustain their milch animals of 6-7 litres per day potential with minimum use of concentrates, thus producing milk at cheaper cost.

Crop	*Varieties*
Hybrid napier	IGFRI No. 6/IGFRI No. 10
Berseem	Wardan
Mustard	Bundel Bersem-2
Maize	Japanese rape/Chinease cabbage
Cowpea	African tall
Sorghum	Bundel Lobia 1/Bundel Lobia 2
	PC-6/HC 136

The country is further endowed with the rich heritage of traditional know-how of raising, maintaining and utilizing forage, feed and livestock resources. Wherever there are opportunities to produce fodder on arable lands, intensive forage production system may be practiced, aiming at achieving maximum sustainable harvest of nutritive herbage per unit area and time. In multiple cropping systems, 3 or 4 high yielding forage crops are grown on a particular piece of land in a calendar year. The crop sequences are tailored with an objective of achieving high yields of green nutritious forage and at the same maintaining the soil fertility. The system assures regular supply of green forage when staggered sowing and harvesting schedules are followed. Under assured irrigation multiple crop sequences like MP Chari - turnip, Berseem + mustard - maize + cowpea - MP Chari + cowpea and Berseem+ mustard - MP Chari + cowpea has been found better. These systems are better suited to well-managed mechanized farms. The concept of overlapping cropping systems may also be practiced. In this, sowing of succeeding crop is done while the preceding crop is still in the field. This practice reduces or even avoids the slack period of forage availability. The system involves growing of combination of appropriate perennial and annual forage species that are not competitive and ensure regular supply of green forage throughout the year to meet the requirement of large dairy farms and also of small farmers having limited land holding to grow food as well as forages.

Successful integration of forage crops in the existing cropping systems (rice-wheat or jowar/bajra/maize-wheat) may be profitable also good from the viewpoint of soil health. There are reports of more profitable forage crop sequences *viz.*, rice-berseem and cowpea-wheat when compared to rice-wheat and jowar-wheat systems, respectively. Similarly, integration of Sesbania on bunds (accounting 5-7 per cent of field area) may provide nutritious forage during summers.

Process

The various steps involved in the green fodder production: sowing, irrigation, recommended fertilizers and pesticides application and harvesting.

Scientific management of water is crucial for year round supply of green fodder from an intensive fodder production system. It aims to provide ideal moisture regime to the crops for realizing their yield potential commensurate with the maximum economy in irrigation water and maintenance of soil productivity. Indian subcontinent is characterized by tropical monsoon climate, active growth of grazing land occurring only during monsoon months. Since ancient times cattle breeding and milk production has been the second.

The fertilizers management strategies in fodder crops aim at increasing the herbage production per unit area and time along with improvement in forage quality parameters and maintenance of soil health. The requirement of fodder crops for nutrients particularly nitrogen is comparatively higher. Chemical fertilizers have played a significant role in increasing crop productivity. But, for sustainable production from arable lands it is important to prepare a balance sheet of nutrients depleted and nutrients supplemented. Majority of the soils at present are rich in potassium but continuous cropping without K application over the years may turn to be deficient. Bio-fertilizers, the products containing living cells of different types of micro-organisms, play important role in enhancing fodder production and also cutting down the usage of chemical fertilizers. Studies have shown that a saving of 20 kg N/ha may be achieved with application of *Azotobacter/Azospirillum* in cereal fodder crops. Similarly, increase in forage yield due to *Rhizobium* inoculation to legume forages ranged between 14-46 per cent. Seed inoculation of berseem with phosphate solubilizer significantly increased the green fodder (103.6 t/ha), dry matter (16.19 t/ha) and crude protein (3.20 t/ha) yields over the control. Integrated use of organic, inorganic and bio-fertilizer sources of nitrogen in sorghum + cowpea–berseem cropping system led to over 25 per cent saving in N through use of *Rhizobium* and/or *Azotobacter*. However, a reliable system of quality control and efficient system of storage, transportation and management of bio-fertilizers is required for its wider applicability. Organic manure-induced improvement in soil physical, chemical and biological properties is well established. Build up of secondary and micronutrients, counteracting deleterious effects of soil acidity, salinity and alkalinity and sustenance of soil health are the key beneficial effects associated with organic manure application. Use efficiency of N fertilizers is improved in the presence of FYM.

Integrated pest management is emphasized these days in order to reduce the use of chemicals. The population of the three major pests namely leaf hoppers, lucerne weevil and aphids can be managed effectively by growing of least susceptible variety, IGFRI-450 in first week of July with fertilizer application of 30 kg N, 100 kg P_2O_5, 80 kg K_2O per hectare and if required application of endosulfan @ 0.08 per cent or berliner @ 0.84 kg/ha may be done. In cowpea, the damage due to major pests like, leafhoppers, semilooper, tobacco caterpillar and grasshopper can be managed without using insecticides by planting the least susceptible variety in the first week of July, using an optimal fertilizer combination of 30 kg N, 100 kg P_2O_5, 40 kg K_2O per hectare with two weeding at 15 and 30 days old crop.

In most of the situations forage cultivation is practiced even on degraded and marginal lands with very less inputs, realizing only low production.

Since ancient times, Indians have practiced mixed farming where livestock formed an integral part of agriculture. Rich genetic diversity exists for cultivated and rangeland species including tree, browse species, and herbaceous grasses and legumes. Major forage genera exhibiting forage biodiversity includes legumes like *Desmodium, Lablab, Stylosanthes, Vigna, Macroptelium, Centrosema* etc.; grasses like

Bothriochloa, Dichanthium, Cynodon, Panicum, Pennisetum, Cenchrus, Lasiurus, etc; browse plants such as *Leucaena, Sesbania, Albizia, Bauhinia, Cassia, Grewia* etc. These genera besides many others form an integral part of feed and fodder resources of the country.

Since ancient times cattle breeding and milk production has been the second most important profession in India after agriculture. Free grazing was practiced and became a way of life. Presently also, livestock production is primarily based on rangeland grazing. The grazing activity is mainly dependent on the availability of the grazing resources from pastures and other grazing lands *viz.* forests, miscellaneous tree crops and groves, cultivable wastelands and fallow land. Such lands are about 40 percent of the total geographical area of the country. The grazing intensity in the country is as high as 12.6 ACU/ha as against 0.8 ACU/ha in developed countries. Therefore, our task is two fold *viz.* improvement of pasture and judicious implementation of grazing management.

There is a lot of difference in the extent of grazing lands in various states. In some states *viz.* Himachal Pradesh, Jammu and Kashmir, Meghalaya, Nagaland and Arunachal Pradesh the grazing land availability is as high as 70 per cent. Pasturelands constitute the main grazing resources of the country, available over an area of 12 m ha (3.94 per cent of the geographical area). The distribution of pasture lands are mostly noticed in the state like Himachal Pradesh (36.44 per cent), Sikkim (13.31 per cent), Karnataka (6.54 per cent), Madhya Pradesh (6.35 per cent),Rajasthan (5.39 per cent), Maharashtra (5.11 per cent) and Gujarat (4.49 per cent). The northernregion has pasturelands in Jammu and Kashmir, Himachal Pradesh and Uttaranchal. This region has a potential resource in the form of green meadows and pasture, which at some places are mixed with the forests. The alpine meadow has an important economic value, providing pasture for the sheep and goats. The western India includes the pasturelands of Rajasthan and Gujarat. In Rajasthan the available grazing land is over 40 per cent and in Gujarat the area is about 30 per cent. The land under permanent pasture is about 5.4 per cent in Rajasthan and 3.5 per cent in Gujarat, providing good quality fodder for livestock. Nearly 30 pastoral communities in northern and western parts of the country depend on grazing based livestock production. Based on the practice followed by these pastoral communities in various regions, the grazing systems may be categorized either on the basis of methods of grazing or patterns of migration.

Considering methods of grazing the pastoral communities change the site of grazing after its utilization. The example of such type of grazing system is Kharak in Uttaranchal and Goals in the desert area of Rajasthan. Based on migratory habits the nomadic tribes are classified in 4 groups *viz.* (i) total nomadism; (ii) seminomadism; (iii) transhumance; (iv) partial nomadism. The success behind any grazing management practice depends on the accurate stocking rates considering the condition of the range. Moderate and conservative stocking rates sustain returns on a long term basis when compared to heavy stocking rate. Under different pasture utilization systems *viz.* rotational, deferred rotational, continuous and cut and carry; highest run-off and soil loss was recorded in continuous system while minimum run-off was recorded in rotational system. However, minimum soil loss was recorded in cut and carry system. The observations showed that improved practices of pasture establishment, contour bunding, grazing management reduced soil loss, made more water available and improved soil conditions. Out of the 4 systems of grazing management, relatively higher values of organic carbon and available nutrients were observed in deferred rotational system, indicating its superiority over other pasture utilization systems. The results indicate that for improvement of the existing rangelands techniques like bush cleaning, application of fertilizers, controlled burning coupled with proper grazing management should be employed.

Alternate Land Use System: Agroforestry

Considering the continuous pressure on cultivable lands and deterioration already set in on the grazinglands, it is imperative to look for alternate land use systems that integrate the concerns for productivity, conservation of resources and environment and profitability. Agroforestry technologies such as silvopasture, hortipasture etc. hold promise not only for bioremediation of degraded habitats but also forage production.Silvopastures integrate pasture and/or animals with trees. Woody perennials, preferably of fodder value, are introduced deliberately and systematically and managed scientifically. Under poor soil, water and nutrient situations where cropping is not possible such systems can serve the twin purposes of forage and firewood production and ecosystem conservation. It has been possible to increase land productivity from 0.5-1.5 t/ha/year to > 10 t/ha/yr (10-year rotation) by developing silvopastures. Now, concept of hortipasture is also finding applicability with the farmers for utilizing their degraded lands. The additional forage availabilitythrough such systems is likely to reduce grazing pressure and thus have important environmental implications. Efforts to design silvopasture systems to produce > 15 t/ha/year through species introduction, planting geometry, canopy manipulation and sustainable management through in situ grazing or cut and carry system are continuing. The grasses have a lot of conservation benefits in reducing soil and water loss to a great extent.Average soil loss from deforested land is reported to be 12-43 t/ha in black soil and 4-10 t/ha in red soils whereas soil loss from natural grassland has been only 3.2 t/ha from a protected site. Silvopastures are still better for soil conservation, soil loss from these areas coming down to 0.9 t/ha. Grasses and legumes are primarily valued as forage and their value in checking soil erosion is often less viewed. There are specific recommendations for a particular type of soil and rainfall. For example, *Dichanthium annulatum* is best in reducing soil loss from 20.5 t/ha to 1.5 t/ha on sandy loam soil (2 per cent slope) in 790 mm rainfall situation.

Forage Seed Availability

One of the main reasons for slow pick up of forage production technologies is unavailability of proper seeds of forage crops and range species. According to an estimate, hardly 20 per cent of forage seed requirement is met. With the development of a number of improved and high yielding varieties in forages, it has become important that quality seed is readily available for supply to the farmers. There is increasing demand of seeds of *Cenchrus ciliaris, Chrysopogon fulvus, Pennisetum pedicellatum, Panicum maximum*, cowpea, berseem, sorghum, and oat. By and large forage crops are shy seed yielders. There are various causes of low seed yield in forage crops *viz.*, indeterminate growth, uneven maturity, blank seed, seed dormancy, site effects etc. Also, the nature of forage seed production is much more complex and a number of environmental and physiological factors have a significant impact on seed production potential of a crop. In majority of forage crops, seed production depends on photoperiod, thermo period, humidity, soil texture, soil structure, soil reaction and moisture. Each forage crop is, therefore, suited to only specific area for forage and/or seed production such as berseem in the northern plains and lucerne in the northwest India. Similarly pasture grasses like seven is best adapted and productive under low rainfall situations of western Rajasthan whereas Congo Signal grass and guinea grass produce better seeds in the highly humid regions of Kerala.

Future Outlook

Time has come to take a stock of present situation and re-examine our research priorities to make them need based. The following points are suggested in this direction:

The current emphasis on forages is to be focused on regional problems. Considering the continuous pressure on cultivable lands it is imperative to look for alternate forage production sites. The main emphasis should now be on degraded lands/problem soils where nothing can be grown and cost of chemical/physical amelioration is very high.

The alarming situation of the fragile ecosystem of the desert and also in the alpine pastures and meadows on the high hills of the Great Himalayas require adequate attention. The lesson that "Forests precede civilization and deserts follow them" should always be kept in mind. Restoration of degraded rangelands and re-vegetation/rehabilitation of wastelands must form a part of the overall development plans. Landscape be the unit of development and resource management. Inventory of degraded lands in all the agro-ecological regions of India should get a high priority.

GIS should be effectively used for making an inventory of biophysical sources and monitoring of development systems. There is an urgent need to have a sincere effort to collect the genetic diversity of various plants from problem soils and their proper evaluation in both field and laboratory conditions in order to identify the suitable germplasm for a particular problem condition. Identification of the traits/genes will be the next step for genetic improvement programme. Identification of such genes/traits can then be attempted for transfer through conventional or biotechnological methods. Efforts should be made to develop crop based feeding system for different agricultural crops grown in different zones of the country. The regional deficits are more important than the national deficit, especially for forage as often is not economical to transport over long distances. Mixed farming system should be supported, as the economic contribution of animals becomes even more significant in the context of beneficial effects like a low input integrated and sustainable system, nutrient recycling, alleviation of rural poverty and improved food security besides positive effects on the environment. The improvement in fodder production on saline soils can be augmented by increasing salt tolerance through exploitation of natural variation in the germplasm or by creating novel variations. Genetic improvement for drought tolerance is an important component in stablizing fodder production in the country. Kharif crops (sorghum, millets, grasses and legumes) are sown at the onset of the rains. Probabilities of moisture deficit are highest at start and the ends of rains. Similarly for Rabi crops dry winter is followed by a hot spring and water deficits are common near flowering stage of the crop. The need is to accelerate the search for unique drought adoptive mechanisms using the full range of diversity available. Bloat causes farmers a lot of concern and results in significant losses in production. Bloat is due to feeding on rich legume forages, where plant lacks protein and precipitates tannin. The transgenic technology could relieve a lot of animal suffering and restore the grazing system to full productivity. Transgenic berseem and lucerne crop based around Bacillus thuringiensis gene giving protection against Heliothis damage, which is a principal seed crop pest in these crops. Identification/building of a resistant gene that provide stable resistance against stem and root rot complex of berseem. Improvement of the amino acid balance of the forage plant species used in intensive feeding and production systems. Improvement in the efficiency of uptake of essential nutrients such as nitrogen and phosphorus through understanding of molecular and cellular mechanism of uptake of nutrients by plant roots from the soil. Formalize the community gene bank existing at the village level. The appropriate sites for undertaking forage seed production may be identified apart from the traditional areas for growing these crops. Adequate emphasis is also required on seed processing and handling.

Sowing Calendar for Supply of green fodder around the year. Name of fodder Sowing time: Sorghum (Jowar) March to May, Pearl Millet (Bajra) March-April and June to July Makchari July, Cowpea (Lobia) March and July, Maize March – August, Winter Maize October Berseem September – October, Lucerne October – November Anjana grass, June – July, Guar June – July, Oats October – Mid December, Mustard October. Development of superior fodder crops rich in proteins and essential amino acids QPM Maize (Rich in tryptophane and lysine) Genetic variability in paddy and wheat straws enhanced palatability and digestibility.

Pasture Development/Management Pasture and grazing lands less than 4 per cent of cultivable area Seeding/Reseeding of grasses Pelleting of grass seeds. Rotational grazing avoid over grazing. Superior fodder, grass and legumes, Use of saline and wastelands, Mixed farming for grains and fodder Perennial grasses + Fodder crops and Multicut varieties (chari, oats) should be given importance.

Suitable Area for Agroforestry

The following type of land can be assigned for agroforestry

1. Cultivable land
2. Field boundaries
3. Along with farm roads and *nallah* sides affected by erosion.
4. Pockets within cultivated holding where cultivation is not possible
5. Old fallows
6. Cultivable waste
7. Other areas like community or panchayat land etc. in which agro-forestry can be incorporated.

System or Practice Agro-ecological Adaptation

On the basis of nature of components the following are common agro-forestry systems prevailing in different agro-ecological regions of India :

1. Agri-silviculture (trees+crops)
2. Boundary plantation (tree on boundary + crops)
3. Block plantation (block of tree+ block of crops)
4. Energy plantation (trees+crops during initial years)
5. Alley cropping (hedges+crops)
6. Agri-horticulture (fruit trees+crops)
7. Agri-silvi-horticulture (trees+fruit trees+crops)
8. Agri-silvipasture (trees+crops+pasture or animals)
9. Silvi-olericulture (tree + vegetables)
10. Horti-pasture (fruit trees+pasture or animals)
11. Horti-olericulture (fruit tree + vegetables)
12. Silvi-pasture (trees+pasture/animals)
13. Forage forestry (forage trees+pasture)
14. Shelter-belts (trees+crops)
15. Wind-breaks (trees+crops)
16. Live fence (shrubs and under- trees on boundary)
17. Silvi or Horti-sericulture (trees or fruit trees+sericulture)
18. Horti-apiculture (fruit trees + honeybee)
19. Aqua-forestry (trees + fishes)
20. Homestead (multiple combinations of trees, fruit trees, vegetable etc).

Besides these common agro-forestry systems, there are many more component combinations followed in different agro ecological regions of India.

References

Farooki, S.H., Kumar, S. and Singh, D.N. 2009. Chara Anusandhan Ke Naye Ayam. Kheti. 62 (7): 28-31.

Gangwar, K.S., Sharma, S.K. and Murari, K. 2004. Alley cropping of rice-wheat with *Leucaena leucocephala* for sustainable food-fodder-fuel security. Indian Farming 54 (3): 33-35.

Gill, A.S. 2004. Ideal forage crops for rainfed areas. Agricultural Extension Review. 16 (1): 18-19.

Kalloo, G. 2009. Forage Research: New Dimensions. In: 42nd Foundation day lecture, pp. 12.

Rawat, C.R. and Hazra, C.R. 1996. Forage production in farming systems. Indian Farming. 45 (10): 29-33.

Chapter 13

Increasing Productivity of Pasture and Range in India

Sant Kumar Singh

ICAR Resesarch Complex for Eastern Rregion, Patna – 800 014

Email: santpatna@yahoo.co.in

Good pasture management strives to keep overall productivity of pastures without sacrificing quality. The most basic principle of grassland management is the need for sustainability. The management measures must be to obtain the maximum herbage yield with the highest possible nutritive value throughout the year at the lowest possible cost. Protection from Uncontrolled Grazing, Grass-legume Balance, Reseeding and Planting with better Forage Species, Over-sowing and Sod Seeding, Planting Tree Legumes in the Pastures, Manure and Fertilizer Application, Control of Weeds and Bushes, Soil and Water Conservation Measures, Grazing or Harvesting Height, Assessment of Forage Yield and their utilization and stocking rate are some of the important points of consideration.

Grasses are capable of supporting or converting solar energy into incredibly huge amounts of biomass. They also support a rich and diverse variety of flora and fauna. They are efficient in absorbing rain water and play vital role in water retention and hydrology of an area. It is imperative to recognize the ecological, hydrological, economical and sociological role of grasslands as a source of survival for millions of livestock and rural people, as protector of soil and water, of rare wild life species and biodiversity conservation in general.

Let us have a look at the term *Pasture* which is grassland, usually with improved species, carefully managed with inputs such as irrigation and manures for grazing livestock. Pastures can be considered as "domesticated grasslands". It may be either natural or sown pasture. Whereas, the term *Range* is a natural grassland in which the climax vegetation or the potential plant community consists principally of native grasses, forbs (broadleaf herbs) and shrubs that are valuable as forage and are in sufficient quantity to justify its use for grazing. Although there is often a tendency to consider native and sown pastures as separate entities, this is an artificial division once a sown pasture is established. Both are usually grass dominant used for feeding livestock, and respond to the same basic principles of management.

Good pasture management strives to keep overall productivity of pastures without sacrificing quality. The most basic principle of grassland management is the need for sustainability. The pasture must be maintained as a sustainable agricultural system. Sustainability can only be achieved if the current enterprise is economically viable, and the basic resource, the soil, is conserved. While ecological stability is an important aspect of sustainable agricultural systems, it must also be socio-economically sustainable. Sustainable agricultural systems are characterized by inputs and outputs. In terms of physical inputs such as water and nutrients necessary to drive the systems, losses and gains to the systems should be in balance. Different agro-eco-systems have limits to inputs and outputs, which if exceeded, lead to non-sustainability. The management measures must be to obtain the maximum herbage yield with the highest possible nutritive value throughout the year at the lowest possible cost.

Pasture and Range Management

In India, grazing based livestock husbandry continues to play an important role in rural economy of the country as around 50 per cent livestock depend on grazing in forests and other grazing areas. Total area available for grazing in the country is in the range of about 40 per cent of the land area. In states like Himachal Pradesh, Uttarakhand, Jammu and Kashmir, Meghalaya, Nagaland and Arunachal Pradesh over 70 per cent of land area is utilized as grazing ground. Since India is characterized by tropical monsoon climate and active growths in grazing land occur only during monsoon months and deficits of various levels in other months. Thus, there is already growing emphasis on animal feed security systems and fodder banks to overcome such problems. The surplus production from grasslands during rainy season needs to be carefully preserved in various forms to meet the forage requirements of the lean periods. The post harvest technologies such as biomass processing, enrichment and densification appear to be the key for better animal husbandry in the deficit zone.

Good pasture management strives to keep overall productivity of pastures without sacrificing quality. The most basic principle of grassland management is the need for sustainability. The pasture must be maintained as a sustainable agricultural system. Sustainability can only be achieved if the current enterprise is economically viable, and the basic resource, the soil, is conserved. While ecological stability is an important aspect of sustainable agricultural systems, it must also be socio-economically sustainable. Sustainable agricultural systems are characterized by inputs and outputs. In terms of physical inputs such as water and nutrients necessary to drive the systems, losses and gains to the systems should be in balance. Different agro-eco-systems have limits to inputs and outputs, which if exceeded, lead to non-sustainability. A stable pasture system is able to meet the nutritional requirements of pasture plants and grazing livestock without depleting the soil resource in the longer term. In other words, the management measures must be to obtain the maximum herbage yield with the highest possible nutritive value throughout the year at the lowest possible cost.

Protection from Uncontrolled Grazing

Uncontrolled grazing will prove fatal to the grasslands and pastures. Grazing must be regulated. The botanical composition and the overall percentage of grasses present in the grasslands must be assessed first. Based on the assessment, decisions regarding protection against uncontrolled grazing or complete renovation through reseeding can be taken. Protection to such areas would encourage quick growth and coverage of the area within two to three years through auto-reseeding. Fencing the area is a prerequisite for giving protection. Materials such as angle iron poles with barbed wire/woven wire, cement or stone posts with wire fencing, stone or brick walls, and live hedges can be used successfully. Plants such as *Lawsonia alba, Inga dulci, Zizyphus nummularia,* and *Parkinsonia* are very ideal for establishing live-hedges. Hedges can also be made with fodder shrubs such as hedge lucerne, gliricidia or subabul. Grazing or cutting interval depends on forage species, soil fertility, season, climate and method of harvesting. The yield of pasture is substantially greater, if the interval between cutting or grazing is long. Nevertheless, the higher yield of herbage produced under a lenient cutting or grazing system is often of lower nutritive value than the younger material. On the contrary, severe and frequent defoliation reduces the vigour and survival of trailing legumes because of their reliance on rooted stolon production. In rainy season, forage grasses are capable of producing high dry matter yields of moderate quality with reasonable persistence, if cut at about six week intervals. This interval appeared to give the best compromise between dry matter yield, forage quality and persistency. However, depending upon the pasture status, the intervals must be long or none at all in winter and summer season.

Grass-legume Balance

Nutritive value of tropical grasses is relatively low. On an average, they contain 6.0 per cent crude protein at the time of their optimal utilization. Overseeding them with leguminous forage species increases the overall nutritive value and palatability of these grasses. It is suggested that good pasture management should aim to maintain at least 20-30 per cent legume component. Many legumes are well known for improving the fertility of soil through nitrogen fixation. In a pasture, the most important factors, which are likely to affect pasture legume content are initial management during the first 6-12 months and the intensity and frequency of grazing. In addition, application of large quantities of nitrogen fertilizer can have negative effects on legume percentage. The grass-legume balance is likely to fluctuate and will not remain static. Taking all these into consideration, the inter- and intra-row spaces of the grasses should be utilized by adopting suitable seed-rates of the compatible species of the legumes. Stylo, siratro, centro, atylosia, clitorea, rhyncosia, alyce clover, puero, etc. form good mixtures in grasslands. However, some are affected by severe drought. Siratro, centro and atylosia are drought tolerant. For a successful good grass-legume mixture in the pasture, bunch grasses are preferable to strongly stoloniferous or rhizomatous grasses. Short and leafy grasses reduce competition for light. Soil type also plays a role, for instance, para grass is less vigorous on lateritic soil, while Centro grows quite well, so that a satisfactory grass-legume balance is maintained. Some rhizomatous and stoloniferous grasses such as signal grass and para grass are incompatible with legumes, mainly because of competition for light and nutrients, particularly N and P. The legumes usually persist for one to two years; the N fixed by legumes encourages grass growth, and the legume is eventually suppressed, leading to grass dominant pasture. Grass-legume mixtures are best kept as simple as possible, since management is complex. Good management of mixed pasture should aim to maintain at least 20 to 30 per cent legume component.

Reseeding and Planting with Better Forage Species

If the grassland is in a completely degenerated condition, reseeding or planting with better yielding, adaptable, persistent and aggressive species become essential. Planting slips or cuttings is always more successful than reseeding but costlier and require more skilful operations. Reseeding or planting is done by the beginning of rainy season. The technique of reseeding is quite simple and can be followed successfully. Land must be prepared well after clearing undesirable bushes and weeds. One or two disc harrowing, followed by planking will give necessary tilth. At the time of final land preparation, 10-15 Mg/ha of farmyard manure should be added and thoroughly mixed with the soil. As the grass seeds are very small, light and with appendages, chances of being washed by water or blown away wind are high. Therefore, the seed is processed into small pellets. These pelleted seeds are easy to handle, and are less vulnerable to water and wind currents. The seed is worked into a homogeneous thick paste (by mixing three parts of sand, one part of clay, and one part of cowdung) and pellets of convenient size are prepared in such a way that each pellet contains two or three seeds. Pellets prepared in this way are either placed in shallow furrows opened either with a bullock or tractor drawn cultivator at a recommended distance for each species or cultivar or are broadcast with the onset of the monsoon. Like grasses, the legume seeds are also turned into pellets after rhizobial inoculation and lime pelletting. It is not possible to take up the entire area at a stretch for re-seeding or planting because the grazing livestock cannot be left starving. A detailed plan, therefore, has to be devised whereby certain areas are open for grazing and certain areas closed and improved simultaneously.

Over-sowing and Sod Seeding

In the tropical grasslands, legumes may be planted along with the existing pasture for improving the pasture quality. Overseeding refers to seeding of a grass or legume into an existing pasture or grassland to improve the overall productivity and quality of herbage. Sowing or planting a legume into existing native tropical grasslands is a low cost pasture improvement method. Often, existing grasses are temporarily set back by heavy grazing, burning, ploughing or improves the feeding value of the pastures.Legumes can be successfully established in grasslands or pastures either by sowing into ploughed or disced strips or by sod seeding into rows using a disc seeder (The practice of mechanically placing seeds of grasses and legumes in to an existing grass sod is called sod seeding). When seeds are used, spraying along the rows with a non-residual, non-selective herbicide such as paraquat, taking care not to touch any existing pasture species, controls many weeds. The feeding value of the pasture is thus improved. However, efforts must be taken to create conditions that favour the germination and establishment of the legume seedlings. This means that competition from the established root system of the existing grass must be reduced to ensure good soil-seed/vegetative propagule contact for the introduced species. This can be done as follows: Heavily graze the pasture to a height of 5-10 cm to reduce the leaf canopy and the accumulated layer of litter. If this is done initially, then subsequent seeding operations achieve adequate soil contact. Usually, overseeding with vegetative propagules into a heavily grazed existing pasture will not offer much competition for light especially in the early stages of establishment.

Planting Tree Legumes in the Pastures

Tropical and subtropical grasses provide good grazing for the livestock in the monsoon and post-monsoon periods. They, however, become dormant with the advent of summer. In the subtropical regions, winter also make the grasslands dormant. Grazing cannot be done until the beginning of monsoons. However, during the lean periods of spring and summer, treetops come to the rescue of the livestock owners. Some of the important trees with good coppicing abilities capable of giving good lopping yields are Subabul (*Leucaena leucocephala*), Gliricidia (*Gliricidia sepium*) and Shevri (*Sesbania aegyptica*). These trees may be planted along the boundaries of the fields, in the cattle-yards, etc., to serve as shade-cum-fodder producing plants. Besides the planting of trees on the farm for various purposes, trees can also be planted in the pastures as companion species with grasses as in a silvipastoral system. Under this system, compatible fodder-trees are planted 5-7 metres apart both ways during the monsoon. The fodder from the trees is available after 4-5 years.

Manure and Fertilizer Application

Although small amounts of urine and manure are being added while grazing, these are insufficient to make good the losses by way of nutrient removal through forages cut or grazed. To replace the soil nutrients thus removed, regular fertilizer application is necessary unless the soil is very fertile. Often, a small dose of nitrogen is used as a 'starter' dose even for legume establishment. However, the application of large quantities of nitrogen fertilizer can have negative effects on legumes. Presence of high levels of nitrate or ammonium ions will inhibit nodulation, and in plants already nodulated, nitrogen fixation may be reduced. Tropical grasses show good response to the application of nitrogen. When a legume is growing with a grass, the latter usually competes strongly for available nitrogen, takes up most of that applied, and grows rapidly at the expense of the legume. This may result in a reduction in the legume population in the pasture. Regular application of phosphorus fertilizers has been shown to have a significant and positive effect on legume percentage, pasture production and live weight gain. Organic manures must also be used copiously in the form of compost, farmyard manure, sheep or goat manure or green manures. If good testing facilities are available, then periodic soil and foliage tests can assist when decisions are made on fertilizer requirements. Fertilizers should be applied once a year or at least once in 2-3 years, to keep up the grass species in optimum botanical composition and at a high production level.

Control of Weeds and Bushes

In the tropics, it is a challenge to manage any crop production systems without weed growth.

Natural grasslands may become heavily infested with undesirable weeds and bushes. These bushes and weeds compete with pasture species for space, light, nutrients, besides being poisonous to the grazing animals and harbouring some carnivorous animals. These are detrimental to livestock production and have to be managed. Common problem bushes in the tropics like *Chromolaena odorata*, (Siam weed), *Sida acuta, Sida rhombifolia, Zizyphus, Butea monosperma* (Flame of Forest), *Carissa, Lantana*, and *Mimosa*, should be cut manually at the ground level and removed. Wherever possible, 2, 4-D or paraquat can be used for bush clearance. The maintenance of good pasture is primarily a matter of good management, as poor grazing management may lead to weed reinfestation. With consistent overgrazing and the opening of the sward, weeds may become reestablished. It is all the more important to take early action to prevent the weeds from seeding. Regular checking of the pastures by the farmer is a must for this.

Soil and Water Conservation Measures

Undulating topography of certain grasslands may pose serious problems in their improvement.

Even though grass sods give protection to soil from erosion; in steeper soils, mechanical measures are also needed. In such areas, adequate soil and water-conservation measures, *viz.*, contour bunding, terracing, pitting, contour trenching, etc. should be adopted. In sloppy areas, contour planting system must be adopted for planting grasses and trees. Various management measures needed for the grassland or pasture must also be planned across the contour. Low-cost vegetative measures should get preference over costly mechanical measures. Grasses with good soil binding and drought tolerance ability must be used for adopting measures such as contour grass strips and hedge or barrier strips with grass. Hedgerow intercropping or alley cropping with fodder trees and grasses must be tried wherever feasible. Rainwater harvesting by plugging gullies or streams, especially in areas with undulating topography, or by erecting check dams at several points on a stream, is done to harness water for utilization during the periods of moisture stress for rejuvenating the grasses. Contour staggered trenches or infiltration pits also do the same purpose. Similarly, the low-lying areas vulnerable to frequent floods are drained of excess water and utilized during the periods of moisture stress for increasing fodder yields.

Growth Stages and Pasture Quality

Quality of pasture is affected by the stage of plant growth at which grazing or cutting is done. As the plants mature, feed digestibility and quality decreases but the quantity of dry matter increases. Young leafy plants are usually high in protein and the cell walls contain readily digestible cellulose. Upon ageing, protein levels and digestibility decline with grasses producing stems and inflorescences. This increasing proportion of stem in pasture reduces the overall nutritive value of the available dry matter. Grasses tend to decline more rapidly in protein content with increasing maturity than legumes. However, some grasses have different rates of decline in crude protein levels. The critical level of crude protein in tropical grasses is about 7.0 per cent below which dry matter intake is depressed. Legumes in a pasture have an important role to play in maintaining adequate protein levels in the feed. Good nutrition is important not only in terms of live weight gain but also in terms of reproductive efficiency, calving percentage and other factors. To achieve high production from tropical pastures, animals must be given good quality feed. Grazing management, thus aims to maintain pastures in a young leafy stage as long as possible. Usually, a compromise has to be made between quality (high protein content of young re growth)and quantity (high dry matter content of mature pastures). Overgrazing can result in reduction in growth rate and weakening of plants. Undergrazing, on the other hand, fails to maximize the plant's potential to produce feed thereby reducing the quantity available to the livestock.

Grazing or Harvesting Height

Frequency of grazing and cutting height can have a very serious effect on the pasture's ability to regrow. Severe defoliation can stop root growth. It causes slow root extension for a period of 6-18 days, reducing nutrient uptake and thus plant vigour. In these circumstances, chances are that more vigorous weed species invade the pasture. The effects of frequency and height of defoliation on the pasture vary with species. Plants with buds close to the ground are more able to withstand heavy grazing pressure than those with exposed grazing points. It is suggested that the grazing practice should be guided by the need to maintain a certain critical leaf area to ensure maximum net photosynthesis. For most of the grasses, average grazing height is about 15 cm. However, grasses like Napier should not be cut or grazed below about 20-30 cm and Guinea below about 15-20 cm. Some grasses like Toco grass and Carpet grass are able to tolerate hard grazing down to 5 cm. Grasses like Napier should not be allowed to grow taller than about 1-1.5 m before grazing (although as Napier grass is usually harvested by cut-and-carry methods, a height of 2.0-2.5 m may be reached before cutting), while most grasses should be grazed before they reach the 1.0 m mark. Nevertheless, height is not the only criterion; maturity, protein content and stage of growth are also important. In addition, pastures may be closed out to allow seeds to set and drop to thicken the stand.

Assessment of Forage Yield

In a grassland or pasture, forage availability may not be uniform throughout the year. In the tropics, if the crop is under irrigation, forage availability may not be affected much by seasons. However, if the crop is rainfed, then the amount of feed available during the rainy and summer seasons show great variations. For many purposes, such as deciding optimum grazing intensity, herbage allowance, etc., forage availability has to be estimated. Estimation of forage availability is essential to keep the livestock under optimal production levels. Even for a small dairy farmer, a good approximation of the amount of forage available on his own farm will be of great use to take decisions on feeding. Several methods are available to quantify the availability of fodder or forages. It is not easy to estimate yield by observing each individual plant in the farm. Instead of the whole farm, only sample area is studied. The simplest method to estimate fodder yield of grasses and annual legumes is by using a quadrat. Sampling units, which are areas of definite size, are called quadrats. Farmers can make a quadrat using wood or iron rods. Usually, a quadrat is of one-meter square (1m x 1m). The fodder inside the quadrate is cut and collected. Sample as many number as possible and mean yield per quadrat is arrived. Sampling should be done at the center, diagonal crossing and edges of the field. Using this mean yield per square meter, estimate the yield per hectare by multiplying with 10,000.

Forage Utilization

In a grassland or pasture, there are three possible ways of utilization. Most important is to allow the animal to graze the standing pasture through various grazing or stocking systems. Others are cut-and-carry the forage to the animal (cut-and-carry system or zero graving) and cut, store and feed the forage to the animal at some future date (forage conservation). Although grazing is the usual method of harvesting pastures, it is rather inefficient in that only about 50 per cent of the forage produced is actually consumed. The rest is knocked down, trampled and soiled.

Grazing System

Depending on situations, grazing methods may be continuous or rotational or some of their variants such as deferred, strip or creep grazing. In both continuous and rotational systems, the number of animals can be kept constant or varied according to pasture growth. Animals select palatable species while grazing, and rotational and deferred grazing are important to save these species.

Continuous Grazing

Continuous grazing is a type of grazing system in which grazing is confined within a enclosed area over a period of time, which may last for a grazing season or an entire year. In this system, the pastures may be set stocked (number of animals kept constant), variably stocked (numbers varied according to pasture growth.) or put and take (some animals are present on the pasture at all the time). In the 'put and take' system of grazing, the pastures are grazed continuously for the whole of the grazing season by a fixed number of animals, the 'testers'; and to graze excess forage beyond that needed for them, a variable number of 'put and take' animals, also called 'regulators' or 'grazers', are also allowed in and removed when the quantity of herbage declines. A put and take system ensures a constant grazing pressure. The advantages of continuous grazing are: selective grazing by animals, which choose the most nutritious forage, and lower input required for fencing and watering points. However, selective grazing may prove disadvantages as it may lead to deterioration of pasture quality and production capacity. Trampling and fouling of forages can take place in poorly grazed patches and the forage become rank and coarse.

Rotational Grazing

The pasture area is subdivided into a number of enclosures or paddocks with at least one more enclosure than the group of animals. Animals are grazed in a regular sequence before returning to graze again the first paddock after a regrowth period of about 2-6 weeks. This is also called paddock grazing. Like continuous grazing, here also the system may be set stocked, variably stocked or put and take. In a rotational system, another variable-time interval is also involved for grazing each unit before moving to the next, which may be short or long or relative to the stocking rate and pasture production. In the humid tropics, the grazing interval usually ranges from 30-60 days.The main advantage of the system is more flexibility. It is useful for practical operations such as fertilizer application, weed control, harvesting and control of grass-legume balance plus increased farmer-animal contact. It may also allow certain legumes to flower and set seeds and may allow defoliated legumes such as subabul to regrow. However, the disadvantages are the requirement of extra sub-divisions and water points; the need to have more attention to retain the pasture status; increased labour requirement to move cattle; and the requirement of better management skills.

Deferred Grazing

Rotational system operates on a set rotation where each unit is successively grazed in sequence or on a deferred rotational grazing system. This recognizes critical periods in the phenology of the pasture species such as seed germination and seedling establishment, fruiting, seed set, etc.

Strip Grazing

In strip grazing, animals are confined for grazing in a limited area for a short period. Sometimes, it is also called ration grazing. Strip grazing is important to make the animals feed on less palatable grasses as well. The system allows the animal access to a limited area of fresh crop either twice daily or daily or for longer intervals. The area once grazed is immediately fenced off.

Creep Grazing

Creep grazing is usually followed for goats and sheep. Growing animals need palatable and nutritious herbage. When paddocks are set up, a special arrangement in the fence is made to make the lambs creep into and eat the lush growth of herbage followed by ewes who may need maintenance ration only. This is creep grazing.

Tethering

Tethering of cattle using ropes to some trees or some movable pillars is a common practice in the villages. The animals graze on a limited area around the tree or the pillar. Tethering is particularly suitable for small holders who keep only a few cattle.

Grazing in situations such as natural pastures under tree crops, small patches of pasture in the households, premises of buildings, and roadsides is possible only through tethering. Tethering restrict the movement of the animals thus avoiding the possibility of damaging other crops and properties.

Cut and Carry System/Zero Grazing

Zero grazing or the feeding of fresh cut grass to livestock is an age-old practice followed in India. It is also called green soiling or stall-feeding or cut-and-carry feeding systems. In the tropics, zero grazing is widely used because of several reasons. Small size of holdings with limited grazing area, fragmentation of land holdings, lack of fencing in pasture areas and cheap labour encourage zero grazing. Cut-and-carry system is suitable for plantation crop-forage system, particularly when the trees are young and vulnerable to damage from grazing animals. Animals do not graze the pastures contaminated with their excreta. Under such conditions also, zero grazing is practiced to reduce selectivity. Herbage is cut by sickles or by mowers, transported, and fed to animals in stalls.

For small dairy· farmers, cut-and carry system is the most ideal. Small holders can develop backyard pastures as a realistic approach to forage development Small plots of land available near the house can be converted to productive pasture with little effort.

Stocking Rate

Stocking rate is defined as the number of animals grazing a unit area at a particular time. It is usually expressed as animals/ha or more precisely Adult Cattle Units per hectare (ACU/ha). Stocking rate is also expressed in terms of its reciprocal, ha/ACU. In India, one cow weighing 350 kg is taken as the basic unit, equivalent to one AG Therefore, ACU = (Weight x Livestock No.)/350.

A cow of over 2.5/3years weighs about 350 kg is equivalent to one ACU. A cow below 2 to 3 years weighs about 200 kg. A buffalo over 2.5/3 years weighs 400 kg and those below 2.5 years weigh about 250 kg. A goat or sheep is equivalent to 0.2 ACU and an equine/camel is equivalent to 1 ACU (Tyagi *et al.*, 1995). Stocking rate is one of the most important variables influencing both the productivity per animal and per hectare. This is the 'instantaneous stocking rate'. The long-term stocking rate integrating short-term fluctuations is referred to as the grazing capacity or carrying capacity implying an 'optimum-stocking rate', which can, be sustained. Grazing capacity or carrying capacity is defined as the number of animals that can graze in a unit area of pasture without over- or under-grazing for a given period.

Stocking rate relates animal numbers to land area. However, the most important aspect is not the land area, but the quantity of forage produced. Grazing pressure takes care of this. It refers to the number of unit animals per unit of available forage. It is the stocking rate at which animals fully utilize a pre-determined amount of pasture herbage. Grazing pressure gives us an idea on the balance between the quantity of forage on offer and livestock requirement. The terms 'grazing pressure' and 'grazing intensity' are used synonymously. Grazing pressure may be expressed in terms of ACU/ha, if actual pasture area is considered; or ACU/Mg, if available forage in terms of quantity of forage is considered. The reciprocal of grazing pressure is the herbage allowance, the weight of herbage available in kilogram or Megagram of dry matter per ACU (kg/ACU or Mg/ACU). As stocking rate increases, the grazing pressure also increases, herbage allowance falls and competition between animals increases. The greatest influence on animal performance and sustainable forage production is the choice of stocking rate. The grazing capacity or stocking rate of pastures may vary according to the differences in botanical composition; soil fertility and amount of fertilizer used; type and age of animals; grazing system and pasture condition; availability of supplementary feeds; and management. Climatic factors and season exert a greater influence, for instance, stocking rate will be lower in very dry periods. There are no general recommendations for optimum carrying capacity because of the complexity of factors, which governs stocking rate.

In nutshell, an optimum stocking rate is one: that maintains pasture stability and botanical composition that maintains a good grass-legume balance, saleable products such as meat and milk. A stocking rate of about 80 per cent of the average carrying capacity maintains the condition grassland, except in excessively dry years (Crowder and Chheda, 1982). Unless the pastures are efficiently utilized and converted to animal products, then the farmer will not get the good returns on his investment. Optimum carrying capacity is often measured in terms of maximum sustainable live weight gain per hectare. Nevertheless, the most important criterion for farmers selling animals to the abattoir or local butcher, who require a quick turn off, is rather monetary gain per head.

Under poor soil, water and nutrient situations (where cropping is not possible), silvipasture systems, integrating woody perennials and pasture species, can serve the twin purpose of forage and firewood production and ecosystem conservation. It has been possible to increase land productivity on a rotation of 10 years through such interventions from 0.5-1.5 t dry matter/ha/year to about 10 t dry matter/ha/year. On good lands under rainfed situation, horti-pasture is emerging as an economically viable option for sustainable income generation.

High priority on generation of information on temperate grasslands with emphasis on (i) low input, clover based sheep grazing system; (ii) ideal pasture for mixed grazing systems; (iii) inventory of grazing routes and grazing systems; (iv) designing of suitable production system for migratory graziers of Himalaya and the Thar desert. There is a need to develop a policy of regulated grazing that is managed on scientific principles so that desirable vegetation development could be ensured. As the grazing policy alone can not mitigate the problems of forage availability in the country, a matching approach on fodder production, agro-waste-use and fodder trees should be brought under one umbrella in form of a national fodder mission.

In arid and semiarid regions of the country, large blocks of lands away from human habitations could be developed as grass reserve and their production may be preserved in form of hay in fodder banks. The technology of densified bales or enrichment in form of feed blocks may be practiced for ease in transport and enhancing forage quality.

The establishment of a nodal agency is required to coordinate production and marketing of quality range seeds, both as regional and national levels, involving commercial seed companies, NGOs and farmers' cooperatives etc. There is a need of capacity building at various levels for the rangeland development and seed production of range species activities with the objective of restoring range health. IGFRI and CAZRI exclusively deal with grassland and desert ecosystems, respectively. They have adequate capacities and capabilities to address the respective issues. However, their geographical coverage and overall approach (ecosystem services and intrinsic values) needs to be broadened particularly a long term ecological and monitoring studies on grassland in representative zones. The traditional Kangayam grassland model offers a time tested grassland management technique which can be replicated elsewhere under similar rainfall and climatic situation.

In the context of development and suitable use of grasslands, some of worth noticing recommendations of there is an urgent need for a National Grazing Policy to ensure the sustainable use of grasslands and biodiversity conservation. As early as 1937, cattle conference held in Simla gave resolution that in all provinces standing fodder and grazing committee should be established. However, we still do not have a grazing or grassland policy on ground. Government of India has formulated 'Draft Grazing and Livestock Management Policy (1994)', and Draft National Policy for Common Property Resource Lands (CPRLs). These policies have not been implemented effectively in the field. Secure tenure for pastoralists (both resident and nomadic) over pastures, and genetic improvement of livestock (using indigenous breeds, not exotics ones) have not been taken into consideration in animal husbandry programmes of the country. In our country, only livestock is considered as wealth, not the grasslands on which the livestock depends. Forest department is mainly interested in trees, Agriculture Department is interested in agriculture crops and veterinary Department is concerned with livestock, but not the grass on which the livestock depends. Grasslands are the common lands of community and while there used to be robust traditional institutions in the past ensuring their sustainable management, today due to take-over by government or breakdown of traditional institutions they are the responsibility of none. Indeed they are often looked at as 'wastelands' on which tree plantations have to be done, or which can be easily diverted for other uses, resulting in a cascading chain of degradation. The Forest policy of 1894 was the most elaborate of all the policies in explaining the modalities of grazing in protected forests. The Forest policy of 1954 was extremely critical of unrestricted and uncontrolled grazing and refuted it as contrary to scientific management of forests. However, it also admitted that in some forest/grassland types, limited grazing does not do much harm, and may actually improve the grassland/forests. Dhebar Commission (1966) recommended that the Forest Department should promote growth of improved varieties of grasses in forest areas and grazing fees should be regulated. The National Commission on Agriculture (NCA) (1976) also recommended strict control on grazing and regulation on grazing. It also recommended the promulgation of grazing rules by each state specifying the grazing rules by each state specifying the grazing rates and providing for the manner in which grazing should be permitted. The National Forest Policy (1988) is in consonance with the previous policy on the issue of grazing, except for an important qualifier that grazing in forest areas should be regulated with the involvement of the local community and a committee under the chairmanship of Mr. C.D. Pandya, IGF (Retd.), also recommended that " A National Grazing Policy should come into effect at the earliest".

References

Singh, K.A. 2007. Grasslands and grazing. IGFRI Newsl. 13 (2): 4-6.

Thomas, C.G. 2003. Forage crop production in the tropics. Kalyani Publication. Ludhiana, India, pp. 259.

Chapter 14
Organic Forage Production

Rajeew Kumar, Santosh Chaudhary and Priyanka Shukla

Department of Agronomy, G.B. Pant University of Agriculture and Technology, Pantnagar – 263 145

Global trend in animal production indicates a rapid and massive increase in the consumption of livestock products. It is predicted that meat and milk consumption will grow at 2.8 and 3.3 per cent per annum respectively in developing countries like India. At present, the country faces a net deficit of 61.1 per cent green fodder, 21.9 per cent in dry crop residues and 64 per cent in feeds. To achieve the current level of livestock production and its annual increment, the deficit in all components of fodder, dry crop residues and feed has to be met from either increasing productivity, utilizing untapped feed resources and/or increasing land area. Sizeable amount of fodder demand is fulfilled through vast grasslands and rangelands. Any positive or negative change in its position has impact on several environmental issues. Similarly, the increase in livestock population will also affect the availability of organic waste which in turn can boost the agricultural production. Hence, eco-friendly fodder production system is of prime importance. Opportunities exist through strengthening research and development.

Need of Organic Farming

Need of Organic farming is a holistic production management system which promotes and enhances agro-ecosystem health, including biodiversity, biological cycles, and soil biological activity. It emphasizes the use of management practices in preference to the use of off-farm inputs, taking into account that regional conditions require locally adapted systems. This is accomplished by using, where possible, cultural, biological and mechanical methods, as opposed to using synthetic materials, to fulfill any specific function within the system. An organic production system is designed to enhance biological diversity within the whole system increase soil biological activity, maintain long-term soil fertility recycle wastes of plant and animal origin in order to return nutrients to the land, thus minimizing the use of non-renewable resources rely on renewable resources in locally organized agricultural systems promote the healthy use of soil, water and air as well as minimize all forms of pollution thereto that may result from agricultural practices handle agricultural products with emphasis on careful processing methods in order to maintain the organic integrity and vital qualities of the product at all stages and become established on any existing farm through a period of conversion, the appropriate length of which is determined by site-specific factors such as the history of the land, and type of crops and livestock to be produced. In recent years, global awareness of health and environmental issues has increased strongly, and sustainability has become the key word in discussions in relation to developing countries. The ever-growing number of health and environmentally concerned consumers, until now mainly in the industrialized countries, is at the root of this development. The international community is becoming more and more conscious of these issues, and government policies in industrialized as well as developing countries increasingly encourage organic and other forms of sustainable agriculture.

Organic Farming Scenario

Indian farmers were basically organic farmers before the advent of inorganic fertilizers and chemical pesticides. Over time the use of these synthetic inputs has come to the level of causing a concern to the environment and human health. Consequently, it is felt necessary to advocate the use of the age-old practice of organic farming not only to ensure uncontaminated food production but also to sustain the agriculture by keeping the land in a healthy condition. In the recent past, this has become a major concern where the consumers started demanding produce grown organically by not using chemicals. To make organic farming successful, it is essential that eco-friendly technologies, which can maintain or increase the agricultural productivity, have to be developed and made available to the farmers. Organic food products, especially the fruits and vegetables are slowly gaining momentum in the foreign markets like USA, Europe and Japan and fetching premium prices. The areas under organic farming are slowly increasing due to the awareness of the impact of chemicals on the environment and human health. World trade in organic products for 2000 was estimated at US $17.5 billion which includes 7.0, 8.0 and 2.5 billion US $ in Europe, USA and Japan, respectively. Trends indicate that the organic food market would grow substantially in most of the European countries, USA and Japan. Currently, the demand out places the supply. In the UK, demand for organic food increased by 55 per cent in 2000. The organic fruits and vegetables sector represents between 15 and 25 per cent of the total organic world market.

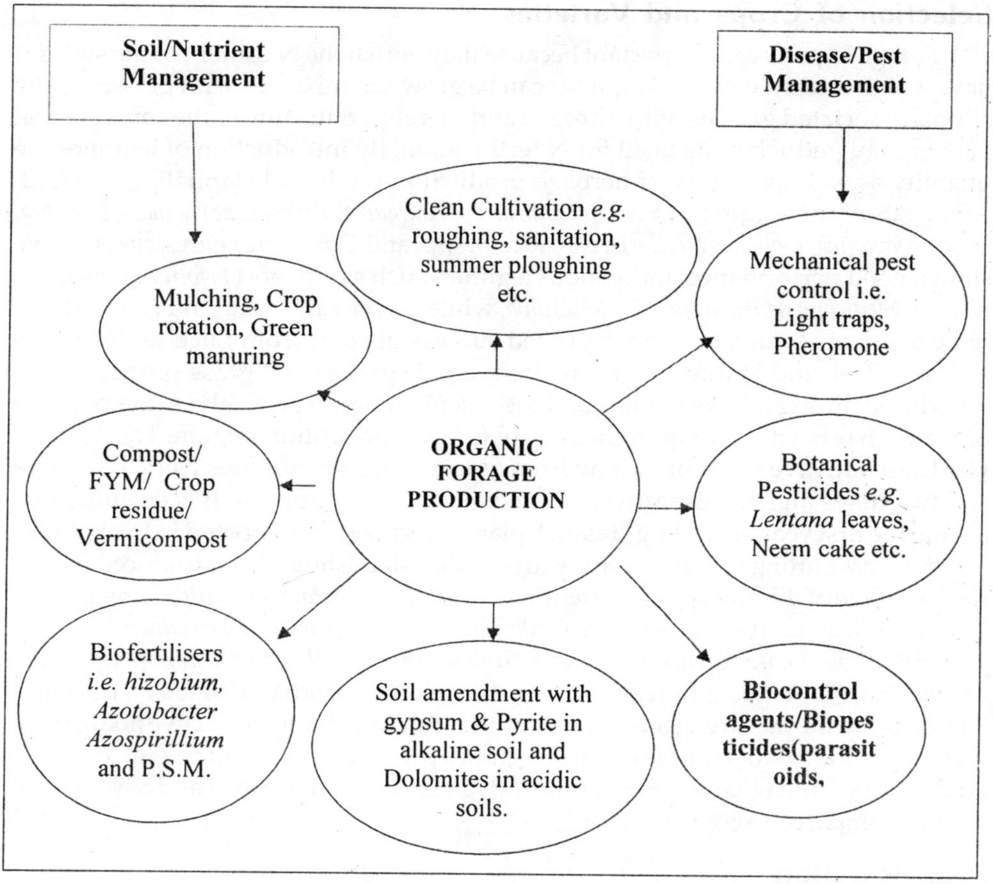

Figure 14.1: Organic Production System

Organic Forage Production and Management

Urgent attention is required to produce organic forage because of three things i.e high chemical residue retention in livestock products, environmental threat due to chemical farming and loss of sustainability of agro-ecosystem including soil. The following things should be consider during organic forage production

Selection of Area for Organic Forage Production

The following points should be kept in mind during area selection

- ✰ Total area should be as per requirement of feed and fodder
- ✰ Land should have deep well or residue less irrigation facilities
- ✰ Land should contain minimum synthetic residue
- ✰ Land should be a part of natural grass land.

Selection of Crops and Varieties

Forage legumes are important because they enrich the N content of the soil and have a high nutritive value. Legumes can be grown in mixtures with grasses. They supply associated grasses with nitrogen and thereby contribute to the conservation of energy by reducing the need for N fertilization. By introduction of legumes the quantity as well as quality of herbage production can be substantially increased. Among the legumes, *Siratro* (*Macroptelium atropurpureum*), *Stylosanthes hamata, S. scabra, Glycine javanica, Dolichos auxilaris, Desmodium* spp. and *Centrosema pubescens* etc. have shown good performance. Indigenous legumes such as clovers (*Trfolium pratens, T. repens*), *Medicago denticulata, Melilotus alba,* white clover var. *Ladino* and *Lousiana* and red clover var. *Montgomery* have proved successful apart from Lucerne (*Medicago sativa* cv. T- 9, and Hunter river) and berseem. Legumes and grass species can be introduced during July by seeding and tussock planting, respectively. A combination of Siratro has been found quite successful for the mid altitude region. The herbage yield and nutritive value of the hay from grasses-legume mixtures were found five and two times higher, respectively than the hay of local species. It is essential that during the first year of seeding/tussock planting, grazing is restricted in treated sites and the grass cutting is done carefully to help the establishment of introduced fodder species. Some of the grasses; *Cenchrus ciliaris, Dactylis glomerata, Dicanthium annulatum, Festuca* sp., *Lolium* sp., *Pennisetum pedicellatum,* etc. and legumes; *Desmodium intortum, Dolichos lablab, Phaseolus artopurpureus, Stylosanthes humilis, Trifolium* sp. etc. have been found adapted to different agro-climatic regions of Indian Himalaya. Legumes introduced in the pastures generally do not establish well due to ineffective nodulation. Rhizobium inoculation of the pasture legumes provides synergistic effect for better establishment and obtained 59 per cent and 72 per cent higher green and dry herbage yield as compared to control.

Crop Rotation

The following rotations are found suitable in organic farming system

- ✩ Maize-soybean/sarson+cowpea –barseem/oat/barley
- ✩ Mustard-sorghum+cowpea
- ✩ Interculture with guinea +cowpea/guniea+ berseem is found suitable in organic farming system

Nutrient Management

The present poor production potential of forage could be attributed to poor fertility of soils. To raise the fertility status and rectify the deficiencies, soil testing coupled with field trials need to be conducted to work out the fertilizer requirement of different forage. Generally, forage crops are by default organic because of no fertilizer is added to forage crop. Judicious use of fertilizer can boost the vegetative growth and is also economically feasible. Organic manure is the main source of nutrient in the forage production system. The following points should be keep in mind during nutrient management:

- ✩ Use of FYM on the basis of nitrogen requirement for fodder produce from organically adopted animal

- ✰ Liquid manure through fermentation of green leafy materials, cattle urine etc
- ✰ Use of composting dung, vermicomposting, vermiwash locally adopted earth worm will be utilized for nutrient vermicomposting
- ✰ Biologically derived nutrients, mulching composting etc should be used

Management of Crop Residues

Organic nutrient sources include plant residues, leguminous cover crops, mulches, green manure, animal manure, and household wastes. Under continuous cropping, recycling and reusing nutrients from organic sources may not be sufficient to sustain forage yields. Nutrients exported from the soil through harvested biomass or lost from soil by gaseous loss, leaching, or erosion must be replaced with nutrients from external sources. The judicious use of chemical fertilizer is essential to maintain soil fertility. The beneficial effects of organic matter are well known. Physically, it improves soil structure and increases water holding capacity. Chemically, it increases the capacity of the soil to buffer changes in pH, increases the cation retention capacity (CEC), reduces phosphate fixation, and serves as a reservoir of secondary nutrients and micronutrients. Biologically, organic matter is the energy source for soil fauna and microorganisms, which are the primary agents that manipulate the decomposition and release of mineral nutrients in soil ecosystems. Organic matter in soil exists as partially decomposed plant and animal residues, living and dead microorganisms, and humidified organic matter or humus. Stable humus constitutes 50 to 75 per cent of the total soil carbon and is little affected by management. The labile soil organic matter pool, which is important for nutrient release during the growing season, can be manipulated through various soil management practices. In general, more than 95 per cent of the total N and S and up to 75 per cent of the P in surface soils are in organic forms.Rates of decomposition of both fresh plant residues and humidifed soil organic matter are three to five times greater in the humid tropical environment than under temperate conditions. Therefore, in cultivated fields in the humid tropics, frequent application and larger quantities of organic materials are required to maintain adequate soil organic matter levels than in temperate regions. Strategies and practices for soil organic matter management include:

- ✰ Returning organic materials to the soil, to replenish soil organic carbon lost through decomposition (recycling of plant and animal residues, green manuring, cover crop rotation);
- ✰ Ensuring minimum disturbance of the soil surface (residue mulch, conservation tillage) to reduce the rate of decomposition;
- ✰ Reducing soil temperature and water evaporation by mulching the soil surface with plant residues; and
- ✰ Integration of multipurpose trees and perennials into cropping systems to increase the production of organic materials.

Integrated Nutrient Management

Sustainable soil nutrient-enhancing strategies involve the wise use and management of inorganic and organic nutrient sources in ecologically sound production systems. The primary goal of integrated nutrient management (INM) is to combine old and new methods of nutrient management into ecologically sound and economically viable farming systems that utilize available organic and inorganic sources of nutrients in a judicious and efficient way. Integrated nutrient management optimizes all aspects of nutrient cycling. It attempts to achieve tight nutrient cycling with synchrony between nutrient demand by the crop and nutrient release in the soil, while minimizing losses through leaching, runoff, volatilization and immobilization.

Cutting and Grazing Management

The response to cutting of a forage plant depends upon its seasonal yield of carbohydrate storage, its growth habit and extent of inflorescence development. Frequency of cutting also significantly influences the yield and quality of herbage produced. The areas with high temperatures may require larger interval and low intensity of cutting to build up sufficient carbohydrate storage for regrowth. Tall fescue (Festuca arundinacea) produced highest dry matter, when it was cut at 30 days interval during second year. Cutting grasses twice from natural grasslands recorded higher fresh forage yield (14.54 t/ha)than one cut (12.08 t/ha) and three cuts(13.30 t/ha). The crude protein content was higher with two cuts compared to one cut

Round Year Forages Production Plan Under Organic System

Under organic farming forage production should be planned in advance considering following recommendations:

- ✰ Combination of leguminous and non leguminous forage crops should be include in the system
- ✰ Scheduled crop rotation considering local weather condition should be adopted
- ✰ Soil conditions preferably minimum pesticide residue
- ✰ Under assured recommendation, selection of forage suitable under irrigation, however land should be at least 50 per cent irrigated for one production.
- ✰ Ability of sufficient bio-fertilizer, FYM, vermicompost green manuring etc.
- ✰ Use of locol varieties of forage under rain-fed conditions and improved varieties under irrigated conditions.
- ✰ Minimum two year conversion period of land from conventional soil to organic soil.
- ✰ Use of integrated forage production system
- ✰ Preferably forage seed should be produced organically.

Control of Pesticide Residue in Forage

Through green revolution agriculture production has been increased many fold, but use of insecticide and pesticide has also been increased in crop as well as in forage production. Synthetic organic insecticide including Organochloro, Organophosphate and Organocarbamate exhibit a high degree of persistence in environment as compared to other classes of pesticide. The organpchloro is not only persistent but also lipophylic, therefore accumulates in plant and animal tissue. Therefore there is a need to control use of pesticide which must be in safe limit. For this use of bio pesticide and pest resistant varieties should be preferred.

References

Aamlid, T.S. Yield and Quality of Mixed Seed Crops of timothy (Phleum pratense) and Clovers (*Trifollium* spp.) in an organic cropping system. Acta Agriculturae Scandinavica 52:18-24.

Agboola, A.A., and A.A. Fayemi. 1972. Fixation and excretion of nitrogen by tropical legumes. Agron. Jour. 64: 409-412.

Ayodele, O.J. 1986. Effect of continuous maize-cropping on yield, organic carbon mineralization and phosphorus supply of savannah soils in western Nigeria. Biol. Fertil. Soil. 2: 151-155.

Bationo, A., C. B. Christianson and M.C. Kalij. 1993. Effect of crop residue and fertilizer use on pearl millet yields in Niger. Fertilizer Research 34: 251-258.

Boelt, B., Deleuran, L. C. and Gislum, R. 2002. Organic forage seed production in Denmark. Newsletter, The International Herbage Seed Production Research Group no. 34:3-5.

Borm, G.E.L. 1995. Possibilities of Mechanical Weed Control at Different Row Distances in Perennial Ryegrass grown for Seed. Proceedings of the third International Herbage Seed conference, Germany, 1995:271-275.

Fore, R.E. and B.N. Okogbo. 1974. Yield response of maize to various fertilizers and lime on Nkpologu sandy loam. Nigerian Agricultural Journal 9: 124-127.

Hauser, H. and B.T. Kang. 1993. Nutrient dynamics, maize yield and soil organic matter in alley cropping with Leucaena Leucocephala. In: Soil Organic Matter Dynamics in Sustainable Tropical Agriculture, K. Mulongoy and R. Merck (Eds.). John Wiley, Chichester, U.K. pp. 215-230.

Hossner, L.R. and D.W. Dibb. 1995. Reassessing the role of agrochemical inputs in developing country agriculture. In: Agriculture and Environment: Bridging Food Production and Environmental Production in Developing Countries, A.S.R. Juo and R.D. Freed (Eds.). Special Publication No. 60. American Society of Agronomy, Madison, Wisconsin, USA, pp. 17-33.

Janssen, B.H. 1993. Integrated nutrient management: the use of organic and mineral fertilizers. In: The Role of Plant Nutrients for Sustainable Crop Production in Sub-Saharan Africa, H. van Reuler and W.H. Prins (Eds.). Ponsen and Looijen, Wageningen, The Netherlands, pp. 89-105.

Kang, B.T. 1986. Cropping systems and soil fertility management in the humid and subhumid tropics with special reference to West Africa. In: Management of

Nitrogen and Phosphorus Fertilizers in Sub-Saharan Africa, A.U. Mokwunye and P.L.G. Vlek (Eds.). Proceedings of a Symposium, Lome, Togo, March 25-28, 1985. Martinus Nijhoff Publishers, Boston, USA, pp. 83-94.

Kang, B.T., G.F. Wilson, and L. Sipkens. 1981. Alley cropping maize (*Zea mays* L.) and leucaena (*Leucaena leucocephala* Lam.) in Southern Nigeria. Plant and Soil 63: 165-179.

Lal, R., A.S.R. Juo and B.T. Kang. 1984. Chemical approaches towards increasing water availability to crops including minimum tillage. In: Chemistry and World Food Supply, T.W. Schemil (Ed.). Pergamon Press, New York, USA, pp. 55-77.

Manu, A., A. Bationo, and S.S. Geiger. 1991. Fertility status of millet producing soils of west Africa with emphasis on phosphorus. Soil Science 152: 315-320.

Marten, G.G. 1986. Traditional Agriculture in Southeast Asia: A Human Ecology Perspective. Westview Press, Boulder, Colorado, USA.

Martin, W.S. 1935. Mulching with grass and plantain trash and its effects on crop and soil conditions. East African Agricultural Jour. 1:140-144.

Payne, W.A., R.J. Lascano, L.R. Hossner, C.W. Wendt, and A.B. Onken. 1991. Pearl millet growth as affected by phosphorus and water. Agron. Jour. 83: 942-948.

Swaminathan, M.S. 1983. Our greatest challenge: Feeding the hungry world. In: Chemistry and the World Food Supplies: The New Frontiers, G. Bixlet and L.W. Shemit (Eds.) CHEMRAWN II. Perspectives and recommendations. IRRI, Los Baños, Philippines, pp.25-31.

Young, A. 1990. Agroforestry for the management of soil organic matter. p. 285-303. In: Organic Matter Management and Tillage in Humid and Subhumid Africa, E. Pushparajah and M. Latham (Eds.). International Board for Soil Research and Management, Bangkok, Thailand. IBSRAM Proceedings No. 10.

Chapter 15

Nutrient Management in Fodder/Forage Crops

Anil Kumar Singh[1], *Lal Singh*[2], *Sushil Dimree*[3] *and Ajit Kumar Pandey*[4]

[1]*Senior Scientist, Agronomy, Division of Land And Water Management, ICAR Research Complex for Eastern Region, ICAR Parisar (P.O.- B.V.College), Patna – 800 014, Bihar*
E-mail: aksingh_14k@yahoo.co.in

[2]*Assistant Professor, Division of Agronomy, SKUAST-K, Shalimar, Srinagar – 191 121, J&K*
E-mail: drlalsingh@rediffmail.com, drlalsingh72@rediffmail.com

[3]*Assistant Professor, Department of Soil Science and Agriculture Chemistry, CSAUA&T, Kanpur – 208 002, U.P.*
E-mail : dimri.astro.gmail.com

[4]*Division of Land and Water Management, ICAR Research Complex for Eastern Region, ICAR Parisar (P.O. - B.V.College), Patna – 800 014, Bihar*

Introduction

Livestock is the integral component of Indian agriculture, since time immemorial and its contribution to national economy through milk, meat, wool as well as farmyard manure (FYM) is enormous. We have approximately 20 per cent of world's cattle, 50 per cent of buffaloes, more than 120 million goats and 60 million sheep (Deb Roy, 1993). Due to religious belief the population of unproductive cattle is increasing. This huge population and poor fodder availability has widened the gap between demand

and supply of forage crops. Moore (1974) maintained that considerable fodder resources were wasted on the maintenance of an excessive number of poorly fed and low yielding animals, which contributed to the process of pasture destruction. At present land holdings are very small and the farmer is always biased in the choice of the crops. Due to these reasons agricultural land ratio does not permit diversion of land from food production to cultivated fodder. Thus the area under fodder crops is meager and is not more than one per cent of the total cultivated land. The management practices play an important role in determining the productivity of cultivated fodders/ forages and grasslands, yet this has been the last priority of the farmers in the country. Presence of inferior and unproductive grass species, lack of balance fertilization, and absence of legume component, improper cutting and indiscriminate grazing are some of the factors responsible for poor productivity of the grasslands. Among these nutrient management will play a pivotal role to enhance the productivity per unit area and total production without expansion of the area. There exists a wealth of indigenous knowledge for the proper utilization and management of the natural resource base but the farmers because of increasing population pressure and declining land productivity are not using it.

Tree flora of Indian sub-continent is one of the richest on this globe. The over exploitation and unscientific management of fodder trees has depleted this resource at huge environmental cost. Keeping in view the constraints in fodder production and in order to overcome the gap between demand and supply, the emphasis need to be given on several steps for augmenting the fodder production. Existing resource utilization pattern needs to be studied in totality according to a system approach. Fodder production is a component of the farming system and efforts need to be made for increasing the forage production in a farming system approach. The holistic approach of integrated resource management will be based on maintaining the fragile balance between productivity functions and conservation practices for ecological sustainability. For maintaining fat percentage in buffalo or cow milk, the important factor is nutrients in the diet of animal. Whatever fodder provided by you to animal it converts in to milk, dung and also used for daily maintenance and production activity. In this relation if soil where you are producing fodder lacking important nutrients, then the fodder/forage also lacking in that nutrients which are deficient in the soil. So there is relation between milk and soil. Most of the time farmers are not giving balanced fertilizers (Organic and inorganic) to the fodder crops and ultimately there were deficiencies in the fodder which we produced for animals. So it is better to do soil sample analysis prior to sowing/planting fodder species and also provide balanced nutrient in the field to attain good quality of milk, meat, wool production and animal health.

Nutrient Accumulation

The quantity of each nutrient absorbed by a field grown crop varies considerably; it is depend on fertilization practices, residual level of nutrient in the soil, plant population per ha, rainfall and other environmental factors. Higher levels of nitrogen application, generally, increase the yield and quality of forage and fodder crops especially non legume crop. The essential elements exist as structural components of a cell, maintain cellular organizations, and function in energy transformations and enzyme reactions. Based on their relative requirement by plants these elements may be classified as:

Macro-nutrients

Those nutrients which are required by plants in concentrations exceeding 1 ppm are termed as major or macro-nutrients. These are further classified into two categories as:

Primary Nutrients

C, H, O, N, P, K are the primary elements which are essential right from seed germination. Out of these C, H, O are found abundantly in water and atmosphere and they arc supplied whereas N, P and K are either obtained from soil or supplied through chemical fertilizers.

Secondary Nutrients

They are secondary in the sense that they are needed only when the plants have started growing they have grown for some period. These are Ca, Mg and sulphur.

Minor or Micro-nutrients

The elements which are required by plants in concentration less than 1 ppm are put in this category. They arc also called as trace elements, oligo elements, supreme elements, etc. Nicholas (1963) used "Ultra micro-nutrient" for molybdenum and cobalt which arc required as low as one part per billion by most of the plants.

Carbon, hydrogen and oxygen form about 94 per cent of the dry weight of plants. These are the major components of carbohydrates, proteins and fats. Besides their structural role, they provide energy required for the growth and development of plants by oxidative breakdown of carbohydrates, proteins, and fats during cellular respiration. The very fact that the micronutrient elements are required by plants in very low concentration suggests that they all function as catalysts or are at least closely linked with some catalytic processes in plants. Evidences are there that at least copper; zinc and manganese are components of certain biological oxidation-reduction systems. Some of the micronutrient elements are essential for and related to the activity of some enzyme systems. The interactions and antagonisms of the elements influence their role in plant metabolism. Such antagonistic effects, between manganese and iron, calcium, copper and also between calcium and boron are amongst many. Manganese toxicity is associated with calcium deficiency and vice versa. A close antagonism between calcium and boron is well known.

Role of Plant Nutrients

Every nutrient plays a specific role in nutrition, growth and development of plants. These roles may be described as under:

Carbon

It is available in abundance from air. CO_2 is used by green plants for photosynthetic activities. It is also required for cell formation in plants. About 45 per cent or more parts of the plant tissues are made up of carbon.

Hydrogen

It is essential for cell and tissue formation in plants. This is obtained from water and is required for energy reactions. It forms about 6 per cent parts of the plant tissues.

Oxygen

Plants take oxygen from air and water. It forms about 43 per cent parts of the plant structure. It is required for photosynthetic and respiratory activities. It helps in formation of tissues and cells.

Mineral Elements

Nitrogen

Among the elements essential for the commercial production of forages and fodder, none has as pronounced an effect nor requires the degree of attention in fertility practices as nitrogen because it is a major structural part of the cell. Nitrogen is an essential constituent of metabolically active compounds such as amino acids, proteins, enzymes and some non-portentous compounds. It plays a vital role in various metabolic activities of plants and is a constituent part of amino acids, proteins, nucleic acids, porphyries, flavins, purine and pyrimidine nucleotides, enzymes, co-enzymes and alkaloids. When nitrogen is a limiting factor, the rate and extent of protein synthesis are depressed and as a result plant growth is affected. The plant gets stunted and develops chlorosis. Cytoplasm and the particulate fractions of the cell organelles contain nitrogen in varying amounts which exist in combination with C, H, O, P and S. Primary cells are found to have about 5 per cent of nitrogen. It helps in harvesting solar energy through chlorophyll, in energy transformation through phosphorylated compounds, in transfer of genetic information through nucleic acids. Besides, it is essential in cellular and protein metabolism and, acts as biological catalyst. From the seedling stag through final harvest, the soil nitrogen regime affects the process of plant development more than any other mineral element. With respect to time of absorption, form in which absorbed, concentration in the leaf at various stages of growth, and in numerous other aspects, the role of nitrogen in the development and properties of the tobacco leaf is of major importance. The influence of variations in nitrogen supply on growth of the plant and properties of the cured leaf has been investigated more extensively than have the effects of any other essential element.

Role of Nitrogen

Green Leaves Quality

Total number of leaves produced by a plant is not appreciably influenced by the level of available nitrogen, but results in an increase in the area of the leaf but a decrease in the weight per unit area, the latter effect being due primarily to a decrease in the thickness of the leaf.

Deficiency Symptom

Although apparent nitrogen deficiency symptoms do not generally occur when the concentration in a particular leaf is above 1.5 to 2.0 per cent, a slight yellow color indicated reduction in chlorophyll content and length and width measurements showed a reduction in leaf area. The reduction in area, however, seldom was accompanied by a reduction in the total dry weight of the leaves; instead there was a greater weight per unit area with the stress than with the nonstress condition.

Phosphorus

Although phosphorus has received major attention in fertilizing forages and fodder, primarily with studies on general growth effects from rates of applications with the principal attention given to yield and quality of the cured leaf. It was observed that a relatively high application of fertilizer phosphorus resulted in a marked increase in the early growth of plants. Phosphorus plays a vital role as a structural component' of cell constituents and metabolically active compounds. Phosphorus is a structural component of cell membranes, chloroplasts, and mitochondria and a constituent sugar phosphates, *viz.* ADP, ATP, nucleic acid, etc., phospholipids and phosphatides. The most essential constituents of plant cells like esters, phosphatides and phospholipids are synthesized by phosphorus when it combines with different organic acids. It also plays an important role in energy transformations and various metabolic activities of plants. Phosphorus plays an important role in energy transformation and metabolic processes in plants. It' stimulates root growth. Being a constituent of adenosine phosphate, phosphoglyceraldehyde and rebulose phosphate, it helps in basic reactions of photosynthesis and activates several enzymes participating in dark reactions in photosynthesis.

Role of Phosphorus in Plant Growth

The responses to phosphorus fertilization are more frequent in the early growing season than are in later stages of final yield and quality of the cured leaf because the rapid and vigorous development of young plants is stimulated by a high level of available phosphorus. An early-maturing normal plant develops more desirable properties than one which matures late.

Deficiency Symptoms

There are, however, reports that a relatively low rate of phosphorus resulted in higher leaf yields than did heavier applications.These effects may be related to the influence of phosphorus on the absorption of other ions. The major symptoms of phosphorus deficiency in tobacco are retardation in growth, particularly during the first month after transplanting, and a delay in maturity of the plant.The leaves tend to be narrower than normal, high in nitrogen, and abnormally green. Occasionally there may be numerous small white spots on the lower leaves. The leaves from plants that are deficient in phosphorus do not mature normally and, therefore, are generally of low quality. The cured leaves tend to be dark brown or, because of their immaturity, may be dark greenish in color and lack the luster of leaves from normal plants. It is generally considered that no adverse effects on the growth and development of tobacco will occur from excessive amounts of phosphorus.

Phosphorus Absorption

Absorption of phosphorus occurs at a fairly constant rate through the growing season although speed is slightly higher during early phase of growth and development. Sources of nitrogen and potassium used in fertilizers are quite soluble in the soil solution and thus readily available for absorption. Phosphorus is only slightly soluble in the soil solution, but as phosphorus is absorbed by the plant a replenishment of the level in solution is likely to occur throughout the growing season. The smaller absorption is a genetically controlled factor, whereas difference in the absorption is primarily due to difference in the relative availability of the elements in the soil. The uptake of phosphorus was inhibited by the hydrogen ion in the very acid region and by the hydroxyl ion in the alkaline region.

Phosphorus Application

A response in earlier growth and higher phosphorus content in the plant has occurred however, by the end of the season there may be no differences in the uptake of fertilizer phosphorus between different placements. The major source of phosphorus in the plant in the early stages of growth is from the fertilizer only.

Potassium

Potassium helps in the maintenance of cellular organization by regulating the permeability of cellular membranes and keeping the protoplasm in a proper degree of hydration by stabilizing the emulsions of colloidal particles. Its salts stabilize various enzyme systems. It plays a catalytic role in activating several enzymes as incorporation of amino-acids in proteins synthesis of peptide bonds etc. Presence of potassium is essential for optimal activation of aldehyde dehydrogenase etc. Potassium increases resistance in plants against drought, heat frost and various diseases caused by fungi nematode and other microorganism. It helps in formation of mechanical tissues in resulting into resistance to lodging.

Deficiency Symptoms

Owing to the richness in potassium content in general all most all Indian soils due to its parent material or heavy applications of potassium fertilizer, deficiency symptoms are rarely observed in commercial production. Non application of potassium fertilizer, however, will, on most soils, result in observable deficiency symptoms but some time potassium deficiency may be accentuated by excess nitrogen, particularly in the ammonium form, and by high levels of magnesium or sulfur. It has been observed that potassium deficiency symptoms appear first and are most severe on rapidly growing plants and spread progressively to the upper leaves. During early phase of growth period, potassium deficiency is manifested by a slight mottling and the appearance of brownish yellow specks, especially near the tips of the leaves. In the later stages of plant growth (near topping), the abnormalities may appear first on the upper leaves. At acute deficient condition brownish yellow spots appear along the margins of the leaves. The tips of leaves may curl downward and the margins curve inward until they are almost at right angles to the plane of the leaf.

Potassium Absorption

The high plant requirements for potassium exist not only after transplanting, but also in the seedling stage. Potassium in combination with either sulfate or nitrate appeared to be the most beneficial of elements on root development. Uptake of potassium by the plant generally increases linearly over a wide range of application rates. The potassium content of the leaf of field-grown burley tobacco ranged from a low of 2.0 to a high of 3.2 per cent. The rate of potassium absorption is very high during the early stages of growth and diminishes rapidly during the later phase. It is assumed that the reduction in absorption is due primarily to an exhaustion of the supply of available potassium in the soil.

Potassium Sources

The response of forages and fodders to various sources of potassium *viz.*, the potassium sulphate, potassium chloride, potassium carbonate, and potassium nitrate salts were tested. In these studies there was a greater difference among the sources on leaf burn than there was on yield or quality score. When alkalinity of the water-soluble ash was used as an indicator of fire-holding capacity, the sulfate source gave lower values than did the nitrate or carbonate source. It was suggested that these differences were due to effects of sources on the sulfur content of the leaf.

Calcium

The requirements for calcium are relatively high because it constitutes one of the principal inorganic constituents of tobacco and next to potassium is the mineral element absorbed in the largest quantity. Calcium is a constituent of the cell wall, is an activator of different plant enzymes and is essential for the stability of cell membranes. It is a structural part of the chromosomes in which it binds the DNA with protein. It is required by a number of enzymes for their proper functioning *viz.* lipase, phosphatase D, α-amylase and pyrase. It makes the stems stiff and thereby reduces lodging in plants. It also neutralizes the organic acids formed within the plant 'body and eliminates their toxic effects. Values of 1.5 to 2.0 per cent calcium in the cured leaves are common. The amount of calcium necessary for the normal growth of tobacco leaves is in excess of 1 percent of the dry weight.

Deficiency Symptoms

Deficiency symptoms occur in the upper portion of the plant indicates that during stress there is little calcium transportation from the older tissues to the growing points. Leaves of calcium-deficient plants are dark green colour and in the advanced stage the terminal bud dies. When this condition occurs early in the development of the plant, the top leaves may become much thickened. Hooking down word at the tips of the bud leaves and ultimately a breakdown of the tissue at the tips and margins of the leaves are some prominent symptoms in the early stages of calcium deficiency. Complete breakdown and death does not occur and growth takes place later. Older leaves on calcium-deficient plants generally retain their normal shape.

Magnesium

Magnesium is a constituent of chlorophyll and chromosome. Being constituent part of polyrioosemes, it helps in protein synthesis in the plants. Mg is also a constituent part of chromosomes. It is known to play a catalytic role as an activator of a number of enzymes, most of which are concerned with carbohydrate metabolism, phosphate transfer and de carboxilations. It is involved in photosynthesis and organic acid metabolism. Mg helps in synthesis of fat and increases oil content in oilseed crops when it combines with sulphur.

Deficiency Symptoms

Magnesium is a constituent of the chlorophyll molecule, loss of green color would be expected to be one of the first symptoms of magnesium deficiency. Discolourations start first on leaves at the base of the plant, and on an individual leaf it begins at the tip and margins and proceeds toward the base and center. The area around the veins of the leaf generally remains green after the rest of the leaf has been decolorized. In extreme cases the leaf may become practically white, suggesting a breakdown of the yellow pigments also.

Sulphur

Sulphur is required for synthesis of sulphur containing amino acids and proteins like cystein, methionine vitamins (thiamine and biotine), lipoic acid, acetyl coenzyme A ferredoxin and glutathione activity of proteolytic enzymes, and increases oil content in oil bearing plants.

Sulphur Deficiency Symptoms

Deficiencies are most likely to occur in the early stages of growth or during dry weather. When they do occur, the first evident symptom is a lighter color of the young leaves compared with the older ones. A distinguishing difference between and nitrogen deficiency is that in the sulfur the plants do not lose their leaves by "firing." Flower emergence is delayed when sulfur deficiency occurs early in plant development.

Iron

Iron is a constituent of cytochromes, haem, nonhaem enzymes and metalloproteinase like ferredoxin and hemoglobin in· plants. These cytochromes play a vital role in oxidative and photophosphorylations during respiratory electron transport and photosynthesis respectively. The ferredoxin helps in reduction of carbon dioxide, sulphate and of atmospheric nitrogen. It synthesises chlorophyll precursor (protoporphyrin-9) which forms chlorophyll in green pigments. Its specific requirement has been identified in synthesis of enzymes like oxidoreductase, sulphate oxidase; catalase~ peroxidase and aconitase etc. Being a constituent part of metabolically active compounds, iron' is responsible for all major metabolic processes in plants.

Manganese

Being a part of nitrite reductase and hydroxylamine reductase, it helps; in the nitrogen assimilation. Manganese performs some function in photosynthesis, acts as regulator of the intake and state of oxidation of certain elements. It activates several enzymes related to' oxidation-reductions (oxidoreductase), hydrolysis (hydrolases), breakdown of phosphates bonds in ATP or ligases. It activates photosynthesis and, nitrogen metabolism. It also accelerates enzyme participating in' Calvin cycle, helps in chlorophyll and chloroplast synthesis for boosting photosynthetic rates.

Manganese Deficiency Symptoms

Manganese deficiency is most likely to occur where the soil reaction is neutral to alkaline; a condition that seldom exists in soils used for tobacco production. The first apparent symptom is chlorosis of the young leaves. The tissue between the veins becomes light green to almost white, giving the plant a checkered appearance. In the advanced stages, chlorosis may be followed by necrotic spotting and the affected tissue may drop out, producing a ragged appearance.

Zinc

Zinc is concerned with the functioning of sulphydryl compounds such as cysteine, in the regulation of oxidation-reduction potential within the cells. It regulates the auxin concentration in plants and helps in synthesis of protean carotene and chl1lrophyll etc. The average content of zinc in various commercial tobaccos is reported to be between 51 and 84 ppm. The minimum requirement for zinc is less than 10 ppm in the tissue.

Zinc Deficiency Symptoms

Leaf spot disease that occurs during wet periods and is attributed to bacteria is reported to correspond in many respects to symptoms produced experimentally by a shortage of zinc.

Copper

Copper is an essential component of the oxidizing enzyme catechol oxidase or polyphenol oxidase. The compounds containing copper like plastoquinones and plastocyanins help in, electron transport from chlorophyll to NADP and from water to chlorophyll during photosynthesis. The copper content of tobacco leaves varied from 12 to 60 ppm; the stalks contained about 10 ppm. An abnormality that resulted in a breakdown of the tissue in tobacco leaves as they approached maturity was corrected by the addition of copper sulfate. The affected leaves were high in total and protein nitrogen and low in sugars. Use of copper sulfate reduced the nitrogen and increased the sugar content. The addition of copper sulphate to field-grown tobacco has been reported to give a more even ripening and an increase in the body of the leaf.

Molybdenum

Molybdenum is a constituent of nitrate reductase and nitrogenase enzyme and is associated with nitrogen utilization and in nitrogen fixation. It helps in protein and amino-acids synthesis. It accelerates nitrogen fixing efficiency of aerobic (*Azotobacter*), anaerobic (Clostridium), blue green algae, *Azolla* and symbiotic bacteria. It regulates the carbohydrate metabolism in plants.

Molybdenum Deficiency Symptoms

Molybdenum deficiency symptoms were associated with high nitrate and low protein nitrogen in the leaves. The deficiency molybdenum would be expected to occur more commonalty on acid soils since its availability is lower under acidic condition. The first indication of molybdenum deficiency was mottling of the lamina in leaves at the mid stalk position. Bending and twisting of the leaf lamina were usually present and were followed by the appearance of small necrotic areas which gradually enlarged until the entire leaf was withered. These symptoms gradually spread to other leaves. Deficiency caused a delay in blooming and in extreme cases a loss of flowers.

Chlorine

During photosynthesis it helps in evolution of oxygen. It is a part of anthocyanin and affects protein synthesis. It increases turgor pressure. Chlorine stimulates the activity of some enzymes and influences carbohydrate metabolism. The uptake and accumulation of chloride ion by plants increases linearly over a wide range of concentration in the substrate. Concentrations in the leaf as high as 10 per cent of the dry weight have been measured in plants grown in the green house.

Chlorine Toxicity

Under high concentrations of chlorine, the green leaf becomes greatly thickened and exceedingly brittle, the leaf margins curl upward, and the leaf presents a distinctive sleek, glabrous appearance. When ammonium was the principal form, however, at chloride levels as low as 0.95 per cent the leaves possessed the color and configuration normally associated with chloride toxicity. Much of the leaf abnormality historically attributed to excess chloride may have been due either to ammonium toxicity or to an ammonium-chloride interaction. An increase in pH would promote nitrification, thus minimizing the possibility of an ammonium-chloride interaction leading to toxicity.

Effects on Chemical Properties of Leaves

The use of excessive amounts may cause adverse effects on combustibility of the leaf more than any other inorganic constituent. The causal mechanism through which chloride reduces the burning qualities of the leaf is not understood, though the Potassium salts or organic acids were the main factor determining the rate of burn. A decrease in total nitrogen content of the plant has been associated with high chloride nutrition. Nitrogen metabolism also appears to be influenced by chloride. The chloride effect on carbohydrate metabolism was on the amylytic enzyme in the leaf which resulted in the marked accumulation of starch when the chloride content was high.

Boron

It regulates development and differentiation of vascular tissues, formation and lignifications of cell-wall. It is associated with reproductive phase in plants and under imbalanced nutrition it causes sterility and malformation in reproductive organs. It is involved in carbohydrate metabolism, particularly in translocation of photosynthesis. Influences cell development by its influence on polysaccharide formation. It regulates translocation of sugars across membranes and polyphenolase activity. It boosts nodulation in legumes, 'regulates water absorption and is essential for synthesis of A IT, DNA, RNA and pectin. Under normal production practices. Deficiencies of this element are rare. They are most likely to occur at a reaction in the neutral to alkaline range which renders the soil boron unavailable or on coarse sandy soils during periods of high rainfall. The development of deficiency symptoms is also favored by rapid plant growth.

Boron Deficiency Symptoms

Deficiency symptoms have been observed in plants with a concentration of 5.5 ppm of boron. Because minute amounts of boron are needed, and slight excesses have a deleterious effect on the plant. Boron deficiency affects primarily the youngest leaves and the terminal bud and occurs during periods of rapid growth indicates that little or no transfer occurs from the older plant parts to the younger growing points. The bud leaves become light green in colour and distorted, In advance stages, breakdown of the tissue and death of the terminal bud occurs and the remaining leaves become normally thickened. Application of more than 400g of boron per acre caused boron toxicity on young plants. Prominent symptoms of toxicity are slight stunting of the plant and a slight specking and marginal burning of the leaves, however, did not affect the final yield and value of the leaf.

Sodium

It maintains the osmotic pressure. It also regulates water uptake by plants. Deficiency symptoms appear if it falls below 0.3 per cent in the whole plant (Gamer 1934). Potassium (K_2O) content of leaves varies from 2.5-7.0 per cent and that of stalk from 2.5-5.0 per cent. Deficiency symptoms appear when its content in the plant falls below 2 per cent. Deficiency symptoms of calcium are expected if its content in leaf fall below 1.0 per cent. When magnesium content of leaf falls below 0.4 per cent, deficiency symptoms become visible. Deficiency symptoms of sulphur are expected if its content in leaf is below 0.1 per cent.

Sodium partially supplemented potassium in the nutrition of the plant. In a greenhouse experiment, the yield was increased and potassium deficiency symptoms were decreased by the addition of sodium to the nutrient media. Burn rate was increased by sodium application in absence of potassium. Sodium ion could partially substitute in some plant process or processes when potassium is low.

Organic Manures

Since organic manure is known to have residual effect on succeeding crop up to 2-3 seasons, the beneficial effect of organic manure on current and succeeding crops might be due to its contribution in supplying additional plant nutrients and its capacity to improving solubility of native soil nutrients. Adequate availability these efficient and great partitioning of metabolites and grater translocation of synthesized food material to the different organs. Organic manure is rich in organic matter and can be supplemented with the chemical fertilizers. Although it is costlier than chemical fertilizers on unit nutrient basis, the other beneficial effect that has on soil can compensate for the added cost. Brady (1996) generalized that half of nitrogen and one-fifth of phosphorus of the applied organic manure may be recovered by the crop. Thereafter rest nutrients are available at slower rates to the subsequent crops. A minimum nutrient loss due to slow release of nutrients from organic manures is an added advantage. Therefore, the residual nutrients can be utilized by next crop in a crop rotation. Organic manures offer an opportunity to cut down the use of chemical fertilizers. Application of organic materials will add a good amount of macro and micro-nutrients in addition to organic carbon to the soil. The beneficial effect of organic sources of manures on crops could be attributed to the fact that after proper decomposition and mineralization, the manure supplied available plant nutrients directly to the plants and also had solubilizing effect on fixed forms of nutrients in soil. Therefore, increase in dry matter production can be ascribed due to overall improvement in plant organs associated with higher growth of crop under residual effect of organic sources (Satyajeet *et al.*, 2006; Mandal and Sinha, 2002)

Inclusion of Legumes with Grasses in Fodder/Forage Production

Forage legumes are important because they enrich the N content of the soil and have a high nutritive value. Legumes can be grown in mixtures with grasses in cultivated fodders/forages and grasslands. They supply associated grasses with nitrogen and thereby contribute to the conservation of energy by reducing the need for N fertilization. By introduction of legumes the quantity as well as quality of herbage production can be substantially increased. Among the legumes, Siratro (*Macroptelium atropurpureum*), Stylo (*Stylosanthes hamata, S. scabra*), *Glycine javanica, Dolichos auxilaris, Desmodium* spp and *Centrosema pubescens* etc. have shown good performance (Melkania, 1995). Indigenous legumes such as clovers (*Trifolium pratens, T. repens*), *Medicago denticulata, Melilotus alba*. White clover (*Trifolium repens*) var. Ladino and Lousiana and red clover (*Trifolium pratens*) var. Montgomery, Lucerne (*Medicago sativa* cv. T- 9, and Hunter river) and berseem (*Trifolium alexandrinum*) var. Meskavi have proved successful in Kashmir valley apart (Gupta, 1977). A combination of Siratro has been found quite successful for the mid altitude region (Melkania, 1987). The herbage yield and nutritive value of the hay from grasses-legume mixtures were found five and two times higher, respectively than the hay of local species. Legumes introduced in the pastures generally do not establish well due to ineffective nodulation. Hazra (1998) observed that the Rhizobium inoculation of the pasture legumes provides synergistic effect for better establishment and obtained 59 per cent and 72 per cent higher green and dry herbage yield as compared to control.

Some of the grasses; *Cenchrus ciliaris, Dactylis glomerata, Dicanthium annulatum, Festuca* sp., *Lolium* sp., *Pennisetum pedicellatum*, etc. and *legumes; Desmodium intortum, Dolichos lablab, Phaseolus artopurpureus, Stylosanthes humilis, Trifolium* sp. etc. have been found adapted to different agro-climatic regions of India (Shastry and Patnaik, 1990).

Fertilizer Management

The present poor production potential of pastures could also be attributed to poor fertility of soils. To raise the fertility status and rectify the deficiencies, soil testing coupled with field trials need to be conducted to work out the fertilizer requirement of different fodder/forage crops and pastures. Generally, fodder/forage crops grown in under nutrition and no fertilizer is added to rangelands except the dropped excreta by animals. Judicious use manures and fertilizers in fodder/forage crops and fertilizers for pasture can boost the vegetative growth and is also economically feasible. Application of nitrogen fertilizer must be given in split doses for better utilization, whereas phosphorus and potash should be supplied as basal dose in case of grasses. In legumes the full dose of nitrogen, phosphorus and potash should be given as a basal dose in furrows or by broadcasting at the time of sowing. Dogra *et al.* (1997) found 120 kg N/ha and 40 kg P/ha as the most economical dose. Herbage yield increased significantly with the application of nitrogen @ 60 kg/ha and phosphorus @ 30 kg/ha (Sood and Sharma, 1996). Nitrogen @ 40 kg/ha and Phosphorus @ 30 kg/ha applied as basal and two splits (onset of Monsoon and 45 days after first application in natural grassland) increased the forage yield significantly. Two splits were significantly superior to single application (Singh, 1995). The experiments on N and P requirement in Himachal Pradesh reveal that application of 80 kg/ha each of nitrogen and phosphorus was found to be the best (Sood and Bhandari, 1992).

Soil Test Based Recommendations

The imbalance use of inorganic fertilizer is one of the causes of low productivity in this region. For efficient use of inorganic fertilizers, soil test based recommendations needs to be followed. The soil test is possible only in soil testing laboratory. In country some of the state departmental level and state agricultural universities soil-testing laboratories are in operation. Due to lack of enough information and unawareness, farmers are reluctant to soil testing and do not follow soil test based recommendations for fertilizer application. In view of this problem, long-term experiments under Soil Test Crop Response (STCR) project in different crops based on series of experiments on soil test crop response, regression equations for application of fertilizers in different crops have been made. With these equations, soil test based recommendations can be made. For soil test, a portable rapid soil testing kit is available. With this kit, soil pH, nitrogen, phosphorus, potassium and organic carbon can be estimated by farmers or extension workers in the field itself. The results obtained with the help of this kit are comparable with the results provided by field soil testing labs. With this kit, soil can be classified as fairly low, medium and high categories for the purpose of fertilizer recommendations.

Use of VAM for Efficient Utilization of Nutrient as well as Soil Water

Since the crop is grown on residual soil moisture under rainfed conditions. Irrigation management is difficult, as it causes mortality in crops. In these situations, use of vesicular arbuscular mycorrhiza (VAM) inoculum under low cost input technology is useful to protect the host against the moisture stress conditions to some extent. It was observed that the symbiotic association of VAM fungus can increase water absorbing capacity of plant root not only from rhizobsphere but also from mycorrhizosphere, both under wet and dry seasons, which is related to larger root absorbing area of mycorrhizal plants. To identify suitable strains of VAM capable of protecting host against water stress and its consequence on performance of crops and crop sequence, a series of experiments in lab, glass house and field study were conducted. The VAM strains were identified for their capabilities to increase relative water content of leaf (RWC), water use (WU) and water use efficiency (WUE) including nutrient mobilization. All ten VAM strains tested under controlled conditions, varied in their capabilities. The selected composite culture of VAM strains? *Glomus aggrocarpum, G. etunicatum,* and *G. deserticola* under natural field conditions were found to protect host against water stress by increasing RWC, WU (Water use) and WUE in mycorhizal plants over uninoculated non mycorrhizal control plants.

Manures and Fertilizer Management in Grasses

Oat (*Avena sativa*)

Fertility requirement of oat as forage crop some what is higher than grown as a seed crop. 70 kg of nitrogen per hectare should be broadcast and mixed with the last harrowing before sowing. Another 80 kg of nitrogen should be broadcast in two splits, 40 kg nitrogen at first irrigation (25-30 days after sowing) and rest 40 kg after first cut. 60 kg phosphorus and 20 kg potash should be applied at the time of sowing as basal. Phosphorus has a favourable effect on the height and tiller number in earlier stage of growth. In addition to chemical fertilizers, 10-20 t/ha of well decomposed FYM is recommended.

Fodder Sorghum (*Sorghum bicolor*) and Pearl Millet (*Pennisetum typhoides*)

Farm Yard Manure @ 5-10 t/ha should be incorporated in the field 20-30 days before sowing of the crop. For dry crop apply 30-40 kg N, 40 kg P_2O_5 and 30 kg K_2O/ha as basal and 30-40 kg N/ha at 40 DAS. For irrigated crop, 50-60 kg N, 60 kg P_2O_5 and 40 kg K_2O/ha applied as basal followed by 50-60 kg N/ha at 40 DAS and 40 kg N/ha should be applied after each cut.

Fodder Maize (*Zea mays*)

Farm yard manure @ 10 t/ha may be applied at the time of preparation of land as basal dressing. N, P_2O_5 and K_2O at the rate of 120, 60 and 40 kg/ha respectively, may be given as top dressing.

Perenial Grasses

Hybrid Napier (*Pennisetum typhoides* x *P. purpureum*)

Hybrid Napier is a quick growing crop and responds to high fertility. Add 20 tnnes of farm yard manure or compost to the soil at least one month before planting. Apply phosphorus and potash @ 50 kg/ha each according to soil tests, in deficient soils at the time of land preparation. It should be followed by top dressing of nitrogenous fertilizers at the rate of 200 kg nitrogen per hectare. The topdressing should be done twice in a year, 115 days after planting at the rate of 100 kg nitrogen per hectare and the other towards the end of winter season. If convenient, divide the whole amount of nitrogen into three or four split equal doses and apply after each cutting followed by gentle raking, if possible for quick growth.

Perennial Ryegrass (*Lolium perenne*)

For geting good fodder production @ 10 t FYM/ha should be incorporated before the sowing of ryegrass and mixed well. In addition to FYM chemical fertilizer @ 100 – 120 kg N/ha/yr, 50 – 60 kg P_2O_5/ha/yr and 20 kg K_2O/ha/yr also be apply.

Guinea Grass (*Panicum maximum*) Gamba Grass (*Andropogon gayanus*)

In the prepared field, trenches of 10 cm width and 20 cm depth are made. In these trenches, FYM should be applied along with phosphorus and potassium fertilizers. Mix with soil and cover the trenches and form ridges of 15 cm height for planting slips. In acid soils, application of lime @ 500 kg/ha in alternate years is desirable. A basal dose of 10 tonnes of FYM, 50 kg P_2O_5 and 50 kg K_2O per ha (applied in trenches) is recommended. For topdressing, use 200 kg N per ha in two split doses, the first dose immediately after-first cutting and the second dose during the northeast monsoon period. If irrigation facilities are available, topdressing can be given in more splits. The fertilizer may be applied on either side of the plants, along the row and earthed up.

Setaria Grass (*Setaria anceps*)

Organic manure, either FYM or compost @ 10 t/ha may be applied at the time of land preparation. The crop responds well to application of fertilizers especially N. The fertilizer requirement depends on the initial nutrient status of the soil.

Orchard Grass (*Dactylis glomerata*) and Tall Fescue (*Festuca arundinaceae*)

The field may be fertilized with 100 – 120 kg N/ha/yr, 40 – 50 kg P_2O_5/ha/yr and 20 kg K_2O/ha/yr. 10 t FYM/ha should be incorporated before the sowing and mixed well for better performance.

Para Grass (*Brachiaria mutica*)

The crop is highly responsive to irrigation with cattle-shed washing or sewage water. Forty tonnes of FYM or compost along with 30 kg P_2O_5 and 30 kg K_2O per ha is to be given as basal dose. Top dressing N 40 kg/ha after each harvest is found to enhance the forage production.

Congosignal Grass (*Brachiaria ruziziensis*)

Basal application of 5 t/ha of FYM along with 50 kg/ha each of P_2O_5 and K_2O is recommended. Nitrogen @ 100-150 kg/ha may be applied in two or three splits.

Manures and Fertilizer Management in Legumes

Berseem (*Trifolium alexandrinum*)

Berseem is a forage legume crop which fixes the atmospheric nitrogen in soil through symbiotic bacteria. However, a light does of nitrogen is needed at the initial stage which helps in better and quick growth of berseem seedlings. For obtaining a good yield add about 15 tonnes of farmyard manure or compost along with 20-25 kg nitrogen and 80 kg phosphorus per hectare. Farm yard manure should be added about 20-30 days before sowing and mixed well in the field. Full does of nitrogenous and phosphetic fertilizers should be applied at the time of sowing as basal application. Addition of farmyard manures helps in improving the physical condition of the soil for better working of symbiotic bacteria.

Lucerne (*Medicago sativa* L.)

Lucerne being a leguminous crop fulfils its major part of nitrogen requirement through the process of symbiotic nitrogen fixation which works effectively from 3-4 weeks after sowing. However, soils with low organic matter and poor nitrogen supply may require 20-25 kg nitrogen per hectare as a starter dose which can meet plant requirement before the formation of effective nodules. An adequate supply of phosphorus is also necessary for the proper functioning of the nodules. Therefore, 80 kg P_2O_5 per hectare should be applied in lucerne crop. Lucerne responds well to farm yard manure on sandy loam soils. Being a perennial crop, it is advantageous to apply every year 25-30 tonnes of farm yard manure per hectare per year. A dose of 20 kg K_2O/ha is also beneficial.

Lucerne is a heavy feeder on soil nutrient. Some of the essential micronutrients such as boron, iron, zinc and manganese etc are also likely to become deficient in certain soils. Proper care should be taken to supply these nutrients in different soils. Boron deficiency is generally observed in leached and coarse textured soils. The leaves of Lucerne plant develop numerous pale yellow spots leading to disorder known as "Lucerne yellow". Spray of 0.2 per cent borax could be done safely to remove this deficiency. Iron deficiency, leading to chlorosis is fairly common in poorly drained alkaline soils. Liming the soil well in advance of sowing is helpful in areas where soil is acidic. Potash may be required in sandy soils.

Clover

Red Clover (*Trifolium pretense*), White Clover (*Trifolium repense*), Sainfoin (*Onobraychis viciaefolia*) and Vetch (*Vicia sativa*). A basal dose of 20 kg nitrogen as starter dose and 80 kg phosphorus and 20 kg potash per hectare should be applied. In poor fertility soil farm yard manure @ 5-10 t/ha may be incorporated. Seed inoculation with *Rhizobium* culture is a beneficial practice.

Guar (*Cyamopsis tetragonoloba* L.)

Being a leguminous crop, guar does not require additional nitrogen. Only on extremely poor soil (sandy with very low organic matter content) 20 kg nitrogen should be used as starter dose at the time of planting. It is desirable to apply 40-60 kg P_2O_5 per hectare to ensure good yields from grain crops. All the fertilizers should be applied at the time of sowing in furrows 4-5 centimetre below the seed. When guar is sown on poor soils after an exhausting crop, it is desirable to apply about 10 tonnes of farmyard manure or compost about one month before sowing.

Rice Bean (*Vigna umbellate*)

Rice bean, in spite of its grain yield potential comparable to major pulse crops and excellent nutritional qualities, failed to emerge as an important pulse crop in

India. But it is potential leguminous forage crop of the country. The crop can be raised in a variety of soils, but not under ill-drained conditions. Before sowing apply fertilizers at the rate of 20 kg nitrogen, 40 kg phosphorus and 30 kg potassium per ha and incorporate in the soil by raking. Inoculation of biofertilizers is also beneficial.

Cowpea (*Vigna unguiculata*)

For rainfed crop, at the time of land preparation, FYM @ 10 t/ha is applied and basal application of N, P_2O_5 and K_2O @ 25, 60 and 30 kg/ha is recommended. For irrigated crop in addition to 10 tones FYM and the basal dose of 40: 80: 30 kg N: P_2O_5: K_2O/ha, topdressing of N and K_2O each at 10 kg/ha after each cut is to be given.

Stylo (*Stylosanthes* spp.)

Recommended dose of N, P_2O_5 and K_2O for both annual and perennial stylosanthes are 20, 80 and 30 kg per ha, respectively. For perennial crops, phosphorus @ 80 kg/ha and potash @ 30 kg/ha may be applied in subsequent years. Application of lime @ 375 kg/ha is also recommended in acid soils. Application of phosphorus @ 120 kg and lime @ 375 kg per ha is beneficial for getting maximum yield.

Fodder Trees

Subabul (*Leucaena leucocephala*)

A basal application of N:P_2O_5:K_2O at the rate of 20:50:30 kg/ha is recommended.

References

Brady, N.C.1995.The nature and properties of soils. Edn. 10 501-502.

Deb Roy (1993). Reap more biomass through diversity in forestry. Intensive Agriculture. XXXI (5-8): 23-26.

Dogra, K.K., Katoch, B.S., Sood, B.R. and Singh, Gurudev. (1997). Production potential and quality of silviherbage systems viz-a-viz natural grasslands in the humid sub tropics of H.P. Range Management and Agroforestry 16(2): 165 –168.

Gupta, R.K. (1977). Energy forests- Farm and Community lands. Indian Farming.

Hazra, C.R. (1998). Integrated nutrient management for sustainable forage production in pasture and silvipastures. Fertilizer. News. 43(3): 33-45.

Mandal, K.G. and Sinha, A.C. 2002. Effect of integrated nutrient management of yield and nutrient uptake by Indian mustard (Brassica juncea) in foothills soils of eastern India. Indian Journal of Agronomy. 47 : 109-113.

Melkania, N.P. (1987). Search for an ideal grass-legume introduction for mid-Himalayan grasslands. Indian Journal of Range Management. 8:51-53.

Melkania, N.P. (1995). Forage resources and forage production systems in Central Himalaya. In: New Vistas in Forage production (ed. Hazra, C.R and Misri, Bimal). AICRPF (IGFRI). Publication Information Directorate, New Delhi. 203 –211.

Moore, M.P. (1974). Secular aspects of the sacred cow: The productivity of Indian farm animals. IDS discussion paper, Institute of Development Studies, University of Sussex.

Shastry, G and Patnaik, U.S. (1990). Grassland management. In: Watershed Management (ed. V.V. Dhruranarayana, G. Shastry and U.S Patnaik). Central Soil and Water Conservation Research and Training Institute (CSWCR and TI), Dehradun, ICAR Publication. pp. 68-73.

Satyajeet; Nanwal, R.K. and Yadav, V.K. 2006. Effect of integrated supply on the crop productivity and soil health under pearl millet-mustard cropping sequence. Farming System Research and Development. 12 (1 and 2) : 26-30.

Singh Virendra (1995). Technology for forage production in Hills of Kumaon. In : New Vistas in Forage Production (ed. Harzra, C.R and Misri Bimal). AICRPF (IGFRI). Publication Information Directorate, New Delhi. pp. 197 -202.

Sood, B.R. and Bhandari, J.C. (1992). Response of *Setaria anceps* cv. Narok to nitrogen and phosphorus in natural grasslands of Kangra valley. Range Management and Agroforestry. 13(2): 139-141.

Sood, B.R. and Sharma, V.K. (1996). Effect of nitrogen and phosphorus levels on the productivity of natural grassland. Haryana Journal of Agronomy. 12 (1) : 68-74.

Source : http://www.thehindu.com/thehindu/seta/2002/04/25/stories/2002042500020400.htm

Chapter 16

Production of Forages and Grasses with Poor Quality Waters

Chunchun Kumar* and Anil Kumar Singh**

ICAR-RCER, Patna – 800 014, Bihar
**E-mail: ckvidyarthi@rediffmail.com*
***E-mail: aksingh_14k@yahoo.co.in*

The additional area required to meet the projected fodder shortage of 40 per cent is about 10 million ha while presently India has only 6.9 million ha land under fodder production, which is about 4.7 per cent of the total cultivated area. The possibility of increasing deficit fodder supply by adopting latest forage production technologies on the available cultivated area is very limited. In Haryana state, ground water quality of 55 per cent of the tube wells is poor. Due to scarcity of good quality waters these poor quality (including alkali and saline) waters have to be used for irrigation even though adversely affecting yield and quality of the crops grown.

The productivity of crops depends on the type and concentration of salts in the soil and in the irrigation water at germination and other active phases. The crops that can withstand saline and alkali soil conditions are generally more suitable for saline and alkali irrigation conditions as well. As the two soils and problem waters contain different types of salts, their effect on crops and soil is also likely to be different so they require different agronomic management practices for better crop production. The following discussions therefore narrate how crops are affected by the soil salinity, saline irrigation water, soil alkalinity and alkaline irrigation water, respectively.

Forages in Saline Soils

Plants differ widely in their ability to tolerate salts in the soil. Forage grasses and legumes are generally more tolerant to saline condition (Magistad and Christiansen, 1944). The plants/crops, which can restrict evaporation, provide better root penetration and enhance organic matter are more suitable from the reclamation point of view for these soils and better forage production.

In saline soils, the volunteer plant species like *Cresa cretica, Cyperus rotundus, Chloris pallida, Sporobolus pallidus, Sessuvium portulacastrum, Haloxylon salicornicum, Aeluropus lagopoides, Zygophyllum simplex, Dicanthium annulatum, Suaeda fruticosa and Butea monosperma* can be seen growing. Rana (1983) identified four species, *Alhagi pseudealhagi, Salsola baryosma, Suaeda fractiosa and Salvadora oleoides* as indicator plants of saline soils.

Forage Crops for Saline Conditions

Richards (1954) reported Sorghum (*Sorghum bicolor* (L.) Moench) as summer (*kharif*) crop as medium tolerant to salinity, but Kalippan and Rajgopal (1969) reported it to be fairly tolerant with no significant effects on germination and early vigor up to 6.3 dSm^{-1}. However, Mehrotra and Das (1973) exhibited 50 per cent decrease in sorghum at EC_2 8 dSm^{-1}. In a sand culture experiment, Kumar and Gill (1994) reported that germination of *Sorghum sudanense* was delayed with the rise of salinity and crop failed to germinate at 18 dSm^{-1} but survived at ECe 6 dSm^{-1}.

Amongst the forage crops grown during winter (*rabi*), oats (*Avena sativa* L.) has been found most tolerant to saline conditions. According to Ashok Kumar and Sharma (1995) the yields of five forage crops *viz.*, lucerne (*Medicago sativa* L.), berseem (Egyptian clover) (*Trifolium alexandrinum* L.), shaftal (Persian clover) (*Trifolium resupinatum* L.), chinese cabbage (*Brassica pekinensis* (Lour.) Rupr. and oats are not affected significantly when irrigated with saline water of 5 dSm^{-1} but beyond this there could be significant decrease. The yields of berseem, shaftal and Chinese cabbage were significantly reduced at $EC_2$7.5 dSm^{-1}, however, Lucerne showed significant decline only when saline water of 10 dSm^{-1} was applied while that of oats was unaffected at this level also.

The order of tolerance of the forage crops to salinity was: oats> Chinese cabbage> lucerne> berseem> shaftal. Under field conditions, Yadav and Ashok Kumar (1997) recorded observations on sorghum, pearl millet (*Pennisetum americanum* (L)), maize (*Zea mays* (L)), barley (*Hordeum vulgare* (L)), oats and berseem and found that germination of all the crops except barley and oats reduced drastically where soil salinity exceeds EC_2 3.0 dSm^{-1}. Berseem, pearl millet and maize recorded decline in tillering/branching and height to a great extent with increase in soil salinity. The results suggested that where soil salinity exceeds $EC_2 > 4.5$ dSm^{-1}, only barley and oats give satisfactory green forage yields. The forage crops like berseem, sorghum and pearl millet produce satisfactory yields only up to EC_2 of 4.5 dSm^{-1}. Maize fails to produce satisfactory green forage yield beyond a salinity level of 3 dSm–1. Yadav *et al.* (1997) reported that on saline soils (EC_2 3–4 dSm^{-1}), the mixtures of berseem with mustard and/or oats out performed their pure stand in green yield, dry matter yield, protein and mineral matter yield.

Salt Tolerance of Grasses

Grasses tolerate salinity more than forage crops. Bermuda grass, Karnal grass and *Panicum coloratum* can be regarded most tolerant grasses as these show less than 50 per cent yield reduction at the EC 14-19 dsm^{-1}. Malik *et al.* (1986) also reported that Karnal grass (*Kallar* grass) can survive up to an EC 40 dsm^{-1} but is economical to grow up to EC 22 dsm^{-1}. Anjan grass is sensitive to soil salinity.

Forages and Grasses as Affected by Saline Irrigation

In most parts of the arid and semi-arid regions in the country, saline ground waters pose a threat to crop production. In areas where this problem exists, generally good water is not available for irrigation purpose. A few sporadic attempts on the use of saline water for growing forage grasses and crops have been made and the results are presented as under.

Ashok Kumar (1988) from field trials at Sampla in Rohtak district inflicted by the problems of high salinity, water table and high salinity of the ground water (EC, 22 dsm^{-1}) reported that when good water was used for irrigation in saline soils, Para-grass yielded the best, but under saline water irrigation, Karnal grass showed best results. When the average yield of 3 years was considered, Gatton panic, Karnal grass and Blue panic were better under saline environment. Ashok Kumar and Gill (1994) while examining the effect of high salinity waters under sand culture condition on several forage grasses reported that the grasses, in general, showed poor survival when sown by seeds but established well when transplanted by the rooted slips. *Panicum coloratum*, Karnal grass, Gatton panic and Rhodes grass were relatively less affected owing to irrigation with high saline waters. When the water of 19 dSm^{-1} EC was used for irrigation, *Panicum coloratum* showed 50 per cent *Leptochloa fusca* and *Chloris gayana* about 65 per cent Gatton panic 75 per cent while hybrid Napier 100 per cent yield reductions. *Cenchrus ciliaris* and *Panicum antidotale'* could not cope up even with saline water of 12 dsm^{-1}.

Under hydroponics' conditions, Ashok Kumar *et al.* (1991) found that the salinity levels reducing 50 per cent shoot and root weights were 80 and 92 mol^{-3} for sudan grass; 47 and 65 mol m^{-3} for maize; and 29 and 36 mol m^{-3} for teosinte thus indicating that sudan grass was more tolerant to salinity. In a sand culture experiment conducted with saline waters *viz.* BAW (best available water) 2.5, 5.0, 7.5 and 10 dsm^{-1} on five *rabi* forage crops namely, berseem, shaftal, lucerne, oats and Chinese cabbage, Ashok Kumar and Sharma (1995) reported that oats produced the highest green forage yield among the crops tried and did not depict any yield reduction up to saline water of 10 dsm^{-1}. Lucerne and Chinese cabbage performed better than other crops under saline water conditions.

Yadav *et al.* (2004) reported that alternate irrigation with saline drainage water increase the yields of all forage crops compared with using saline drainage water only. They further stated that alternate irrigation starting with canal water was superior to alternate irrigation starting with saline drainage water because less salt was added in total. Oat produced the largest green forage yield (32.3 t ha^{-1}) in the first year while rye grass gave maximum in the second (34.6 t ha^{-1}) and third year (37.0 t ha^{-1}). Persian clover performed better than Egyptian clover in all the 3 years. Interaction between species and irrigation treatments was significant. In comparison to canal irrigation water, there were 36.4 per cent, 42.1 per cent, 53.5 per cent, 67.5 per cent, and 85.3 per cent yield reduction in rye grass, oat, Persian clover, Egyptian clover, and senji, respectively when only saline drainage water was used for irrigation purpose. Thus, the tolerance of winter forage crops to salinity was in the order of rye grass> oat> Persian clover> Egyptian clover> senji.

Alkali Soils

For reclamation of alkali soils use of an amendment, usually gypsum is recommended. The quantity of gypsum applied depends mainly on the pH of soils and the crops to be grown.

Forages for Alkali Soil

Sporobolus marginatus, S. diander, Desmostachya bipinnata, Kochia indica and Suaeda maritima. However, other species such as *Dicanthium annulatum, Capparis aphylla, Cynodon dactylon, Andropogon squarrosus, Sporobolus coromandelianus* can also be noticed. On low lying areas where water is liable to accumulate for any length of time, *Leptochloa fusca*, commonly called Karnal grass in India is found in abundance.

Most of the volunteer grasses found in alkali soil are very coarse with very little nutritive value excepting a few like Karnal grass, Bermuda grass, *Dicanthium annulatum* etc. As pointed out earlier, Karnal grass grows well where water accumulates, so the land should be shaped in such a way to accumulate maximum rainwater. For this, it is suggested that alkali land should be bunded to make smaller plots. Simple bunding or closure of alkali lands will encourage the palatable grasses to grow and to control the coarse grasses.

The relative salt tolerance of grasses and forage crops grown in the northern India under irrigated conditions based on the results by Abhichandani and Bhatt, 1965; Ashok Kumar, 1987; Ashok Kumar, 1988; Ashok Kumar, 1990; Ashok Kumar and Abrol, 1986; Ashok Kumar and Gill, 1994; Ashok Kumar *et al.*, 1991; and Maas, 1986.

Grasses are in general, more tolerant to alkali soils than the other forage crops. Karnal grass can be rated as most tolerant grass to alkali soil conditions as it did not show any yield reduction even at the highest pH of 10.4; rather it gave more yields on alkali soil compared to on normal soil (Ashok Kumar and Abrol, 1983 a; Ashok Kumar, 1986). This is followed by Rhodes grass that did not show reduction up to a pH of 10,0 and beyond this also there was a small yield reduction. In terms of green forage yield, Rhodes grass gave higher yield than Karnal grass at all the pH levels including at pH 10.4 because of its greater yield potential. Gatton panic, Bermuda grass, Coastal Bermuda and Para grass are also tolerant grasses as these showed less than 50 per cent yield reduction up to a pH of 10. Blue panic, Setaria grass and hybrid Napier can be regarded as moderately tolerant grasses as these showed good yield up to a pH 9.6, the former two grasses showing less than 50 per cent yield reduction. Guinea grass, Anjan grass and Deenanath grass are sensitive to soil alkalinity, almost failing even at pH 9.2.

Amongst the forage crops, *rabi* crops seem to be relatively more tolerant than *kharif* crops. Sorghum, Bajra (Pearl millet) and Dhaincha (*Sesbania*) are moderately tolerant forage crops as these showed less than 50 per cent yield reduction at pH 9.6; teosinte and maize as moderately sensitive showing less than 50 per cent yield reduction at a pH of 9.2 while cowpea and Guar as sensitive. All the *rabi* forage crops are moderately tolerant; oats and shaftal being more tolerant than berseem (Ashok Kumar, 1987). All the *rabi* forage crops tested appears to be more tolerant in comparison to wheat (Chhabra and Ashok Kumar, 1990-91). Growing of shaftal (*Trifolium resupinatum* Linn.) as a forage crop in *rabi* season has been found more profitable than wheat (*Trifolium aestivum* Linn.) (Sharma *et al.*, 1983).

Kharif Forage Crops during Initial Reclamation

In highly alkali soil, forage crops like sorghum [*Sorghum bicolor* (L.) Moench.], guar [*Cyamopsis tetrgonaloba* (L.) Taub.], cowpea [*Vigna unguiculata* (L.) Walp.], Deenanath grass (*Pennisetum pedicellatum* Trin.) and sudan grass [*Sorghum sudanense* (Piper) stapf.] either fail to grow or give very poor yields even after application of 100 per cent of gypsum requirement (Chillar and Bhumbla, 1970). Similar findings were recorded by Kumar (1993) wherein it was observed that *kharif* forage crops singly or as a cereal-legume combination grown on alkali soils with gypsum at 50 per cent of requirement yielded very much lower (Table 16.1) than that obtained from normal soils. Maize (*Zea mays* L.), guar, and maize + guar failed completely giving no forage yield. These results suggest that *kharif* forage crops should not be grown in the beginning of reclamation of alkali soils.

Kharif Forage Crops After the Rice-Wheat Rotation

After following rice-wheat rotation for two years, Chillar (1982) applied gypsum and pyrite at 0, 25, 50, 75 and 100 per cent of their requirements and observed the performance of sorghum, oats and maize. In control plots (pH 9.8), even after two crops of rice and wheat, forage yields of sorghum and maize were extremely low and the yields of both the crops increased with increasing dose of amendments. Sorghum was more tolerant than maize.

Rabi Forage Crops during Initial Reclamation

Among the *rabi* forage crops shaftal has been reported to be relatively more tolerant (Ashok Kumar, 1987). From the results of a field experiment, it was evident that the performance of berseem during first year was extremely poor even after application of 25 t ha^{-1} gypsum, whereas shaftal gave good yields even with 10 t ha^{-1} gypsum application In the second year also similar results were recorded. However, in the third year berseem showed significant increase in green forage yield with each increasing level of gypsum up to 15 t ha^{-1} whereas shaftal gave significantly greater yield at 10 t ha^{-1} gypsum. From the same experiment a yield reduction of 25 and 50 per cent in berseem at an ESP of 32 and 41, respectively was recorded while for shaftal such reductions were observed at the ESP of 37 and 53 respectively, indicating

Table 16.1: Green Forage Yield (t ha^{-1}) of Grasses and Forage Crops on Normal Soil and Relative Yield on Alkali Soil at Different pH/ESP Levels

Botanical Name	Grass/Crop	Yield (t ha^{-1})					
			Alkali Soil (pH/ESP)				
		Normal Soil	8.8	9.2	9.6	10.0	10.4
			15	30	50	70	90
Grass							
Leptochloa fusca	Karnal grass	35	40	42	45	45	40
Chloris gayana	Rhodes grass	60	63	65	63	60	50
Panicum maximum	Gatton panic	60	60	65	60	45	35
Cynodon dactylon	Bermuda grass	40	40	40	40	30	25
Cynodon dactylon	Coastal Bermuda	40	40	40	35	28	20
Brachiaria mutica	Para grass	100	100	90	70	50	40
Panicum antidotale	Blue Panic	75	75	75	60	30	20
Setaria anceps	Setaria grass	75	75	60	40	15	0
Pennisetum purpureum	Hybrid Napier	100	100	75	40	15	0
Panicum maximum	Guinea grass	75	60	20	8	0	0
Cenchrus ciliaris	Anjan grass	25	10	0	0	0	0
Pennisetum pedicellatum	Deenanath grass	60	10	0	0	0	0
Forage crops							
Kharif							
Sorghum bicolor	Sorghum	40	40	35	25	0	0
Pennisetum americanum	Pearlmillet	30	30	25	20	0	0
Euchlaena mexicana	Teosinte	50	45	30	10	0	0
Zea mays	Maize	40	35	30	10	0	0
Sesbania aculeata	Dhaincha	30	30	25	20	5	0
Vigna unguiculata	Cowpea	30	20	10	0	0	0
Cyamposis tetragonaloba	Guar	25	15	10	0	0	0
Rabi							
Avena sativa	Oats	50	50	45	35	15	0
Trifolium resupinatum	Shaftal	80	80	70	45	25	0
Trifolium alexandrinum	Berseem	90	80	65	35	0	0

Table 16.2: Green Forage Yield (t ha^{-1}) of Grasses and Forage Crops on Normal Soil and at Different EC Levels on Saline Soil

Botanical Name	*Grass/Crop*	*Yield (t ha^{-1})*				
		Normal Soil	*EC Level*			
			4	*9*	*14*	*19*
		Grass				
Cynodon dactylon	Bermuda grass	40	40	35	28	20
Leptochloa fusca	Karnal grass	40	40	30	25	20
Panicum coloratum	–	35	35	30	25	20
Brachiaria mutica	Para grass	100	90	70	40	10
Panicum maximum	Gatton panic	60	60	35	25	20
Panicum antidotale	Blue panic	75	68	45	30	15
Chloris gayana	Rhodes grass	60	60	40	30	15
Pennisetum purpureum	Napier hybrid	100	80	55	35	0
Cenchrus ciliaris	Anjan grass	40	20	0	0	0
		Forage Crop				
		Rabi				
Avena sativa	Oats	50	50	40	35	10
Trifolium alexandrinum	Berseem	90	75	60	35	0
Trifolium resupinatum	Shaftal	80	65	55	40	0
Medicago sativa	Lucerne	90	75	60	45	0
Brassica pekinensis	Chinese cabbage	30 25	22	20	0	
		Kharif				
Sorghum sudanense	Sudan grass	60	55	30	15	0
Sorghum bicolor	Sorghum	40	35	20	10	0
Zea mays	Maize	40	20	10	0	0
Euchlaena mexicana	Teosinte	50	20	5	0	0

higher tolerance of shaftal over berseem. Sharma *et al.* (1983) reported that shaftal was more profitable as fodder crop than wheat in the initial year of reclamation itself. The results of a field experiment conducted at CSSRI on highly alkali soil (pH 10.5) with the application of 50 per cent GR to evaluate the effect of zinc application on rice crop and its residual effect on the yield and chemical composition of following *rabi* forage crops revealed that all the forage crops (berseem, shaftal, oats and barley) were relatively more tolerant than wheat. Amongst *rabi* forage crops, shaftal was found to be most tolerant. Barley produced higher seed yield than oats. Chillar (1982) also noticed that oats for forage grow well after following rice-wheat rotation.

Forage Production with Alkali Irrigation

For reclamation of alkali soils and to mitigate the adverse effect of alkali water irrigation, use of gypsum application is generally recommended. The quantity of gypsum required depends on several factors such as soil characteristics (including initial pH) of the soil, residual sodium carbonate (RSC of water), tolerance of the crop to be grown, climate and water requirement etc. Relatively very small work has been done on the forage production in relation to alkali irrigation. The results of the work done are described below under different heads.

Rabi Forages

On a moderately alkaline soil (pH 9.29), with alkaline water irrigation [EC_{iw} 1.8 dS/m, residual sodium carbonate (RSC) 10.5 and sodium adsorption ratio (SAR) 8], oats, shaftal and berseem were more suitable than other crops tested. However, berseem gave comparatively lower yield in the first year of cultivation. Application of gypsum (2 t ha^{-1}) increased the yields of forage crops in the second and third year of experimentation. In comparison with their yields in normal soil, oats showed minimum yield reduction under alkali soil irrigated with alkali water, hence was more tolerant among the forage crops studied.

In a separate field experiment with alkali water irrigation it was found that in the first harvest recommended seed rate of berseem at 25 kg ha^{-1} + 2 kg ha^{-1} mustard gave significantly greater yield than the remaining mixture treatments. Application of gypsum at 2 and 4 tha^{-1} increased the yield of berseem population treatments in all the cuts as well as for total yield.

Kharif Forages

Sudan grass and sorghum were found to be more suitable for the situations where soils were moderately alkaline and irrigation waters is also alkaline. Maize, teosinte and pearl millet should not be grown with alkali water irrigation. Application of gypsum at 2 t ha^{-1} in the first year improved the yield of kharif forage crops by 15 per cent over the control but during the third year there was 17.5 per cent and 35 per cent improvement in the yield owing to gypsum application at 2 and to 4 t ha^{-1}, respectively.

Conclusions

Since kharif forage crops are sensitive to high alkali soil conditions, even with the ample availability of amendments, like gypsum, these should not be grown in the beginning. However, Shaftal (*Trifolium resupinatum*) yields successfully in first year itself. Grasses, in general, are more tolerant to alkali conditions because of their better root system, are, therefore, recommended. The basic difference in the reclamation strategies under two conditions is that in alkali soils where good quality water is available, gypsum has to be applied only once in the beginning and with cropping and time space the soils get improved and the soils go round from bad to good while in case of alkali water irrigation, the soil turn from good to bad. For successful growth of crops including forage crops, gypsum needs to be applied to each crop and the quantity of gypsum depends upon the tolerance of the crop and their water requirement.

References

Abhichandani, C.T. and Bhatt, P.N., 1965. Salt tolerance at germination of bajra (*Pennisetum typhoides*) and Jowar (*Sorghum vulgare*) varieties. *Ann. Arid zone*. 4: 36-42.

Ashok Kumar, 1987. Relative performance at Egyptian and Persian clovers at different levels of gypsum application in a barren alkali soil. *Indian J. Agric. Sci.* 57 : 157-162.

Ashok Kumar,1988. Performance of forage grasses in saline soils. *Indian J. Agron.* 33: 26-30.

Ashok Kumar, 1988. Long term forage yields of five tropical grasses on an extremely sodic soil and resultant soil amelioration. *Expl. Agric.* 24 : 89-96.

Ashok Kumar, 1990. Effect of gypsum compared with that of grasses on the yield of forage crops on a highly sodic soil. *Expl. Agric.* 26 : 185-188.

Ashok Kumar and Abrol, I.P. 1979. Performance of five perennial forage grasses as influenced by gypsum levels in a highly sodic soil. *Indian J. Agric. Sci.* 49 : 473-477.

Ashok Kumar and Abrol, I.P. 1983. Effect of gypsum on tropical grasses grown in normal and exremely sodic soil. *Expl. Agric.* 19: 167-177.

Ashok Kumar and Abrol, I.P. 1986. *Grasses in alkali soils.* Bull. No. 11, Central Soil Salinity Research Institute, Karnal.

Ashok Kumar and Gill, K.S. 1994. Effect of varying salinity of irrigation waters on the yield and chemical composition of forage grasses. *Forage Res.* 20: 93-101.

Ashok Kumar and Sharma,P.C. 1995. Effect of salinity on the performance and ionic concentration in *rabi* forage crops. Forage Res. 21: 87-90.

Ashok Kumar, Batra, L. and Chhabra, R. 1994. Forage yield of sorghum and winter clovers as affected by biological and chemical reclamation of a highly alkali soil. *Expl. Agric.* 30: 343-348.

Ashok Kumar; Datta, K.S., Gorham, J., Wyn Jones, R.G. and Hollington, P.A. 1991. Influence of salinity levels at two growing stages on the performance of three tropical forage crops. *J. Indian Soc. Coastal Agric. Res.* 4 : 493-512.

Chillar, R.K. 1982. *Ph. D. Thesis*, Kurukshetra University, Kurukshetra.

Chillar, R.K. and Bhumbla, D.R. 1970. *Annual Report*, Central Soil salinity Reserch Institute, Karnal. P35.

Chatterjee, B.N. and Das, P.K. 1991. *Forage crop production, principles and practices,* oxford and IBH Publishing Co. Pvt. Ltd., New Delhi.

Chhabra, R. and Ashok Kumar, 1990-91. Effect of zinc applied to rice on the yield and chemical composition of following *rabi* fodder crops. *Annual Report*, CSSRI, Karnal. P. 41.

Kalipan, R. and Rajgopal, A. 1969. Salinity effect on germination and early vigour of 5 sorghum varieties. *Madras Agricultural Journal*. 56: 282-5.

Maas, E.V., 1986. Salt tolerance of plants. *Applied Agriculture Research* 1(1) : 12-26.

Magistad, O.C. and Christansen, J. E. 1944. *U.S. Department of Agriculture Circular*, 707.

Malik, K.A., Aslam, Z. and Nagvi, M, 1986. *Kallar grass- a plant of Saline soils*, Nuclear Institute for Agriculture and Biology, Faislabad, Pakistan.

Mehrotra, C.L. and Das, S.K. 1973. Influence of exchangeable sodium on the chemical properties of important crops at different satges of crop growth. *J. Indian Soc. Soil Sci.* 21: 355-365.

Rana, R.S. 1983. Plant indicators of soil problems. *Annual Report*, CSSRI, Karnal PP 93-95.

Richards, L. A. 1954. *Diagnosis and improvement of saline and alkali soils*. Hand Book No. 60, U.S. Dept. Agric., Washington.

Sharma, D.P. Mehta, K.K., Yadav, J.S.P. and Singh, G. 1983. Effect of crop growth and soil properties in sodic soil on farmers field. *Indian J. Agric. Sci.* 53- 681-685.

Yadav, R.K. and Ashok Kumar 1997. Feasibility of cultivating different forage crops on saline soil. *Crop Research*. 13: 45-49.

Yadav, R.K., Ashok Kumar, Lal, D. and Batra, L. 2004. Yield responses of winter (rabi) forage crops to irrigation with saline drainage water. *Experimental Agriculture* 40: 65-75.

Chapter 17

Agrotechniques of Production of Non-Leguminous Forages/Fodder Crops in India

Sant Kumar Singh

ICAR Resesarch Complex for Eastern Rregion, Patna – 800 014, Bihar
E-mail: santpatna@yahoo.co.in

Serious limitation in fully utilizing the production potential of improved animals is availability of quality forages. The National Commission on Agriculture (1976) recommended that a minimum 10 per cent of the arable area in the country (about 16.5 million ha) should be under improved forage crops to meet the green forage needs of the livestock population. Despite this, the area under forage crop is only 4.4 per cent now, leaving a wide gap between the supply and demand. Careful attention is needed in various aspects of agro-techniques of major forages and fodder crops to ensure maximum productivity. This chapter is a humble collection of agrotechniques of major forages and fodder crops.The term forage is used broadly to mean all the plant materials that are eaten by herbivorous animals. It includes fodder crops. Fodder crops is defined as any plant that is cut before being fed to animals in green state (soilage) or after converting to hay or silage such as Guinea grass, Napier grass, Fodder sorghum and Fodder maize etc.

Non-Leguminous Cereal Forage Crops

Cereal forage crops are some cultivars of cereal plants and are particularly able to produce huge vegetative growth, which may not be efficient in grain production.

They are seasonal crops, but produces substantial quantity of fodder in a very short period. According to the season in which they are grown, cereal fodder crops are usually referred as Kharif or Rabi crops. In the rabi season of North and Central India, winter annuals can be grown except in the high altitude regions. Cereal crops are propagated through seeds only. They belong to the family poacea (Gramineae). Growing of cereals and legumes together either in mixed or in any intercropping system is protective way to get quality forage in appreciable quantity with added advantage of soil enrichment.

Fodder Maize (*Zea mays* L.)

Maize is the most nutritious, succulent and palatable fodder. Fodder maize is usually grown in mixed cropping system or in rotation with several compatible grasses and legumes.

Climate and Soil

Maize can be grown in all seasons *viz.*, summer, rainy and winter. It is cultivated from the sea level to about 2500 m altitude. It cannot tolerate heavy rains and dry hot winds. Although it can grow in areas with a rainfall of more than 500 mm, for the best yields, regions with 1200-1500 mm are preferred. Nevertheless, it is often grown with irrigation, and good yield can be obtained with irrigation only. Although it is fairly drought tolerant up to five weeks, thereafter, is very susceptible. Maize can tolerate a minimum temperature of about 10°C and maximum temperature of 45 °C. However, the most suitable temperature for germination is 21°C and for growth 32°C. Maize prefers fertile, well drained alluvial loams and clay loams.

In warmer climates, maize can be grown throughout the year. Maize can be sown as rainfed crop with the break of monsoon, actual dates vary from region to region-usually from April to July. If irrigation facilities are available, rabi (August-September) and a spring crop (January-February) can also be raised. With irrigation, the crop can be raised throughout the year. In South India four to five crops are possible.

Cultivars

Although most of the cultivars can be grown for fodder purposes, hybrid cultivars that give high herbage yield are preferred. 'African tall', 'Vijay composite', 'Moti composite', 'Manjari Composite', 'Jawahar', etc. are some of the cultivars recommended for the whole country.

Establishment

Maize plant is propagated through seeds only. A deep, friable seedbed should be prepared, as maize is comparatively shallow rooted. The land is prepared by two to three ploughing, and ridges and furrows are formed. For irrigated crops, level beds are prepared. Seeds are broadcast or dibbled at a spacing of 30 cm between rows and 15 cm between plants. In North India, drilling by *pora* or *kera* method is widely used. When grown for forage, leaf: stem ratio is important. To reduce the stem size and increase the proportion of leaf, closer than normal spacing should be used. When broadcasting is done, use a seed rate of 80 kg/ha; while for dibbling, seed requirement is less, and the seed rate recommended is 40-60 kg/ha.

Manures and Fertilizers

The crop requires heavy manuring and respond well to fertilizers. At the time of sowing, farmyard manure or compost at the rate of 10 t/ha is required. Nitrogen, phosphorus and potassium shall be given at 90:30:30 kg/ha. Two-third of nitrogen (60 kg/ha), full phosphorus (30 kg/ha), and potassium (30 kg/ha) shall be given as a basal dose. The remaining one-third nitrogen shall be given at 30 days after sowing.

Weeding

Fodder maize has poor competitive ability until the crop canopy is closed. For the initial weed control, pre-emergence herbicides, simazine or atrazine at the rate of 1.0 kg/ha may be applied. In a mixed cropping system, herbicide application may not be possible. In such a situation, one hand weeding or interculture may be done at three to four weeks after sowing.

Water Management

Fodder maize is fairly drought tolerant, and it can withstand a drought period of up to five weeks, but thereafter is very susceptible. In the summer season, it is often grown with irrigation. Irrigation is necessary on the day of sowing. Subsequent irrigations shall be given at 10-15 day intervals. Maize is susceptible to water logging; and hence, provisions for drainage must be done during the rainy season especially in low-lying fields.

Harvesting and Uses

Fodder maize is a fast growing crop. The crop can be utilized in many ways such as grazing, soilage and for silage. However, it is best utilized for silage making. Fodder maize makes the best silage of the grass family, with heavy yields and high acceptability and without the need of additives. It is also fed directly to cattle. Grazing is possible only if planted dense. When maize is grown for grain purpose, the stover and the stalks left after husking can also be used as an important feed for cattle.

Harvesting of herbage is done after 60 days of planting or at the milk stage of the crop. If there is sufficient moisture, a second cut can also be taken. The stages of harvest differ according to the purposes. It should be cut at tasseling to wax-ripe stage and falls sharply after the milk stage. However, for silage making, it may be cut when the grain is full and glazed-in the medium dough stage. The herbage contains 7-10 per cent crude protein and 25-35 per cent crude fibre. The herbage is highly palatable.

Fodder maize produces heavy herbage yields from a very short period of growth. Green fodder yield ranges from 35-55 t/ha depending upon management.

Fodder Sorghum (*Sorgum bicolor* (L.) Moench. (*Sorghum vulgare* Pers.)

Jowar is the important fodder crop of low rainfall. It withstand heat and drought. It also tolerates water logging better than maize.

Climate and Soil

Sorghum is an annual, which can be grown with a rainfall of 300-750 mm. It is cultivated in areas too dry for maize. It is often grown as a rainfed crop and is sown from May to July depending upon the onset of monsoons. The crop can be grown with irrigation too, and with irrigation, sowing can be done any time of the year except cold season. Sorghum has adapted to a wide range of soils except very sandy soils. Good drainage, however, is necessary in heavy soils. It tolerates poor soils and can withstand moderate salinity. Its preferred soil pH range lies between 5 and 8.5.

Cultivars

Most of the sorghum cultivars suitable for growing as fodder are single cut types. However, double cut and multi-cut cultivars are also seen.

Single cut types- PC-6, PC-9, PC-23, SL-44, UP Chari-1, UP Chari-2, Raj Chari 1, Raj Chari 2, HC-136,HC 171, HC 260, MPK V-136, JS-20, S-1049 and JS-3.

Double-cut types-Co 37

Multi-cut types-MP Chari, Meethi sudan (SSG-59-3), Pioneer-988, X-988, X-988, MFSH-3, Hara Sona.

Establishment

Sorghum is propagated through seeds. As it is a shallow rooted crop, deep cultivation is not required. Land is prepared by 2-3 ploughings. A seed rate of 40-50 kg/ha is followed for fodder purposes. Seeds are sown in line by *Pora* method or by *Kera* method. Some times, broadcast in the rainy season. Spacing usually followed is 25-30 cm (between rows) and 10-15 cm (between plants).

Manures and Fertilizers

For a rainfed crop, farmyard manure at 10 t/ha is to be applied at the time of sowing. However, for an irrigated crop, a higher amount of farmyard manure at 25 t/ha must be given. Fertilizers need to be applied based on the number of harvests. Irrigation should be followed after top dressing of the fertilizers. The general pattern is as follows.

Single cut types Total requirement of N:P:K (kg/ha)-90:30:0. Basal dose: nitrogen, two- third (60 kg/ha); phosphorus, full (30 kg/ha); remaining one-third nitrogen at 30 days after sowing.

Double cut types Total requirement of N:P:K (kg/ha)-120:30:0. Basal dose: nitrogen, half (60 kg/ha); and phosphorus, full (30 kg/ha); remaining one-third nitrogen at 30 days after sowing.

Multicut types Total requirement of N:P:K (kg/ha)-210:60:60. Basal dose: nitrogen 60 kg/ha; phosphorus, full (60 kg/ha); and potassium full (60 kg/ha). The remaining nitrogen is to be applied in three equal splits at the rate of 50 kg/ha each after first, second and third cuts.

Weed Management

Weeds may be a problem in the early stages, and can be controlled by herbicides or shallow inter cultivation with a harrow or hoe. Later, when the rows close in, weeds would be suppressed by the shade of the crop canopy. For early season weed control, apply atrazine 1.0 kg/ha, pre-emergence. For broad leaf weed control, 2,4-D can be applied post-emergence by 4-5 weeks at 0.75-1.0 kg/ha when the parasitic weed *striga* is a problem, follow rotation of crops.

Water Management

Often fodder sorghum is grown as a rainfed crop. However, with irrigation, the crop can be sown any time of the year excepting winter season. Irrigated crop yields very heavily. During the hot dry months, the irrigated crop must be given water at least every fortnight. Usually, five to seven irrigations are required. For rainfed crop too, if it is planted late, irrigation may become necessary. In the post-monsoon period, irrigation may be provided once in every three weeks, if sufficient soil moisture is not present. Water stagnation must be prevented in the rainy season and drainage facilities need be provided to drain out excess water.

Harvesting and Uses

Fodder sorghum are very palatable, especially in the young and flowering stages. At the flowering stage, the herbage contains low amounts of HCN. The crop at this stage can be safely grazed or cut and fed. Therefore, harvesting is normally done at 50 per cent flowering stage. Harvesting beyond this stage is not advisable, because the crude protein content and digestibility of nutrients decline sharply after this stage, while crude fibre content increases. Palatability also suffers, as the stems become hard and woody. In multi-cut types, the first harvest is taken after two months, and subsequent cuts at an interval of 35-40 days. The crop produces 30-50 t/ha from a single cut crop depending upon the receipt of rainfall and land and crop management from the multi-cut types, yields may be in the range of 50-85 t/ha. The fodder contains about 7.8 per cent crude protein and 32.3 per cent crude fibre.

The crop is mainly utilized as silage and hay as damages from HCN injury are reduced. Sorghum is one of the best crops for silage because of its high yields, sugar content and juiciness of its stalk. For silage making, sorghum is harvested at the milk or soft dough stage, as silage from mature plants keep better.

Fodder Bajra (*Pennisetum typhoides*)

It is an important millet crop suited to adverse conditions of weather, and provides staple food for the poor in the arid and semi-arid tracts of the country. Bajra hybridizes freely with napier grass (*Pennisetum purpuream*). Bajra-Napier hybrid popularly called hybrid Napier, having the good characters of both plants, it is an important fodder grass in the country.

Climate and Soil

Fodder bajra tolerates drought as well, and is grown extensively in the semi-arid tracts of the world without irrigation. It is grown even in areas with a low rainfall of 250 mm, but performs better with a rainfall of 400-750 mm. Some of the roots may penetrate to 360 cm depth, although 80 per cent of the root weight is in the top 10 cm. It can be grown on a wide variety of soils, but performs best in well-drained fertile sandy loams and loams. It cannot tolerate water logging.

It is grown as a rainfed crop in Kharif or irrigated crop during the summer season. In India it is mainly a Kharif crop sown during June-July.

Cultivars

Cultivars of bajra include 'Giant bajra', 'L-72', 'L-74', 'TNSC-1' and 'Rajko'.

Establishment

Simple land preparation with a country plough or tractor drawn plough is enough as deep cultivation is not required. Normal seed rate followed is 8-10 kg/ha. Bajra can be sown broadcasted, drilled or sown behind the plough. When sown behind the plough or drilled, a spacing of 30-40 cm between rows is given. When grown irrigated, three irrigations are usually necessary.

Manures and Fertilizers

Fodder bajra requires a fertilizer dose of 40:30:20 kg/ha of N,P and K in addition to 10-12 t/ha of farm yard manure.

Weeding

Normally, weeds may not be a problem, as the crop develops rapidly with vigour. However, if heavy weed growth is expected in the early stages of establishment, herbicides can be applied. Pre-emergence application of Atrazine at 0.5 kg/ha on the same day of seeding will take care of germinating weeds.

Water Management

Often fodder bajra is grown as a rainfed crop during kharif. Water stagnation must be prevented in the rainy season and drainage facilities need be provided to drain out excess water. However, when grown irrigated during summer three irrigations are usually necessary.

Harvesting and Uses

The crop is ready for harvest by 60-75 days. It is harvested at the boot stage or at 50 per cent flowering stage. A mean green fodder yield of 25-30 t/ha is expected. Fodder bajra is usually cut and fed in the stalls as green fodder. Hay and silage are also made. The herbage contains crude protein in the range of 6.8-12.8 per cent and crude fibre in the range of 29-34 per cent.

Teosinte (*Euchlaena mexicana*)

Teosinte crosses freely with maize and the hybrids are fertile with normal meiosis.

Climate and Soil

Teosinte requires a hot humid climate with an annual rainfall of about 1000 mm or more. The crop prefers well drained alluvial loams and clay loam soils.

Cultivars

'Sirsa', 'Rahuri', 'Maizente', 'TL-1' and 'Improved teosinte' are some of the cultivars recommended for cultivation.

Establishment

The land is prepared by two or three ploughing and beds and channels are formed. Seeds are broadcast or dibbled at a spacing of 30 cm between rows and 15 cm between plants. Drilling by *pora* or *kera* method is widely used. The crop can be sown in May-June or September-October. Normally a seed rate of 40 kg/ha is used. Transplanting is also sometimes practiced.

Manures and Fertilizers

At the time of land preparation, farmyard manure at the rate of 10t/ha may be added. In addition, a basal dose of N,P and K 30:30:25 kg/ha is also required. An additional dose of nitrogen at the rate of 35 kg/ha shall also be given as top dressing.

Weeding

Weeds may be a problem in the early stages, which can be kept under control by applying pre-emergence herbicide, Simazine or Atrazine at 1.0 kg/ha combined with hand weeding.

Harvesting and Uses

Teosinte has quick regrowth ability after cutting. Early sowing provides two cuttings. The best stage of harvesting is 7-10 days before the appearance of the tassel. Herbage yield will be in the range of 40-45 t/ha. The nutritive value of the forage is reported to be in the range of 4.5-12.0 per cent crude protein and 19.6-32.2 per cent crude fibre, depending upon the stages of cut and growth conditions. Teosinte is often cut and fed green in the stalls, but makes good silage as well. Hay can also be made. For silage, harvesting must be done when the tassels are fully formed and the seeds are at watery to milk stages. From teosinte, wilted silage is usually made. For this, the moisture content of the succulent herbage has to be brought down to 65 per cent before ensiling.

Fodder Oat (*Avena sativa* L.)

It is the most important and ideal annual (winter) cereal fodder crop due to its wide adaptability, excellent growth habit, quick regrowth, better nutritive value, more palatability and digestibility and higher yield. It is quick growing, palatable, succulent fodder crop. Paired row planting (20/40 cm) of oat intercropped with cowpea (*Vigna sinensis* L.) (1 row) has been reported to give very good green forage yield (59.73 t/ha). Oat is sown with pea, mustard also.

Climate and Soil

It is a crop of temperate region, and in India, it is grown during the Rabi season. The crop can be grown under rainfed as well as irrigated conditions.Oat requires loamy to clay loam soil.

Cultivars

'Kent', 'OS-6', 'OS-7', 'OL-9' 'OL-125', 'UPO-212 and 'JHO-822'(single/double cut); and 'UPO-94', 'JHO-86', 'JHO-851', 'PO-3' (multi cut) and RO-19 (multicut).

Establishment

The land is prepared by two or three ploughing and beds and channels are formed. Sowing is done from the first week of October to December to supply green fodder from December to end of April. The early crop provides two cuttings and fodder during lean period in December-January. Sowing is done by drilling or in furrows. Sowing is done at distance of 25-30 cm from row to row. Seeds are sown at the rate of 100-125 kg/ha.

Manures and Fertilizers

The crop is manured with 20 tonnes of FYM. The crop responds well to fertilizers, nitrogen 80 kg/ha (2 split doses, 40 kg basal and 40 kg, 40 days after sowing), phosphorus 30 kg/ha and potash 20 kg/ha is generally applied. For double cut crop another dose of 40 kg urea is applied after the first cut.

Weeding

Weeds may be a problem in the early stages, which can be kept under control by applying pre-emergence herbicide, Simazine or Atrazine at 1.0 kg/ha combined with hand weeding.

Harvesting and Uses

Teosinte has quick regrowth ability after cutting. Early sowing provides two cuttings. The best stage of harvesting is 7-10 days before the appearance of the tassel. Herbage yield will be in the range of 40-45 t/ha. The nutritive value of the forage is reported to be in the range of 4.5-12.0 per cent crude protein and 19.6-32.2 per cent crude fibre, depending upon the stages of cut and growth conditions. Teosinte is often cut and fed green in the stalls, but makes good silage as well. Hay can also be made. For silage, harvesting must be done when the tassels are fully formed and the seeds are at watery to milk stages. From teosinte, wilted silage is usually made. For this, the moisture content of the succulent herbage has to be brought down to 65 per cent before ensiling.

Cultivated Grasses

Grasses are cultivated for grazing as well as for cut-and –carry systems. In India, several grasses have been introduced from tropical countries for cultivation and now it is an important class of forage crops. Cultivated grasses include both annual and perennial types. In general, annual types are propagated by seeds only. However, an addition to seeds, perennial types are usually amenable to vegetative propagation too. Most of the exotic grasses introduced are heavy yielders with good herbage quality. Nevertheless, proper land and grazing management is essential to exhibit their full production potential. Cultivated grasses also belong to the same family as that of cereals, Poaceae (Gramineae).

Guinea Grass (*Panicum maximum*)

Guinea grass is the most popular fodder grass grown under irrigated conditions in India, because of its wide adaptation, quick growth, ease of establishment, palatability, herbage yield, good persistence and good response to fertilizers. Guinea grass can be conveniently and profitably grown as a component of agroforestry systems and comes up well under Subabul. It can be grown together with leguminous plants such as cowpea and stylo.

Climate and Soil

Guinea grass comes up well under humid tropical conditions. In the tropics, it can be grown in regions where the temperature ranges between 15°-38°C. It does not tolerate severe drought and requires a rainfall in excess of 1000 mm per year. Irrigation may become necessary, if dry period persists. Although it can tolerate partial shade, it does not come up well under continuous shade. It grows best on well-drained, light textured soils-preferably sandy loam or loams with medium to high fertility. It can not tolerate heavy clays or prolonged water logging. Normally, it will not tolerate acid or saline conditions too. Nonetheless, wherever, drainage is satisfactory, it can tolerate acid conditions.

Cultivars

Altogether, there are four groups of cultivars.

Common Guineas

These are tall tufted cultivars growing to about 2.0 m when not grazed, eg., 'Risversdale'. Common guineas are most suitable for tropical low lands with high rainfall, especially more than 1300 mm per annum. Common guineas are suitable for soilage, silage, hay and some times grazing. Some cultivars can be grown under shade especially under coconuts but may compete with the trees unless well fertilized and managed.

Coarse Guineas

Coarse guineas include tall and robust cultivars. They grow taller than guineas, *e.g.*, 'Hamil'. Like common guineas, these are also suited to wetter areas, but best used in non-grazing situations and in cut-and-carry systems.

Creeping Guineas

These are fine-stemmed short guineas with a prostrate creeping habit, eg., Embu'. However, it is not recommended for general use because it requires very good management and is unlikely to persist under heavy grazing and poor management.

Green panic (*P. maximum* var. *trichoglume*). It is a shorter, fine stemmed botanical variety of guinea grass. Green panics are small or low growing types with fine stems up to 1.0 m high and short leaves suitable for grazing. The 'panics' are, in general, suitable for the sub-tropics and elevated moist tropics with moderate tolerance to drought, frost and shade. They are mostly established from seeds and include cultivars such as 'Petrie', 'Sabi' and 'Gatton'. Although more suited for drier areas, it is not recommended for general cultivation because of cost and difficulty.

Some of the of the most important cultivars of guinea grass cultivated in India are:

'Makueni', 'Hamil', 'Riversdale', 'Coloniao' and 'Marathakom'

Establishment

Under natural conditions, guinea grass spreads slowly by seeds. However, when cultivated, propagation can be by slips or by stem cuttings in addition to seeds. Rooted slips are preferred for summer planting as the stem cuttings may dry up in the dry weather. To obtain slips, old clumps are uprooted and slips with roots are separated. Normally, 1.25 lakh of slips are required for one hectare. One hectare will provide material for five hectares of planting.

When seeds are' used, a seed rate of 3-6 kg/ha is required depending on the seed quality. However, 'Hamil' normally requires only 1-2 kg seeds per hectare. Drilling or dibbling on the contour in small drill furrows and pressing in with press wheels gives an excellent stand. Sowing depth or coverage should not be more than 1.5 cm. Germination may be better by using a straw mulch to cover the surface-sown seeds. Transplanting of guinea seedlings is more reliable. When transplanting is done, a seed rate of 3.0 kg/ha shall be enough. Nevertheless, as the ears ripen unevenly, seeds shed early and viability is low, seed production in Guinea grass is a difficult task and vegetative propagation is more reliable.

Guinea grass requires a weed free seedbed for establishment. Land preparation involves two or three ploughing and leveling. Planting is done on ridges for better drainage and for better irrigation. Firstly, small trenches of 10 cm width and 20 cm depth are made. In these trenches, farmyard manure is applied along with phosphorus and potassium fertilizers and mixed well with the soil. Cover the trenches and form ridges of 15-cm height. In acid soils, it is desirable to apply lime at 500 kg/ha. Seedlings or slips are planted on the ridges at three numbers per hill. Give a spacing of 60 x 30 cm when grown as a pure crop or as Guinea grass-legume mixture. Although the grass is a perennial, one-third replacement once in three years should be followed to maintain the vigour of sward.

Manures and Fertilizers

The crop responds to heavy manuring. Organic manures, either farm yard manure or compost, at 10 t/ha and $N:P_2O_5:K_2O$ at 200:50:50 kg/ha per year is recommended. The entire quantity of organic manure must be applied basally in the trenches along with full P_2O_5 and K_2O and incorporated. For the rainfed crop, apply N at 200 kg/ha in two splits, immediately after first cutting and during the Northeast monsoon period. Apply it on either side of the plants along the row and earthed up. However, for the irrigated crop, more splits of nitrogen should be given. For the irrigated crop, N at 40 kg/ha shall be applied basal, followed by top dressing of N at 30 kg/ha after each harvest.

Guinea grass can be grown with sewage irrigation or barnyard washings; and in such cases, separate addition of organic manures is not needed. However, for the decomposition of the accumulated organic debris from the sewage, frequent interculture is necessary.

Weeding

Initially, the fields may be kept weed free for about two months. During this period, two inter cultivations may become necessary for weed removal. Pre-emergence application of atrazine at 1.0 kg/ha is also effective for early season weed control. After three or four cuttings, inter cultivations are again required for loosening the soil and for weed removal. Spraying 2,4-D at the rate of 1.0 kg/ha about 5-6 weeks after planting is effective for the control of broadleaf weeds.

Water Management

Guinea grass is usually planted in May-June along with the receipt of pre-monsoon rains. If sufficient rains are not received within 7-10 days after planting or if the crop is being raised as an irrigated crop, two irrigations are required for quick establishment. Subsequently, irrigations may be given depending upon rainfall and soil type. Usually, irrigation once in 7-10 days is required. In soils with good water holding capacity, irrigation once in 15 days may be enough. Guinea grass can be grown with sewage irrigation or barnyard washings. However, when the grass is grown with sewage irrigation, frequent interculture is necessary for the decomposition of the accumulated organic debris from the sewage.

Harvesting and Uses

Guinea grass is mainly utilized in a cut-and-carry system, though limited grazing is also practiced. Although it tolerates a good deal of defoliation, it should not be grazed or cut below about 15-20 cm. Otherwise, it recovers slowly only. Rotational grazing gives better control of pasture growth rather than continuous grazing. Under extremely wet conditions, grazing is not feasible as trampling may damage the pastures.

The crop may become ready for harvest 9-10 weeks after planting; and at this stage, the plant may have reached a height of 100-150 cm. After a harvest at this stage, subsequent cuts are made in every 45-60 days. The grass should not be allowed to seed before harvesting. About 7-9 harvests are possible in a year. To maintain yield, one-third or one-fourth of the plants should be replanted every year. Total forage yield may range based on factors such as cultivars and soil conditions. Nevertheless, a mean yield of about 80-100 t/ha of green fodder per yea: can be expected. Certain cultivars are capable of giving herbage yield of 140 t/ha.

Guinea grass is palatable to all kinds of livestock, and it is usually fed directly as cut fodder It also makes good hay and silage. It is nutritious and free from oxalates, and contains 8 -14 per cent crude protein and 28-36 per cent crude fibre.

Napier and Hybrid Napier (*Pennisetum* spp.)

Napier grass (*Pennisetum purpureum* Schum.) is a native of Zimbabwe in tropical Africa. It is a tall clumped grass with thick growth, which gives the name *elephant grass.* Its peculiarity is its high herbage yield. The name Napier grass was given in honour of one Col. Napier of the Rhodesian Department of Agriculture, who was responsible for its development. It was introduced to India in 1912. With the popularisation of the interspecific hybrid, Bajra x Napier (*Pennisetum americanum* x *Pennisetum purpureum]*, Napier grass lost its significance as a fodder grass in India.

Hybrid Napier can be intercropped with a number of crops. Legume intercropping improves the quality of forage. It mixes well with calopo, centro and glycine. In North

India, when the crop becomes dormant during winter, it can be intercropped with berseem, Indian clover, peas, fodder oats or Chinese cabbage. In the summer, with irrigation, cowpea, rice bean, velvet bean or cluster beans may be grown. In the Kharif season, it may be intercropped with maize or cowpea.

Climate and Soil

In the tropics, Napier grass grows through out the year. Light showers alternated with bright sunshine are very congenial to the crop. Planting can be done throughout the year, provided irrigation facilities are present. However, the usual time of planting is by the onset of South West monsoon.

Napier grass can grow on a variety of soils. However, sandy loams to clay loams are preferred. *Deep fertile* soil, rich in organic matter and nutrients are ideal. Water logged and flood prone areas are not suitable. It can tolerate a pH ranging from 5 to 8.

Cultivars

Seyeral hybrid cultivars were released for cultivation in India. Pusa Giant, Co1, Co-2, Co-3, IGFRI-3, IGFRI-6, IGFRI-7, IGFRI-I0, NB-5, NB-17, NB-21, NB-35, Gajraj and BN-6 are some of the cultivars recommended.

Establishment

Only vegetative propagation is possible, as the plant is a sterile hybrid. Rooted slips or stem cuttings (setts) are used for propagation. Cuttings from moderately mature (about three months old) stems, and from the lower two-third of the stem length sprout better than the older stem. Stem pieces for planting are made with at least three nodes.

The crop requires a deep, weed free, compact seedbed. The land will be ready with three to four ploughing followed by a harrowing. Incorporate organic manures at the time of land preparation. Phosphorus and potassium fertilizers are also applied basal at the time of land preparation. Provision must be given for drainage during rainy season, as the crop cannot withstand water logging. Hybrid Napier is planted in the same way as sugarcane. Planting is done by the onset of South West monsoon. The setts are planted in furrows made by a plough at a spacing of 60 cm x 60 cm, or 50 cm x 50cm. In situations where plough is not used, it can be planted in small pits taken in rows. In intercropping systems, spacing must be adjusted to accommodate the companion crops; and usually 200 cm x 50 cm, or 100 cm x 50 cm is given depending upon the cropping scheme. Total requirement of setts under 60 cm x 60 cm spacing is 27,800 per hectare. The stem pieces are planted either vertically or at an angle, burying two nodes inside the soil, exposing the third above the soil surface. The cuttings may also be planted in furrows about 15 cm deep, initially covering with about 7.5 cm of soil. The furrow gets filled gradually as the plant grows.

Manures and Fertilizers

The crop must be manured heavily with organic manures. Farmyard manure or compost at the rate of 25 t/ha is recommended at the time of land preparation. However, the organic manure application can be dispensed with, if sewage irrigation or irrigation with barnyard washings is possible. As the grass is a heavy yielder, it responds well to nitrogen application. For rainfed crops, nitrogen may be applied in two or three splits at 200 kg/ha followed by gentle ranking. Under irrigated conditions, apply nitrogen at 40 kg/ha as a basal dose. Subsequently, nitrogen must be given at 30 kg/ha after each herbage harvest. However, phosphorus and potassium must be given as basal dose (50 kg/ha each) at the time of land preparation.

Weeding

Weeding and inter cultivation is required in the early stages. Initially, inter-cultivation once or twice is necessary for better establishment. Later, inter-cultivation may be given as and when necessary after harvests to stir the soil and to control weeds, if any. After the establishment of full canopy coverage, normally it does not require weeding. Occasionally, decayed tillers from the clumps must be removed, and the field must be maintained clean and healthy. Herbicides have limited applicability. If broadleaf weeds are a problem, 2,4-D can be applied at 1.0 kg/ha.

Water Management

Hybrid Napier requires irrigation to express its full biological potential. If sufficient rains are not received, about two irrigations are required for the establishment. During the summer period, regular irrigation at fortnightly intervals is required. In the rainy season, the field should be provided with drainage facilities by way of channels or furrows as the crop cannot withstand water logging.

Harvesting and Uses

Hybrid Napier is commonly used in a cut-and-carry system. The crop will be ready for the first harvest by 75 days after planting. Subsequent cuts can be done at an interval of 30-45 days, or when the plants attain a height of 1.5 m. The cultivar'Co-3' can be harvested at a lesser interval, say 25-30 days. About 7-10 cuts are possible in a year. In the winterless South India, with good management, not less than 10 cuts are usually made. The cut should be made retaining at least 10-15 cm stubble. However, cutting in rainy periods requires caution. If the rainwater gets accumulated at the cut ends, it may rot. Avoid harvesting in days of heavy downpour. Usually, an annual mean green fodder yield of 200-250 t/ha is obtained.

Hybrid Napier is superior in quality than Napier grass, and usually contains about 10.2 per cent crude protein and 30.5 per cent crude fibre'. It is also more palatable than the Napier. Some cultivars may contain high oxalate content, which reduces the quality of the fodder. However, the oxalate content gets reduced by harvesting at longer intervals, usually 45 to 60 days. The grass is good for soilage, silage and hay. Before making silage or hay, the fodder is usually chopped and made into small pieces.

Para Grass

Other Names

Buffalo grass, Water grass, Angola grass, Mauritius signal grass) : *Brachiaria mutica* (Forsk) Stapf.

Para grass is a perennial grass widely grown throughout the tropics and sub-tropics. It is a water loving grass growing well in moist damp, well developed soils, where other grasses are not suited. Para grass is popular because of its ease of propagation from cuttings, easy and rapid establishment, high yields, good quality and palatability, and tolerance to waterlogged and flooded.

Climate and Soil

Para grass grows best in the hot and humid climate of the tropics and sub-tropics where rainfall is above 1000 mm. It does not tolerate shade. It is suited for growing on lower terraces where water stagnates in the rainy season and grows luxuriantly in warm moist condition. In winter it remains dormant. It is drought resistant and survives also during the cold winter and hot summer season on uplands.

It prefers alluvial and hydromorphic soils. Nevertheless, it grows on a wide range of moist soil types excluding highly wet soils. It thrives better on fertile clay and clay loam textured soils with high moisture retention capacity. It can grow even on sandy soils with good irrigation facility. It tolerates slightly acid to alkaline soils, and highly tolerant to saline or sodic soil conditions. Hence, it is used as an excellent grass in soil reclamation. Apart from these, it grows well on field bunds, banks of streams and canals, lowlands, and soils too wet for normal farm crops.

Establishment

It is propagated vegetatively by stem cuttings or rooted slips. Stem cuttings or rooted slips with three joints are prepared and fixed in the soil after land preparation in a slanting condition to establish easily. Two buds are covered in the soil and one bud is allowed above the ground.

Manures and Fertilizers

It gives high yield under organic manures. 30 tonnes of F. Y. M. are applied along with 30 kg P_2O_5 and 30 kg K_2O to be given as basal application. In addition, top dressing of nitrogen at the rate of 20 kg/ha must be given after each harvest.

Weeding

Para grass competes successfully with weeds. Being an aggressive creeping grass, once established, it can suppress weeds very well. However, for the initial two months after planting, the land should be kept weed-free by way of intercultivation once or twice.

Water Management

The crop responds well to sewage irrigation. Para grass is grown in many sewage farms set up near cities. Normally, two to three light irrigations are required for the initial establishment of the crop. In the later stages, irrigation may be given once in 10-15 days. If the crop is grown in a moist environment, irrigation is not normally required.

Harvesting and Uses

The grass becomes suitable for harvesting in 8-10 weeks after planting. Subsequent cuts are taken at 4 - 6 weeks interval during rains. The crop yields about

120 t/ha with 10 cuttings in south India. However, in North India, because of winter, only 4-5 cuttings are possible yielding about 70 t/ha. Under irrigated condition yields are obtained from spring season to autumn. After every cutting, 20-25 kg. N is given when cattle yard washing or sewage irrigations are not available. In the pre-bloom stages, the crude protein content ranges from 5.9 to 15.4 per cent and crude fibre from 27.2 to 41.5 per cent.

Para grass stems and leaves are very palatable when young. It is mainly used for hay making. Silage making is also rare. However, in a "cut-and carry" system, hay and silage can be made. Grazing and cutting can seriously deplete para grass stands.

Deenanath Grass (*Pennisetum pedicellatum Trin.*)

Deenanath grass grows on a wide range of soil types ranging from unfertile, undulated hill slopes to fertile level lands in warm humid rainy and autumn seasons. Under irrigation it also grows luxuriantly during the spring and summer seasons. This grass is very leafy, thin stemmed and platable to animals at all stages of growth. In dry lands, this grass gives higher yields than different Jowar varieties grown for fodder purpose.

Climate and Soil

The grass requires a warm habitat and is fairly drought tolerant. They thrive well in relatively dry and moderately humid areas with a mean rainfall of 500-1500 mm. It is able to grow in marginal wastelands with moderate biomass yield. The grass can tolerate both acidic and alkaline soils. It is also tolerant of soil salinity, and reported to tolerate an electrical conductivity value of up to 12 DS/m. As a rainfed crop, seeds are sown with the onset of South West monsoon in May-July. With irrigation, seeds may be sown in early March to have 2-3 cuttings.

Cultivars

The cultivars identified in India for general cultivation include: PS-3, IGFRI-4-2-1, IGFRI-43-1, IGFRI-3808, and Pusa Deenanath.

Establishment

This grass is easily established by sowing seeds in the beginning of monsoon.Two or three shallow ploughing by an indigenous plough or digging by a hoe to a depth of 10-15 cm is enough to prepare a seed bed. A population of 30-40 no./m^2 is retained under broadcast sowing. Sowing is done by broadcasting or drilled in lines 20 cm apart. It is mixed with slightly moist, powdered soil for uniform distribution. Usually, the seed rate is 30-40 kg/ha seeds with fluffs (about 7.5 kg kernel). Pasture under deenanath grass, though dry up in summer, revives with the first monsoon showers and reseeding must be done after every three years.

Manures and Fertilizers

For higher fodder yields 10 tonnes of F. Y. M. per hectare is applied at the time of land preparation. It responds to N,P and K application. Nitrogen shall be applied at 60 kg/ha (half basal + half top dressed after 6-8 weeks growth) and P_2O_5 and K_2O may be applied at 30 kg/ha each as basal dose. First harvesting is done after 80 - 90 days in the end of September when sown in June or early July. Next harvest is done after two months.

Weeding

As the crop growth is not fast in the early seedling stage, weeding is needed in the initial two months of sowing. Application of Atrazine at 1.0 kg/ha as pre-emergence is recommended for weed control.

Water Management

It is grown as monsoon season crop and rarely irrigated. During the rainy season, sufficient drainage must be given to drain excess water in the soil. The soil should remain moist but not wet. If the crop is sown in early March, irrigation may become necessary for the establishment depending upon the receipt of rainfall.

Harvesting and Uses

The crop flowers by September in South India and start senescence by mid November. In North India, it flowers by October-November. The first cut gives good vegetative yield at 70-100 days and subsequent cut coincides with flowering stage. Clipping should be done about 10 cm above the ground level to encourage better regeneration. Usual interval for cutting is 35-45 days. For an earlier sown crop, three cuttings are possible. Average yield from a single cut is in the range of 40-50 t/ha. The crop may give a total green fodder yield of 120 t/ha. It is generally used as green forage. However, high quality hay and silage can also be made. For making hay, harvest at the boot stage, dry in thin layers in the sun for a day and then in shade of trees to preserve colour. Wilted silage is usually made with deenanath grass. For this, harvested fodder is first wilted up to 65 per cent moisture content; and if possible, mixed with legumes before ensiling.

Congosignal Grass (*Brachiaria ruziziensis* Germain and Everard)

It is a perennial grass with stiff erect bottle green leaves. It gives a bushy growth and completely covers the ground, growing to a height of 2-3 ft. This grass is found to remain green long after rainy season, although in dormant condition. Congosignal plus *Stylosanthes* is a good combination.

Climate and soil

Congosignal prefers a warm and moist tropical climate receiving rainfall above 1000 mm. Nonetheless, it has good drought tolerance. It can be grown in almost all types of soils, but requires high soil fertility conditions. It tolerates acidity, but cannot tolerate waterlogging and needs drainage. It can tolerate shade to an extent.

Establishment

Land is prepared by ploughing and cross ploughing. The ideal time for planting congosignal grass is May-June and September-October with the onset of rains. Both seeds and slips can be used. A seed rate of 2-5 kg/ha is recommended. The seeds are sown at 1-2 cm depth. To protect the seeds from ants, dusting carbaryl 5 per cent DP or methyl parathion 2 per cent DP in the soil at the time of sowing is effective. When slips are used, they are planted at a spacing of 40 cm x 20 cm.

Manures and Fertilizers

Organic manures such as FYM may be applied at 5-10 t/ha at the time of planting. At the time of planting, basal application of 50 kg/ha each of P_2O_5 and K_2O is also recommended. Nitrogen at 100-150 kg/ha must be applied in two or three splits after defoliation followed by gentle raking for rapid growth and high biomass yield.

Weeding

In the early stages, weeds may compete with the crop and may delay its establishment. One or two inter cultivation during early growth stage is advisable to check weed growth. After establishment, being a tufted creeping perennial, it covers the soil, and forms a dense mat under grazing.

Water Management

It is seldom grown with irrigation. It is a rainfed crop. Nonetheless, if irrigation is given, yield will be doubled. In summer months, irrigation may be given at fortnightly intervals.

Harvesting and Uses

It will be ready for grazing for the first time 50 days after planting and subsequently at 30-40 days interval. Occasionally, silage and hay are also made. It has been reported that, it makes very good silage with Stylo with 1.0 per cent molasses and without any other additives. Under average management, a rainfed crop yields about 35-45 t/ha of green fodder, whereas, the yield may increase to about 50-100 t/ha under irrigated conditions. The herbage is very palatable and contains about 13 per cent crude protein and 27 per cent crude fibre. Congosignal grass is also utilized for soil erosion control in sloppy areas, as it forms a dense mat, and prevents the soil from getting eroded away. Usually, it is grown in strips for this purpose.

Buffel Grass (*Cenchrus ciliaris* L.)

Buffel grass is a perennial grass and is also known as Anjan grass, African foxtail, Dhaman grass and Kusa grass. It has a deep and clustered root system, which holds soil in its place. Because of its soil binding property, buffel grass is usually used for rehabilitating sand dunes and highly eroded areas.

Climate and Soil

It is suited for low rainfall areas especially below 1000 mm. It is able to survive prolonged drought and grows vigorously when favourable conditions set in. It is usually grown as rainfed crop suited to pasture and range lands. It is one of the dominant grass species of hot deserts of India. It can be grown in almost all types of soils.

Cultivars

Often, buffel grass cultivars are divided into three groups.

Tall Rhizomatous Types

These are tall, grow up to 1.5 m high, vigourous, rhizomatous, hard stemmed, and suitable for cattle grazing, *e.g.*, Biloela, Molopo and Lawes.

Medium Non-Rhizomatous Types

These are medium in height, grow upto 1.0 m, non-stoloniferous, with finer and softer stems, and suitable for sheep grazing, *e.g.*, Gayandah, American

Short Non-Rhizomatous Types

These are short, non-rhizomatous types, *e.g.* West Australian

Nevertheless, cultivars having characteristics of all the above types are available. The cultivars recommended for cultivation in India include Bundel Anjan (IGFRI-3108), Marwar Anjan (CAZRI75) and Blue Anjan (Co-I). Some of these are evolved from crosses between the above types.

Establishment

Buffel grass is propagated mainly through seeds. The fluffy seeds are broadcast after mixing with moist soil on a fine seedbed with the break of monsoon in May-June. It can also be sown shallow in furrows by *Kera* method. Seed rate shall be 5-6 kg/ha in sandy soils and 7-10 kg/ha in loam to clay loam soils. It is surface sown and lightly harrowed or rolled wherever possible. Rolling, trampling by man or cattle after sowing improves germination. Usual spacing ranges from 60 cm x 30 cm to 75 cm x 75 cm depending upon soil fertility, soil type and rainfall.

The seeds may be soaked in water for about eight hours before sowing to remove the germination inhibitors. For smaller areas, transplantation of seedlings produced in nursery or direct vegetative propagation from turf splits or rhizome can also be done.

Manures and Fertilizers

In semi-arid environment, fertilizers are not normally applied. However, both organic manures and fertilizers improve the productivity of the grass. Organic manures, either farm yard manure or compost at 10 Mg/ha is given at the time of land preparation. Nitrogen fertilizer at 40 kg/ha is given as a basal dose and at 20 kg/ha after every harvest. Phosphorus and potassium fertilizers are applied at 20 kg/ha each at the time of sowing and repeated every year.

Weeding

In the initial stages, one or two weeding or interculture may be given for easy establishment.

Subsequently, an intercultural operation may be done during summer to stir the soil, to stimulate the growth of the grass and to control the established weeds.

Water Management

Buffel grass is usually raised as a rainfed crop. It does not tolerate waterlogged conditions and hence, good drainage is essential during the rainy months. However, if possible, light irrigations can be given during summer months, which facilitate quick recovery and faster growth.

Harvesting and Uses

Buffel grass withstands considerable grazing pressure once established. The grass will be ready for utilisation about 9-12 months after sowing. However, it normally takes about two years for full establishment. No grazing shall be allowed in the first year. However, light grazing can be done in the second year. From third year onwards, rotational grazing shall be practised. In a cut and-carry system, only one cut is allowed during the first year. In the second year, three cuts can be taken. Six cuts are, however, possible in subsequent years. Height of cutting is also important. Close cutting or grazing at 5-10 cm above ground level gives higher yields and better herbage quality than cutting at higher levels.

The grass is very palatable when young, and remains fairly palatable at maturity. The crude protein content ranges from 7-11 per cent and crude fibre from 31-42 per cent. A mean yield of 20 to 30 Mg/ha is expected from third year onwards. Buffel grass is ideal for soilage, grazing, ensiling and hay.

Setaria Grass (*Setaria sphacelata*)

Other Names

Golden Thimothy, Golden bristle grass) *etaria sphacelata* (Sch.) Stapf. and Hubb (*S. anceps* Stapf.)

It was first introduced in India in 1950 through the Indo- Swiss Project. It is well adapted to subtropical regions with annual rainfall exceeding 750 mm. Setaria is palatable, utritious, establishes easily from seeds, and persists under grazing on a wide range of soils. Although some cultivars contain high oxalate content, it usually gives little trouble than similar asses. Setaria can be grown either alone or in mixtures with perennial grasses and legumes.

Climate and Soil

Setaria grow vigorously under high annual rainfall ranging from 1000 to 1500 mm. It can also survive long, hot and dry seasons. The grass grows well at 20° to 25°C. Setaria shows some cold tolerance and gives early spring growth. It is more cold tolerant than most other tropical and subtropical grasses. It, therefore, establishes well in the high ranges. It comes up in a variety of soils. It is fairly drought tolerant and withstands water logging. It does not tolerate shade.

Cultivars

Nandi, Kazungula, Narok and Splenda are the major cultivars. Kazungula and Splenda are adapted to the wet tropics.

The cv. Nandi, originated in the highland Nandi district in Kenya, has relatively low oxalate content, while cv. Kazungala is high in it. 'Nandi' is superior to most others for the high rainfall areas at high altitude. It tolerates waterlogging and flowers earlier than cv. Kazungula, and if ungrazed, flowers through the season and becomes stemmy.

Ungrazed, flowers through the season and becomes stemmy. It is cross-pollinated and seeds we It establishes less readily than cv. Narok and cv. Kazungula.

The cv. Splenda is a tall robust perennial, which has broader leaves and lo[illegible] spikelets thar other cultivars, with flowering stems reaching upto 2.8 m height, [illegible] adapted to the higlrainfall tropics and sub-tropics.

'Kazungula' is from Zambia developed for grazing, and is more vigorus, very tall and leafy with thick stems and broad bluish green leaves. 'Kazrmgala' is coarser and more robust than 'Nandi', and is a tetraploid. It has high sodium content. Cattle accept stubble grazing with 'Kazrmgula' more readily than with 'Nandi'. The seed-heads are lighter than 'Nandi', and tend to bend. The sheath of basal leaves have blue-green coloration, and the stigmas are purple. Seeds are slightly smaller than that of 'Nandi'. 'Kazungula' is the most tolerant of poor sandy and stony soils, which tolerate drought conditions as well.

The cv. Narok was collected from the Aberdare Mormtains, Kenya. It is more robust than 'Nandi' but less so than 'Kazungula'. It is greener in colour than both, and some plants lack the red pigmentation at the base common to 'Nandi' and 'Kazrmgula'. The inflorescence is rustcoloured and the seeds larger than 'Nandi'. It is also a tetraploid. 'Nandi' and 'Narok' prefer medium-textured, fertile soils.

Establishment

Propagation in setaria grass is usually by rooted cuttings or divided root stocks or by seeds.

Mainly, rooted slips (single tillers with roots) are used. Seedlings can also be raised in a nursery and transplanted during the rainy season. Land is prepared for planting by two to three ploughing or digging followed by a levelling. Weeds should be removed and the land should be free of stubbles. If irrigation facilities are available, planting can be done anytime of the year except winter. As the seed is small, it should be sown no deeper than 1.8 cm for cv. Kazungula and 2.5 cm for cv. Nandi; then it is lightly covered and rolled.

When grown as a pure crop, it is planted at 50 x 30 cm spacing. For intercropping with legumes, 100 x 30 cm spacing is followed. For seed propagation, a seed rate of 3.5 to 4.0 kg/ha is used. In the case of rooted slips, the number varies between 33,500 to 67,000 per hectare.

Manures and Fertilisers

Organic manures either farm yard manure or compost at 25 t/ha is given at the time of land preparation. The crop responds well to fertilizers especially nitrogen. Nitrogen fertilizer may be applied at the rate of 40 kg/ha as a basal dose and subsequently at the rate of 20 kg/ha after every harvest. Phosphorus fertilizer is applied as basal dose at the rate of 40 kg/ha. Potassium application is essential when the K content in the plant falls below one per cent. In such situations, K may be applied at 40 kg/ha.

Weeding

The crop requires one or two weeding and inter culturing in the first two to three months after planting. Subsequently, to encourage fresh sprouts and to control weeds, one or two intercultivation is necessary every year after the cold season is over.

Water Management

It is difficult to grow setaria grass without irrigation. The grass grows luxuriously in moist soils. However, the fields should be well drained during the rainy season.

The crop requires two successive light irrigations at 7 to 10 days interval for establishment. Subsequent irrigations may be given as and when necessary depending up on the receipt of rainfall.

Harvesting and Uses

Setaria grows luxuriously with irrigation and manuring. It is able to withstand heavy grazing and forms a robust crown. However, early grazing may cause plants to be pulled up by the roots especially when the soil is moist. The grass should be lightly grazed until established, then heavily grazed to prevent it becoming stemmy and coarse. Under grazing causes the plants to become coarse, and it shades companion legumes. This is a usual problem with cv. Kazungula. Because of its erect growth habit, it may be more suited to cut-and-carry feeding systems. Under the cut-and carry system, first harvest may be possible by 70 to 80 days after planting. Subsequent cuts can be taken after every 40 to 50 days depending on crop growth. About 7-8 cuts are possible in a year. While harvesting, a stubble height of about 10 to 15 cm is left for good regeneration. The crop yields about 75-150 t/ha of green fodder per year.

Apart from grazing, the grass can be fed directly as green cut fodder or preserved as silage or hay. The grass is very palatable when young but less so as they approach maturity. The crude protein and crude fibre content of the grass ranges from 4.8 to 18.4 per cent and 24 to 34 per cent respectively.

An important drawback of Setaria is the high acidity of fresh herbage (pH 4.8). The high acidity is linked to high content of organic acids especially oxalic acid. The amount of oxalate varies with the cultivar and stage of growth. Young plants contain more oxalate content than older plants, and strains higher in nitrogen are also higher in oxalate. Oxalate content ranging from 3.7 per cent in cv. Nandi to 7.8 per cent in cv. Kazungula has been reported. Lactating cows and horses are more liable to be affected. Affected cattle may develop staggering gait and diarrhoea. Chances of poisoning are rare, but animals, especially lactating cows, should not be fed on young, luscious Setaria especially after a period of starvation.

References

Farooki, S.H., Kumar, S. and Singh, D.N. 2009. Chara Anusandhan Ke Naye Ayam. Kheti. 62 (7): 28-31.

Hazra, C.R. 1996. Improved cultivars for increased forage production. Indian Farming. 45 (10): 3-8.

Sarkar, R.K. and Mahasin. 2007. Productivity, quality and profitability of oat (*Avena sativa*)-based inter-cropping system under graded fertility levels. Indian J. of Agril. Sc. 77 (12): 862-5.

Thomas, C.G. 2003. Forage crop production in the tropics. Kalyani Publication. Ludhiana, India. Pp. 259.

Tripathi, S.B. and Hazra, C.R. 1996. Forage production on problem soils. Indian Farming, 45(10): 9-13.

Chapter 18

Improved Production Technology for Major Leguminous Fodders/ Forages

Aditya Kumar Singh[1], *Anil Kumar Singh*[2] *and Bishram Ram*[1]

[1]*College of Post Graduate studies, Central Agricultural University, Barapani Umiyam, Shillong*

[2]*ICAR Research Complex for Eastern Region, Patna – 800 014, Bihar*

E-mail: aksingh_14k@yahoo.co.in

Livestock is an integral part of rural life in India. After green revolution, it was the turn of white revolution and milk production in the country increased tremendously. This trend must be continued and sustained. A serious limitation in fully utilizing the production potential of improved animals is availability of quality forages. Forage, in broad sense, includes all the feed consumed by the livestock. However, principal forages are grasses in the family poaceae and legumes in the family leguminaceae. The importance of forage crop cannot be overlooked and every possible effort is needed to increase its availability in the country.

In dairy sector, cost of feed alone constitutes about 60-65 per cent, of the total cost of milk production. This cost must be brought down by 30-40 per cent, if quality roughages are available. Because of several constraints, most farmers are unable to feed their animals properly with good quality roughages. Green forages, consisting of both natural and cultivated forages, are the cheapest source of roughages. However, production of forage or fodder, which forms bulk of roughages, is still a sideline activity in India that is integrated with other areas of agricultural production.

Improved Agro-technique for Producing Ground Leguminous Fodder

The word legume is derived from a Latin word *Legere* which means to gather because the pods have to be gathered or picked by hands as distinct from reaping the cereal crop *viz.*, poaceae family. Legume plants produces and supply the major portion of world's plant protein. This is utilized directly by man through eating or indirectly by livestock feeding and subsequent consumption of livestock products. Fodder legumes are also refered as masals and have immense value in animal nutrition as they are nearly equal to concentrates and are likely to be substituted for the later.Legumes grown for forage purposes are usually classified into ground legumes and tree legumes. All the herbaceous types of legumes, which grow in close proximity with the ground, are called ground legumes.

Both annual and perennial types are present in ground legumes, and include erect, spreading, climbing or decumbent types. They contain high amount of protein and improving the quality of feed. Thus, they have a direct bearing on the level of animal production. Therefore, attempt should be made to include sufficient leguminous fodder in feed rations. In brief, importance of forage legumes are summarized below as:

1. Short duration in nature.
2. Can be raised as catch crop.
3. Improves soil fertility by way of BNF.
4. Multipurpose in uses.
5. Increase intake of fodder by improving fodder availability.
6. Capable of replacing concentrates in animal rations and save feeding costs.

Legumes can be grown as pure crop srparately, cut and fed along with grasses in mixtures or can be grown with grasses or cereal fodders in a right portion to get high yields coupled with high quality. The production technology of some of the important ground leguminous fodder cultivated for forage purposes on commercial scale is given below:

Berseem or Egyptian Clover (*Trifolium alexandrium* L.)

Because of its introduction from Egypt, Berseem is also known as Egyption Clover. It is a most important leguminous green fodder of winter (Rabi) season specially of northern India. It provides fodder with high tonnage over a long period from November to May in 5-7 cuttings. It is a most digestible and palatable green fodder to the cattle and milch cattle are specially benifitted with its consumption. It contains about 20-24 percent protein, 3 per cent calcium and 0.46 percent phosphate at the green stage along with 70 percent drymatter digestibility. It has very good soil restorer capacity and add about 0.4 to 0.45 per cent of organic carbon. It also leaves about 45-50 kg of available nitrogen and about 15-16 kg of available phosphorus in a hectare of land. Cultivation of berseem also improves physical condition of soil viz loosening of compact soil, better soil aggregation and a decrease in bulk density. An increase in population and activity of soil microorganisms is also reported that helps in improved growth of succeeding crop.Due to its smothering effects on weeds, the crop also minimizes intensity of weed flora. During glut period, it can be preserved as silage with all nutritive values intact, to feed the cattle in lean months. However, it must be mixed with straw or other dry fodders before feeding to the cattle to avoid bloating to them.

Origin and History

It is believed to be originated in Asia minor from their it was introduced to Egypt. It was brought to India in1904 by Sir Fletcher probably from Egypt and first grown at Mirpur Khas farm of then Sind Province, presently at Pakistan. From Sind, it was brought to Punjab and then to other north Indian states. It is now grown in many southern parts of the country too.

Area and Distribution

Major growing countries- Egypt, Seria, Iran, Pakistan and India.

Major growing states(India)- In irrigated areas of Punjab, Haryana, Uttar Pradesh, Rajasthan, Bihar and some parts of Gujarat, Maharashtra and Andhra Pradesh.

Plant Description

It is a small shrubby annual, growing upto a height of 60-80 cm with a rooting depth of about 40-50cm. The stems are hollow and succulent but become fibrous after flowering stage. Leaves are small, tender, slightly hairy on upper surface and trifoliate. Flowers are yellowish white with round heads. The seeds are egg shaped, small (2mm in length)and greenish yellow in colour.

Ecological Requirements

Climate: It requires dry and cool climate for proper growth and therefore, grown well in Rabi season. The crop is sown only when average daily temp reaches to 13-15.5°C.It requires about 25-30°C temperature for germination and approximately 18-21°C for best branching and vegetative growth. However, at flowering, slightly higher temp. of about 35-37°C is needed. A temp. of lower than 12°C with a frosty night or more than 40°C are detrimental for crop growth and development. This crop can be grown successfully in areas which receive an annual rainfall of 150-250 cm or even lower but irrigation must be ensured.

Soil: The crop grows best in well drained, medium loam soils rich in calcium and phosphorus having good water holding capacity and free from excess soluble salts.The crop can also tolerate to mild soil acidity.

Improved Varieties

Improved Varieties of the berseem can be grouped in two groups-

1. Diploid- most promising varieties are BL1,Miscavi, Khadravi, Chhindwara and IGFRI-99-1.

Tetraploid- Pusa Giant and T-678.

Seeds of tetraploid varieties is required to be produced every year by proper treatment because they have a tendency to become diploid in successive generations.Tetraploid vars are winter hardy, quick growing, very leafy and succulent. However, re-growth after harvesting is not possible, if day temp beyond 27°C in the month of April. Diploid vars. Are susceptible to extremely low and very high temp but still they perform better during April-May when temp. is above 27°C. Other latest released varieties of berseem are- BL42, BL180, Jawahar Berseem 1, 2 and 3, Bundel Berseem 2 and 3, UPB110 and Hisar Berseem1.

Field Preparation

As seeds of berseem are very small,it requires very fine,well pulverized and leveled seedbed. A fine seedbed ensures better contact of seeds and soil particles for higher germination. For ensuring a good seedbed, one deep ploughing with mould board or disc plough and then three to four harrowing followed by planking are required to break all the soil clods and leveling of the field. If soil does not have sufficient moisture for germination, a pre sowing light irrigation must be given. The field may be laid out into smaller beds of convenient sizes according to land topography and source of irrigation water.

Integrated Plant Nutrient Supply System

Being as a leguminous crop, berseem has a capacity to meet most of its nitrogen requirement through the process of symbiotic biological nitrogen fixation in its nodules. However, about 15-20 kg of nitrogen is needed as starter dose at the initial stage which helps in better and quick growth of berseem seedlings. For good yield, spread approximately 15-20 tones of well rotten FYM in a ha. of land about 20-30 days before sowing and mixed well in soil. Approximately 50-60 kg of P_2O_5 must be given for better yield. However, application of 60-80 kg of K_2O/ha may require depending on potassium supplying status of soil. Full dose of nitrogen, phosphorus and the potassium (if required on soil test basis)fertilizers should be applied as basal application at the time of sowing. Addition of FYM or compost will help in improving the physical condition of soil for better working of symbiotic bacteria along with meeting their food requirements.

Seed and Sowing

Sowing Time

After arrest of rains, sowing of berseem can be done from last week of September to first week of December in North West to Eastern and Central India. The ideal time for sowing of berseem is when mean day temp is around 25.C, which is recorded mostly in the first to third week of October in North India.

Seed Rate

The optimum seed rate for timely planted crop is 25-30 kg/ha that may be increased up to 35 kg/ha in case of early and late sowing conditions. For best growth and yield, tetraploid and diploid should be mixed in 2:1 ratio. The initial harvest, specially first cut, are usually poor. Therefore, for yield compensation in first cutting, about 1-2 kg/ha of fodder mustard seed should be mixed with normal seed rate of berseem.

Sowing Method

There are two methods for sowing of berseem:

1. *Dry Method:* Seed is broadcast, mixed and covered with half to one cm fine soil. Sufficient soil moisture should be ensured for better seed germination. After sowing, the field is divided into convenient sized beds and later the beds are flooded with water. Seedling will establish within a fortnight when seeds are sown by this method. First irrigation should be given only after proper germination.
2. *Wet Method:* Nicely prepared beds are flooded with 5-7 cm water and planking is done in order to level the surface. After 2-3 hours, the seeds are uniformly spread in standing water. About 24 hours after seeding the water is drained out and re-irrigated after 3-5 days of sowing when germination is complete. Before sowing, the water in beds should be stirred thoroughly so as to make a soil water suspension either by puddler or by planking in water. This would help in proper setting of seeds.

Seed Treatment

For higher germination and best growth, select only healthy, bold and yellow seeds free from weeds. In general, Kasani (*Chicorium intibuis*) seeds are found mixed with berseem. These Kasani seed can be separated from berseem seed before sowing by putting it into 5 per cent common salt solution. Kasani seeds being lighter, float on the surface of the solution, can be skimmed, washed three to four times with clean water so as to remove salt crust from seed surface before sowing.

When berseem is sown first time in a field, seed should be dual inoculated with *Rhizobium trifolii* and any other phosphorus solublising microbial culture which helps in BNF after crop establishment as well as increased availability of soil P. Before treating the seeds, they should be first soaked in fresh water for about 8-12 hours. For better sticking of culture with seeds, a solution is prepared with jaggery. About 100-150 g of jiggery or Gur is mixed with 1-1.5 liters of water and boiled. After cooling, 2.5 packets each of berseem, *Rhizobium* and PSB culture are mixed with it and then seed is well mixed and dried in a cool shady place. In case, the Rhizobium culture is not available, about 60kg of powdered soil from thr field where berseem was grown earlier should be applied in the field before sowing of the crop.

Berseem in Mixed Cropping

Berseem can be grown as an intercrop with sugarcane and napier etc. Both these crop remain dormant during winter when berseem can grow well and gives very good yield besides enriching the soil with nitrogen.

Water Management

The depth and frequency of irrigation is decided by soil type, number of cutting and nature of crop *i.e.*, sole or mixed. The first irrigation should be given after the seedling emergence and subsequent two or three light irrigation (4-6cm) should be given at weekly intervals. Subsequent irrigation may be given at an interval of 10 days in October, 12-15 days in November to January, 10-12 days in Feburary to

March and 8-10 days in April-May. Thus, about 12-15 irrigations will be needed during the entire crop season. Normally, the crop should be irrigated after each cutting.

Weed Management

For higher herbage and high quality fodder, chicory seeds must be eliminated as per the procedure given earler. Application of Fluchloralin @1.2 kg a.i./ha as pre-sowing also control chicory and other weeds effectively. In forage crops, herbicide is generally not used fearing residual toxicity on the animals.

Harvesting:

The first cutting should be taken when the crop is about 50-55 day sold. Subsequent cutting may be taken at 25-35 days intervals and number of cuts depending on the rate of the growth and temperature during the life cycle. Each cutting should be done at about 5-7 cm height above ground level for better and quick growth.

Yield

A good berseem crop, on an average, can give about 800-1000 quintals of green fodder with 18-20 per cent dry matter,if seed is not taken. The seed crop gives about 500-600 quintals of forage and 4-6 quintals of seed per hectare.

Seed Production

In case, the crop is to be left for seed, no cutting should be taken after middle of March. Normally the crop is left for seed production after three or four cuttings. More cutting may not only decrease the seed yield but also produces the seeds with poor viability. Provide light irrigation upto flowering to check excessive vegetative growth. At flowering and seed setting stage, irrigation should be given frequently. Seed crop mature in the end of May when seed bolls turn yellow tom brown in colour. The crop should be harvested and threshed either by beating with sticks or by trampling with bullocks.

Lucerne (*Medico sativa*)

Lucerne or alfalfa is one of the oldest cultivated fodder crop of the world and is also known as queen of forage crop.Alfalfa is an Arabic word. Being a deep rooted crop it can extract water from deeper soil layers and that is why it is generally grown in deep water table areas where water supply is inadequate for berseem. It can be grown both an annual as well as perennial crop and may supply green fodder continuously for 3-4 years from the same field. Its nutritious and palatable fodder contains about 15-205 crude protein with 725 digestibility, 1.5 per cent calcium, 0.2 per cent phosphorus on dry weight basis along with high amount of vitamins A,B and D. Like berseem, an excess of Lucerne fodder may causes bloat in cattle. Some animals are more prone to this complaint than others. Therefore, start with a small quantity of its green feed in the beginning, and then gradually increase the quantity. Feeding of Lucerne hay that is particularly rich in protein, calcium and vitamins may replace grains and concentrates. It supply green fodder for a longer period (November to June)than the berseem (December to April).

Origin and History

Lucerne is generally believed to be originated in south-west Asia and first cultivated in Persia (Iran). It was known to Greek and Romans in about 470 B.C.This crop was inbtroduced in India from North West sometimes in 1900 and now become quite a popular green fodder.

Botanical Description

Lucerne is a herbaceous annual as well as perennial plant, growing upto a height of 60-150cm. It has a deep root system consisting of a strong tap root and number of lateral roots making it a strong drought resistant. Stem is erect and branches arise from the crown, a woody base on stem near ground level. The number of branches may be as many as forty. The leaves are trifoliate; the middle leaflet possesses a short petiole, a characteristic which distinguishes it from berseem. Flower colour is usually purple, but it may be blue, yellow or white and they are fertilized specially by bees. Seeds are kidney shaped,very light in weight and yellowish brown with a shiny surface.

Climatic Requirements

Although a native of temperate region of south-west Asia, it is grown successfully in most of the tropical countries. Plants can withstand fairly low temperature. This crop can be grown from below sea level to as high as 2500m from mean sea level.

Soil

This crop does best on well drained fertile deep loam soils. It can also be grown in acidic soils with liberal application of lime.

Improved Varieties

Chetak, RL88,Sirsa No.8, Sirsa No.9, moopa, Rambler, Anand-2, Anand-3, NDRI Sl.1, IGFRI-S-244, IGFRI-S-244, LLC-3 and LLC-5 are some of the varieties can be grown successfully in most of the Lucerne growing tracts.

Field Preparation

The crop needs a fine well levelled seedbed with adequate soil moisture. A fine seedbed ensures better contact of seeds with soil particles and facilitates higher germination. Field should be prepared thouroughly and all the weeds and crop stubbles should be removed from the field before sowing. For ensuring a good seedbed, one deep ploughing with mould board or disc plough and then three to four harrowing followed by planking are required to break all the soil clods and leveling of the field. If soil does not have sufficient moisture for germination, a pre sowing light irrigation must be given.

Seed and Sowing

For higher germination and best growth, select only healthy, bold and yellow seeds free from weeds.

Sowing Time

Lucerne should be planted from end of September to end of October; however, middle of October is the best planting time. Any delay in sowing results in a lower germination, poor growth and subsequently low fodder yield due to very low temperature in the beginning and high temperature in the month of February / March.

Seed Treatment

When lucerne is sown first time in a field, seed should be l inoculated with *Rhizobium melilotii* culture which helps in BNF after crop establishment. Before treating the seeds, they should be first soaked in fresh water for about 8-12 hours. For better sticking of culture with seeds, a solution is prepared with jaggery. About 100-150 g of jiggery or Gur is mixed with 1-1.5 liters of water and boiled. After cooling, 2.5 packets *Rhizobium* culture is mixed with it and then seed is well mixed and dried in a cool shady place. In case Rhizobium culture is not available, about 350 kg of well powdered soil from the field where lucerne was grown earlier should be collected and evenly spread / broadcasted in the field before sowing of the crop.

Seed Rate

The seed rate depends upon method of sowing, type of crop *i.e.* wether it is intercropped with sugarcane or nappier, or grown as a pure crop. The broadcasting obviously needs ahigher seedrate ranging between 17-22 kg / ha whereas, line sowing at 30cm distance needs only 11-15 kg / ha. Howevr, intercropping required only 6-12 kg / ha seedrate.

Sowing Method

Line sowing is a superior sowing method over the broadcasting. In line sowing, the water soaked seeds is sown in shallow furrows and then field is flooded from across the furrow direction so that the seed is not collected at one place towards slope.

In Maharashtra, the crop is sown on ridges at 45-60cm apart and this method could be adopted in some areas of north India too where growth conditions are very favorable.

Plant Nutrient Supply System

Being as a leguminous crop, after three to four weeks of sowin, lucerne has a capacity to meet most of its nitrogen requirement through the process of symbiotic biological nitrogen fixation from air in its nodules. However, about 15-20 kg of nitrogen is needed as starter dose at the initial stage in the soils with low organic matter and poor nitrogen supply which helps in better and quick growth of the crop seedlings. Being a perennial crop,it is advantageous to apply every year 25-30 tones of well rotten FYM/ha and mixed well in soil. An adequate supply of phosphorus is also necessary for proper functioning of root nodules. Therefore, approximately 50-60 kg of P_2O_5 should be given for better yield. However, application of 60-80 kg of K_2O/ha may require depending on potassium supplying status of soil. Full dose of nitrogen, phosphorus and the potassium (if required on soil test basis)fertilizers should be applied as basal application at the time of sowing. Addition of FYM or compost will help in improving the physical condition of soil for better working of symbiotic bacteria along with meeting their food requirements. Boron deficiency is quite common in coarse textured leached soils where leaves of Lucerne may develop numerous pale yellow spots leading to a disorder known as Lucerne Yellow.This disorder can be rectified with a spray of 0.2 per cent borex.

Water Management

The depth and frequency of irrigation is decided by soil type, number of cutting and nature of crop *i.e.* sole or mixed. To attend good germination, one pre sowing irrigation is must if stored soil moisture is not sufficient. The crop needs very frequent irrigation at an interval of one week during early growth periodbut once plants are established and grown little taller, they need subsequent irrigation at an interval of 15-20 days during winter and 10-12 days during spring and summer seasons. Normally, the crop should be irrigated after each cutting but if weather is too hot, then two irrigations may be given between two subsequent cuttings,

Weed Management

Lucerne takes along time to establish itself and gives ample scope for weed growth upto the time of first cutting. If the crop is sown in lines, weeding and hoeing becomes easier. First weeding should be done about 20-25 days after sowing. Trifuralin, @ 4kg/ha should be applied before sowing for good harvest.For seed production, weeding is must. Dodder or akasbell (*Cuscuta*) is a most harmful weed of this crop. This weed should be carefully removed along with the host plant and burnt. Never allow to fodder plants to set seeds because their seeds can remain viable in soil for so many years. This weed can also be controlled by spray of crude oil in localized patches at the time of flowering of dodder plants. In forage crop, herbicide is generally not used fearing residual toxicity on the animals.

Seed Production

For seed production, the crop is allowed to flower after taking a cutting in the end of January. The crop should be irrigated frequently during vegetative phase and no stress should be given during flowering. However, irrigation should be withheld after fruiting as this may resut in regeneration which affect the seed yield adversely. Being a cross pollinated crop, the prospects of better seed setting are improved if few hives of honeybees are kept in or near the field. Seed crop mature in the end of May when seed bolls turn yellow tom brown in colour.The crop should be harvested and threshed either by beating with sticks or by trampling with bullocks.

Harvesting

The October sown crop becomes ready for first harvest in second or third week of when the crop is about 50-55 days old. Subsequent cutting may be taken at 25-35 days intervals when crop attains the height of 60 cm from ground uptil May after which it may be left for setting of seeds as the plants start flowering.

Yield

A good annual lucerne crop, on an average, can give around 6-8 cutting which gives a total of about 800-1000 quintals of green fodder and 2-3 quintals of seed yield/ha while perennial crop provides about 80-1100 tonnes/ha.

Cowpea (*Vigna unguiculata*)

Cowpea is also known as lobia, barbati, southern pea and black eyed pea etc. and has multiple uses like pulse, vegetables, feed, fodder and green manuring etc. Its seed is a nutritious component in human diet and cheap live stock feed as well.Being rich in protein and containing many other nutrients, it is also known as vegetable meat. On dry weight basis, seed contains about 23 per cent protein, 1.8 per cent fat and 60 per cent carbohydrate along with a rich source of calcium and iron. This crop forms excellent forage because of very heavy vegetative growth and is quit comparable with Lucerne. Such a heavy growth cover the ground so well that it checks the soil srosion in problem areas and can be later ploughed in as green manure. It also has considerable promise as an alternate pulse crop in dryland areas.

Origin and History

It is a native of south Africaand has been cultivated since very ancient times in the Medditerranean region of Greeks, Romans and Spaniards. It is also claimed to be indigenous to be India.

Area and Distribution

About 90 per cent of the world acreage is in Africa and to some extent, it is also grown in Asia, North and South America, Australia and central and southern parts of Europe. In India, it is mainly grown in central and peninsular regions. In northern India, it is grown in U.P., Punjab, Delhi and Haryana in irrigated areas during summer as a dual crop for fodder as well as green pods.

Botanical Description

The common cowpea is atwining annual herbaceous plant. Root system consists of a well developed tap root along with numerous laterals.Most of the root portion located in upper 40cm soil. Stem is slighltly rigid and almost glabrous. Leaves are trifoliate, alternate andwith scattered short hairs. Flowers are white, yellow or pink and usually self pollinated. Pods are long and cylindrical and constricted between the seeds. The seeds are bean shaped and spotted with different colours such as brown, white and motteled.

Climate

Being a warm weather crop, it can withstand a considerable degree of drought and can be grown in all tropical and sub tropical regions. Germination is better in between 12-15°C temperature and crop thrives well in between 27-35°C. It is more tolerant to heavy rainfall than any other pulsed but suffers heavily from water stagnation and heavy drought.

Soil

It can be grown in a variety of soil but well drained loam or slightly heavy soil is best suited for its cultivation. This crop can also be grown on sloppy lands in hilly tracts and heavy loam soils.

Improved Varieties

Some important dual.purpose(for grain as well as for fodder) are- Kohinoor, HFC42-1 (Hara Lobia) GFC1, GFC 4 UPC5286, UPC-5287,Sweta, UPC-9202, Bundel lobia1,bundle Lobia2, Pusa Phaguni, Pusa Rituraj, Pusa Dofasli, Pusa Barsati, and Cowpea-74.

Field Preparation

Make the field clean after removing the stubbles and other residues of previous crop. Two or three cross harrowing followed by planking is sufficient for preparing a friable, smooth seedbed. Give a pre-sowing irrigation before starting preparation for summer, irrigated fodder cowpea.

Seed and Sowing

For higher germination and best growth, select only healthy and bold seeds free from weeds and procured from a reliable source.

Sowing Time

Summer fodder–March–April

Kharif–June–July

Hills–April–May.

Seed Treatment

When cowpea is sown first time in a field, seed should be inoculated with suitable *Rhizobium* culture which helps in BNF after crop establishment. Before treating the seeds, they should be first soaked in fresh water for about 8-12 hours. For better sticking of culture with seeds, a solution is prepared with jaggery. About 100-150 g of jiggery or Gur is mixed with 1-1.5 litres of water and boiled. After cooling, 2.5 packets *Rhizobium* culture is mixed with it and then seed is well mixed and dried in a cool shady place. Before culture treatment, seeds should also be treated with bavistin @ 2g/kg of seeds to save the crop from soil borne diseases.

Seed Rate

The seedrate depends upon time, purpose and method of sowing. The broadcasting obviously needs a higher seedrate than the line sowing. For fodder purpose, seed rate should be 35-45 kg/ha for a pure crop that can be reduced proportionally in case of inter or mixed cropping. Higher seeds rate should be used in summer because of poor growth due to high temperature and hot desiccating winds.

Spacing

Kharif–Row spacing 35–45 cm

Plant spacing: 8–10 cm

Spring/summer–Row spacing 25–30 cm

Sowing Method

Line sowing is a superior over the broadcasting. In line sowing, the water soaked seeds is sown in shallow furrows and then one light planking is done to cover the seeds.

Plant Nutrient Supply System

Being as a leguminous crop, after three to four weeks of sowin, lucerne has a capacity to meet most of its nitrogen requirement through the process of symbiotic biological nitrogen fixation from air in its nodules. However, about 15-20 kg of nitrogen is needed as starter dose at the initial stage in the soils with low organic matter and poor nitrogen supply which helps in better and quick growth of the crop seedlings. An adequate supply of phosphorus is also necessary for proper functioning of root nodules. Therefore, approximately 50- irrigation 60 kg of P_2O_5 should be given for better yield. Addition of FYM or compost will help in improving the physical condition of soil for better working of symbiotic bacteria along with meeting their food requirements. However, application of K_2O may depend on potassium supplying status of soil. Full dose of nitrogen, phosphorus and the potassium (if required on soil test basis)fertilizers should be applied as basal application at the time of sowing and is placed about 5-7 cm below the seeds.

Water Management

For a rainy season, rainfed crop, drainage is most important than irrigation. An early sown crop may require one or two irrigation during pre-monsoon period.But for a summer crop,5-6 irrigations are needed at an regular interval of 10-15 days because of high temperature and low relative humidity. Actual number of irrigation in summer crop may vary according to soil type prevailing weather during crop growth.

Weed Management

Effective weed control is required during first 25-35 days. One hand weeding at 25-30 days after sowing oran application of Pendimethalin @ 1 kg/ha just after sowing as pre emergence is recommended to maintain the quality of the fodder. In forage crop, herbicide is generally not used fearing residual toxicity on the animals.

Seed Production

For seed production, the crop is allowed to flower after taking a cutting in the end of January. The crop should be irrigated frequently during vegetative phase and no stress should be given during flowering. However, irrigation should be withheld after fruiting as this may resut in regeneration which affect the seed yield adversely. Being a cross pollinated crop, the prospects of better seed setting are improved if few hives of honeybees are kept in or near the field. Seed crop mature in the end of May when seed bolls turn yellow tom brown in colour.The crop should be harvested and threshed either by beating with sticks or by trampling with bullocks.

Harvesting

For fodder purposes, crop should be cut when it attains the age of 40-45 days. Green pods for use as vegetable can be harvested at 45-90 days after sowing depending on the variety.

Yield

A good crop of fodder cowpea can yields about 300-400 quintals of green fodder from one hectare of a pure crop.

Cluster Bean or Guar (*Cyamopsis tetragonoloba* L.)

Among all leguminous fodder, guar is comparatively more drought crop. It is grown for multi-purposes *viz.* feed, fodder and vegetables. It also occupies an important place in national economy because of its industrial importance due to presence of gums in its seeds. The fodder of the guar as well as its grain is highly nutritive and they are rich in protein, fat and minerals.

Origin and History

Tropical Africa is considered as its primary centre of origin. However, it is being grown in India since very ancient time for fodder and vegetable purposes. Therefore, some peoples think that it as indigenous to India.

Area and Distribution

In India, Rajasthan is the major guar growing state contributing around 60 per cent of total national production. Gujrat, Haryana and Punjab are other guar growing states where it is grown for fodder cum pods purpose.

Botanical Description

Guar plant is a robust, erect annual usually grows upto a height of 90-180 cm. It has well developed tap roots and trifoliate and toothed leaves. Flowers are purplish in color. Pods are somewhat flattened, 3-13cm long containing 5-12 seeds and borne in cluster (that's why plant is known as cluster bean).

Climate

Being a warm weather crop, it can withstand a considerable degree of drought and can be grown in all tropical and sub tropical regions. Germination is better in between 12-15°C temperature and crop thrives well in between 27-35°C.

Soil

It can be grown in a variety of soil but well drained medium to light soil with pH range of 7-8.5 is best suited for its cultivation.

Improved Varieties

Durgapur Safed, Ageta Guara-111, HFG-119, HFG-156, Guara-80, Bundel guar1, Bundel guar2, AGFRI-212-9 are some important dual purpose (for grain as well as for fodder) etc.

Field Preparation

It does not require much field preparation. Two- three ploughing with a local plough or two cross harrowing and planking is sufficient. In Kharif, preparation start after soil soaking rains but in summer, give one pre sowing irrigation to ensure sufficient soil moisture for proper germination.

Seed and Sowing

For higher germination and best growth, select only healthy and bold seeds free from weeds and procured from a reliable source.

Sowing Time

Summer fodder–March–April (North India)

Kharif (main season)–June–July, depending on occurrence of monsoon

The early sown crop for fodder makes luxuriant growth under irrigated conditions and gives higher yield of fodder.

Seed Treatment

When cowpea is sown first time in a field, seed should be inoculated with suitable *Rhizobium* culture which helps in BNF after crop establishment. Before treating the seeds, they should be first soaked in fresh water for about 8-12 hours. For better sticking of culture with seeds, a solution is prepared with jaggery. About 100-150 g of jiggery or Gur is mixed with 1-1.5 liters of water and boiled. After cooling, 2.5 packets *Rhizobium* culture is mixed with it and then seed is well mixed and dried in a cool shady place. Before culture treatment, seeds should also be treated with bavistin @ 2g/kg of seeds to save the crop from soil borne diseases.

Seed Rate

The seedrate depends upon time, purpose and method of sowing. The broadcasting obviously needs a higher seedrate than the line sowing. For fodder purpose, seed rate should be 35-45 kg/ha for a pure crop that can be reduced proportionally in case of inter or mixed cropping. Higher seed rate should be used in summer because of poor growth due to high temperature and hot desiccating winds.

Row Spacing

Kharif: Timely planted 35–45cm

Late planted: 25–30cm

Spring/summer: 25–30cm

Sowing Method

Line sowing is a superior over the broadcasting. In line sowing, the water soaked seeds is sown in shallow furrows and then one light planking should be done to cover the seeds.

Plant Nutrient Management

Being as a leguminous crop, after three to four weeks of sowin, lucerne has a capacity to meet most of its nitrogen requirement through the process of symbiotic biological nitrogen fixation from air in its nodules. However, about 15-20 kg of nitrogen is needed as starter dose at the initial stage in extremely poor soils with low organic matter and poor nitrogen supply which helps in better and quick growth of the crop seedlings. An adequate supply of phosphorus is also necessary for proper functioning of root nodules. Therefore, approximately 50- 60 kg of P_2O_5 should be given for better yield. When guar is grown after an exhaustive crop, addition of 15-20 tones of FYM or compost one month before sowing will help in improving the physical condition of soil for better working of symbiotic bacteria along with meeting their food requirements. Full dose of nitrogen and phosphorus fertilizers should be applied as basal application in the furrow at the time of sowing and is placed about 4-5 cm below the seeds.

Water Management

For a rainy season, rain fed crop, drainage is most important than irrigation. One or two irrigation may be required in case of delayed monsoon or a very long gap in two consecutive rains. But for a summer crop, 5-6 irrigations are needed at a regular interval of 10-15 days because of high temperature and low relative humidity. Actual number of irrigation in summer crop may vary according to soil type prevailing weather during crop growth.

Weed Management

Effective weed control is required during first 25-35 days in Kharif season crop. One hand weeding at 25-30 days after sowing or an application of Flucloralin(Basalin) @ 1 kg/ha before sowing as pre emergence is recommended to maintain the quality of the fodder. In forage crop, herbicide is generally not used fearing residual toxicity on the animals.

Harvesting

For fodder purposes, the crop should be cut when it start flowering or the pods just start to form during the age of 50-85 days after sowing.

Yield

A good crop of fodder guar can yields about 300-400 quintals of green fodder from one hectare of a pure crop.

Other spp. of forage legumes are:

Stylosanthes

Stylosanthes is a genus of many fodder-cum-cover crops a nd most of them are native from tropical region of south and Central America and Carribean Islands. These are adopted wellin warm and humid climates. After establishment in the field it can compete very successfully with weeds and can invade even natural grasslands.It can tolerate water logging and grows well on acid soils. It is an exotic spp. and is also known as Muyal masal. Recently introduced and acclimatized in manyparts of the south India. It's a a hardy type and extremely drought resistant and comes up well under dryland condition. It is a perennial herb attaining a height of 60-75 cm in about six month period. It is relatively unpalatable during early growth stages but cattle readily eat it once they become accustomed to it. It produces many semi erect branches with lot of secondary and tertiary branches assuming a bushy appearance. The most important spp. grown for forage purposes are-Common stylo or Brazilian Lucerne, Caribbean Stylo, or pencil flower, Townsville Stylo or Magsaysay/wild Lucerneetc.The seeds have hooks for dispersal through animals. They may also pass through digestive system of animals.

These group of crops have been grown successfully with a a number of grasses *viz.* molasses grass, Guinea grass, Rhodes grass and setaria grass etc. Once established, it grows very fast and puts good canopy cover made by tri pinnately trifoliate leaves. First cut can be made around six months and subsequent cuts at every three months afterwards. If grown under irrigated condition, it can be cut at every 45 days at a height of 45Cm from ground level. Stylo is propagated by seed but can also be established by stem cuttings. Seeds normally require scarification for improving germination, but rhizobium inoculation is not required,It can also be over sownwithout land preparation or minimal cultivation. in general. It has got quick regeneration capacity and yields upto 30-35 tones of green fodder as pure crop with a protein content of 19 per cent.

It is best recommended as an inter crop with rainfed BLOU BUFFEL OR KOLUKATTI grass as one row of it after every three rows of the grasses to get a balanced and nutritious fodder.

Desmodium (*Desmodium toruosum and D. unisineatum*)

It is anutritious and highly palatable fodder contains about 21 per cent pritien with 27 per cent dry matter and is also called as Alfalfa of the Tropics or Tropical Medic. Two tropical spp. *i.e.* Green leaf desmodium (*D. tortusum*) and Silver leaf desmodium (*D. unisineatum*) are very poupular in many tropical parts of southern India. Green leaf desmodium is a erect, perennial growing upto a height of 2-2.5m and is distinguished by the brown flecking on the upper surface of hairy leaves. Its pods are in chain like form and maturation starts from top downwards. It is used as protein supplement in poultry feeding. It comes up well in acid soil. Silver leaf spp. has an irregular silver band along the margin of the leaf. Livestock relish both on the leaves and shoots.

It is an irrigated forage legume and perennial habit. It is a shade tolerant crop, which grow well under the shades of coconut, tamarind and eucalyptus trees. It can be grown upto a height of 30 cm four times in a year. It is a self sown crop and shattered seed on the ground germinate at optimum soil and moisture conditions.

Mothbean (*Phaseolus acontifolius*)

It is the most drought resistant crop of arid regions of north western parts of the India among all kharif pulses and is largely grown as a dry crop either alone or as subsidiary to millets or as a components in crop mixture. On account of its spreading habits, it forms a mat like structure on the ground and is highly useful against the wind erosion in sandy areas. It can be used both as green as well as dry fodder. Seeds are broadcasted @ 10-15 kg/ha when grown alone or 5-10 kg/ha when grown mixed

mixed with other crops. Since it is grown in typical arid parts with no irrigation, in general, it is a low yielder and gives green fodder yield about 10 tones/ha.

Rice Bean (*Phaseolus calcaratus*)

Rice bean is a native of South East Asia and occurs in India in wild and cultivated forms.Its grainsresemble de-husked rice grains, hence the name rice bean. However, the trifoliate leaves and stem resembles like green gram. The plants are semi erect to twining in habit with several small branches growing upto a height of 3m. It is mainly grown as a rainfed crop between May to August. A seed rate of 30-35 kg/ha is normally used and is usually sown in lines in row spacing of 30-40 cm. about 20kg of N and 40 kg of P/ha, respectively is applied as basal at the time of sowing. It is grown for dual purposes for fodder as well as pulse crop. Important varieties recommended for fodder purposes are K-1, K-16, BC11, BC15, BC17 and BC18. On an average, a good crop may yield about 20-30 tones of green fodder.It requires and warm and humid tropical climate. It is fairly resistant to drought but susceptible to water logging. It is a annual herb of 70-90 days growth and prefers sandy loam to clay loam soils.

Lab-lab or Field Bean (*Lablab niger*)

It is a quick growing, profusely branching and twining in habit. It is grown as inter crop with rainfed cholam in kharif. It yields about 30t/ha of green fodder in 65-70 days.

Soybean (*Glycine max*)

Though it is basically a oilseed crop but creeping type old varieties *viz.* Pusa Selection yellow and Chocolate are also used fodder purposes with 13-16 tones of green fodder yioeld with in 60-70 days.The seed is sown in row spacing of 45-60cm with the seed rate of 25-30kg/ha.

Forage Legume for Temperate Regions

Different spp. of clovers such as red clover (*T. pratense*), white clover (*T. repense*) crimson clover (*T. incarnatum*) and subterranean clover (*T. subterraneum*) centrosema, calapogonium and vetches (*Vicia* spp) are the important forage legumes suitable for cold temperate places and high elevations. These are usually grown in mixture with grasses at the seed arte of 8-10 kg/ha and maintained as perennials. For pure crop, 40-50 kg/ha seeds are required. These contain 15-16 per cent of crude protein and all of them are resistant to frost. On an average, they yield about 10-15 tones of green fodder/ha.

References

Bhagmal; Singh, K.A.; Roy, A.K.; Ahmed, S and Malviya,D.R.2009. Forage crops and grasses. In: Handbook of Agriculture. ICAR, New Delhi, India. pp1617.

Brain, F.B. 1971. Crop production-cereals and legumes. Academic Press, London and New york.

Jayanthi, C. ; Devasenapathy, P. and Vennila,C. 2008. Farming Systems:Principles and practices. Satish Serial Publishing House, New Delhi, India. pp. 330.

Hazra,C.R. 1995. Advances in Forage Production Technology. IGFRI, Jhansi. pp. 51.

ICAR. 1989. Technology for Increasing Forage Production in India. Coordinating unit, IGFRI, Jhansi.

Menhi, Lal, Shukla, N.P., Tripathi, S.N.; Sinsinwar, B.S.; Niranjan, K.P.; Gupta,S.D.;and Arya,R.L. 1994. Forage Production Technology. IGFRI, Jhansi. pp. 72.

Miller, D.A. 1984. Forage Crops. Mc Graw Hill Co. New York. pp. 530.

Narayanan, T.R.; and Dabadghao, P.M. 1972. Forage Crops of India.ICAR, New Delhi, pp. 373.

Sankaranarayan, K.A. and Vinod,S. 1984. Grasses and Legumes for Forage and Soil Conservation. M. 1972. ICAR, New Delhi, pp. 155.

Sharma, N.K.; Singh, R.P.; Yadav, M.S. and Singh, K.C. 1999. Forage Production in Drylands. Scientific Publishers, Jodhpur. pp. 152.

Singh, R.P.(ed.). 1995. Forage Production and Utilization. IGFRI, Jhansi. pp. 7367.

Singh,C.; Singh,P.; and Singh,R. 2003. Modern techniques of raising field crops. Oxford and IBH Publishing house, New delhi, India. pp. 583.

Singh, S.S. 2005. Crop Management under irrigated and rainfed conditions. Kalyani Publishers, Ludhiana, India. pp. 538.

Thomas, C.G.2003. Forage crop production in the tropics. Kalyani Publishers, Ludhiana, India. pp. 259.

Wilson, P.N. 1998. Production of animal feed. In: Webster, C.C. and Wilson, P.N. (eds.) Agriculture in the Tropics. pp. 294-312.

Yellamanda Reddy, T. and Sankara Reddi, G.H. 1995. Principles of Agronomy. Second edn. Kalyani Publishers, Ludhiana. pp. 515.

Chapter 19
Tree Legumes Fodder

Nitu Kumari[1], S.K. Singh[2] and Ratna Rai[3]

[1]*Krishi Vigyan Kendra, Buxar*
ICAR Research Complex for Eastern Region, – Patna 800 014, Bihar
[2]*Krishi Vigyan Kendra, Auraiya*
[3]*G.B.P.U.A.T., Pantnagar, Uttaranchal*

Legumes are a large group of plants that have double-seamed pods, containing a single row of seeds; depending on the variety, the seeds, pod and seeds together, or the dried seeds, are eaten. Nutritionally they are 2-3 times richer in protein than cereal grains. Tree legumes are usually long-lived and require low maintenance, and therefore enhance the sustainability of farming systems. Leaves of many legume trees, such as *Leucaena, Calliandra, Gliricidia, Acacia, Albizia, Caliandra, Dalbergia, Erythina* and *Sesbania*, are palatable to livestock. Because of high protein content (from 15 to 25 percent), legume tree leaves make good fodder or supplements to other feeds. Along with naturally occurring stands, tree legumes are often planted specifically for forage both in extensive grazing systems and in association with crops. There is a wide range of ecological adaptation among tree legumes, although there are no single species suited to the entire range of conditions. Managers must select carefully to ensure successful growth of tree species in their environment. The tall stature and deep-rooted habit of tree legumes effectively insulate them from the most serious effects of competition. Consequently, in pasture systems, management of tree legumes to maintain grass/legume balance is simpler than with herbaceous legumes. The hard or waxy coat on the seed of many tree legumes inhibits the absorption of water and prevents uniform germination. The seed coat must be broken or scarified before germination will occur. Forage tree legume plants which are not effectively nodulated will be pale in colour and will not grow vigorously due to inadequate nitrogen fixation leading to nitrogen deficiency. Planting of tree legumes by seed to form hedgerows is the most common method for broad-acre sowings. More extensive research is immediately warranted to harvest the fruit of *"sustainable quality fodder"*.

Legumes are a large group of plants that have double-seamed pods, containing a single row of seeds; depending on the variety, the seeds, pod and seeds together, or the dried seeds, are eaten. Legumes have been used in agriculture since ancient times. Legume seeds or pulses were among the first sources of human food and their domestication and cultivation in many areas occurred at the same time as that of the major cereals. Nutritionally they are 2-3 times richer in protein than cereal grains and many also contain oil. Leguminous mulches have always been used as a source of nutrient-rich organic matter and nitrogen for crops. In more recent times, legumes have become important as high quality fodder for livestock both in cultivated pastures and in naturally occurring associations.

The legumes are the third largest group of flowering plants comprising over 18,000 species in 650 genera which are well distributed in most environments throughout the world. Taxonomists have divided the legumes into three families:

- ✰ The Caesalpiniaceae contains about 2,800 species, most of which are shrubs and trees. Among them are Brazilwood, Royal Poinciana, Honey locust, and the Redbuds.
- ✰ Mimosaceae also contains about 2,800 species. These are predominantly small trees and shrubs of semiarid tropical regions of Africa, the America and Australia. *Acacia* species are the best known examples of this family.
- ✰ Fabaceae contains over 12,000 species; mainly herbs and small shrubs distributed worldwide, and include the well-known food crops such as beans and peas.

Legumes in Agriculture

Nitrogen is the most limiting element in agricultural production, and deficiency reduces the productivity of crops, pastures and animals. There are several potential sources of nitrogen to overcome this shortfall, namely:

- ✰ N from the mineralization of soil organic matter,
- ✰ N from artificial fertilizers,
- ✰ N from biological nitrogen fixation in legumes, and
- ✰ N from organisms associated with tropical grasses.

Of these, N from soil is often insufficient for plant growth especially in most tropical soils which are low in organic matter. N from organisms associated with grasses is a minor source. Fertilizer N and N fixed by legumes are the largest potential sources with the latter being the cheapest source. Biologically fixed N is transformed into leguminous protein and this may be consumed directly by animals to meet their protein requirements and the excess returned to the soil via animal wastes. Alternatively, N may be returned directly to the soil as organic mulch.

Since few other plant families include species with a nitrogen fixing ability, legumes produce most biologically fixed nitrogen and are therefore crucial to maintaining the N-balance in nature. In Australia, Steele and Vallis (1988) estimated annual use of 35,000 t of artificial fertilizer N on pastures compared with 1.2 Mt of N derived from biological nitrogen fixation. Very high yielding leguminous crops can add up to 500 kg of nitrogen to the soil per hectare per year (NAS, 1979) although inputs of 100-300 kg N/ha/year from good quality legume-based pasture would be a more realistic expectation (Steele and Vallis 1988). Legume associations are therefore vital to sustaining soil nitrogen fertility over long periods. The practice of shifting cultivation, traditional in many countries, is heavily dependent on the leguminous component of the primary and secondary forest cover for fertility restoration.

Another advantage of legumes is their high quality for animal production. The nutritive value of legumes is measured in terms of the potential intake of digestible dry herbage and, in general, legumes have both higher digestibility and higher intake than grasses and their nutritive value tends to remain higher as plants mature.

Tree Legumes

Until recently, tree legumes were largely neglected by researchers because their utilization and management fell between the disciplines of forestry and pasture agronomy. They are now receiving increased research attention because of their multipurpose value and some distinctive features which set them apart from herbaceous legumes. Their special characteristics may be summarized as follows:

Tree Legumes

- ☆ Are usually long-lived and require low maintenance, and therefore enhance the sustainability of farming systems,
- ☆ Provide high quality forage for feeding of livestock,
- ☆ Stabilize sloping lands against erosion because of their deep-rooted habit,
- ☆ Supply N-rich mulch for cropping systems,
- ☆ Can be used to colonise and rehabilitate adverse environments, *e.g.* saline or arid locations,
- ☆ Provide a source of timber and firewood for either domestic or industrial use,
- ☆ Are used in farming systems as living fences, as shade trees for plantation crops, and as living trellises for climbing crops, and
- ☆ Are a source of fruit and vegetables for human consumption.

Tree legumes can therefore be regarded as truly multipurpose trees for agriculture.

Tree Legumes as Fodder for Animals

Leaves of many legume trees, such as *Leucaena, Calliandra, Gliricidia, Acacia, Albizia, Caliandra, Dalbergia, Erythina* and *Sesbania* are palatable to livestock. Because of high protein content (from 15 to 25 per cent), legume tree leaves make good fodder or supplements to other feeds. Thus, legume trees in agroforestry systems could be used to produce fodder. Care must be taken, however, to establish tree species that are compatible with both the livestock and the site.

Fodder may be produced from legume trees in two possible ways: (1) by cutting off the branches and leaves and feeding them to animals located elsewhere; and (2) by allowing the animals to graze directly among the trees. Because the trees are integrated into an agroforestry system, and the inter-planted agricultural crops could be damaged by grazing animals, the first system, cut-and-carry, would seem more suitable to agroforestry fodder production.

On the other hand, where short-duration crops, such as upland rice or corn, are inter-planted with fodder-yielding legume trees, it may be possible to allow animals to graze in the field periodically, perhaps immediately after the cereal crop is harvested and prior to planting the succeeding food crop. Three advantages of this system are: (1) the animals may feed on the cereal stalks as well as on the tree legumes, thus increasing the amount of forage available per hectare; (2) the labour of cutting and carrying the fodder to the animal pens is eliminated; and (3) the manure produced by the grazing animals is recycled directly into the field.

Legume trees are raised principally for fodder in agroforestry systems will at the same time stabilise and, to some degree, fertilise the soil to the benefit of the food crops. They will not produce fuelwood, however, because, as in the case of green manuring, the tops and branches are pruned regularly and used as feed on site or carried to the penned animals. This frequent cutting demands that the tree species chosen must sprout, or coppice, rapidly and profusely. Legume-tree species such as *Leucaena* and *Gliricidia*, the leaves of which are palatable to livestock, have this coppicing capability.

When lopping off the branches and tops, as in the case of producing green manure, stumps must be left at heights that are convenient for farmers to cut or easy for animals to graze on the coppices. In general, from 30 cm to waist-high stumps are adequate, although as the higher the stump, the greater the fodder yield. The highest yield comes from trees where, instead of pruning, the leaves are simply plucked from the twigs. The added labor needs for this slow process must be balanced against the added yield when deciding what methods of forage-harvesting are used.

If the trees are pruned for forage at appropriate intervals, a few trees may need to be left uncut, as shade trees for the livestock. This is especially important to prevent livestock dehydration in hot regions.

The most complex agroforestry combination is one that carries all three components: food crops, trees, and animals, as described above. The most usual forms consist only of two components, however, such as a crop-tree combination or a livestock-tree combination. Thus, a farmer may sometimes grow legume trees alone for fodder with no food crops inter-planted. In this case, the animals may graze directly on the sprouts at the tree stumps. Tree spacing in this case would depend on the need for animals to move about rather than on the fertilizer and shade needs of food crops. Spacing between trees could be wider, 5 to 10m, with 200 to 100 trees per hectare. An advantage of wide spacing is the possible availabity of additional forage

In another combination of products, livestock and timber can be produced simultaneously if the trees are left unpruned and are spaced widely enough to allow sufficient amounts of grass to grow on the forest floor to fill forage need. With this system, however, the high-protein legume tree foliage no longer serves as forage. Only the grasses growing under the trees and fertilised by nitrogen - rich litter from the legume will feed the livestock.

The nutritional quality of tree legumes varies from excellent (*Leucaena leucocephala*) *to* quite poor (most Australian *Acacia* species). Poor quality can be due to tannins which reduce the digestibility of both herbage and protein. The presence of tannins is often evident as brownish, reddish tinges in juvenile growth. Another reason for poor quality is that some species have phyllodes (expanded and flattened leaf petioles) instead of compound pinnate or bipinnate leaves which are very high in fibre and therefore of low digestibility, *e.g.* the Australian Acacias.

Forage from tree legumes is often used as a buffer to overcome feed gaps that arise from seasonal fluctuations in the productivity of other feed sources. For example, grasses and other herbs may die when upper soil layers lose their moisture but the deep-rooted trees exploit moisture at depth and continue to grow. During the dry season or in times of drought, trees provide green forage rich in protein, minerals and vitamins while the herbaceous cover provides only poor quality straw.

Under natural conditions, a large proportion of the foliage of tree species will be out of reach of grazing animals so utilisation can be manipulated by cutting or lopping to make it available when needed. Sometimes natural leaf fall through senescence is an important day-to-day component of the diet of some grazing animals. In Africa, goats thrive on the leaf fall of *Acacia melliflora* (Dougall and Bogden, 1958).

Tree Legumes as Planted Fodder in Cropping and Grazing Systems

Along with naturally occurring stands, tree legumes are often planted specifically for forage both in extensive grazing systems and in association with crops.

In many of the more intensive agricultural areas of Asia and Africa, where livestock are raised in small numbers by smallholder farmers, tree legumes are planted as 'forage banks' on unused land along field borders or fence lines, on rice paddy bunds or in home gardens. These areas are usually harvested under a 'cut-and-carry' system and are a principal source of high quality forage used to supplement low quality roughages such as crop residues. Productivity from these areas can be quite high. In the Batangas region of the Philippines, a 2 ha area of *Leucaena leucocephala* grown in association with the fruit tree *Anona squamosa* was able to supply the forage requirements of 20 growing cattle over a 6 month period (Moog 1985). At Ibadan in Nigeria, Reynolds and Atta-Krah (1986) suggested that the surplus foliage produced over a year from 1 ha of *Leucaena leucocephala* and *Gliricidia sepium* planted at 4 m intervals in an alley cropping system could be used as a supplement to provide half the daily forage requirements for 29 goats.

In many of these intensive cropping areas, tree legumes are planted not only for their forage but also for firewood, green manure and other uses.

In the more extensive grazing areas of Australia, Southern Africa and South America, tree legumes are increasingly being planted in association with improved grasses to increase carrying capacity and productivity of grazing cattle. In Central Queensland, over 20,000 ha have been sown to *Leucaena leucocephala* in the past 10 years. The leucaena is sown in wide spaced rows 4-10 m apart and an improved grass such as green panic (*Panicum maximum* var. *trichoglume*), Rhodes grass (*Chloris gayana*), buffer grass (*Cenchrus ciliaris*) or signal grass (*Brachiaria decumbens*) sown between the leucaena rows. A high stocking rate (up to 3-4 animals/ha) and liveweight gain (up to 1 kg/head/day) can be achieved with this system. A record liveweight gain of 1,442 kg/ha for cattle grazing a grass/legume pasture was achieved on an irrigated leucaena/pangola grass mixture in the Ord River District of North Western Australia (Jones 1986).

Other tree legume species that are being investigated for use in extensive grazing systems include *Calliandra calothyrsus, Albizia chinensis, Cajanus cajan, Gliricidia sepium* and *Sesbania sesban.*

The Nutritive Value of Tree Legumes

Tree leaves and pods form a natural part of the diet of many ruminant species and have been used traditionally as sources of forage for domesticated livestock in Asia, Africa and the Pacific (Skerman 1977, NAS 1979, Le Houerou 1980a). Although not all forage trees are legumes, more than 200 species of leguminous trees are reported to be used for forage, with most species being tropical or subtropical in origin. The most commonly used species come from the genera *Acacia, Albizia, Calliandra, Desmanthus, Desmodium, Gliricidia, Leucaena, Prosopis* and *Sesbania* (Brewbaker 1986). Compared with herbaceous legumes, tree and shrub legumes have received relatively little attention in the search for productive and persistent forage sources for the tropics (NAS 1979).

Tree legumes must have both desirable agronomic characteristics and high nutritive value to be useful as forages. The nutritive value of a feed is determined by its ability to provide the nutrients required by an animal for its maintenance, growth and reproduction. Tree legumes have been mostly used as feeds for ruminants, although there are some reports of their inclusion in the diets of non-ruminants (pigs and poultry). The leaves, stems and fruits may be used either as a complete feed or as a supplement to other feeds. In some species, a major limitation to the use of one or more of these components is the presence of toxic and/or anti-nutritive factors.

Nutritive Value

Nutritive value is a function of the feed intake (FI) and the efficiency of extraction of nutrients from the feed during digestion (digestibility). Feeds of high nutritive value promote high levels of production (live weight gain). Feed intake in ruminants consuming fibrous forages is primarily determined by the level of rumen fill, which in turn, is directly related to the rate of digestion and passage of fibrous particles from the rumen. Voluntary consumption of feed may also be modified by animal preference, some feeds being eaten in smaller or larger amounts than predicted by digestibility (D). The acceptance or edibility (palatability) of a feed has been related to both physical characteristics (hairiness and bulk density) and the presence of compounds which may affect taste and appetite (volatile oils, tannins and soluble carbohydrates).

The productivity of ruminants is closely associated with the capacity of a feed to promote effective microbial fermentation in the rumen and to supply the quantities and balances of nutrients required by the animal tissues for different productive states.

There is no simple parameter of the quality of tree legume foliage. Chemical composition alone is an inadequate indicator of nutritive value. Modern concepts of feed evaluation require that quality be assessed in terms of the capacity of a feed to supply nutrients in proportions balanced to meet particular productive functions (Leng, 1986). The nutritive value of feeds should be ranked on the following characteristics:

- ☆ Voluntary consumption potential,
- ☆ Potential digestibility and ability to support high rates of fermentative digestion,
- ☆ High rates of microbial protein synthesis in the rumen relative to volatile fatty acids (VFA) produced (fermentation protein/energy (P/E) ratio),
- ☆ High rates of propionic acid synthesis (glucogenic) relative to total VFA synthesis (fermentation glucogenic/energy (G/E) ratio), and
- ☆ Ability to provide bypass nutrients (protein, starch and lipid) for absorption from the small intestine (absorbed P/E and G/E ratios).

Chemical Composition of Forage Tree Legumes

Much of the considerable information now available on the chemical composition of tree foliage (Gohl 1981, Le Houerou 1980a) is from proximate analysis (total N. ether extract, crude fibre, and nitrogen free extractives) and is of limited value as a predictor of nutritive value. Analyses based on detergent extraction are more useful since plant dry matter is separated into a completely digestible fraction (neutral detergent solubles (NDS)) representing cell contents, and a partially digestible fraction (neutral detergent fibre (NDF)) representing plant cell walls.

Elements (N, P, S, etc.) composition provides values which can be compared with animal requirements. However, whilst values less than predicted requirements are indicative of deficiency, values greater than prescribed are not necessarily indicative of sufficiency. Not all elements are fully available for use by the microbial population in the rumen or for absorption in the intestines.

Proteins and Tannins

The protein content of forage tree legume leaves (12-30 per cent) is usually high compared with that of mature grasses (3-10 per cent). The proteins are digested in the rumen to provide ammonia and amino acids for microbial protein synthesis. Microbial cells then pass to the small intestine, providing the major source of absorbed amino acids for the ruminant. In some cases, feed proteins may escape digestion (bypass proteins) in the rumen and provide additional protein for absorption in the small intestine. The microbial population in the rumen requires a minimum level of ammonia

(70 mg N/l) to support optimum activity; lower values are associated with decreased microbial activity (digestion) and are indicative of nitrogen deficiency. Feeds containing less than 1.3 per cent N (8 per cent crude protein) are considered deficient as they cannot provide the minimum ammonia levels required. All forage tree legumes have N contents higher than this value, and may be judged adequate in protein. However, tannins found in some tree legume leaves form complexes with plant proteins which decrease their rate of digestion (degradability) in the rumen, thereby decreasing rumen ammonia concentration and increasing the amount of plant protein by passing the rumen. Where the tannin-protein complexes are dissociated in the low pH of the abomasum, an additional source of protein is made available for absorption by the animal. In other cases, the tannins protect the proteins from digestion even in the small intestine.

Tannins may therefore have a beneficial effect (increasing bypass protein or decreasing ammonia loss) or a detrimental effect (depressing palatability, decreasing rumen ammonia, decreasing post-ruminal protein absorption) on protein availability. It is clear that the interpretation of the nutritional value of protein in forage trees requires information on the nature and action of tannins. The proteins in the leaves of species which do not have tannins (*Albizia lebbeck, Enterlobium cyclocarpum, Albizia saman* and *Sesbania* spp.) will be rapidly degraded in the rumen, providing high levels of rumen ammonia, much of which will ultimately be wasted by excretion as urinary urea. Such feeds provide N in a similar way to urea. Species which contain some tannins will therefore provide both degradable and undegraded rumen N and will be more effective sources of supplemental N for ruminants. Nevertheless, the significance of tannins in tree legume forage is poorly understood, with low concentrations being beneficial and high concentrations detrimental. It is also likely that not all tannins act similarly; this area requires further study.

Macro and Trace Element Content

There is little information available on the trees trace elements (Cu. Mn, Zn, Co, I) and only fragmentary data on the macro-elements. Sulphur (S) in plant material is mainly found in the form of sulphur amino acids and is required, together with N. for microbial protein synthesis in the rumen. Concentrations greater than 1.5 g/kg dry matter or N: S ratios less than 15:1 are considered adequate. However, where protein digestion in the rumen is limited by complexing with tannins, S rather than N may become the limiting factor in microbial protein synthesis. From the data available, most species appear to meet the S requirements of ruminants, with the possible exception of the *Acacia* spp.

The minimum requirement of ruminants for phosphorus (P) varies from 1.2 to 2.4 g/kg feed dry matter depending on physiological function. Forage tree leaves generally have high P concentration. Mulga (*Acacia aneura*) is an exception with low P contents and responses of sheep to P (+ molasses) supplementation have been reported. Calcium (Ca) is closely associated with P metabolism in the formation of bone, and a Ca:P ratio of 2:1 is usually recommended for ruminant diets. Ca is rarely limiting in forage diets and the same is true for forage trees. However, high concentrations of oxalic acid in leaves may decrease the availability of Ca during digestion. Gartner and Hurwood (1976) have suggested that high oxalate levels in mulga affect Ca metabolism in sheep. Magnesium (Mg) and potassium (K) are found in excess of requirements in tree-leaves and are seldom a limiting dietary factor in ruminants.

Although sodium (Na) deficiency has been recorded in cattle grazing tropical pastures, short term deficiencies are rare. Ruminants effectively conserve tissue Na by recycling it through the rumen. The recommended requirement for Na in ruminant diets is 0.7 g/kg dry matter. Some tree species appear to be marginal in Na, but deficiencies are probably not common as forage tree leaves usually form only part of a ruminant's diet. Deficiencies of minerals other than S and P appear to be unlikely, although leucaena is reported to be low in both Na and I.

Limitations to Nutritive Value of Fodder Trees as a Sole Feed

Some tree legumes contain anti-nutritive factors which adversely affect nutritive value. For this reason, depending on the species, tree legume foliage may be of lower nutritive value as a sole feed than as a supplement to other feeds. The significance of secondary plant compounds becomes more evident when tree foliage is the only feed consumed.

Acacia species are generally of low nutritive value, and as a sole feed are little better than a maintenance feed for stock. Mulga (*Acacia aneura*) has received considerable research attention in Australia. Its nutritive value may be greatly increased by the provision of specific supplements. McMeniman and Little (1974) first demonstrated that supplementation of mulga with P in molasses increased wool growth in sheep.

Research with leucaena has resulted in the discovery of rumen bacteria capable of degrading 3 hydroxy-4 (1H)-pyridone (DHP). Inoculation of cattle with the bacteria increases the intake and productivity of cattle grazing this tree legume in Australia. There are prospects for isolating other bacteria with beneficial functions from the rumen which may be used as an inoculum to animals to offset the detrimental effects of tannins and other secondary plant compounds.

New problems will arise with each new species that shows agronomic promise, and careful evaluation of the nutritive value of each introduction needs to be made.

Agro-ecological Range of Species

Agroforesters require forage tree legumes adapted to a wide range of environments. A summary of the general ecological adaptation of key species is given in Table 19.1.

There is a wide range of ecological adaptation among tree legumes, although there are no single species suited to the entire range of conditions. Managers must select carefully to ensure successful growth of tree species in their environment.

Establishment of Forage Tree Legumes

Some of the great advantages of many tree legumes, once established, are their longevity, vigorous growth and apparent immunity against competition from lower-growing herbaceous species. The tall stature and deep-rooted habit of tree legumes effectively insulate them from the most serious effects of competition. Consequently, in pasture systems, management of tree legumes to maintain grass/legume balance is simpler than with herbaceous legumes.

Table 19.1: Adaptation of some Fodder Tree Legumes to Various Environments

Species	*Acid Soils (pH<5.5)*	*Cool Temp. (15–25°C)*	*Low Rainfall (<500 mm)*	*Medium Rainfall (5–1000 mm)*	*High Rainfall (>1000 mm)*	*Poor Drainage*	*High Salinity*
Acacia aneura	T	T	T	NT	NT	NT	NT
Acacia angustissima	T	NT	NT	NT	T	NT	NT
Acacia nilotica	NT	NT	NT	NT	NT	T	T
Acacia tortilis	NT	NT	T	NT	NT	NT	NT
Albizia chinensis	T	T	NT	T	T	NT	NT
Albizia lebbeck	T	T	T	T	T	NT	T
Albizia saman	T	NT	NT	NT	T	T	NT
Calliandra calothrysus	T	NT	NT	NT	T	NT	NT
Chamaecytisus palmensis	NT	T	NT	T	NT	NT	NT
Cratylia agentea	T	NT	NT	NT	T	NT	NT
Desmodium rensonii	T	NT	NT	NT	T	NT	NT
Desmodium virgatus	NT	NT	NT	T	T	NT	NT
Erythrina spp.	T	NT	NT	NT	T	NT	NT
Faidherbia albida	NT	NT		NT	NT	T	NT
Flemingia macrophylla	T	NT	NT	NT	T	T	NT
Gliricidia sepium	T	NT	NT	NT	T	NT	NT
Leucaena diversifolia	NT	T	NT	T	T	NT	NT
Leucaena KX2 hybrid	NT	T	NT	T	T	NT	NT
Leucaena leucocephala	NT	NT	NT	T	T	NT	NT
Leucaena pallida	NT	T	NT	T	T	NT	NT
Leucaena trichandra	NT	T	NT	T	T	NT	NT
Prosopis juliflora	NT	NT	T	NT	NT	NT	NT
Sesbania grandiflora	NT	NT	NT	NT	T	T	T
Sesbania sesban	NT	T	NT	NT	T	T	T

T: Tolerant; NT: Not tolerant.

Source: Roshetko *et al.* (1996), Shelton (1994)

However, most tree legumes exhibit very slow growth as seedlings and at this stage of growth are vulnerable to competition from weeds and predation from wildlife. Long-lived species are often particularly slow as seedlings as much of their early growth is directed to establishing a strong root system. The poor establishment record of *Leucaena leucocephala* in subhumid Queensland is currently the major impediment to its more rapid adoption. This often results in an extended period before first grazing can occur and in a high rate of establishment failure (Lesleighter and Shelton 1986).

Methods for establishment of tree legumes are required which are quick and more reliable. Slow and/or unreliable establishment increases the time and level of risk before economic returns are feasible, and farmers are much less likely to adopt new technology when establishments risks are significant. This section summarizes what is known about establishment of tree legume species for utilization in agricultural systems.

Planting Methods

Direct Seeding

Planting of tree legumes by seed to form hedgerows is the most common method for broad-acre sowings. Appropriate row spacings for leucaena in pasture sowings vary from 3 to 10 m with wider spacings used in drier environments. The current recommendation in central Queensland is single or double rows 1 m apart with approximately 4-5 m between centres. If plant spacing within rows is 30-50 cm, this gives a population of 13,000-33,000 plants/ha, vastly less than the 75,000-140,000 plants/ha reported to be necessary to achieve peak forage yield (Brewbaker *et al.*, 1985). Smaller plant populations, in wider rows, may provide better rationing of limited water supply and an opportunity to intercrop the rows with grass.

In Australia, tree legumes (primarily leucaena are normally directly seeded using a seed drill, into fully prepared and clean cultivated seedbed..

Planting Seedlings

Most tree legumes are readily established from transplanted seedlings. Seedlings are first grown in greenhouse nurseries in polythene bags or in small plastic dibble tubes until they reach a height of 30-50 cm. After a short period of 'hardening' in the open air, seedlings are directly transplanted into the field into moist soil. Weeds need to have been previously controlled either mechanically or chemically. Watering and protection of seedlings from predators will be necessary until trees become well established.

Some nitrogen fixing trees can be planted from stump cuttings which are easier to transport into the field.

Stump cuttings can be made from seedlings which reach 60-90 cm in height and 10-20 mm in diameter in nursery seedbeds. They are first carefully removed when the seedbed is thoroughly wet and stems and roots cut 15-20 cm above and below the crown. Gliricidia stump cuttings can survive several weeks if kept moist although survival is best when transplanted promptly.

A recent comparison of planting methods shows the superiority of transplanted seedlings over direct seeding methods especially when weed growth is not adequately controlled. Survival of transplanted seedlings was also higher than that of plants from direct seeding methods.

Vegetative Propagation

Vegetative propagation of tree legumes is commonly practiced with some species. Its advantage is more rapid establishment of new stands which are genetically identical to the parent lines without the need for seed collection. Disadvantages are that it requires more hand labour and the root development of cuttings may be shallow and devoid of a strong taproot compared with seedling grown trees. Shallow rooted trees are more susceptible to drought and wind damage.

Gliricidia sepium is commonly planted vegetatively and a full description of propagation methods is given in Glover (1989). Gliricidia establishes readily from cuttings or 'quick sticks' and is ideal for shade trees, support trees or 'living fences'. Cuttings should be mature branches >7 cm in diameter which are brownish-green in bark colour. The cutting is normally cut obliquely at both ends, discarding the younger tips, and the base inserted 20-50 cm into the soil depending on the length of the cutting. Cuttings for living fences may be up to 200 cm long whilst those for hedgerows may be 30-50 cm in length. In Indonesia, cuttings are sometimes planted as close as 10 cm apart with alternate cuttings bent sideways at 45° and plaited onto upright cuttings. This makes a surprisingly strong fence. In other areas, barbed wire is strained along the line of rooted cuttings and anchored on supported comer posts to make an equally strong fence. The fences can be periodically pruned to provide fodder, green manure, fuelwood or stakes for new fences. Frequency of pruning depends on the environmental conditions for growth and the end use of prunings. Living fences around agricultural fields need to be pruned regularly to reduce shading.

The *Sesbania* species seed prolifically and are normally planted from seed, although research suggests that some *Sesbanias* can be established from cuttings (Evans and Macklin 1990). *Sesbania* species can also be propagated using *in vitro* methods.

Similar techniques may be used for other species of tree legumes.

Root Systems of Tree Legumes

The poor establishment characteristics of tree legumes can be partly related to their rooting characteristics. Trees have evolved for long-term survival rather than quick early productivity. Their root systems have a high component of permanent structural roots as well as a system of fine roots responsible for nutrient and water uptake. A greater proportion of assimilates is translocated to non-productive structural development in tree roots.

Trees also have a much lower root length density than grasses (Bowen 1985). Atkinson (1980) reported root length densities ranging from 0.8 to 69 cm/cm^3 for horticultural trees compared with 1004,000 cm/cm^3 for grasses. Swasdiphanich (1993) measured root length densities in the surface 50 cm of soil of 0.5 cm/cm^3 for *L. leucocephala* and 2 cm/cm^3 for *Calliandra calothyrsus*. Low root densities make it difficult for roots to access poorly mobile nutrients ($H_2PO_4^-$, K^+ and NH_4^+) or immobile nutrients (Cu^{2+} and Zn^{2+}) in the soil (Bowen 1985). Since the major concentration of tree roots occurs in the surface 15-30 cm of soil (Bowen 1985) it is clear that grasses will have a competitive advantage over trees especially in the seedling stage. Weeds, and particularly grass weeds, must be carefully controlled during the seedling phase of tree legumes. *Eucalyptus* species, which have adapted to grow in low nutrient soils, have quite fine roots (Bowen 1980) and this may make *Eucalyptus* seedlings more competitive with grasses.

There are a number of factors which can be manipulated to improve seedling growth rates and these are now considered.

Factors Affecting Seedling Growth Rates

Seed Treatment

Scarification

The hard or waxy coats on the seed of many tree legumes inhibit the absorption of water and prevents uniform germination. The seed coat must be broken or scarified before germination will occur. Without scarification, the germination percentage may be <10 per cent. Anon (1989) provides a summary of scarification methods for a range of tree legumes (Table 19.2). The most common method is hot water treatment, but sulphuric acid or mechanical scarification methods are also used.

Table 19.2: Pre-germination Treatment for Some Nitrogen Fixing Trees (Anon. 1989)

Sl.No.	Species	Treatment*
1.	*Acacia acuminate*	C, D
2.	*Acacia aneura*	A, C
3.	*Acacia angustissima*	A, C
4.	*Acacia auriculiformis*	A for 30 s, B for 15 min. C
5.	*Acacia crassicarpa*	A for 30 8, C
6.	*Acacia holosericea*	A for 1 min. C
7.	*Acacia mangium*	A for 30 s, C
8.	*Acacia mearnsii*	A, C
9.	*Acacia melanoxylon*	A, B for 15 min. C
10.	*Acacia nilotica*	A, C, D
11.	*Acacia polyacantha*	D
12.	*Acacia saligna*	A, C
13.	*Acacia senegal*	C, D
14.	*Acacia tortilis*	A, C, D
15.	*Albizia lebbeck*	A, C, D
16.	*Albizia procera*	A
17.	*Albizia saman* (syn. *Samanea saman*)	A, C
18.	*Alnus species*	no treatment needed
19.	*Cajanus cajan*	no treatment needed

Contd...

Table 19.2–Contd...

Sl.No.	Species	Treatment*
20.	*Calliandra calothyrsus*	A, C, D
21.	*Casuarina species*	no treatment needed
22.	*Chamaecytisus palmensis*	A for 4 min
23.	*Dalbergia spp.*	D
24.	*Enterolobium cyclocarpum*	C, D
25.	*Erythrina poeppigiana*	D
26.	*Faidherbia albida* (syn. *Acacia albida*)	A, B for 20 min. C, D
27.	*Flemingia macrophylla*	no treatment needed
28.	*Gliricidia sepium*	no treatment needed
29.	*Leucaena* spp.	A, B for 5-10 min. C
30.	*Mimosa scabrella*	C, D
31.	*Paraserianthes falcataria* (syn. *Albizia falcataria*)	A, B for 10 min. C
32.	*Pithocellobium dulce*	no treatment needed
33.	*Prosopis spp.*	A, C
34.	*Pterocarpus indicus*	no treatment needed
35.	*Robinia pseudoacacia*	A, B for 20-60 min. C
36.	*Sesbania grandiflora*	C, D
37.	*Sesbania sesban*	C, D
38.	*Tipuana tipu*	no treatment needed

* Treatments:

A. Pour boiling water over seeds, about 1 litre water per 250 9 of seeds or about five times as much water as seed, stir gently, pour off after 2 min (or as specified), replace with tap water and soak overnight.

B. Cover seeds with concentrated sulphuric acid, stir gently for recommended soaking time, pour off acid and rinse well in water.

C. Scratch or nick the round end of each seed with a file, knife or nail clipper. Do not cut the cotyledon.

D. Soak in cold/tepid water for 24 h.

Rhizobium Inoculation

Forage tree legume plants which are not effectively nodulated will be pale in colour and will not grow vigorously due to inadequate nitrogen fixation leading to nitrogen deficiency. In preliminary evaluation trials it is possible to confuse this problem with poor adaptation to the environment.

Quite a number of tree legume species have specific *Rhizobium* requirements for effective nodulation and nitrogen fixation. However, in many cases, the specificity of the species is not known, and the most effective *Rhizobium* strain has yet to be identified.

The *Rhizobium* requirements of leucaena are now well known and it is possible to obtain peat cultures of effective *Rhizobium* from seed suppliers, when ordering seed. This is not the case with most other species.

Gliricidia sepium and *Sesbania sesban* are known to require specific *Rhizobium*. In the latter species, there is a host-strain interaction and different accessions of *S. sesban* require different strains of bacteria.

Until more is known about host plant - bacterial strain specificity, care should be exercised when evaluating new varieties of tree legumes. In some cases, it may be wise to apply nitrogen fertiliser, or to use soil from around the roots of well-grown nodulated trees.

Site Selection

Successful establishment of tree legume species will only be achieved if the characteristics of the proposed planting site are matched against the climatic and edaphic requirements of the species. If the establishment requirements are not fully met, growth of seedlings will be poor unless the soil is amended or an alternative site is found.

Fertilizer Application

The responsiveness of the tree legumes to applications of fertilizer depends on their external nutrient requirements for maximum growth, and the fertility of the soil being fertilized. There are many experiments providing information on the response of tree species to fertilizer application but these tend to be site specific and therefore of limited value for predicting fertilizer requirements elsewhere. More work is required to elucidate the nutrient requirements of the commonly used species.

Vesicular Arbuscular Mycorrhizae

Most tree legume species form symbiotic associations with naturally occurring soil fungi called vesicular arbuscular mycorrhizae (VAM). This association assists the roots to exploit more fully the soil volume and to gain improved access to available nutrients especially phosphorus. Nutrient ions are transferred to the roots via the hyphae. VAM therefore compensate for the low root length densities of trees. VAM can produce 80 cm of hyphae per cm of fine root infected (Sanders and Tinker 1973) with a length/weight ratio of 500 times that of fine roots. Mycorrhizal infection is therefore an important strategy complementary to, and sometimes replacing, fine root production (Bowen, 1985).

Both *Calliandra calothyrsus* and *L. leucocephala* are known to form mycorrhizal associations. Young leucaena seedlings are very dependent on rapid early mycorrhizal infection of the roots for adequate phosphorus supply.

Weeds

The slow seedling growth of many tree legumes makes them susceptible to competition from fast growing weeds which may slow or completely dominate their growth. It is therefore vital that young seedlings be protected from weed competition until they are well established. This can be performed by hand, using hand-held or tractor drawn machinery, or by the use of herbicides.

Chemical weed control with leucaena is now well understood. Pre-emergence control of weeds can be obtained with Trifluralin (0.5 kg a.i./ha) and Alachlor (3 kg a.i./ha) when incorporated, or The post-emergence herbicides Fluazifop (2 kg a.i./ ha) and Bentazone (2 kg a.i./ha) are effective against grass and broad leaved weeds respectively without being excessively phytotoxic to leucaena.

Little is known about herbicides for use with other tree legumes although Glover (1986) reported that Glyphosate (1 kg a.i./ha) and Simazine (1 kg a.i./ha) were effective and non-phytotoxic pre-emergent herbicides for control of grass and broad leaved weeds in *Gliricidia sepium.*

Conclusion

Tree legumes are multipurpose in nature. They provide high quality forage for feeding of livestock besides other advantages. Tree legumes have been neglected in past in Agriculture as feed for animals. In an era of climate change they offer true potential to colonise and rehabilitate adverse environment. *Leucaena, Sesbania, Erythrina, Acacia, Calliandra, Albizia,Dalberzia, Desmodium, Prosopis* offer a wide range of tree legumes having potential for utilization as fodder for animals. Though a lot of work has already been done in India and abroad; we need to standardize agronomic practices for each legume fodder separately. In other words, more extensive research is immediately warranted to harvest the fruit of "sustainable quality fodder".

References

Atkinson, D. (1980) The distribution and effectiveness of roots of tree crops. *Horticultural Review* 2, 424-490.

Bowen, G.D. (1980) Coping with low nutrients. In: Pate, J.S. and McComb, A.J. (eds), *The Biology of Australian Native Plants.* University of Western Australia Press, Perth, pp. 33-64.

Bowen. G.D. (1985) Roots as a component of tree productivity. In: Cannell, M.G.R. and Jackson, J.E. (eds), *Attributes of Trees as Crop Plants.* Institute of Terrestial Ecology, Natural Environment Research Council, Abbots Ripton, United Kingdom, pp. 303-315.

Brewbaker, J.L. (1986) Leguminous trees and shrubs for Southeast Asia and the South Pacific. In: Blair, G.J., Ivory, D.A. and Evans, T.R. (eds), *Forages in Southeast Asian and South Pacific Agriculture.* ACIAR Proceedings No. 12, ACIAR, Canberra, pp. 43-50.

Brewbaker, J.L., Hegde, N., Hutton, E.M., Jones, R.J., Lowry, J.B., Moog, F. and van den Beldt, R. (1985) *Leucaena Forage Production and Use.* NFTA, Hawaii, 39 pp.

Dougall, H.W. and Bogdan, A.V. (1958) Browse plants in Kenya. *East African Agricultural Journal,* April 1958, 236-246.

Gartner, R.J.W. and Hurwood, I.S. (1976) The tannin and oxalic acid contents of *Acacia aneura* (mulga) and their possible effects on sulphur and calcium availability. *Australian Veterinary Journal* 52, 194-196.

Glover, Nancy. (1986) Herbicide screening for *Gliricidia sepium. Nitrogen Fixing Tree Research Reports* 4, 59-61.

Gohl, B. (1981) Tropical Feeds. *FAO Animal Production and Health Series* No. 12. FAO, Rome, 529 pp.

Jones, R.J. (1986) Overcoming the leucaena toxicity problem to realise the potential of leucaena Australian Institute of Agricultural Science, Queensland Branch, Bulletin No. 289, pp. 4-9.

Le Houerou, H.N. (ed.) (1980a) *Browse in Africa*. ILCA, Addis Ababa, Ethiopia, 421 pp.

Leng, R.A. (1986) Determining the nutritive value of forage. In: Blair, G.J., Ivory, D.A. and Evans, T.R. (eds), *Forages in Southeast Asian and South Pacific Agriculture.* ACIAR Proceedings No. 12, ACIAR, Canberra, pp. 111-123.

Lesleighter, L.C. and Shelton, H.M. (1986) Adoption of the shrub legume *Leucaena leucocephala* in Central and Southeast Queensland. *Tropical Grasslands* 20, 97-106.

McMeniman, N.P. and Little, D.A. (1974) Studies on the supplementary feeding of sheep consuming mulga (*Acacia aneura*) 1. The provision of phosphorus and molasses supplements under grazing conditions. *Australian Journal of Experimental Agriculture and Animal Husbandry* 14, 316-321.

Moloney, R.A., Aitken, R.L. and Gutteridge, R.C. (1986) The effect of phosphorus and nitrogen applications on the early growth of *Adenanthera pavonina, Albizia falcataria* and *Schleinitzia insularum. Nitrogen Fixing Tree Research Reports* 4, 34.

Moog, F.A. (1985) Forages in integrated food cropping systems. In: Blair, G.J., Ivory, D.A. and Evans, T.R. (eds), *Forages in Southeast Asian and South Pacific Agriculture.* Proceedings of an international workshop held at Cisarua, Indonesia. ACIAR Proceedings Series No. 12, Canberra, pp. 152-156.

NAS (1979) *Tropical Legumes Resources for the Future.* National Academy Press, Washington DC, 331 pp.

NAS (1979) *Tropical Legumes Resources for the Future.* National Academy Press, Washington DC, 331 pp.

Reynolds, C. and Atta-Krah, A.N. (1986) Alley farming with livestock. In: Kang, B.T. and Reynolds, L. (eds), *Alley Farming in the Humid and Subhumid Tropics.* Proceedings of an international workshop held at Ibadan, Nigeria pp. 27-36.

Robinson, P.J. (1985) Trees as fodder crops. In: Cannell, M.G.R. and Jackson, J.E. (eds), *Attributes of Trees as Crop Plants.* Institute of Terrestrial Ecology, Huntingdon, UK, pp. 281-300.

Roshetko, J.M., Dagar, J.C., Puri, S., Khandale, D.Y., Takawale, P.S., Bheemaiah, G. and Basak, N.C. (1996). Selecting species of nitrogen fixing trees. In: Roshetko, J.M. and Gutteridge, R.C. (eds), Nitrogen Fixing Trees for Fodder Production - A Field Manuel. Winrock International, Morrilton (AR), USA. pp.23-23.

Sanders, F.E. and Tinker, P.B. (1973) Phosphate flow into mycorrhizal roots. *Pesticide Science* 4, 385-395.

Shelton, H.M. (1994a). Environmental adaptation of forage tree legumes. In: Gutteridge, R.C. and Shelton, H.M.(eds), Forage Tree Legumes in Tropical Agriculture. CAB International, pp. 120-131.

Shelton, H.M. (1994b). Establishment of forage tree legumes. In: Gutteridge, R.C. and Shelton, H.M.(eds), Forage Tree Legumes in Tropical Agriculture. CAB International, pp. 132-142.

Skerman, P.J. (1977) *Tropical Forage Legumes.* FAO Plant Production and Protection Series No. 2. FAO, Rome, 609 pp.

Skerman, P.J. (1977) *Tropical Forage Legumes.* FAO Plant Production and Protection Series No. 2. FAO, Rome, 609 pp.

Steele, K.W. and Vallis, I. (1988) The nitrogen cycle in pastures. In: Wilson, J.R. (ed.), *Advances in Nitrogen Cycling in Agricultural Ecosystems.* CAB International, Wallingford, UK, pp. 274-291.

Swasdiphanich, S. (1993) Environmental adaptation of some forage tree legumes. PhD thesis, The University of Queensland.

Chapter 20

Cow Pea (*Vigna sinensis* L.) Legume Forage for Enhancing Quality Fodder

Deokaran

Krishi Vigyan Kendra, Buxar
ICAR Research Complex for Eastern Region, Patna – 800 014, Bihar

Cow pea (*Vigna sinensis* L.) commonly known as 'lobia' is used as a pulse, a fodder and a green manure crop. Being rich in protein and containing many other nutrients. It is used for both human consumption and as a concentrate feed for the cattle. The crop forms excellent forage. The feeding value of cowpea forage is high and quite comparable to lucerne. On dry weight basis cowpea grains contains 23.4 per cent protein, 1.8 per cent fat and 60.3 per cent carbohydrate. It is also a rich source of calcium and iron. The crop gives vegetative growth and covers the ground so well that it checks the soil erosion in problem area and can later be ploughed in a green manure.

Origin and History

Cowpea is probably a native of central Africa where almost all the wild forms are found. It has been cultivated since very ancient time in the Mediterranean region by the Greeks, Romans and Spaniards. It has now been introduced in many countries throughout the world. It is also claimed to be indigenous to India.

Area and Distribution

In India it is mainly grown in central and peninsular regions. In northern India it is grown in Uttar Pradesh, Bihar, Madhya Pradesh, Punjab, Haryana and Delhi.

Climate

It is a warm season crop and can be grown in all tropical and subtropical areas. It can tolerate drought to some extent but cannot tolerate water logging. The germination is better at 12–15°C temperatures and the crop thrives best between 21°-35°C temperature. Frost is harmful for this crop. Partial shade can be tolerated. The varieties show varying response to temperature and day length. It can also grow under shade of trees but cannot tolerate cold or frost.

Soil

Cowpea can be grown in almost all type of soil but grows better in well-drained loam or slightly heavy soils. Saline or alkaline soils are not good. It can be grown on 7.0 to 7.5 pH.

Varieties

The criteria in the development of cowpea varieties for fodder purpose have been primarily the fodder yield and its quality. However, the same variety may be raised for more than one purpose. For instance, variety meant for grain may be used for vegetable and fodder purposes. Similarly fodder variety may be used for green manuring purpose. Several cow pea varieties are grown in various parts of our country. The promising ones are as below:

Type-2

It is a late maturing (125-130 days) variety suitable for growing in plains of Uttar Pradesh and Bihar. Plants are spreading type with dark green leaves. Flowers are bluish-purple. Its yield potential is 300-325 quintals of green fodder per hectare and green yield is 12-18 quintals per hectare.

Cowpea-74

This variety is suitable for cultivation as a fodder crop as well as a grain crop. It has erect, dark green and non-twining plants. It is resistant to virus diseases. Its grains are medium, white and attractive. It gives about 300 quintals of green fodder and about 15 quintals of grain per hectare. It is suitable for growing in Punjab and Haryana.

C-152

This variety takes about 105-110 days to mature. It is suitable for Kharif season planting. Seeds are brownish and medium bold. It is a grain variety but sometimes used for fodder purpose also. It is also to grow in Punjab, Jammu and Kashmir, Delhi, Haryana, U.P., Bihar and M.P. It is good for pure and mixed cropping.

UPC-5286, PC-287, IFC-8401

These varieties are suitable for Kharif and summer crop and yield potential is 350-400 quintal per hectare.

K-397, K-585, Sirsa-10, EC-4216, FOS-1, FS-145, Co-2, S-450 and S-457

These varieties are exclusively recommended for high yielding fodder purposes.

Rotations and Mixed Cropping

Cowpea can be grown in rotations with many crops. The most common rotations is as below:

1. Jowar+cowpea-berseem-maize+cowpea (one year)
2. Maize-oat-maize+cowpea (one year)
3. Maize-berseem-maize+cowpea (one year)
4. Jowar-berseem-maize+cowpea (one year)
5. Sudan grass-berseem-maize+cowpea (one year)

Field Preparation

The stubbles and other residues of previous crop should be picked up as far as possible to have a clean and smooth. Field should be prepared by giving two or three cross harrowings followed by planking.

Seed and Sowing

A seed rate of 30-35 kg per hectare for rainy season crop, while 35-40 kg is required in summer. Variation in seed rate depends on the seed size of a variety. For green manuring crop sown by broadcasting, the seed requirement is 35-40 kg per hectare.

In most of the areas cowpea is grown during the rainy and summer seasons. Sowing in June-July is common for rainy season crop, but it could be extended to August for early-maturing varieties. Similarly sowing in February-March is common for summer crop, particularly in northern plains, which could be extended upto mid-April.

Manures and Fertilisers

Depending upon availability, 15-20 tonnes/ha of farmyard manure may be incorporated in the soil at the time of field preparation. Being a nodule forming crop, cowpea does not require heavy nitrogen fertilization. Thus, only 20-25kg/ha N along with 50-70kg/ha of P_2O_5 and K_2O may be applied as basal dose before sowing. The doses of P and K may be based on the soil test value to economize on fertilization. Seed inoculation with *Rhizobium* culture is beneficial. In zinc-deficient areas zinc sulphate @ 10-15kg/ha may be applied in the soil.

Water Management

The rainy season crop do not requires irrigation but good drainage is essential. The early sown rainy season crop may need one or two irrigations in the pre-monsoon period. For raising summer crop, five to six irrigations may be given. During summer, because of high temperature and low relative humidity more irrigation are needed as compared to Kharif crop. The number and frequency of irrigation depend upon the soil type and weather prevailing during the growth period. Generally the crop should get irrigation at an interval of 10-15 days.

Weed Control

Effective control of weeds in the first 20-25 days of the crop season is essential. At least two weedings and hoeings are required to check the weeds. During rainy season weeds can be controlled by the use of chemicals too. Use Basalin 1 kg a.i. per hectare in 800-1000 litres of water as pre-planting spray. It should be well incorporated in the soil before sowing.

Diseases

Bacterial Blight of Cowpea

It is caused by *Xanthomonas vignicola*. Symptoms of this disease first appear on the cotyledons, primary and the new trifoliate leaves. The affected cotyledons are red and shrivelled. Necrotic spots are found on the margins of the primary leaves. Thereafter, the casual bacterium affects the stem and finally covers other parts of the plant.

Control Measures

1. Grow resistant varieties.
2. Use healthy seed from disease free field.
3. In case of severe infection, crop may be sprayed with 0.2 per cent Fytolan.

Cowpea Mosaic

It is a disease caused by a virus transmitted by aphids. The affected leaves become pale yellow and exhibit mosaic, vein banding symptoms. The affected leaves become reduced in size and show puckering. Pods are also reduced and become twisted.

Control Measures

1. Use healthy seed from healthy crop.
2. For controlling aphids spray Metasystox 0.1 per cent or any other systemic fungicide.

Powdery Mildew

This disease is caused by a fungus *Erysiphe polygoni*. Powdery mildew symptoms are visible on all the aerial parts of the affected plants. Symptoms first start from leaves and then spread to stem, branches and pods. Symptoms start with white powdery growth on leaves which may coalesce and cover the whole leaf with the white powdery growth.

Control Measures

1. After harvest, collect the plants left in the field and burn them.
2. The disease can be controlled by spray or wettable sulphur like Sulfex, Elosal or Hexasual at the rate of 2-3 kg per hectare in 800-1000 litres of water.

Rust

This disease is caused by a fungus, *Urontyces appendiculatus*. This disease affects the leaves, pods and sometimes new shoots. Symptoms on the leaves are very clearly visible and start from the lower surface of the leaf where small white pustules are found. These pustules contain uredia of the fungus. Brown coloured urediospores come out of these uredia.

Control Measures

Spray the crop with Dithane M-45 at the rate of 2 kg per hectare in 1000 litres of water.

Insect Pests

Hairy Caterpillar

It is one of the most important pests of cowpea crop. It causes severe damage to the crop, by eating away all the green matter of the leaves. The adult moth of this caterpillar lays eggs in large clusters and the young larvae are also congregated. They may damage the crop at seedling stage. Damage can be so severe that sometimes re-sowing may be necessary.

Control Measures

1. Collect and destroy the eggs and young larvae.
2. The young caterpillars can be killed by dusting 10 per cent BHC dust at the rate of 25-30 kg per hectare. For full grown caterpillars spray 1.5 litre Endosulfan 35 EC in 1000 litres of water per hectare.

Leaf Hoppers, Jassids and Aphids

The adults and nymphs of these pests suck the juice from the leaves and the damage is more severe when the plants are young. As a result of sucking the sap, the leaves turn brown and crumpled and the plants look sick.

Control Measures

1. Spray the crop with 0.1 per cent Metasystox or 0.04 per cent Monocrotophos 40 EC.
2. Give basal application of Thimet 10 per cent granules at the rate of 10 kg per hectare.

Harvesting

For fodder, the cutting of the crop depends upon the need and the stage of growth of the component crop soon with it. In general the crop should be cut when it attains the age of 40-45 days.

Yield

A good crop for fodder is obtained 250-350 quintals of green fodder yield per hectare.

References

Arvindham, S. and Das, Vijendra, L.D. 1996. Heterosis and combining ability in fodder cowpea. *Indian Journal of Pulses Reseach* 9(1):68-70.

Schoo, M.S., Bharadwaj,B.L. and Beru S.M.1990.Cowpea-88.A new dual purpose variety of cowpea. *Journal of Research.* Punjab Agricultural Universit. 271-383.

Vikrant, Singh Harbir, Malik C.V.S. and Singh B.P.2005, Grain yield and protein content of cowpea as influenced by farmyard manure and phosphorous application.*Indian Journal of Pulses Research.* 18(2):250-251.

Chapter 21
Weed Control in Forages and Fodder Crop

Lal Singh[1], *Aditya Kumar Singh*[2] *and Anil Kumar Singh*[3]

[1]*Division of Agronomy, Sher-e-Kashmir University of Agricultural Sciences and Technology of Kashmir, Shalimar, Srinagar – 191 121 (J&K)*

[2]*College of Post Graduate studies, Central Agricultural University, Barapani, Umiyam, Shillong*

[3]*ICAR Research Complex for Eastern Region, Patna – 800 014, Bihar*

E-mail : drlalsingh@rediffmail.com, drlalsingh72@rediffmail.com

Weed control on farmland is the process of limiting any given weed infestation to the extent that it permits economic crop production. The objective here is to limit the growth of unwanted plants without any attempt to eliminate them from the fodder/forage crop fields. The extent to which a given weed growth should be limited will depend upon the cost involved in the operation and the benefits anticipated from the operation. Prevention of weed is a prerequisite weed control measures. In this process all measures to deny the entry and establishment of new weeds in an area, large or small. However, in practice, weed prevention is also includes measures to check the every year spread of even existing weed species on the farm. Cultural and physical methods are also a good practice to overcome the weed infestation in cultivated as well as pastures and range lands with assurance of good quality feed. If there are necessary then herbicide should be applied to control the weeds effectively and economically.

Introduction

The importance of weed control in forage production should not be overlooked, especially when you consider the high investment cost associated with their production. Weeds reduce forage yield by competing for water, sunlight, and nutrients. In addition to yield losses, weeds can also lower forage quality, increase the incidence of disease and insect problems, cause premature stand loss, and create harvesting problems. Some weeds are unpalatable to livestock or, in some cases, may be poisonous. The chemical analyses of the weedy and weed free forages may some times not show differing nutritive values, particularly when there are leguminous weeds in the mixture, but in feeding tests the animals exhibit clear performance for the weed free forage. When some hungry animals are forced to feed upon weedy forage, their meat and milk are odd-flavoured and tainted. Weeds like *Cichorium intybus*, *Allium vineale*, *Argemone mexicana*, *Brassca kaber*, *Anthemis cotula*, *Thalapsi* sp. *Ambrosia* sp. *Helenium* sp. and *Oxalis acetosella* are particularly notorious in this respect. Certain weeds cause sickness in animals while others may prove fatal due to high level of specific alkaloids, tainnins, oxalates, glucosides, or nitrates. Halogeton (*H. glomeratus*), a weed of arid and semi-arid regions, has brought death to herds of sheep with its high oxalate content. Johnsongrass (*Sorghum halepense*) at its tillering stage and *Xanthium pungens* at its cotyledon stage are poisonous to animals due to their high prussic acid content. Corncockle (*Agrostemma githago*) seeds have been found poisonous to horses, cattle and poultry. Their toxic principle is githagin or agrostemine. Puncturvine (*Tribullus terrestris*), a weed of dry lands, induces in sheep extra-sensitivity to light. Also its thorny fruits cause sores in the hooves of animals. The spiny fruits of *Xanthium strumarium*, *Achyranthus aspera*, *Cenchrus settigerous*, *Cirsium arvense*, and *Cenchrus incertus* stick on to the mouth, tail and body of animals and annoy them badly. The leaves of lantana (*Lantana camara*) induce acute photosensitivity jaundice in animals due to their toxic principle 'Lantra dene-A'. Carrotgrass (*Parthenium hysterophorus*) causes contact dermatitis in livestock and it is reported to be poisonous to sheep. *Ageratum* spp. growing in abundance in Himalayas and Nilgiris, are often hazardous to livestock. A weed golden crownbread (*Verbesina enceliodes*) is poisonous to sheep and goat. Locoweeds (*Astragalus* and *Oxytropis* spp.) inflict abortive and tetratogenic effects on sheep and cattle. Such weeds possess the toxic alkaloid swainsoine. *Lupinus sericeus* induces crooked calf disease in cattle. Rhododendron species has been found responsible for causing diarrhoea in milch animals in Kashmir. It also stains the milk of affected animals with blood. Leafy spurge (*Euphorbia esula*) causes scours and weakness in cattle, and it is fatal to sheep. Sweet clover (*Melilotus alba*) contains a dicumarin which act as an anti-blood coagulant. *Heliotropium amplexicaule*, *Helenium* spp. and *Datura stramonium* are other sickening weeds for animals. Under drought conditions weeds like *Chenopodium*, *Amaranthus*, *Cirsium* and *Polygonum* spp. develop nitrate levels as high as 1000ppm or more, which cause asployxia in animals. Spines of *Tribullus terestris* can puncture animal skin. When ingested it can also injure the stomach. The hair of neetle (*Utrica urens*) cause severe itching and inflammation in animals. Due to these above facts weeds should be removed from cultivated fodder/forage crops as well as pastures and range lands to produce the quality fodders and maintain the health of livestock.

Weed management strategies in forages should focus first on precautions and cultural practices and then on chemical weed control practices. Vigorous, dense-growing forage stands have fewer weed problems. Thus, cultural and management practices that promote a highly competitive forage stand may prevent many weed problems. These practices include:

Prevention of Weeds

Strictly speaking, prevention of weeds embodies all measures to deny the entry and establishment of new weeds in an area, large or small. However, in practice, weed prevention is also includes measures to check the every year spread of even existing weed species on the farm. Weed prevention is a long term planning so that later the weeds could be controlled or managed more effectively than is possible when these are allowed to disperse freely. The important weed prevention measures are as follows :

1. Use clean and weed free seeds for sowing in cultivated field or rejuvenation of pastures/rangelands.
2. Avoid feeding, screenings and other material containing weed seeds to the farm animals.
3. Avoid adding weeds to the manure pits.
4. Clean the farm machinery thoroughly before moving it from one field to another.
5. Keep irrigation channels, fence-lines and uncroped areas clean.
6. Inspect your farm frequently for any strange looking weed seedlings. Destroy such patches of a new weed by digging deep and burning the weed along with its roots. Sterilize the spot with suitable chemical.

Forage Crop–Site Selection

Consider field histories when you select a field for forage production. It might be difficult to establish and maintain a weed-free stand in fields known to be infested with weeds such as thistle, dandelion, and quackgrass. In addition, some herbicides that are applied in previously grown crops have the potential to carry over and cause injury to newly seeded forages. More information on herbicides that have a potential to injure alfalfa and other forages can be obtained from your Manitoba Ag Rep office and directly from the Product label.

Weed Control in New Seedlings

Weed control is more critical during the first year than any other period of forage production. Forage seedlings grow slowly and are easily overcome by rapidly growing weeds. Research has shown that some broadleaf weed seedlings are capable of growing five times more rapidly than certain legume seedlings. Because alfalfa stands gradually decline with age, it is important to start with a good stand. A uniform, dense stand is more likely to survive longer and have fewer weed problems than a thin stand.

Fertilization

Adjusting soil nutrient levels according to soil test recommendations is important during the establishment phase and throughout the life of the forage stand. The objective is to achieve a competitive stand that is capable of suppressing weed emergence and growth. Proper fertility is not effective at eliminating established weeds, especially in areas where the forage stand is poor.

Cutting–New Stands

Clipping or mowing can be an effective option for controlling weeds, such as common cocklebur, in forage stands. This method controls weeds by removing the leaves and lateral buds that develop new growth. Annual broadleaf weeds have buds that develop above the soil surface; they are more easily controlled with clipping or mowing than grasses, which have crown buds near the soil surface. Now as low as possible to be effective. Because alfalfa and other legumes have crown buds, they can tolerate low clipping. When you clip new seedlings, be careful not to smother the forage with heavy residues. Remove clipped vegetation when weed infestations are heavy.

Mainting Established Stands

Established forages are capable of growing rapidly and competing against many weed seedlings during the growing season. However, weeds gradually invade fields where forage stands decline with age. Timely mowing and the use of herbicides may aid in weed control and prolong the life of the stand. If you have a weed problem that occurs in field borders, along fence rows, or in adjacent fields, you should mow or spray to prevent production and spread of weed seed from these areas into forage fields. This is particularly important for such weeds as thistle and dandelion, which are capable of producing a large number of seeds that easily spread to new areas.

Clipping Established Stands

The routine mowing of forages for hay is sometimes effective in controlling some perennial weeds by reducing food reserves and plant vigor. However, in grazed forages, livestock often selectively graze and may leave particular weed species. Mowing soon after livestock have been removed from the field can help control these weeds and prevent seed production and further spread of infestations.

Herbicides for Established Stands

Several herbicide options are available for established forage stands, refer to the Guide to crop Protection 2004 (Forage Crops Chart – Recommended Herbicides) for further information. You can use many of the same herbicides available for new seedlings. Furthermore, the deep root system of established plants such as alfalfa enables them to tolerate certain herbicides that are not suitable for new seedlings. When selecting herbicides for forages, you should consider such factors as: whether the herbicide can be applied as a dormant season, non-dormant, or between cutting treatment; effectiveness on weed species to be controlled; feeding and grazing limitations; rotational crop restrictions; and cost of treatment.

Scounting Methods of Forage Crops

Scouting for weed problems early is an effective tool for identifying and controlling weed problems before they develop into situations that cannot be easily managed. This requires a trained eye and the ability to identify weeds in their early growth stages. Winter annual weeds, such as stinkweed and flixweed, usually germinate in late fall and are present in early spring, whereas, the summer weed complex, which includes weeds such as wild oat and lamb's quarters, will be present after the first harvest through a killing frost in the fall.

Weed infestation levels or weed density should be determined by estimating the percentage of ground cover occupied by weeds. This can be accomplished by randomly selecting one site for every 10 acres within a field. A minimum of three sites should be selected in fields with fewer than 20 acres. At each field site, an area approximately 30 feet by 30 feet should be used to determine the percentage of weeds present. Keep in mind that fields that appear almost weed free could have a 5 per cent weed density. Only in extremely poor forage stands will weed infestations in excess of 50 per cent occur. At each site, record the predominant species and its size at the time of sampling.

Cultural and Physical Methods of Weed Control

Any weed management programme on farmland essentially begins with adaptation of good crop husbandry practices leading to a sturdy crop which could overpower the weeds, and make their subsequent control easy and more economical. On the contrary when fundamental principals of good crop husbandry are ignored, the crop become so weedy that their control is often disappointing. Some important good crop husbandry practices which can bring about effective suppression of weeds in farming systems are as follows :

Proper Crop Stand

Inconsistent and under population crop are prone to heavy weed infestations which become difficult to control later. Therefore, practices like selection of proper seed, right method of sowing, adequate seed rate, protection of seed from soil borne pests and diseases, etc. are very important to obtain proper and uniform crop stand capable of offering initial competition to the young weeds. In this respect it also essential to adopt a sowing method which will ensure emergence of the crop prior to the weeds.

Selected Crop Stimulation

Basal placement of fertilizers in the seed rows often helps in selectively stimulating the crop seedlings which can withstand competition from the weeds much better. Initially healthy crop seedlings also grow fast and close-in early to shade out the subsequent flush of weeds. In this respect, the various plant protection measures adopted to maintain the crop plant healthy should be considered as part of good crop husbandry leading to effective weed management. The top dressing of a fertilizer in the inter-rows should be done in a manner that major portion of it is utilized by crop plants, and the weeds are able to make only limited use of it.

Crop Rotation

Many of our weed problems exist with us because of practicing monocultures *i.e.* growing of same crop year after in the same field. For instance growing of leguminous fodder after leguminous fodder or perennial fodders throughout the year has led to several difficult weed problems. Introduction of some cereal crops after legumes or annual crop after perennial as a break crop has solved this problem. Parasitic weeds as well as the crop associated weeds can be discouraged by adopting well conceived crop rotations.

Summer Tillage

In major part of India there is a clear cut solar energy-rich, dry period of summer available, which should be utilized for desiccation of rhizomes, tubers, and roots of the perennial weeds to death. Initial tillage of the field in summer for this purpose should encourage clod formation. These clods which embody the weed propagules, upon drying desiccate the weed propagules within, much better than a pulverized soil. Subsequent tillage operations should break these clods into smaller units to further expose the shriveled weeds to the hot sun. Scientific studies have been conducted with certain perennial weeds to find the optimum weed growth stages for practicing summer tillage when downward translocation of food from the leaves to the rhizomes shall be minimal. Most studies indicate that an interval of 15-20 days after fresh weed emergence was optimum for the purpose. Dry weather and dry soil are pre-requisites for the success of summer tillage as a weed management practice.

Solarisation

This is another method of utilization of solar energy for the desiccation of weeds. In this method the soil temperature is further raised by 5-10 °C by covering a pre-soaked fallow field with thin transparent plastic sheet. The plastic sheet checks the long wave back radiation from the soil and also prevents loss of energy by hindering moisture evaporation. In some regions the technique raised the soil temperature to as high as 50 °C, which was sufficient to kill several kinds of weed propagules. In many experiments, four weeks or longer treatment reduced the weed seed viability by 80 per cent or more.

Reduction in Area Under Bunds and Channels

It is a matter of common observation that weeds grow profusely on bunds and channels of crop fields. Also, these weeds are often ignored by the farmers since they don't compete directly with crop plants. In western countries, to obviate this plastic sheets are used to bifurcate the land into small fields for purpose of irrigation. In India where plastic sheets are yet costly, reduction in the frequency of soil bunds by proper leveling and grading of the fields should be given due consideration in any good weed management plan.

Stale Seed Bed

A stale seed-bed is one where initial 1-2 flushes of weeds are destroyed before planting of a crop. This is achieved by soaking a well prepared field with either irrigation or rain and allowing the weeds to germinate. At this stage a shallow tillage or a non-residual herbicide like paraquat may be used to destroy the dense flush of young weed seedlings. This may be followed immediately by sowing a desired crop. The technique allows the crop to germinate in almost a weed free environment. Spike-tooth harrow is a very useful implement for destroying the emerging weeds during the preparation of stale seed-beds.

Inter-Cultivation

Inter-row tillage, also called inter-cultivation, used to remove the weeds largely from between the crop rows leaving the intra-row weeds unhurt. To some extent this situation can be improved if the inter-cultivation is practiced when the weeds are very young. At this stage stirring of the soil close to the crop rows, which occur during inter-cultivations will often disturb the intra-row weeds sufficiently to dislodge them. There are two serious limitations with inter-cultivation as a physical method of weed control first, it is not workable when the field is either too dry or too wet. Often, in rainy season the farmers in the heavy soil areas have to wait for day together before the soil attains the right physical condition for cultivation. During all this period the weeds grow fast and over power the crop. Second, the inter-row cultivations cut the surface roots of crops during the operation which may give a set-back to the crop.

In narrow row crops, hand weeding is the only physical method available for removing weeds. Although very effective, under the present hike in labour wages it is often a very costly operation, unless the field area is small enough to be covered by farmer's own family in spare time. On light soils hand hoes of various shapes and size often prove more efficient and economical than hand blades. On heavier soils, on the other hand, kudalies are more effective.

Intercropping

Many time intercropping helps in the suppression of at least the secondary growth of weeds that occurs after the intercrop has fully covered the ground. It means that the initial weed growth in an intercropping system should either be prevented by some good crop husbandry method or controlled later with a suitable soil active herbicide or by hand weeding. Cowpea is a very good cover crop fodder legumes which can be grown in between maize, sorghum, bajra, hybrid napear and other perennial grasses to increase the total crop productivity and suppress the secondary growth of weeds. A fast shading intercrop is also known as smother crop.

Chemical Weed Control

Because of the aggressive nature of some weed species, they can become established despite preventive and cultural efforts. Therefore, herbicide treatment might be necessary to combat some weed problems. Herbicides are commonly used for controlling weeds in cultivated and uncultivated/pastures/rangelands. Numerous forms of application techniques and equipment are available to apply herbicides. The appropriate option will be determined by the size of the infestation, the available resources, access and personal preferences. The specific herbicides and control strategies available for use will depend on the type of forage you grow (alfalfa/grass mixtures, clovers, forage grasses, etc.), whether your stand is a new seeding or an established stand, and the crop growth stage (dormant, non-dormant, between cutting). Before using an herbicide, always read and follow product label directions. An herbicide is a chemical used to kill some target plant (s). In agriculture these target plants are weeds. Thus, when we talk to herbicides, we mean chemical warfare against weeds to the extent of killing or severely stunting, depending upon our objective.

Advantages and Limitation of Herbicides

The major objective of discovering and using herbicides has been to replace the torturous, back-breaking, manual weeding and let the framer use his time thus spared in some other farming operations. The farm children need to go to school instead of wasting their lives in weeding. Ever rising wages and fuel costs have also given impetus to the farmers to switch over to herbicides to control weeds, at least in part. Besides such reasons, there are some additional benefits that frequently from the use of herbicides in agriculture as follows;

1. Herbicides can control weeds even before they emerge form the soil so that crops can germinate and grow in completely weed-free environment during their seedling stage. This is usually not possible with the physical weed control procedures.
2. In broadcast sown and narrow-row crops, herbicides prove very effective in reaching every weed. Mechanical weeding methods cannot be employed in such crops.
3. In wide row crops although inter-cultivation is very commonly practiced to remove the inter-row weeds, but it leaves the intra-row weeds unhurt. Herbicides reach both inter-row and intra-row weeds, equally well.
4. There are certain weeds which resemble during their vegetative phase crop plants with which they are associated. Such weeds, therefore, escape the farmer's hoe. But now herbicides are available which can distinguish between such weeds from the crop plants and control them easily, without any damage to the crop.
5. In the event of incessant rainfall, there is no opportunity for the farmer to use his hoe even though the weeds may be growing by leaps and bounds. Early application of herbicides at the time of sowing can allow the young crops to grow in a weed-free environment.
6. Herbicides withhold the weeds for considerable period after their application. This is in variance with physically uprooted weeds which tend to grow back soon.

Probably one could think of some more advantage of using herbicides in agriculture. But at the some time one should be familiar with certain important limitations of introducing herbicides, particularly in the developing countries. These limitations are as follows:

- ☆ The use of herbicides requires some technical know how on the part of the farmer in respect of the choice of particular herbicides their appropriate time and method of application and precaution required in their strong and use. The success of a herbicide is greatly dependent on the soil type, crop and its variety to be treated, weed flora per cent and the prevailing environmental conditions. A farmer must under stand these implication will before introduction any herbicide of his farm.

- ☆ Over and under-dosing of herbicide can make a marked deference between the success and failure of obtaining selective weed control.
- ☆ Certain herbicides, because of their long term residues in soil may impose limitations on the choice of crop rotations on the farm.

Herbicide drifts to crops growing in the neighborhoods can inhibit unhealthy quarrels.

Table 21.1: Scientfic Weed Management Practice in Different Fodder/Forage Crops

Sl.No.	*Crop*	*Scientfic Weed Management Practice*
1.	Sorghum (*Sorghum bicolor*)	In large areas weeds may be controlled by weedicides. Atrazine, ropazine @1kg a.i./ha one to three days after sorghum sowing may be applied as pre-emergence and 2,4-D as post-emergence at 4-5 weeks stage at the rate of 0.75 -1 kg a.i./ha with 750 liter of water. In small areas or when sorghum is mixed with legumes, one or two inter-cultivation by kudali or handhoe is recommended wherever line sowing was done to remove weeds in between rows. Striga weed parasites cause severe damage and they may be controlled by rotation of crops.
2.	Sudan grass (*Sorghum sudanense*)	One weeding or inter-cultivation may be necessary at 3-4 weeks after sowing. Once the crop canopy covers the field further weed growth is suppressed. In pure stands pre-emergence herbicides like atrazine @ 1 kg a.i./ha or in the case of broad leaf weeds, post-emergence herbicides like 2,4-D @ 0.75 -1 kg a.i./ha with 750 liter of water may be applied 4-5 weeks after sowing.
3.	Maize (*Zea maize*)	In India, 1.5 kg a.i./ha of simazine or atrazine in 700-800 liter of water sprayed on the sown field combined with hand weeding increased dry matter yields from 5.25 t/ha in control plot to 15.69 t/ha in treated plots (Pandey *et al.*, 1969). Broad leaved weeds in the crop can be controlled by 2,4-D amine acids @ 1 kg/ha acid equivalent. For mixed crops one weeding or inter-cultivation may be done at 3-4 weeks after sowing and repeated if necessary.
4.	Teosinte (*Euchlaena maxicana*)	During the first two months the plants grow slowly and thereafter they grow fast. Therefore, weeding at the early stage is necessary. Pandey *et al.* (1969) obtained good results with the application of 1.5 kg a.i. simazine/ha combined with hand weeding.
5.	Bajra (*Pennisetum typhoides*)	Because of the rapid growth of the young plants, weeding is not a major problem. However if weeds come up in numbers and affect the crop, pre-emergence treatment with atrazine at 0.5 kg a.i./ha or 1 kg a.i. 2,4-D/ha about 5-6 after sowing is helpful.
6.	Oat (*Avena sativa*)	In pure stands, pre-emergence or post-emergence herbicides are useful but in legume mixed oat herbicides should be applied carefully. Benazelin spray @ 0.75 kg a.i./ha with 600 to 700 liter water may be applied in oat + senji mixture, after sowing (Relwani,1979). 2,4-DB @ 0.75 kg a.i./ha may be applied in oat + legume mixture, at 2-3 weeks after sowing.

Contd...

Table 21.1–Contd...

Sl.No.	Crop	Scientfic Weed Management Practice
7	Deenanath Grass (*Pennisetum pedicellatum*)	Since growth of the grass is not fast during the first two months of establishment, weeding may be important. Rampant growth of weeds suppresses the main grass crop under heavy fertilization. Application of atrazine @ 1 kg a.i./ha as pre-emergence provides good control of the un wanted weeds in the pure grass crop (Mukherjee *et al.*, 1981).
8	Guinea grass (*Panicum maximum*)	Germination and emergence of seeds are slow and uneven. The delicate seedlings or newly emerged shoots from slips or cuttings require protection from weeds in the first two months. Two inter-cultivations should be given during this period. Later, inter-cultivation may be necessary after three or four cuttings. Post-emergence spray of 2,4-D @ 1 kg a.i./ha about 5-6 weeks after sowing is effective against broad-leaved weeds.
9.	Gamba grass (*Andropogon gayanus*)	Similar to that of guinea grass
10.	Setaria grass (*Setaria anceps*)	One or two weeding or inter-cultivation is given in the first 2 to 3 months. To control weeds and to encourage fresh sprouts, one or two inter-cultivation has to be carried out every year.
11,	Hybrid napier (*Pennisetum purpureum* x *P. typhoides*)	Early inter-cultivation once or twice is necessary before the plants establish and grow vigorously. Subsequently, inter-cultivation should be given as and when necessary. High density crop dose not require weeding after establishing full coverage. Post-emergence spray of 2,4-D @ 1 kg a.i./ha to the standing crop can control the broad-leaved weeds.
12,	Paragrass (*Brachiaria mutica*)	The land should be kept weed free for the first two months. Since it is a sturdy and aggressive grass, once it gets established, the weeds that appear later are suppressed. During this period, the grass planted in lines may be easily inter-cultured once or twice.
13,	Pangola grass (*Digitaria decumbens*)	During the first two months of initial establishment two weedings and inter-culturing will be necessary for the grass to completely cover the ground. Later the grass smothers the weeds. To a certain extent pangola grass is sod bound or root bound. After three years the pasture should be cultivated with a disc harrow or rotovator to renovate the grass and also to kill weeds. The same effects may be had by burning the old stand immediately after the cold season.
14.	Anjan grass (*Cenchrus ciliaris*)	For initial establishment, one or two weeding or inter-culture may be given. In subsequent years, one inter-culture may be given during the summer season to destroy the weeds, stir the soil and stimulate the growth of the grass. In India, one weeding increased the herbage yields in the next year by 86 per cent while two weeding increased the yield slightly.
15.	Black anjan grass (*Cenchrus setigerus*)	As anjan grass
16.	Rhodes grass (*Chloris gayana*)	The seedlings are small and weak, so weed competition is highest at establishment. One or two weeding or inter-culturing are necessary in the early stage to keep down the weed till the grass grows vigorously to cover the field. After the cold season, one ploughing with a country plough or light cultivation with a tractor should be given to thin out the stands, kill the weeds and to promote fresh growth. Burning the old stands also serve the purpose.

Contd...

Table 21.1–Contd...

Sl.No.	Crop	Scientfic Weed Management Practice
17.	Doob grass (*Cynodon dactylon*)	Inter-culturing with desi plough, harrow or hoe every 2-3 weeks is necessary to uproot the weeds in the first 6-8 weeks of establishment. Spraying pre-emergence herbicides like 2,4-D or simazine immediately after planting controls the weeds. Grazing or cutting at the early stages is also recommended for weed control. Later the grass will grow vigorously and aggressively. After 2-3 years, weeds can be destroyed and the grass renovated by light discing, harrowing or ploughing.
18.	Congosignal grass (*Brachiaria ruziziensis*)	The crop is generally planted in May-June and September-October with the onset of rains. Prepare the land by ploughing one or two times, remove weeds and level the land.
19.	Berseem (*Tryfolium alexandrinum*)	Chicory (*Cichorium intybus*) is a notorious weed; it even overruns the entire berseem field. It has low protein but high silica and crud fiber content, Chicory roots and leaves seriously affect the health of cattle and pigs. This weed should, therefore,be regularly and systematically controlled from the beginning. One deep ploughing immediately after the final harvest of berseem effectively buries the weed seed deep into the soil. Chicory seeds can be eliminated from berseem, if the seeds at the time of sowing are dipped in 15-20 per cent common salt solution. The chicory seeds along with immature berseem seeds float on the surface and are removed. Heavy berseem seeds are then collected and thoroughly washed with fresh water before sowing. Chicory seeds can also be separated from berseem seeds by spreading the seeds over an inclined smooth metal sheet having a jental jerking motion. The spherical berseem seeds will role down while the angular chicory seeds will be left on the metal sheet. Contact herbicide like dinoceb acetate, etc. may be tried as post emergence spray. Spot treatment of the weed is more effective. Some other weeds such as Asphodelus, Chenopodium, Convolvulus, etc. do not pose any problem to the berseem crop. These are generally suppressed by vigorous growth of the crop after the first cut.
20.	Lucerne (Alfalfa) (*Medicago sativa*)	Inter-culture in early stages is effective. Frequent cutting of grassy weeds in the monsoon are good checks for competitive grass and growth of lucerne. Pre-sowing soil application of herbicides like trifluralin, EPTC and dalapon @ 4 kg/ha, or pre-emergence treatment of diuron, dinoseb, alachlor and 2,4-D ester and sodium salt and post emergence application of IPC granule are effective measures. Dodder (*Cuscuta reflexa*) is a parasitic weed on lucerne. It propagates vegetatively and through seeds. Spot treatment with 2 per cent dinoseb or diquat spray before the first cut of lucern is helpful. Control of its seed setting is important.
21.	Cowpea (*Vigna unguiculata*)	Mechanical weeding or inter-cultivation by kudali or star weeder if sown in lines and hand weeding at 30 DAS can be done. One or two weeding may be required in the early growth stages to combat weed problems. Usually high seed rates are effective in smothering weeds in forage crops. However, trifluralin @ 0.75-1 kg a.i./ha as pre-plant application (before sowing and final harrowing) and choramben @ 3 kg/ha may also be effective in reducing the damage caused by weeds.

Contd...

Table 21.1–Contd...

Sl.No.	*Crop*	*Scientfic Weed Management Practice*
22.	Rice bean (*Vigna umbellata*)	In the usual course, rice bean requires clean cultivation and one are two weeding or inter-culturing for establishment of the crop. The grassy and broad leaved weeds which are prevalent in rice bean plots can be controlled chemically by herbicides applied as pre-emergence. These herbicides are EPTC @ 3-4 kg a.i./ha (pre-emergence) or trifluralin/ nitralin @ 1-1.5 kg a.i./ha seven days before planting.
23.	Senji (*Melilotus parviflora*)	Due to fast and early growth, senji having high density for fodder crop fully covers the field in three to four weeks. As such, it suppresses weed growth. Only tall and thorny weeds are required to be removed.
24.	Methi (Fenugreek) (*Trigonella foenumgraecum*)	No weeding is usually necessary to the crop. The crop establishes by four to five weeks and spreads out aggressively to suppress the weed.
25.	Moth (*Phaseolus aconitifolius*)	The crop may be weeded if necessary.
26.	Glycine species	One or two weedings may be needed.
27	Velvet bean (*Mucuna pruriens*)	One initial weeding is necessary to promote good crop growth and an early dense cover of foliage. Due to aggressive nature the velvet beans can smother weeds easily.
28	Guar or cluster bean (*Cymopsis tetragonoloba*)	One or two weedings are necessary before the establishment of the crop. Pre-sowing incorporation of trifluralin @ 0.75-1 kg a.i./ha is effective in controlling annual grasses and broad leaf weeds. TCA and dalapan @ 6.5 kg a.i./ha check *Cynodan dactylon.* Vernolate (PPTC) is most effective in controlling *trianthema portulacastrum* and EPTC for broad leaf weeds. Singh and sharma (1967) obtained 1,125 kg/ha forage of guar with vernolate applied @4.5 kg a.i./ha. Dubey and Singh (1977) checked the three predominant weeds-*Cyperus rotundus*, *Pulicaria wightiana* and *Verutrella divericata* with TOK-E 25, lasso trifluralin.
29	Turnip (*Brassica compestries* var. rapa)	One weeding and inter-culturing should be given early. Once the foliage covers the ground, subsequent weed growth is suppressed. Pre-sowing treatment of land with herbicide like trifluralin or nitralin @ 1.5 kg a.i./ha is helpful.
30	Carrot (*Daucus carota* var. sativa)	Weeding and thinning are very important operations during the first 4-6 weeks as otherwise weeds may smother and the yield may suffer a serious set-back. Pre-emergence chemical weed control with linuron, tenoran etc. is effective.
31	Forage beets (*Beta vulgaris* var. Crasea)	Crowded seedling should be thinned and the crop should be given one or two weeding.
32	Stylo (*Stylosanthes* spp.)	As slow initial growth one or two weeding or inter-cultivation is given in the first 1-2 months for establishment of the crop.
33	Subabul (*Leucaena leucocephala*)	Since the early growth of the crop is slow, the tender plants are to be protected from aggressive weeds. Two or three inter-row cultivation is essential to check weeds in early life. Once established, the plants seldom smothered even by vigorous grasses.

Removal of Bushes and Other Species

The bushes and noxious weeds and poor quality grasses may offer severe competition for light and nutrients. The most common obnoxious weeds of the Himalayan grasslands are lantana, ageratum, eupatorium, cactus, barberry etc. These weeds can be controlled by cutting and stems treated with herbicides to prevent regrowth. The herbicides like Weedon 64, Picloram, Paraquat and Glyphosate etc @ 1.0-2.0 Kg/ha could be applied around the bush. Sood and Singh (1986) have found that paraquat spray in the 15 cm band @ 0.6 lit. a.i./ha reduced the weed incidence in the grasslands and the fresh herbage yield increased by 26.8 per cent.

References

Chattarjee, B.N. and Das, P.K. Forage crop production principles and practices. Published by Oxford and IBH Publishing Co. Pvt. Ltd., New Delhi, pp. 433.

Gupta, O.P. 2005. Weed management principals and practices. Second Edition. Published by Agrobios (India), pp. 290.

http://www.ikisan.com/links/ap_fc_sorghum.shtml

http://igfri.ernet.in/crop_profile_berseem.htm

http://www.ikisan.com/links/ap_tc_cowpea.shtml

http://gbpihed.gov.in/envis/HTML/vol92/vol92InderDev.html

http://www.kau.edu/pop/foddercrops.htm

Sood, B.R and Singh C.M. (1986). Effect of method of introduction and fertilizer management on the production potential and quality of natural grassland of Kangra valley. Range Management and Agroforestry. 7(2) : 119-126.

Chapter 22
Insect Pest of Fodder and Forage Crops and their Management

Ajay Kumar Pandey

College of Forestry and Hill Agriculture,
G. B. Pant University of Agriculture and Technology,
Hill Campus Ranichauri – 249 199, Tehri Garhwal, Uttarakhand
E-mail: drajay2002@gmail.com

A cereal crops like corn, wheat, barley and oats belongs to grass family from which the grain or seed is harvested to utilize as feed. Once the seed is harvested the remainder of the plant is either discarded or utilized as fodder to increase the energy of livestock. A forage crop is a plant (except its roots) that is grown for feeding to livestock. Often these plants come from the grass family but not all forage crops are grasses. Forages are a "bulky" food with higher fiber content and lower energy content than a cereal. Thus they are good at filling up an animal and providing a maintenance ration India has a huge livestock population of over 343 millions and the average cultivated area under fodder production is only 4.4 per cent of the total area with 2.51 and 2.1 crore hectares area under forests and grazing, respectively. All these resources are not sufficient to meet the annual forage demand due to which the animals survive on the straw of jowar, bajra, ragi, wheat, barley, etc. The straw or *bhusa* are available mainly from wheat, paddy, bajra, jowar, ragi, etc. However, among the large number of limitations, insect pest are of great importance which reduce the quality and quantity of grain and fodder crops. The insect pests problem in forage crops have been well recognized as a major problem in India.

In India losses due to insect pest in forage crop are very poorly documented. Insect pest cause yield loss, both quantitative and qualitative, depend on many factors like the level of infestation, crop variety, agronomic practices and other related factors. Insect pests induced losses are not only in term of green or dry fodder yield but also quality factors which affect the re-growth period and canopy structure. For example, feeding of Lucerne weevil leads to reduction of dry matter up to 1.7 tonnes/ha and 28.8 per cent losses in lucern have been reported from India.

In India *Jowar* (*Sorghum vulgare*), Maize (*Zea mays*), *Bajra* (*Pennisetum typhoides*), Oat (*Avena sativa*) Cowpea (*Vigna sinensis*), Berseem (*Trifolium alexandrinum*), Lucerne (*Medicago sativa*) are the important fodder and forage crop which is infested by different type of insect pest which reduce the quality and quantity of the fodder. Some of the important insect pests of legumes and non-legumes fodder and forage crop are as under:

Insect Pest of Lucerne and Berseem

Lucerne (*Medicago sativa* Linnaeus) is a deep-rooted, perennial legume plant, well adapted to the Indian environment. It uses water from suitable soils depths *i.e.* two meters and below and provides additional green feed at the start and at the end of the normal winter growing season. It has the ability to respond quickly to summer rainfall to produce high quality green feed. Once established lucerne has good drought tolerance but will go dormant during extended dry periods. The major pests of lucerne are Helicoverpa, Lucerne flea, aphids, lucerne weevil and grasshoppers.

Fruit Borer (*Helicoverpa armigera*)

This pest causes damage to corn, sorghum, soybean, cowpea, lucerne, and berseem (*Trifolium alexandrinum*) besides the vegetables. It is widely distributed in Canada, South America, and USA including India. An adult male is yellow-brown while a female is orange-brown in color. It has a wingspread size of about 3.8cm. Eggs are pinhead-sized and yellow-green in color. Larvae vary in color from bright green, pink, brown to black. Alternating light and dark bands run lengthwise along their bodies, the heads are yellow and the legs are almost black. Mature larvae vary in length about 3-5 cm, feed on leaves and floral part of the lucerne and bean. They remain feeding in the tip areas until they leave to pupate in the soil. It defoliates the leaf and floral part of whole crop.

Management

- ☆ Two to three spraying of Neem seed kernel extract (NSK) of 5 per cent concentration at 10 days intervals starting from initiation of flowering has been found effective against this pest.
- ☆ Alternate row spraying with of Endosulfan @ 0.7 kgha^{-1} or Monocrotophos @ 0.4 kgha^{-1} or Phenthoate @ 0.5 kgha^{-1} has also been found effective to manage this pest.

Aphids (*Acyrthosiphon pisum* and *Therioaphis trifolii f. maculata*)

The spotted alfalfa aphid is the major aphids associated with alfalfa crop. It is pale yellow coloured with six or more rows of black spots on the back. It occurs in all the Lucerne growing areas of India and is particularly severe in the north-west, western Himalayas including Ladakh region. The aphid sucks sap from the leaves and stems and are mostly confined to the lower parts of the plants and underside of the leaves. The nymphs and adults cause the damage by sucking the sap from the foliage due to which plants show yellowness tips, wilting and stunting symptoms. Excessive infestations can cause reduction in leaf size and prevent flowering. The damage is more severe at the seedling stage which can cause mortality of the plant. The most suitable temperature and relative humidity for the development is 20-25°C maximum, 9-10° minimum and 80-85 relative humidity per cent respectively (Faruqui *et al.*, 1986).

Management

- ☆ Some of the natural enemies like *Ahidius* sp., *Aphelinu* sp, *Ephedrus plagiator* and predators like *Coccinellids*, syrphid flies, *Chrysopa*, *Syrphus* sp. keep the population under control. The population of these predators and parasitoids should be encouraged to maintain the pest population under check.
- ☆ A number of chemical controls are available, although chemical use in dryland systems is less economic than in irrigated. Spraying of phosphamidon @ 0.3 $kgha^{-1}$ or endosulfan @ 0.525 $kgha^{-1}$ is effective but it should be applied at least 30 days before the utilization of fodder.

Lucerne Weevil (*Hyper postica*)

This is the most damaging pest of Lucerne occurring in all the Lucerne growing areas of the world and is particularly severe in western Himalayas and central plains of India. Besides Lucerne, it also damages sweet clover, peas *Medicago* and *Melus* sp. Adult weevils are brown snout beetles about 0.5 cm long. Dark strip extends downwards more than half the length of the body. The body becomes darker with age. Adult females lay 10-30 lemon yellow coloured eggs in the holes made by them in the Lucerne's stem. The damage is mostly caused by the larvae that feed within the plant tips on the young leaves. In heavy infestation the lower foliage is also consumed. The adults also feed on the foliage. Incidence of the weevil is higher during the month of February to the first week of March. The beetle development is rapid at a temperature range of 11.5-31.6°C and 63-76 per cent relative humidity (Pandey and Faruqui, 1990).

Management

- ☆ Sowing of crop during the third week of October escape the infestation of lucerne weevils (Shri Ram *et al.*, 1987).
- ☆ Fertilizer combination of 100 kg P_2O_5 and 50 kg K_2O per hectare make the crop less attractive to insect pest.
- ☆ Soil treatment with Phenthoate @ 0.75 $kgha^{-1}$ or spraying of Malathion @ 0.05 per cent or Cypermethrin @ 0.02 per cent manages the pest very effectively.

Leaf Roller

Leaf roller can reduce plant vigour and damage the leaf by rolling and feeding inside on green matter thereby it leads to decreasing the quality of fodder. Infestations are usually most severe in summer. Small dark-green larvae with black heads produce a web in the terminal growth of each lucerne tiller, skeletonise the leaf, roll the tips and exacerbate leaf drop.

Management

- ✰ Cutting or grazing infested crop is the primary management option.
- ✰ Spraying is generally uneconomic. However, insecticide application could be considered when more than 30 per cent of terminal growth is rolled.
- ✰ Spraying of Malathion @ 0.05 per cent or Cypermethrin @ 0.02 per cent suppress the pest population control.

Jassids (*Empoasca* sp.)

The vegetable jassid is yellowish-green insect with a wider host range and is normally present in the fields during summer and winter but more severe in winter. Jassids suck from soft stems and from the leaves which result in puncturing and killing of the cells. The damage soon shows as a small white spot. Continuous feeding results in a stipple pattern on the leaves. Lucerne jassids have a more severe effect on the plant and can have stunt crop growth.

Management

- ✰ Early sowing escapes the crop from the infestation of jassid and aphids.
- ✰ One or two spraying of 0.03 per cent solution of diamathoate or phosphamidon reduces the pest incidence.

Cutworms (*Agrotis* spp)

Cutworms *Agrotis* spp is a serious pest of maize, barley, lucerne, berseem beside the pulse and vegetable crops. Caterpillars cut young plants at ground level in the night. Larvae shelter in trash and soil during the day and feed at night. In older plants, they cut shoots and feed upon leaves leading to reduction in growth. In sever infestation at nursery stage of crop, whole field is covered with cut plants are a moth larvae. Indicators of cutworm activity are thinning in seedling stands and plants that are cut off at ground level. Digging at the base of plants can reveal the large soft-bodied brown larvae.

Management

- ✰ Removal of weeds in and around fields will reduce egg-laying sites and will help in the prevention of cutworm infestation.
- ✰ Plow and harrow fields properly before sowing. This will destroy eggs and expose larvae to ants, birds, and other predators.
- ✰ In rice fields, keep area flooded. This will prevent cutworm population

- ✰ Soil application of Chlorpyriphos 1.5 dust @ 10-15 kg per hectare at the time of land preparation or seed treatment with Chlorpyriphos 20 EC @ 500 ml dissolved in 5 liter water for one quintal of seed.
- ✰ Cutworms are readily controlled with insecticides. Apply the above insecticide at evening or night when the larvae are active.

Lucerne Caterpillars, *Spodoptera exigua*

Lucerne caterpillars, *Spodoptera exigua* Hubner are a polyphagous insect which is also a major pest of lucerne. The dark-brown female moth lays eggs in clusters on the lower portion of young plants. Caterpillars vary in colour from pinkish brown to light green with a narrow dark longitudinal stripe on the back. Pupae of this pest live in the soil. Larvae feed gregariously on leaves and soft twigs thereby it defoliate the plants and ultimately reduce the quality and quantity of fodder production.

Management

- ✰ Collection and destruction of adult using light trap is an effective method.
- ✰ Hoeing in between the rows is also effective method to control the population of this pest.
- ✰ Spraying of Endosulfan @ 0.05 per cent or Carbaryl @ 0.2 per cent reduce the infestation of this pest.

Although in case of barseem the insect pest problem is not so severe, but minor insects like hairy caterpillars (*Euproctis virguncula and euproctis lunata*), semilooper (*Plusia orichalcea*), red pumpkin beetle, (*Rhaphidopalpa foveicollis*), red cotton bug, (*Dysdercus koenigii*), aphid (*Aphis craccivora*), dusky bug (*Oxycarogenus* sp.), leaf miner (*Phytomyza sp.*) etc may cause damage at different stage of crops.

Cowpea (*Vigna unguiculata*)

Cowpea (*Vigna unguiculata* L. Walp) is considered the most important food grain legume which is also a good source of quality fodder for livestock and provides cash income. Many biotic and abiotic factors greatly reduce the cowpea productivity. Among these constraints, some insect pests cause damage to this crop.

Flea Beetles (*Pagria* sp.)

The insect appears in meancing forms in cowpea during July to December. The adult beetle is small with shiny metallic brown elytra having whitish spots. The adults cause damage by feeding on leaves which appeared as shot hole on leaves. Grubs generally live inside the soil and feed upon the roots of the host plant.

Management

- ✰ Application of Phorat 10 CG at the rate of 10 $kgha^{-1}$ followed by one spray of Monocrotophos (0.04 per cent) or Malathion (0.04 per cent) reduces insect pest incidence.

Hairy Caterpillar

There are several hairy caterpillars but two are predominant *viz. Spilosoma* oblique (Walker) and *Amsacta lactinea* Cram which are most important pest of cowpea. It causes severe damage to the wide variety of crops. Young hairy larvae are gregarious. Initially young larvae feed upon green matter of leaves of maize, cowpea and groundnut, etc they by they made the leaves papery white. Mature larvae eat away the leaves leaving behind only the veins. In maize grown for grain purpose, fresh silk is also damaged by the caterpillar resulting in partially grain filled cobs. They may damage the crop at seedling stage which can be so ever that sometimes re-sowing may be necessary.

Management

- ✰ Collect and destroy egg masses and caterpillars of early instars remaining congregated on the undersurface of leaves.
- ✰ The cowpea crops sown between the third week of June and the last week of July generally escape the incidence of leafhoppers and defoliates.
- ✰ The young caterpillars can be killed by dusting Malathion 5 per cent @ 10-15 kgha^{-1}.
- ✰ For full grown larvae, spraying of endosulfan 35EC @ 0.07 per cent reduce the incidence of caterpillars.

Semi Looper (*Plusia orichalsia*)

These caterpillars from a loop while moving so are known as semi loopers. The adults are large stout moths which are active during the night. The females lay eggs on the leaves in batches which hatch in about two to three days. The caterpillars are voracious eaters of leaves resulting in defoliate of plant.

Management

- ✰ Collect and destroy egg masses and caterpillars of early instars.
- ✰ For full grown larvae, spraying of endosulfan 35EC @ 0.05 per cent reduce the incidence of caterpillars

Leaf Hoppers (*Empoasca kerri*)

The jassid is widely distributed in India and is one of the destructive pests in the north-western region. The adults are about 3 mm long and greenish in colour. The winged adults jump at the slightest disturbance and are attracted towards light during the night. The female lays eggs on the underside of leaves. After hatching of eggs, nymphs come out and start sucking the sap from the underside of the leaves and mature in 7-21 days. Besides damaging the plant through sap loss they also inject toxins due to which leaves turn pale and eventually dry up and fall. The pest appears at the early stage of the crop and attains the peak during the third week of August.

Management

Control of leafhopper is difficult because young wingless hoppers hide on the underside of leaves and the adults are quite resistant to insecticides..

- ☆ The cowpea crops sown between the third week of June and the last week of July generally escape the incidence of leafhoppers and defoliates.
- ☆ Intercropping of cowpea with sorghum pearl millet or maize reduces leafhoppers, defoliators and viral infection (Shri Ram and Gupta, 1987, Sharma and Verma, 1984).
- ☆ A higher dose of Potassium (100 kg P_2O_5 per hectare) and potash (50 kg K_2O per hectare) reduces the incidences of leaf hoppers and defoliators like flea beetle, semi-looper, tobacco caterpillar and grass hoppers (Sri Ram and Gupta, 1990).
- ☆ Application of Phorat 10 CG at the rate of 10 kgha^{-1} at the time of sowing followed by two sprays of Monocrotophos (0.04 per cent) or Malathion (0.04 per cent or Metasystox (0.03 per cent) in 40 and 60 days after sowing reduces insect pest incidence.
- ☆ Care should be taken if insecticides are used for controlling the pests. Do not feed the fodder to the animals for a period of 20-25 days after spray.

Thrips (Thysanoptera:Thripidae)

It is also a sporadic pest but some time it becomes key pest of cowi pea, lucerne and alfalfa crops. The adult has a slender small body, yellowish to dark-brown in color, and is cigar-shaped. It is 1-2 mm long with a well-pronounced 5-8 segmented antennae. Thrips have rasping-sucking mouthparts and feed by rasping the surface of the rapidly growing tissues of the leaves and sucking up the released plant fluid. Thrips cause tiny scars on leaves called stippling, which can cause stunted growth. Damaged leaves may become papery and distorted. Infested terminals lose their color, rolled, and drop leaves prematurely. In beans, thrips causes feeding damage on flowers and young pods.

Management

- ☆ Lack of water increases the susceptibility of plants to thrips damage. So keep plants well irrigated
- ☆ Remove heavily infested plant parts.
- ☆ Remove weeds as the thrips population utilize as secondary host for survival.
- ☆ Synthetic pyrethroid like cypermethrin at the rate of 0.01 per cent or neem seed kernel extract at the rate of 3 per cent applied in 15, 30 and 60 days after sowing reduce the pest incidence.
- ☆ Crops raised for seed purpose coerced bug and pod borers can be effectively controlled by application of neem seed kernel extract at the rate of 9 per cent (Jackai *et al.*, 1992).

Some other insects pest like tobacco caterpillar (*Spodoptera litura*), grasshopper (*Colemania sphenarioides, Chrotogonus trachypterus* and *Attractomorpha crenult*), aphids (*Aphis craccivora*), stem fly (*Agromyza phaseolia*), *leaf miner* (*Phytomyza atricornis*) etc also cause minor damage to this crop.

Insect Pest of Sorghum, Maize and Bajra

Modern sorghum is a product of human ingenuity. It has been domesticated to meet human needs. Commonly, sorghum is grown in areas too hot and too dry to produce other crops such as maize. Sorghum is known for its production reliability. But, this crop may be infested by an array of insect.

Some insect pests cause damage to sorghum, maize and bajra crops at any plant growth stage while others cause economic damage only at specific plant growth stages. Most insect species infesting sorghum are not host specific. In sorghum agroecosystems, one or two key insect pests such as sorghum midge, Contarinia *sorghicola* (Coquillett), shoot fly, *Atherigona soccata* (Rondani), and corn earworm, *Helicoverpa zea* (Boddie), are the major pest. Many other insects like wireworms, white grubs, cutworms, aphids, leaf feeding caterpillars, stalk-borers, bugs, etc are occasional pest that cause economic damage only in limited to a small areas. Few of the major pests are as under which damage to sorghum, maize and bajra crops.

The Sorghum Ear Head Bug (*Calocoris angustatus*)

It is one of the most destructive pest of sorghum in southern India. This bug has been recorded on a number of cereals, millets and grasses, but its breeding is mainly restricted to sorghum, on which it attains the status of a pest. The adult is a small, slender, greenish yellow bug, measuring 5-8 mm in length and over 1 mm in width. The adult appear on sorghum crop as soon as the ear emerges from the leaf sheaths. The female lays 100-150 eggs under the glumes or in between anthers of florets, by inserting its ovipositor. Nymphs who come out after egg hatching and adult start feed on sap of developing grains. As a result of feeding by the bugs, the grains remain chaffy or shriveled. When a large army or tiny nymphs feeds, the whole ear may become blacked at first and may eventually dry up, producing no grains.

Management

- ☆ Burning of panicle residue and chaff obtained after threshing of grains to destroy diapausing larvae
- ☆ Uniform date of sowing and preferably grown only one variety is also somewhat effective to reduce the infestation.
- ☆ Spraying of endosulfon @ 0.05 per cent or Malathion @ 0.03 per cent or using dust formulation @ 15-20 kg per hectare is effective to manage this pest.

Shoot Fly (*Altherigona varia soccata*)

The sorghum shootfly also known as the sorghum stemfly is a widely distributed pest in India, which is a serious pest in southern parts. Besides sorghum, it infests maize, wheat and small millets. It causes damage at seedling and early stage of the crops. The adult is like a small housefly about 4-5 mm long. The male is black in colour while the female has pale gray thorax and yellow abdomen with pawed patches. The female fly lay elongated and flatted 20 to 40 eggs singly on the underside of leaf. The eggs hatch in 1-3 days and the tiny maggots creep out and reach in between the sheath and the axis, and bore into the stem. They feed inside the main shoot for 5-10 days and after maturity they my pupate either inside the stem or come out to pupate in the soil. Several generations are completed in a year. In northern India, this pest goes under hibernation at pupal stage. The maggot bore into the stem and cut the main shoot. The high yielding hybrids are more susceptible to the attack of this fly.

Management

- ☆ Uses of glossy lines of sorghum which are less susceptible to the shoot fly (Bapat and Mote, 1982). The presence of trichomes in leaves is associated with reduced shoot fly susceptibility (Matili and Gibson, 1983).
- ☆ Irregular shaped silica bodies are present in the plant tissues of shoot fly resistant lines of sorghum (Ponnaiya, 1951).
- ☆ Maintained the sufficient plant population by minor increase in seed rate.
- ☆ For biological control of release of different strains of *Trichogramma exiguam*, the parasite which is now established in Delhi and Nagpur (Jotwani, 1982).
- ☆ Apply carbofuron 3G @ 25 kgha^{-1} in furrow to manage this pest.

Stem Borer (*Chilo partellus*)

The stem borer (*Chilo partellus*) is also one of the important pests of sorghum and maize. It also causes damage to other fodder crops like bajra and grasses. The male moth is 20-30 mm across the wings. The female has pale fore wings with almost white hind wings. The adult actives at night and the female lays about 300 eggs in overlapping clusters of about 20 eggs. The eggs are flat oval and yellowish in colour. The caterpillars are grayish white with black head and four brownish longitudinal strips on the back. After hatching, young larvae initially feed on leaves and later on start boring downwards through the central whorl. The insect breed actively from March – April to October and for the rest of the year it remains in hibernation as full growing larva in maize and sorghum stubble, stalks or unshelled cobs. The life cycle completed in about 3 weeks and there are probably 5 generations in a year.

In the central and northern districts of the Punjab, it is the most serious pest of maize which destroys 25 to 40 percent of the young plants population. Sorghum and bajra are also attacked by this pest at the time of grain formation. Damage through this insect to sorghum fodder is perhaps equally great as compared to grain sorghum, but it is not generally noticed. After hatching, young larvae starts feeding on the leaves before boring downwards through the central whorl thus reaching the growing point. This results in the production of dead hearts symptoms.

Management

- ☆ Increase in density of plants, this pest cause higher damage. So, maintain proper the plant density to reduce the pest incidence.

- ✩ Destruction of crop residue, alternate host plants and chopping of stems harboring dispausing larva could be very effective in reducing borer population.
- ✩ Intercropping of sorghum and cowpea reduced the numbers of stem borer, *Chilo partellus* in sorghum and thrips, *Megalurothrips sjostedti* in cowpea.
- ✩ Light traps could be used to monitor the field population. The trapped moths can be killed to reduce the insect population in the field.
- ✩ Foliar application of endosulfan 35EC @ 0.1 per cent 15 days after sowing at 15 - 20 days interval has been found effective to manage this pest.

White Grub (*Holotrichia* spp.)

The most obvious and significant damage occurs during the spring soon after sorghum plants emerge from the soil. Damage results from grub feeding on roots which cause seedlings. Infested plants not killed are severely stunted and may never produce grain. The losses occur within seven to ten days after plants emerge in severely infested fields. Another kind of damage to roots also occurs where the roots are pruned out by over wintered and current-season grub. Such injured plants may produce panicles after damage but sufficient roots to prevent lodging.

- ✩ Light trap should be installed in the month of May- June to collect and destroy the adult.
- ✩ Deep ploughing should be done to expose the immature stages to natural enemies.
- ✩ Soil application of Imidacloprid 200 SL (0.048 kg a.i./ha) proved to be most effective insecticide for the management of *Holotrichia* in Soybean followed by Imidacloprid 0.75G applied @ 0.09kg a.i/ha and Chlorpyriphos 20 EC (0.40 kg a.i/ha).
- ✩ Post-sown soil application of Imidacloprid 200 SL (0.481 g a.i./ha) or Imidacloprid 0.75G (0.09 kg a.i./ha) in standing occurred most effective insecticide against the *Holotrichia* followed by Chlorpyriphos 20 EC (0.80 kg a.i./ha).

Cutworms (*Agrotis* spp.)

Different species of cutworms may damage sorghum plants by cut off sorghum plants at or slightly below the surface of the soil, feed on above-ground plant parts and by feeding on underground plant parts including roots of seedlings. Plants with severed stems die. Leaf feeding by cutworms causes ragged leaves, while root-feeding cutworms kill small plants or stunt larger plants.

Management

- ✩ Ploughing and harrowing fields should be done properly to destroy eggs and expose larvae to ants, birds, and other predators.
- ✩ Insecticide can be used against cutworms in sorghum, but effectiveness of control can vary because cutworms hide in the soil during the day, late-afternoon applications sometimes are more effective.

- ☆ Soil application of Imidacloprid 200 SL (0.048 kg a.i./ha) proved most effective for the management of *Agrotis ipsilon* (Hufn.) and wireworm.

Kharif Grass Hopper

There are two common species of grasshopper *i.e. Hieroglyphus banian* Fabr. and *H. nigrorepletus* Bol. besides few another one. Adults are green or dry grass coloured. The Kharif grass hopper generally does not cause much damage to sorghum and maize crop. Both adult and nymph stages of this pest are responsible for the damage. In a favorable season, it may prove very harmful and leaves nothing on the plant except stem and midribs of leaves of maize, rice and sugarcane. Both nymphs and adults feed on the growing tips and young leaves. They may even eat part of the stalk and ears of maize. The period of attack is July to September.

Management

- ☆ Dusting of the crop with Malathion 5 per cent @ 10-15kgha-1 or dusting with quinalphos (1.5 per cent) has been found effective.

Sorghum Midge (*Contarinia sorghicola*)

The adult fly is tiny, fragile, mosquito like insect with a bright orange abdomen and a pair of transparent wings. Full grown maggots are dark orange in colour. After mating, female start laying eggs about 30 to 90 eggs. The female inserts the eggs singly into developing florets. After hatching maggots move down into the ovary and feed inside the developing grains. Pupation takes place within the damaged spikelets which range from 2 to 6 days. Total life cycle from eggs to adult is completed in 14 to 19 days. The maggots feed on the developing seeds often one larva per spikelet which is sufficient to cause complete loss of the grain. The grain head is flattened with tiny shrunken seeds. The severity of infestation has a significant effect on the overall production of grains.

Management

- ☆ Burning of panicle residue and chaff obtained after threshing of grains to destroy imatur stage of pest.
- ☆ Uniform date of sowing and one variety covering whole area is also somewhat effective to reduce the infestation.
- ☆ Spraying of endosulfon @ 0.05 per cent or Malathion @ 0.03 per cent or using dust formulation @ 15-20 kg per hectare is effective to manage this pest.

Aphids (*Rhapalosiphum maides*)

The aphids, dark green coloured with slight white covering, attacks mainly maize and other alternate forage crops like sorghum and grasses. The aphids cover leaves, leaf sheath and inflorescence of the plant. The leaves become mottled and distorted and new growth is dwarfed. Honeydew production is quite prolific which affects the photosynthesis.

Management

- Early sowing of crop escape for the aphid infestation
- Application of proper dosage of Phosphorus and Potash reduce the infestation of aphid.
- Spraying of dimethoate @ $kgha^{-1}$ 0.3 or phosphamidon @ 0.3 $kgha^{-1}$ reduces the aphid population.

Army Worm (*Mythimna separate, Spodoptera exigua*)

These caterpillars are sporadic pests of sorghum and feed on wheat rice maize, millets, oat, barley and various wild grasses. Freshly emerged larvae are dull white and later turn green while adult being pale brown colour. Out break of this army worm occurs after heavy rains and floods. The larvae feed voraciously and migrate from one field to another with the help of threads from which they suspend themselves in the air. The young larvae eat the leaves in the central whorl of the plant. As they grow they are able to feed on older leaves and also skeletonize them totally. In case of severe attack, they defoliate the whole plant. The caterpillars cause severe damage by feeding the tender leaves. These pests attain epidemic status if after high rainfall a dry spell follows.

The sorghum, maize and bajra crops are also attacked by sugarchan leaf fhopper, *Pyrilla perpusilla* Walk. (Hemipter: lophopidae), thrips, *Anaphothrips* sp. Thermits, *Odontotermes obesus* Pambur (Isoptera: Termitidae), the pink borer, *Sesamia inferens* Walker, *Spodopter exempta, Mythimna separate* Walker (Lepdoptera: Noctuidae), red hairy caterpillar, *Amsacta moorei* Butler (Lepidopter: Arctidae), whitegrub, *Hologtrichia conseanguinea* (Blach. (Coleopter: melolonthidae) etc.

Integrated Pest Management for Sorghum, Maize and Bajra Crop

- Early sowing in Kharif and late sowing in rabi season helps in avoiding the insect pest.
- *Crop termination and alternate host elimination* increase the mortality of insect pests by destruction of food sources and overwintering habitats.
- Destroying the crop soon after harvest sorghum crop suppresses insect pest abundance the following year.
- Sorghum stalks should be destroyed soon after harvest to expose and kill insect pests and eliminate their food supply.
- Herbicides can be used to kill sorghum and alternate host plants where reduced tillage is used.
- Destroying food sources and overwintering habitats reduce abundance of cutworms, sorghum midge, green bug, and sorghum midge.
- Crop rotation is most effective against insect pests with a limited host range, long life cycle (one or fewer generations a year) and limited ability to move from one field to another.

- Sorghum varieties should be selected that mature as early and uniformly as practical in a locale. These varieties escape infestation by sorghum midge, corn earworm, fall armyworm, sorghum webworm, and sugarcane borer.
- Sorghum sowing should be as early as practical to reduce the infestation.
- Early and uniform planting of sorghum to avoid a damaging sorghum midge infestation is an excellent example of avoiding high numbers of corn earworm, fall armyworm, sorghum webworm, stalk borers, and panicle-feeding bugs.
- Using too much fertilizer and irrigation can cause sorghum plants to be especially succulent and attractive to insect pests, and may extend the time to maturity, increasing the duration of vulnerability. Therefore, proper dosage of fertilizer and irrigation should be given.
- *Crop rotation* is a cultural management method that involves alternate use of host and non-host crops to reduce insect pest abundance and damage. Growing sorghum in a field planted to a different, non-host crop the previous year significantly reduces the abundance of some insect pests.
- Seed dressing with Carbofuron (4 per cent) or furrow application of Carbaryl 10G or Endosulfan 3G at the rate of 1.5 gm/m row length followed by two sprays of Endosulfan (0.07-0.1 per cent) as per the severity of the pest on crop reduce the infestation of shoot borer and the shoot fly
- Care should be taken if insecticides are used for controlling the pests. Do not feed the fodder to the animals for a period of 20-25 days after spray.

References

Bapat, D. R. and Mote, U. N. 1982. Sources of shoot fly resistance in sorghum. *J. Maha. Agr. Univ.*, pp.238-240.

Faruqui, S. A., Pandey, K. C. and Patil, B. D. 1986. Field population studies and natural control of spotted alfalfa aphid. *Indian J. Ecol.* 13: 120-122.

Hamid, S. F. and Singh S. P. 2000. Handbook of pest management. Kalayani Publishers, New Delhi. 169p

Jackai, L. E. N., Inang, E. E. and Nwdri, P. 1992. The potential for controlling post-flowering pest of cowpea, Vigna unguiculata Walp. using neem *Azadirachta indica. Tropical Pest Management* 38: 56-60.

Jain, R. K. 2001. Pest and diseases of fodder crops and their management. *In: Plant Pathology* (Ed. Trivedi, P. C.). Pinter Publishers, Jaipur. pp 273-289.

Jotwani, M. G. 1982. Factors reducing sorghum yields - insect pests in sorghum. *In the Eighties Proc. of the Int. Symp. on Sorghum,* ICRISAT, Hyderabad, 1981, pp. 251-255.

Kwesi Ampong-Nyarko, K. V. Seshu Reddy, Ruth A. Nyang'or and K. N. Saxena 1994. Reduction of insect pest attack on sorghum and cowpea by intercropping. *Entomologia Experimentalis et Applicata,* 70(2):179-184.

Maiti, R. K. and Gibson, P. T. 1983. Trichome in segregating generations of sorghum matings. II. Association with shootfly resistance. *Crop Sci.* 23: 76-79.

Pandey, K. C. and Faruqui, S. A. 1990. Oviposition and feeding by lucerne weevil *Hypera postica* Gyll. in lucerne genotypes. *Indian J. Genet.* 50: 76-79.

Panwar, V. P. S. 1995. Agricultural Insect pests of crops and their control. Kalayani Publishers, New Delhi. 286p

Ponnaiya, B. W. X. 1951. Studies on the genus sorghum. II. The cause of resistance in sorghum to insect pest *Atherigone indica*. *Madras Univ. J.* 21: 203-217.

Sharma, S. R. and Verma, A. 1984. Effect of cultural practices in virus infection in cowpea. *Zeiy. Fue. Acker. Uad Pflan.* 153: 23-31.

Shri Ram, Gupta, M. P. and Shivankar, V. J. 1987. Integrated pest management of some fodder crops. IGFRI-ICAR, New Delhi.

Singh, C., Singh, P and Singh R. 2004. Modern techniques of raising field crops. Oxford and IBH Publishing Co. Pvt. Ltd, New Delhi. 583p.

Chapter 23
Major Diseases of Fodder Crops and their Management

Bijendra Kumar[1] *and Jameel Akhtar*[2]

[1]*Plant Pathology Section, College of Forestry and Hill Agriculture, Hill Campus, Ranichauri, Tehri Garhwal – 249 199, Uttarakhand*

[2]*Department of Plant Pathology, Birsa Agriculture University, Ranchi, Jharkhand*

Fodders are the plant species cultivated and harvested for feeding the animals in the form of forage (cut green and fed fresh), silage (preserved under anaerobic condition) and hay (dehydrated green fodder). The total area under cultivated fodders is 8.3 million ha (Anon., 2006) on individual crop basis. If we examine the land resources available for growing fodder and forage crops, it is estimated that the average cultivated area devoted to fodder production is only 4.4 per cent of the total area. Similarly, the area under permanent pastures and cultivable wastelands is approximately 13 and 15 million hectares respectively which is declining over the years and the trend could well continue in the future. These resources are able to meet the forage requirements of the animals only during the monsoon season. But for the remaining periods of the year, the animals have to be maintained on the crop residues or straws of *jowar*, *bajra*, *ragi*, wheat, barley, etc. either in the form of whole straw or a *bhusa*, supplemented with some green fodder, or as sole feed. The crop residues are available mainly from wheat, paddy, *bajra*, *jowar*, *ragi*, sugarcane trash, etc.

Among the various constraints, the diseases have always been the major limiting factor for fodder cultivation. Several diseases are causing serious damage to fodder crops. In India, diseases alone can cause losses up to 72 per cent in Lucerne (Ahmad *et al.*, 1977); 74 per cent in cowpea (Chester, 1950); 50 per cent sorghum (Sunderam,

1970); 30 per cent in bajra (Ahmad, 1969; Sunderam, 1970); 75 per cent in clusterbean (Chester, 1950) and 55 per cent in oats (Ahmad, 1969). Besides, these diseases also affect the quality parameters of forages. Keeping this in mind the present article is written with the objective to focus the diseases problems of fodder crops along with their management.

Diseases of Cultivated Non-Leguminous Forages

Sorghum

Smuts

Seven different types of smuts have been reported in sorghum however, four of which occur in India.

Grain Smut (*Sphacelotheca sorghi*)

This is the most destructive of all the smuts, causing extensive damage to grain yield all over the country. It is also known as covered smut, kernel smut or short smut. Grain smut of sorghum is caused by *Sphacelotheca sorghi*. It is reported to cause huge losses in USA, Italy, South America, Sri Lanka, Burma, Manchuria and several other countries. In India, it is known to occur in Tamil Nadu, Karnataka, Andhra Pradesh, Maharashtra, Gujrat and Uttar Pradesh. In certain areas, it is reported to cause up to 25 per cent of grain yield (Rangaswami and Mahadevan, 1999).

Symptoms

The symptoms of the disease become visible only at the time of grain formation stage. The affected grains convert into smut sori. The size of the sori varies with the variety, but generally they are larger than the normal grain. The sori are oval or cylindrical, dirty grey sac, sometimes conical at the tip, and measuring 4-12 mm in length. The sac is surrounded by the unaltered glumes at the base. Sometimes the stamens develop normally, but more commonly they are absent or are involved in sorus, being replaced by 3 conical protrusions from sides of the sorus. These sacs rupture easily than in the unelongated sacs. The interior of the sac is completely filled with the spore powder.

Management

Since, the disease is externally seed-borne; seed treatment with suitable fungicides can easily control the disease. Immersion of seeds in 0.5 per cent formalin for 2 hours and dried quickly. Alternatively, they are treated in 0.5 – 3 per cent copper sulphate solution for 10 – 15 minutes, then dried and sowing had been old and very effective recommendation. The systemic fungicides like carboxin (Vitavax), Bavistin etc. have also been reported to control grain smut disease successfully (Singh, 2000).

Loose Smut (*Sphacelotheca cruenta*)

The loose smut of sorghum caused by *Sphacelotheca cruenta* is reported from China, Iran, Italy, Africa and the USA. In India, it occurs in Tamil Nadu, Karnataka, Andhra Pradesh and Maharashtra. The effects of the smut are not only on the grain but also on the plant growth. Thus, grains as well as fodder yield may be reduced.

Symptoms

The affected plants remain stunted, produce thinner stalks, more tillers, and earlier flowering than the healthy plants. Generally, all the spikelets of the ear are affected and become malformed and hypertrophied. The floral bracts tend to elongate and proliferate. Frequently, the lemma and palea as well as the ovary contain smut sori. The size of the sori ranges from 3-18 mm in length and 2-4 mm in thickness. The affected ear appears like a leafy or leathery structure. The covering membrane of the sori ruptures early releasing the powdery mass of dark coloured spores.

Management

Seed treatment with formalin, Sulpher, copper sulphate, carboxin, Bavistin etc. as recommended for grain smut, is also effective in controlling the loose smut. Where soil survival of spores is possible, crop rotation and field sanitation are recommended (Singh, 2000).

Head Smut (*Sphacelotheca reiliana*)

Head smut caused by *Sphacelotheca reiliana* is reported from many countries in Asia, Southern Europe, Africa and America. In India, it is known to occur in Tamil Nadu, Karnataka, Andhra Pradesh, Maharashtra, Gujrat, Madhya Pradesh, Uttar Pradesh, Punjab and Bihar. This disease causes significant damage to sorghum as it affects the entire earhead, transforming it into a smutted head.

Symptoms

The disease manifests itself only at the time of earhead formation or flowering. The inflorescence is invariably destroyed in the infected plants. In affected plants, the inflorescence is partially or entirely converted into a big sorus fully covered with a thin grayish-white membrane in its early stages but during emergence through the boot leaf the membrane is ruptured and spores are exposed. The sorus is usually 8-10 cm long and 2.5-5.0 cm wide, and is cylindrical in shape. If the wind is blowing at the time of sorus emergence, the air-borne spores resemble a smoky cloud around the head. When the spores are blown off, a network of dark fibers traverses the spore mass and remains adhering even after the spores have been blown away.

Management

The disease can be controlled by a combination of practices such as sanitation, crop rotation and seed treatment (Singh, 2000).since, the smut is only sporadic *i.e.* only few plants are affected in a field it is possible to locate and destroy. Collecting smutted heads in cloth bags and dipping in boiling water to kill the pathogen will reduce the inoculum potential for the following crop (Ranagaswami and Mahadevan, 1999).

Long Smut (*Tolyposporium ehrenbergii*)

Long smut is reported to be prevalent in Egypt, West Africa, Iraq and Pakistan and in India it occurs in Tamil Nadu, Maharashtra, Andhra Pradesh, Karnataka, Madhya Pradesh and Uttar Pradesh. Since, its occurrence is sporadic and confined to a few grains in an ear; it causes little damage to the crop.

Symptoms

The fungus *Tolyposporium ehrenbergii* is known to cause the disease. Since, the disease is confined to a few grains in an ear, only close examination of the ears in the field reveals the presence of long smut. Usually only a few grains are transformed into smut sori which are scattered throughout the ear. The sorus is covered by a thick whitish to dull yellow membrane. The sori are prominent, long, cylindrical, slightly curved, measuring 4 cm in length and 6-8 mm in width. The sori usually remain intact until broken mechanically, releasing the brownish-green spore balls.

Management

Since, the fungus spreads through air-borne spores and sporidia, the disease is difficult to control. However, adjusting the sowing dates seems to help in avoiding the disease (Rangaswami and Mahadevan, 1999).

Rust (*Puccinia pupurea*)

Rust caused by *Puccinia pupurea* is serious disease of sorghum wherever the crop is grown. It is found all over India, causing extensive damage to both irrigated and unirrigated crop. The damage depends on the variety and season. When it occurs late in the season, the loss in grain yield is relatively insignificant, but when infestation starts in the early stages of the plant growth, there is heavy reduction in grain yield.

Symptoms

The fungus affects the crop at all stages of growth, but more often the infection begins when the plants are about 2 month old. The first symptoms of the disease appear in the form of small flecks on the lower leaves (purple, tan or red depending upon the cultivar). Rust pustules (uredosori) appear on both surfaces of the leaf as purplish spots which rupture to release reddish powdery masses of uredospores. As the disease advances, infection spreads to the younger leaves, except for the youngest two or three leaves which are rarely infected. The pustules are elliptical, 1-2 mm in diameter and lie between and parallel with the leaf veins. In highly susceptible cultivars the pustules occurs so densely that almost the entire leaf is destroyed. Reddish brown to black teliospores develop later sometime in the old uredosori or in teliosori, which are darker and longer than the uredosori. The telia develop in linear patterns mostly on the lower surface of the leaves. The pustules may also occur on the leaf sheath and on the stalks of the inflorescence. The rust incidence causes older leaves to dry prematurely and the plants to appear smaller and generally unhealthy, even from distance.

Management

The fungus is known to survive on the alternate host *Oxalis comiculata*; hence, its removal can help in reducing the disease to some extent. Spraying the crop with Mancozeb at 1.25 kg/ha is also recommended.

Downy Mildew (*Peronosclerospora sorghi*)

This disease is reported to be prevalent in many parts of Asia and Africa, Italy and the USA, mostly in a mild form. It occurs in peninsular India, causing much damage to the crop in years when favourable climatic conditions prevail. The damage depends upon the environments and time of infection. The plants may be badly damaged before their full development or, if fully developed, may remain sterile. In a genotype with 100 per cent infection the loss in yield is 74 per cent (Anahosur and Laxman, 1991).

Symptoms

The fungus, *Peronosclerospora sorghi* is the cause of a systemic downy mildew of sorghum, earlier known as *Sclerospora sorghi* Weston and Uppal. It invades the growing points of young plants, either through oospore or conidial infection and as the leaves unfold they show various types of symptoms. The disease manifests itself as a downy, whitish growth on the lower surface of the leaves, consists of conidiophores and conidia. Later, this downy growth spreads over a major portion of the leaf blade, which appears yellowish through upper surface. The first few leaves that show symptoms are only partially infected with green or yellow colouration of the infected portion. As the disease advances, chlorotic streaks develop and turn brown as the leaf tissue die. As the infected bleached leaves mature they become necrotic and the intervenal tissues disintegrate, releasing the resting spores (oospores) and leaving the vascular bundles loosely connected to give the typical shredded leaf symptom. The younger ones remain normal or if infected do not shred and the plants produce healthy ears. However, the green-ear stage though reported (Patel and Kamath, 1950), is not common.

Management

Since, the pathogen is soil-borne in the form of thick walled persistent oospores, cultural control by deep ploughing (30-35 cm), systematic rogueing of diseased plants and crop rotation have been suggested to reduce the oospore population in soil and the loss from downy mildews.

Potassium azide @ 1.12 kg/ha as soil application was found to reduce the incidence of sorghum downy mildew by 23 per cent (Rangaswami and Mahadevan, 1999). Anahosur and Patil (1983) had recommended spraying of metalaxyl 25 WP (Ridomil) @ 2 g a.i. per litre of water at 10 and 40 or 20 and 50 days after emergence for complete control.

Leaf Spot or Leaf Blight (*Exerohilum tercicum*)

Symptoms

The leaf blight caused by *Exerohilum tercicum* (syn: *Helminthosporium tercicum*) also causes seed rot and seedling blight of sorghum. The disease appears in the form of small narrow elongated spots in the early stage but in later stages they extend along with the length of the leaf becoming bigger, spindle shaped, measuring several cm in length and up to one cm in width. The typical symptoms are long elliptical necrotic lesions with dark margins without distinct yellow haloes and bear a faint grey to brown bloom. The straw coloured center becomes darker during sporulation. Many lesions may develop and coalesce on the leaves, destroying large areas of the leaf tissues, giving the crop a distinctly burnt or blasted appearance.

Management

Seed treatment with Thiram @ 2.5 g/kg of seed and spraying crop with Mancozeb @ 2.5 kg/ha in 1000 liters of water are recommended for the management of this disease (Tewari, 2000).

Anthracnose and Red Rot (*Colletotrichum graminicola*)

This disease has a world wide distribution and is prevalent wherever sorghum is grown (Jain, 2001).

Symptoms

The fungus *Colletotrichum graminicola* causes both leaf spot (anthracnose) and stalk rot (red rot) however, in India, leaf spot phase of the disease common. The disease appears as small red, purple or brown spots, with whitish or purple center. The spots are elliptical or spindle shaped, 2-4 mm long and 1-2 mm broad, surrounded by ill-defined margin. The infection is found mostly on the basal and older leaves, the young ones usually being free from infection. With advancement of the disease, the spots may enlarge slightly with characteristic black dots (acervuli) in the center. Red rot phase can be characterized externally by the development of circular cankers, particularly in the inflorescence. Infected stem when split open shows discolouration, which may be continuous over a large area or more generally discontinuous giving the stem a marbled appearance. The stem lasions also show acervuli.

Management

Seed treatment with Captan or Thiram at 4 g/kg seed and spraying of Mancozeb @ 1.25 kg/ha are recommended for the management of disease.

Zonate Leaf Spot (*Gloecercospora sorghi*)

Symptoms

Zonate leaf spot caused by *Gloecercospora sorghi* appear as semi-circular zonate spots with alternate dark and light boundaries on the lamina. The lesions are red brown in colour. As the disease advances the lesions enlarges becoming irregular or semi-circular extending across the lamina with distinct red margins giving characteristic zonatic appearance. The disease is favoured by rain, as severity of the disease is high during the rainy season.

Management

Cultural control by burring the diseased stubbles under the plough after harvesting and spraying the crop with Dithane Z-78 @ 2.5 kg/ha have been suggested for the management this disease (Tewari, 2000).

Sooty Stripe (*Ramulisporia sorghi*)

The disease is reported to be prevalent in Haryana, Delhi, Uttar Pradesh, Andhra Pradesh and Tamil Nadu (Jain, 2001).

Symptoms

The disease appears in the form of elongate elliptical oblong or irregular lesions on the leaves with straw coloured centers of dead tissue and purplish to tan lesion margins, depending on the host cultivar. Later, the surface becomes powdery and sooty coloured as the sclerotic lesions are formed. The lesion is small in the beginning but later several lesions may coalesce to produce large areas of necrotic leaf tissue. The centers of the lesions darken and become grayish when conidia are produced and then blackish or sooty as numerous small black sclerotia are produced. The leaves dry up prematurely, in humid weather when the disease spreads rapidly.

Management

Since, the pathogen is seed and soil borne and survives in the soil with fallen leaves and debris from season to season; the diseased stubbles should be buried under the ground with plough after harvest. Spraying the crop with Mancozeb @ 2.5 kg/ha in 1000 liters of water is also recommended for it control (Tewari, 2000).

Among the other diseases of sorghum the following are of minor importance:

1. Stalk rot, top rot or seedling blight (*Gibberella fujikuroi*)
2. Rough leaf spot (*Ascochyta sorghi*)
3. Cercospora leaf spot (*Cercospora sorghi*)
4. Charcoal rot, stalk rot, blight or hollow stem (*Macrophomina phaseolina*)
5. Ergot or sugary disease (*Sphacelia sorghi*)
6. Bacterial leaf blotch (*Xanthomonas campestris* pv. *rubrisorghi*)
7. Bacterial leaf stripe (*Pseudomonas syringae* pv. *sorghicola*)
8. Phanerogamous parasite *Striga asiatica* and *S. densiflora*

Pearl Millet or Bajra

Green Ear or Downy Mildew (*Sclerospora graminicola*)

Green ear or downy mildew caused *Sclerospora graminicola* is a serious disease of pearl millet (*Pennisetum typhoides*). The disease is known to occur in India, Iran, Israel, China, Japan, Fiji, USA and many African countries. In India, the disease was first reported by Butler (1907) in sporadic form, not causing much damage, except in low lying fields where the loss could be significant. Since, then the disease has become widespread and virulent, causing considerable damage to the grain yield. Loss estimates vary from 6 to 60 per cent. Up to 27 – 30 per cent loss has been estimated in India (Mathur and Dalela, 1971; Nene and Singh, 1976). Up to 60 per cent grain losses have been reported in many African countries.

Symptoms

The symptoms of the disease appear on the leaves and inflorescence. The first symptoms can appear in the seedlings at 3 to 4 leaf stage. The affected leaves show chlorosis in streaks (light green to light yellow colour) on the upper surface. Just below these streaks on the lower surface a fine downy growth of the fungus may appear. As the disease advances, the streaks turn brown and the leaves become shredded along the veins. The downy growth seen on infected leaves consists of *sporangiophres* and sporangia.

The characteristics symptom of the disease is the ear deformities characterized by the transformation of floral parts into twisted leafy structures. This gives the ear an appearance of green leafy mass hence the name "green ear". In most cases the entire ear is transformed into leafy structures, but sometimes only part of the ear is affected, the other part produce normal grains. All the floral parts including glumes, palea, stamens and pistil are converted into green, linear, leafy structures of variable length.

Management

Since, the pathogen is mostly soil-borne it is difficult to control the disease. As the fungus is persistent in soil up to about 5 or more years, crop rotation has little or no value in avoiding the disease. Rogueing of diseased plants within a month of sowing followed by spraying of Dithane M-45 had also been an effective control method (Singh, 2000). The disease is partly disseminated through seeds, seed treatment Ridomil (8 g/kg seed) followed by one spray of 0.1 per cent Ridomil 20 days after planting is recommended for its management (Muthusamy and Narayanaswamy, 1981; Dang *et al.*, 1983). Similarly, Appaji *et al.* (1989) reported that seed dressing with Apron SD-35 (8 g/kg seed) followed by a foliar spray, 25 days after sowing, with Ridomil MZ-72 WP (metalaxyl + mancozeb) at 500 ppm concentration or with Ridomil ZM-280 (metalaxyl + ziram) at 1000 ppm concentration the best combination. Gupta and Verma (1991) also found that seed treatment with Apron SD-35 (2.5 g/kg seed) controlled the disease up to 30 days after sowing.

Rust (*Puccinia penniseti*)

Rust disease caused by *Puccinia penniseti*, is serious disease of pearl millet, occurring wherever the crop is grown. In India, it is prevalent an all the bajra tracts, causing considerable damage from seedling stage to maturity.

Symptoms

The first symptom of the disease appears mostly on the distal half of the lamina. Later, minute, round uredosori occur in groups on both surfaces of the leaves. The uredosori may also occur on leaf sheath, stem and even on peduncles. Later, telial formation takes place on leaf blade, leaf sheath and stem. The telia are black, elliptical and sub-epidermal. While, brownish uredia get exposed at maturity the black telia remain covered by the epidermis for longer duration. In severe infections; the plants appear unhealthy and slightly stunted as compared to healthy ones.

Management

The only effective method of controlling bajra rust to grow resistant varieties. So far most of the varieties tested are susceptible; therefore work is needed to develop resistant varieties. Some control of the rust is reported through preventive sprays with 100 ppm of Cupramar and Dithane S 31 (Rangaswamy and Mahadevan, 1999).

Smut (*Tolyposporium penicillariae*)

Smut caused by *Tolyposporium penicillariae*, is a widespread disease of pearl millet. The disease has been reported from Pakistan, Africa, the USA and India. In

India, it is prevalent in Tamil Nadu, Andhra Pradesh and Maharashtra. Chahal (1986) estimated the loss from this smut by comparing the yield in untreated plots with yield in plots treated with 4 sprays of oxycarboxin (Plantvax). He reported that yield was 20.6 per cent more in the treated plots.

Symptoms

The disease manifests itself at time of grain setting. A few grains scattered on the ear, may be replaced by oval to top-shaped sori, which are generally two to three times of the size of the normal grain. The top of the sorus is bluntly rounded to conical in shape. They are bright green to chocolate brown in the early stages and become dark black on maturity, often projected clearly beyond the glumes. The colour is due to the membrane covering the sorus and often rupture to expose a black spore mass.

Management

The pathogen survive in soil so removal of smutted ears, clean seed, hot weather ploughing, field sanitation etc. can reduce the incidence of disease to some extent (Tewari, 2000). Various non-systemic fungicides including zineb and mancozeb, the systemic fungicides Plantvax, Vitavax and Benlate and antibiotics heptanes and aureofungin have been tried either as seed treatment, foliage and panicle spray. Effective control by foliar and panicle sprays with Plantvax and Vitavax is reported (Chahal, 1986; Chaube and Pundhir, 2005). Four sprays with captafol, zineb and heptanes have also been reported effective (Chaube and Pundhir, 2005).

Ergot (*Claviceps microcephala*)

Ergot caused by *Claviceps microcephala* is an important disease of pearl millet. It was first reported from South India but was not considered a major disease; however, in 1956 it occurred in epidemic form in South Satara area of Maharashtra (Bhide and Hegde, 1957; Shinde and Bhide, 1958). By 1966 the disease had become a major limitation in the cultivation of improved hybrid varieties. Sever epidemics of the disease occurred in Delhi, Uttar Pradesh, Rajasthan, Maharashtra, Karnataka, Tamil Nadu, Andhra Pradesh and Haryana. Natarajan *et al.* (1974) estimated the average incidence to be about 62 per cent with grain loss of about 58 per cent. In Rajasthan, Bansal and Siradhana (1988) observed maximum grain yield loss of 7.64 q/ha.

Symptoms

The disease becomes evident by exudation of small droplets of pinkish or light honey-coloured sticky fluid (the honey dew stage) from the infected spikelets. Under severe infection many such spikelets exude plenty of honey dew which trickles along the earhead. Later, these droplets become darker, coalesce, and cover large areas of the cob. With the advancement of disease, small dark brown sclerotia can be seen projecting from between the glumes. These sclerotia (ergot) contain alkaloids responsible for ergot poisoning in animals.

Management

The sclerotia remain viable for longer time in soil (Kulkarni, 1967) therefore; repeated deep ploughing especially during dry summer, long crop rotation, planting sclerotia free seeds, adjustment of sowing dates, and intercropping may help in avoiding soil-borne inoculum. Many workers have recommended 2-3 sprays of ziram, copper oxychloride + zineb, and wettable Sulpher at 5-7 days intervals starting just before earhead emergence (Singh, 2000). Besides, chemical control Mower *et al.* (1975) has reported that *Fusarium roseum* "Sambucinum" is a highly potential biocontrol agent against *Claviceps purpurea*. Similar parasitism of *C. purpurea* by *Fusarium heterosporum* has also been reported (Cunfer, 1975).

Other diseases of bajra of minor importance are:

1. Twisted top (*Fusarium moniliforme*)
2. Leaf blast (*Pyricularia setariae*)
3. Leaf spot (*Curvularia penniseti*)
4. Leaf spot (*Xanthomonas campestris* pv. *Penniseti*)
5. Leaf blotch (*Xanthomonas campestris* pv. *Annamalaiensis*)
6. Leaf spot (*Drechslera australiense*)

Maize

Downy Mildew (*Peronosclerospora sorghi*)

A number of downy mildew fungi have been reported on Maize in India (Butler and Bisby, 1933; Singh *et al.*, 1967; Payak and Renfro, 1967; Singh and Chaube, 1968; Singh, 1968). However, the most common and dangerous species are *Peronosclerospora sorghi*, *P. philippinensis*, *P. sacchari* and *Sclerophthora rayssiae zeae.*

Symptoms

Varied types of symptoms are produced by downy mildew pathogens:

The most characteristic symptom of downy mildew caused by *P. sacchari* is the development of long, rather broad, chlorotic stripes along almost the entire length of the leaf. When several such stripes coalesce the margins are lost and irregular elongated patches are formed. The affected plants exhibit a stunted and bushy appearance due to the shortening of the internodes. White downy growth can be seen not only on the lower surface of the leaf but also on the chlorotic streaks. The downy growth also occurs on bracts of green unopened male flowers in the tassel. Sometimes miniature to large leaves has been noticed in the tassel. Symptoms of downy mildew caused by *P. philippinensis* and *P. sorghi* on leaves are very similar to those describe above. The colour of chlorotic stripes by *P. philippinensis* is more intense. This fungus produces a characteristic downy growth on leaves followed by yellow discolouration, browning and necrosis of the blade and stunting of plant. The lesions of downy mildew caused by *Sclerophthora rayssiae* var. *zeae* are in the form of chlorotic to yellow streaks. They turn purplish with age.

Proliferation of auxillary buds on the stalk of tassel as well as the cobs is very common (crazy top).

The two types of downy mildews commonly do not occur on the same plant though occurring in the same field. However, a number of plants have been seen in which all the lower leaves carried the symptoms of *Sclerophthora* and the upper leaves those of *P. sacchari*.

Management

The disease can be managed by adopting three approaches such as cultural practices, use of fungicides and planting resistant varieties. Cultural practices include long crop rotation, destruction of collateral hosts and disease crop debris. Several fungicides have been recommended for its management. Nene and Saxena (1970) had suggested 4-6 spraying of Dithane M-45 (@ 0.3 per cent) starting from 10 days after sowing at 10 days interval. Sun (1970) reported Dithane M-22 effective against the disease in Taiwan. Sharma *et al.* (1981) reported good control of Philippine downy mildew of maize with 3 sprays of Dithane Z-78 (0.3 per cent) or 4 sprays of Dithane M-45 (0.3 per cent). Among systemic fungicides, Lal *et al.* (1977) had reported that seed treatment with Demosan in combination with rogueing of diseased plants 20 days after sowing, and one foliar spray of neem oil gives most effective control. Seed treatment with Ridomil (metalaxyl) @ 4 g/kg seed (Lal *et al.*, 1979, 1980) or 0.8 per cent (Sharma *et al.*, 1981) gave best results. Similarly, Figueiredo and Anahousur (1993) reported that Ridomil MZ 75 WP may be used for seed dressing (@ 3 g/kg seed) to control the disease.

Leaf Blight (*Helminthosporium maydis*)

Leaf blight is a common disease of maize, sorghum and other related crops in many parts of the world, including India.

Symptoms

The fungus *Helminthosporium maydis*, also causes leaf spot or stripe symptoms, it affects the maize at young stage. On leaves, the young lesions are small yellowish and diamond in shape. These lesions gradually increase in size; growth is limited by adjacent veins, so finally they assume rectangular shape. Such lesions may coalesce, producing a complete burning and gives blighted appearance. The surface is covered with olive green velvety masses of conidia and conidiophores.

Management

Since, the pathogen is seed-borne, seed treatment with Captan or Thiram @ 4 g/kg seed or spraying the crop with Dithane Z 78 or Sonacol @ 2.5 kg/ha in 1000 liters of water (Tewari, 2000) can reduce the damage caused by the disease.

Smut (*Ustilago maydis*)

Maize smut is not common in India and is confined to Kashmir, less common in the Punjab, west Bengal and rarely seed in the north western parts of the Uttar Pradesh.

Symptoms

The smut caused by *Ustilago maydis* produces gall on the ears, axillary buds, tassels, stalks and rarely on the leaves. When galls are formed on cobs they cause extensive damage to the grain yield. The fungus induces hyperplasia and hypertrophy and excessive development of the phloem elements of the bundles. The epidermal tissues of the gall are dull white or grey in colour, much bulged from the surface and shiny. They rupture to expose the black powdery spore mass. Stem galls result in bending of stalk. Infection of the female flowers gives rise to galls instead of grains. However, the affected seedlings remain stunted and weak.

Management

As, the disease is less common in India not much work has been done, however, crop rotation, field sanitation and seed treatment may help in reducing the incidence of the disease.

Stalk Rot (*Erwinia chrysanthemi* var. *zeae*)

The disease was first reported from Pusa (Bihar) by Prasad in 1930. Later, Hingorani *et al.* (1959) described its occurrence in different states of India. The disease is fairly destructive in Tarai region of Uttarakhand, Bihar, Rajasthan, Andhra Pradesh and Jammu & Kashmir.

Symptoms

The disease occurs in young as well as old plants. The bacteria infects the plants especially at the collar region, at any stage of the growth, as a result basal internodes become soft, discoloured and give a bad fermenting smell. With the advancement of the disease leaves may show yellowing and drying. The infected tissues of the stalk are first soft but later they turn into a dry mass of shredded, easily disjointed fibers. At this stage the plant topples down. In addition to stalk rot, a light to dark brown rotting of the sheath, starting from the point where it is attached to the affected tissues of the stalk, has also been seen during rainy season. The infected ears become blighted, chaffy or rot.

Management

Proper drainage or selection of well drained field, use of potash at recommended dose and application of bleaching powder @ 25 kg/ha in rows (Tewari, 2000; Singh, 2001) may help in reducing the disease to some extent. In India, Sharma *et al.* (1982), Lal and Saxena (1982), Thind and Soni (1983) have reported efficacy of calcium hypochloride in the control of disease. Two applications of calcium hypochloride, containing 22 per cent chlorine, as soil drench @ 25 kg/ha, first before flowering and second 10 days later give control of the disease (Lal and Saxena, 1982).

Besides above-listed important diseases of maize, there are other diseases of minor importance in this crop are: rust (*Puccinia sorghi*), turcicum leaf blight (*Helminthosporium turcicum*), seedling blight or wilt (*Fusarium monilirforme*), Banded leaf and sheath blight (*Rhizoctonia solani* f. sp. *sasakii*), charcoal rot (*Macrophomina phaseolina*), wet rot (*Cephalosporium maydis*) and mosaic diseases.

Oats

Smuts

This crop suffers from two smut diseases, the loose smut and the covered smut. Both smuts are worldwide in distribution.

Loose Smut (*Ustilago avenae*)

The symptoms of loose smut in oats are similar to the loose smut of wheat and barley. All parts of the florets are converted into smut sori which are covered by a thin membrane. This membrane ruptures easily on emergence from the boot to expose the smut spores. Usually, all the eras in a stool and all the grains in a ear are diseased but sometimes a few upper grains of the ear may remain healthy. Sometimes, the smut sori may develop on the upper most leaf blade and in the stalk of the inflorescence.

Covered Smut (*Ustilago kolleri*)

In covered smut the sori replacing the kernels are enclosed in a fairly persistent membrane composed of pericarp ad floral bracts which are not damaged. Thus the spikelets preserve their shape. The spores are exposed at the time of threshing, releasing the black mass of spores over the surface of healthy grains. These sori of covered smut are darker than those of loose smut. The main rachis of the inflorescence may also contain covered smut sori which show spore masses in big cavities in a transverse section.

Management

Since, the disease is seed-borne; seed treatment continues to be the best method of control. Seed treatment with formalin (diluted in equal volume of water) can be used for its management. The liquid is sprinkled and the grains are wetted thoroughly then left covered with moist gunny bags for 19 hours. Pathak *et al.* (1970) had reported control of covered smut of oats by seed dressing with systemic fungicides like Vitavax and Plantvax @ 0.2 and 0.25 per cent, respectively.

Rusts

Stem Rust (*Puccinia graminis* f. sp. *avenae*)

The stem rust caused by *Puccinia graminis* f. sp. *avenae* is common on oats, causing sever damage to the crop. The diseases prevalent in the Nilgiris and at least four different races have been collected from that area. This disease reduce the forage value and yield of the seeds. Like stem rust of wheat, the fungus has the long life cycle and completes its life cycle through Barberis and Mahonia. The uredial and telial stages occurs on culms, leaf sheaths and blades of plant, while pycnial and acial stages on the alternate hosts. Uredial pustules are lorge oblong and brown to dark brown in colour. Telia are black, oblong and exposed. These are formed around the uredia at the advance stages of the growth.

Crown Cust (*Puccinia coronata* var. *avenae*)

It is the most important disease of oats but is confined to the temperate climatic regions. The disease is quite common in North India. Three different races of *Puccinia coronata* have beeen identified. The characteristic symptom is in the form of small orange colour lesions on the leaf surface arranged in lines covering the entire leaf surface. This results in the reduction of the chlorophyll content of the leaves (Ahmad and Yadav, 1977). Certain hosts like; *Phalaris tuberose, P. minor, Avena glauca, A. strigosa* and *Vulpia myuros* aid the perpetuation and annual recurrence of the rust (Randaswamy and Mahadevan, 1999).

Management

The sources of resistance for stem rust lie in the oat varieties lowa-670, Gopher, Kandula, Markton, Curt and Siaia (Misra *et al.*, 1965) and non-cultivated oats for crown rust are *Avena Sterilis, A. barbata, A. magna,* and *A. abyssinica* (Murphy *et al.*, 1968; Simons, 1959; Simons *et al.*, 1962). The disease can be controlled with the use of Zineb or Mancozeb @ 0.25 per cent as foliar spray at 15 days interval (Jain, 2001). Ahmad (1970) had reported a hyper-parasite fungus *Trichothecium roseum* which can be utilized for the management of rust.

Leaf Blotch (*Drechslera (Helminthosporium) avenae*)

This disease is particularly severe in the northern regions of the country (Jain, 2001). The symptoms of leaf blotch disease caused by *Drechslera avenae*, appear as small grayish brown, necrotic elongated spots on the leaf blade and sheath, which coalesce to form larger spots or lesions of irregular shape and size. The lesions are mostly limited by veins, but they spread and coalesce to form elongated large dark brown to grey blotches. As the disease advances, the tissues become necrotic and dry up and the leaves wilt.

Management

For the management of leaf blotch, seed treatment with organo-mercurial fungicides like, Ceresan and Agrosan @ 0.25 per cent completely inhibited the pathogen from the seed (Jain, 2001), however, these fungicides are not recommended now a days. Spraying of Mancozeb @ 0.25 per cent can be used for its management (Jain, 2001).

Besides above-listed diseases sclerotial wilt (*Sclerotium rolfsii*) and red leaf of oat (Viral) are of minor importance diseases known to occur in India.

Diseases of Cultivated Legumes

Berssem or Egyptian CloverCLOVER

Root Rot Complex (*Rhizoctonia solani, Fusarium semitectum, Tylenchrhynchus vulgaris*)

The occurrence of disease is common in the Gangetic and Central plains. Although fungi only can incite the disease but the presence of nematode accelerates the infection rate causing serious damage to the crop (Bhaskar and Ahmad, 1990; Hasan and Bhaskar, 1992). Once the disease established in the field it becomes a permanent source of infection as the pathogen perpetuates in the sopil through their resting structures. Heavy incidence of the disease reduces the plant density and the green fodder yield.

Management

For the control of root disease seed treatment with Thiram (0.25 per cent) and

Bavistin (0.1 per cent) followed by foliar spray of Bavistin (0.1 per cent) can be practiced (Jain, 2001).

Stem Rot (*Sclerotinia trifoliorum*)

This is the major disease of berseem caused by a fungus, *Sclerotinia trifoliorum* present in the soil. This is mainly a disease of temperate zone; however, it is also known to occur in the Gangetic plains of India during January and February when the temperature is around 10°C, which favours the multiplication and spread of the pathogen. The pathogen attacks the basal portion of the stem and causes it rot. Rotting of stem occurs generally 5 cm above the ground. It produces heavy white cottony mycelium in the field, which can be very easily spotted in the field around the wilted patches of the berseem crop. Later, black coloured sclerotial bodies can be seen that helps the pathogen to perpetuate in the field from one season to another making the disease soil-borne.

Management

Use of disease free seed, flooding of infested field during the summer are some the cultural practices which help to reduce the inoculum in the field. Drenching of soil with 0.4 per cent Brasicol, after cutting, can be practiced for its management (Singh, 2001).

Lucerne

Downy Mildew (*Peronospora trifolii*)

In India the disease confined to North India and central India during winter (December to February). The disease causes considerable damage to the ypung plants (Ahmad, 1977) and reduces the stand of the crop and is most conspicuous in the subsequent years. The characteristic symptoms of the disease are light green leaves especially at the apex of the stem and presence of brownish mycelium on the lower surface of the leaves. The mycelium consists of enormous number of conidiophores with hyaline conidia. Heavy infection results in the stunting and defoliation of the plants. The pathogen persists in the soil through oospores lying in the plant debris

Management

Foliar spray of Dithane M-45 (0.25 per cent) and Chlorothalonil (0.1 per cent) were found effective against downy mildew (Jain, 2001).

Rust (*Uromyces striatus*)

Rust caused by *Uromyces striatus*, is a common disease of lucerne in Northwest, Western Himalayas, Gangetic and Western plains of India. The disease is characterized by the presence of reddish brown uredia and telia on the leaves and stems develop late in the season. The pustules usually are single but sometimes arranged in circles around the single pustule. The telia that are black in colour are formed independently or in the same lesions. The rust is heteroecious in nature and forms its sexual stage on *Euphorbia* sp.

Management

Foliar spray of Dithane M-45 (0.25 per cent) has been found effective against this disease (Jain, 2001).

Common Leaf Spot (*Pseudopeziza medicagensis*)

The disease caused by *Pseudopeziza medicagensis* is common to all lucerne growing areas, but more serious in the north-west, Gangetic and central plains of the India. The disease appears in the form of small dark brown to black, circular spots measuring 3-4 mm in diameter. The edge of the spot is toothed with dark brown to black apothecia in the center, a feature that distinguishes it from all other leaf spots. Heavy infection results in the yellowing and subsequently defoliation of leaves.

Management

Early cutting of lucerne can help in reduction of disease to some extent. Foliar spray of Dithane M-45 (0.25 per cent) has been found effective against this disease (Jain, 2001; Singh, 2001).

Bacterial Wilt (*Aplanobacter insidiosum*)

It is the most serious disease of lucerne which is caused by *Aplanobacter insidiosum*. Affected plants show stunting and large numbers of branched stems can be seen in the field. The roots of the affected plants show a brownish-yellow discolouration of the woody tissue.

Management

There is no appropriate control measure for this disease. Avoid growing lucerne in the same field for next 2-3 years.

The other diseases of minor important are anthracnose (*Colletotrichum trifolii*), crown wart (*Physoderme alfalfa*), leaf spot (*Cercospora medicagensis, Stemphylium botryosum, and Alternaria medicagensis*), powdery mildew (*Erysiphe polygoni*), mosaic (Alfalfa mosaic virus), wilt (*Fusarium* sp.) and root rot (*Rhizoctonia* and *Fusarium* sp.).

Guar or Clusterbean

Bacterial blight (*Xanthomonas campestris* pv. *cyamopsidis*)

It is the most serious disease of guar caused by *Xanthomonas campestris* pv. *cyamopsidis*. It occurs mostly in kharif crop. The early symptoms of the disease appear in the form of small intra-veinal spots, which are round and well defined on the dorsal surface of the leaf. They may enlarge or several such spots may coalesce and result in water soaked lesion. These lesions later become necrotic, covering large areas of the leaves. These may appear on flowers and pods also.

Management

For the management of this disease hot water treatment (50°C for 10 minutes) has been recommended (Jain, 2001).

Alternaria Leaf Spot (*Alternaria cyamopsidis*)

The disease is caused by a fungus *Alternaria cyamopsidis*. The disease is characterized by the formation of dark brown, round to irregular spots on leaves, varying from 2 to 10 mm in diameter. These spots are water soaked initially but later turn grayish to dark brown with concentric zonations with light brown lines inside the spots. In severe infection several spots merge together and the leaflets become chlorotic and usually drop off. If plants are infected in the early stages of growth they fail to flower.

Management

Spraying of Dithane M-45 (@0.2 per cent) at 15 days interval has been found effective (Singh, 2001).

There are some other diseases of minor importance such as powdery mildew (*Leveillula taurica*), root rot (*Rhizoctonia solani*), dry root rot (*Macrophomina phaseolina*) and anthracnose (*Colletotrichum lindemuthianum*).

Cowpea

Dry Root Rot (*Macrophomina phaseolina*)

The disease prevalent in all cowpea growing areas of the country and its intensity is influenced by the drought spells during the rainy season.

Symptoms

The primary symptom of the disease is yellowing of leaves. With in a day or two, such leaves droop and in next 2-3 days, they may drop off. Later, the plants may wilt within a week after the first symptoms at the ground level. When the stem is examined closely the dark lesions may be seen on the bark at the ground level. If the plants are pulled from the soil and examined, the basal stem and the main roots may show water soaked lesions, which finally result in the decaying of the entire root system and may show dry rot symptoms. In advanced stage several sclerotial bodies may be seen scattered on the affected tissues.

Management

Since, the disease is soil-borne, it is difficult to control. Nevertheless, deep summer ploughing, field sanitation including cutting and burning of diseased plant may help in reducing the disease to some extent. Seed treatment with Thiram (0.25 per cent) + Bavistin (0.1 per cent) followed by application of bioagents (*Trichoderma viride* or *T. harzianum*) has also been reported effective (Jain, 2001).

Anthracnose (*Colletotrichum lindemuthianum*)

Symptoms: The fungus *Colletotrichum lindemuthianum* infects all plant parts above the ground level and at any stage of plant growth. The most characteristic symptom of the disease is spotting on pods. Firstly, water soaked lesions appear on the pods, later becoming brown. These lesions enlarge to from circular spots of varying size. The spots are usually depressed with dark centers, and bright red, yellow or orange margins. The spots may also be seen on leaves, petioles and stems.

Management

Spraying the plants with Bordeaux mixture at 15 days interval may help in reducing the disease. Seed treatment with organo-mercurial fungicides had also been recommended in early days (Jain, 2001).

Rust (*Uromyces phaseoli typica*)

Though there are different species of rust attacking legumes, the bean rust caused by *Uromyces phaseoli typica* is found on cowpea. It is worldwide in distribution and occurs in almost every state in Inida.

Symptoms

The fungus produces characteristic rust pustules on the leaves, pods and sometimes on new shoots. Symptoms on the leaves are very clearly visible and start from the lower surface of the leaf where small pustules are formed. These pustules contain uredial stage of the fungus. They appear in groups and several pustules coalescing to cover a large area of the leaf blade. As the disease advances, the fungus produces telial stage on the host.

Management

Cultural practices like adjusting sowing dates of the crop can help in minimizing the damage. Although Sulpher dusting can help in checking the spread of rust but it is not economical (Rangaswamy and Mahadevan, 1999). Spraying the crop with Dithane M-45 @ 2kg/ha in 1000 litres of water has been reported to be effective in reducing the damage caused by rust disease (Singh, 2001).

Cowpea Mosaic (Cowpea Mosaic Virus)

It is a disease caused by cowpea mosaic virus transmitted by aphids (*Aphis craccivora, A. gossypi* and *Myzus persicae*). The affected leaves become pale yellow and exhibit mosaic, vein banding symptoms. The size of the affected leaves gets reduced and show puckering. Pods are also reduced and become twisted.

Management

Use of healthy seed and spraying the crop with Metasystox (0.1 per cent) or any other systemic insecticide for controlling aphids, help in reducing the disease (Singh, 2001). It has been reported that viral diseases in the crop are less in summer crop sown during March to May (Sharma and Verma, 1984) so adjustment dates of sowing can be a effective way to reduce the damage by mosaic.

Other diseases of common occurrence in this crop are leaf spot (*Cercospora dolichi*), bacterial blight (*Xanthomonas campestris* pv. *vignicola*), seedling rot (*Rhizoctonia, Pythium, Sclerotium, and Phytophthora* sp.), top necrosis (*Fusarium equiseti*), wilt (*Fusarium* sp.), Die-back (*Colletotrichum capsici*) and powdery mildew (*Erysiphe polygoni*).

References

Ahmad, S.T. 1969. Studies on the crown rust of oats caused by *Puccinia coronata* var. *anaenae* Fraser Led. Ph. D. Thesis, AMU, Aligarh.

Ahmad, S.T. 1970. *Trichothecium roseum*, as hyperparasite of different rusts. *Indian Phytopath.* 23: 634-636.

Ahmad, S.T. 1977. *Sehima nervosum*, a new host record of *Puccinia versicolor*. *Indian Phytopath.* 30: 261.

Ahmad, S.T. and Yadav, R.B.R. 1977. Effect of crown rust infection on chlorophyll content of *Avena* sp. *Forage Research*. 4: 177-179.

Ahmad, S.T., Srinath, P.R. and Gupta, M.P. 1977. Estmation of losses due to downy mildew in lucerne. *Indian Phytopath.* 30: 466-468.

Anahosur, K.H. and Laxman, M. 1991. Estimation of loss in grain yield in sorghum genotypes due to downy mildew. *Indian Phytopath.* 44: 520-522.

Anahosur, K.H. and Patil, S.H. 1983. Effective spray schedule of metalaxyl (25 WP) for sorghum downy mildew therapy. *Indian Phytopath.* 36: 465-468.

Anonymous. 2006. Forage crops and grasses. *In: Handbook of Agriculture*. Directorate of Information and Publications of Agriculture, ICAR, New Delhi. 1346p.

Appaji, S., Thakur, D.P. and Bishnoi, O.P. 1989. Relationship between meteorological factors and the occurrence of pearl millet downy mildew (*Sclerospora graminicola*) and its control with chemicals under field conditions. *Indian J. Mycol. Pl. Pathol.* 19: 68-72.

Bansal,R.K. and Siradhana, B.S. 1988. Estimation of losses due to pearl millet ergot disease. *Indian Phytopath.* 41: 575-577.

Bhide, V.P. and Hegde, R.K. 1957. Ergot of bajra (*Pennisetum typhoides*) in Maharashtra state. *Curr. Sci.* 26: 116.

Butler, E.J. 1907. Some diseases of cereals caused by *Sclerospora graminicola*. *Mem. Dep. Agr. India, Bot. Ser.* 2: 1-24.

Butler, E.J. and Bisby, G.R. 1933. *Fungi of India*. ICAR Monogr. No. 11, New Delhi.

Chahal, S.S. 1986. Yield loss due to smut of pearl millet. *Indian Phytopath.* 39: 292-293.

Chaube, H.S. and Pundhir, V.S. 2005. Crop diseases and their management. Prentice Hall of India Pvt. Ltd. New Delhi. 703p.

Chester, K.S. 1950. Plant disease losses, their appraisal and interpretation. *Plant Dis. Rep. Suppl.* 193: 190-362.

Cunfer, B.M. 1975. Colonization of ergot honey dew by *Fusarium heterosporum*. *Phytopathology*. 65: 250-255.

Dang, J.P., Thakur, D.P. and Grover, R.K. 1983. Control of pearl millet downy mildew caused by *Sclerospora graminicola* with systemic fungicides in an artificially contaminated soil. *Ann. Appl. Biol.* 102: 99-106.

Figueiredo, N.X. and Anahousur, K.H. 1993. Relative efficacy of two formulations of metalxyl in the control of downy mildew of maize. *Indian Phytopath.* 46: 180-181.

Gupta, G.K and Verma, S.K. 1991. Control of downy mildew of pearl millet with Ridomil. *Indian Phytopath.* 44: 448-461.

Hingorani, M.K., Grant, U.J. and Singh, N.J. 1959. *Erwinia carotovora* f. sp. *zeae*, a destructive pathogen of maize in India. *Indian Phytopath.* 12: 151-157.

Jain, R.K. 2001. Pests and diseases of fodder crops and their management. *In: Plant Pathology* (ed. Trivedi, P.C.). Pointer Publishers, Jaipur. 422p.

Kulkarni, U.K. 1967. Viability of sclerotia of *Claviceps microcephala* in relation to their weight and size. *Indian Phytopath.* 20: 139-141.

Lal, S. and Saxena, S.C. 1982. Field evaluation of calcium hypochloride for the control of bacterial stalk rot of maize. *Indian J. Mycol. Pl. Pathol.* 12: 278-282.

Lal, S., Bhargava, S.K. and Upadhyay, R.N. 1979. Control of sugarcane downy mildew of maize with metalaxyl. *Plant Dis. Rep.* 63: 986-989.

Lal, S., Nath, K. and Saxena, S.C. 1977. Integrated control of sugarcane downy mildew of maize. *Indian Phytopath.* 30: 143-144.

Lal, S., Saxena, S.C. and Upadhyay, R.N. 1980. Control of brown stripe downy mildew of maize by metalaxyl. *Plant Dis. Rep.* 64: 874-876.

Mathur, R.L. and Dalela, G.S. 1971. Estimation of losses from green ear disease (*Sclerospora graminicola*) of bajra and grain smut of jowar in Rajasthan. *Indian Phytopath.* 24: 101-104.

Misra, D.P., Singh, S. and Ahmad, S.T. 1965. Performance of oat varieties against *Puccinia avenae* and *P. graminis avanae. Indian Phytopath.* 18: 29-32.

Mower, R.L., Snyder, W.C. and Hancock, J.G. 1975. Biological control of ergot by *Fusarium. Phytopathology.* 65: 5-10.

Murphy, H.C., Sadanaga, K., Zillinksy, F.J., Terril, E.E. and Smith, R.J. 1968. *Avena Magana*-an important tetraploid species of oat. *Science.* 158: 103-104.

Muthusamy, M. and Narayanswamy, P. 1981. Fungicidal control of downy mildew of pearl millet. *Indian J. Agric. Sci.* 51: 511-514.

Natarajan, U.S., Guruswamy Raja, V.D. Selvaraj, S. and Parambaramani, C. 1974. Grain loss due to ergot disease of bajra hybrids. *Indian Phytopath.* 27: 179-182.

Nene, Y.L. and Saxena, S.C. 1970. Studies on the fungicidal control of downy mildew of maize caused by *Sclerophthora rayssiae* var. *zeae*. *Indian Phytopath.* 23: 216-219.

Nene, Y.L. and Singh, S.D. 1976. Downy mildew and ergot of pearl millet. PANS. 22(3): 366-385.

Patel, M.K. and Kamath, M.N. 1950. Occurrence of green ear stage in sorghum. *Curr. Sci.* 19:156.

Pathak, K.D., Joshi, L.M. and Renfro, B.L. 1970. Control of covered smut of oats by systemic fungicides. *Indian Phytopath.* 23: 693-694.

Payak, M.M. and Renfro, B.L. 1967. A new downy mildew disease of maize. *Phytopathology.* 57: 394-397.

Prasad, H.H. 1930. A bacterial stalk rot of maize. *Agric. Jour. India.* 25: 72.

Rangaswami, G. and Mahadevan, A. 1999. Diseases of cereals. *In:* Diseases of crop Plants in India (4th ed.) Prentic Hall of India, Pvt. Ltd. New Delhi, pp.160-264.

Sharma, S.C., Khera, A.S., Bains, S.S. and Malhi, N.S. 1981. Efficacy of fungitoxicant sprays and seed treatment against Philippine downy mildew of maize. *Indian Phytopath.* 34: 498-499.

Sharma, S.C., Randhawa, P.S., Thind, B.S. and Khera, A.S. 1982. Use of Klorocin for the control of bacterial stalk rot and its absorption, translocation, and persistence in maize tissue. *Indian J. Mycol. Pl. Pathol.* 12: 185-189.

Sharma, S.R. and Verma, A. 1984. Effect of cultural practices in virus infection in cowpea. *Zeiy. Fur. Acker. Uad Pflanz.* 153: 23-31.

Shinde, P.A. and Bhide, V.P. 1958. Ergot of bajra (*Pennisetum typhoides*) in Bombay state. *Curr. Sci.* 27: 499-500.

Simons, M.D. 1959. Variability among the strains of non-cultivated species of *Avena* for resistance to races of the crown rust fungus. *Phytopathology*. 49: 598-601.

Simons, M.D., Wahl, I.D. and Silva, A.R. 1962. Strains of cultivated *Avena* sp. resistant to important races of crown rust fungus. *Phytopathology*. 42: 352-360.

Singh, C. 2001. *Modern techniques of raising field crops*. Oxford and IBH Publishing Co. Pvt. Ltd. New Delhi. 523p.

Singh, J.P. 1968. *Sclerospora sacchari* on maize in India. *Indian Phytopath*. 21: 121-122.

Singh, R.S. 2000. Diseases caused by basidiomycotina – ustilaginales. *In: Plant Diseases* (7th Edition). Oxford and IBH Publishing co. pvt. Ltd. New Delhi. pp. 315-370.

Singh, R.S. and Chaube, H.S. 1968. The occurrence of *Sclerospora sacchari* and its oospores on maize in India. *Labdev. J. Sci. Technol.* 6B: 197-200.

Singh, R.S., Khanna, R.N., Chaube, H.S. and Joshi, M.M. 1967. Internally seed borne nature of two downy mildews of corn. *Plant Dis. Rep.* 51: 1010-1012.

Sun, M.H. 1970. Sugarcane downy mildew of maize. Indian Phytopath. 23: 262-269.

Sunderam, N.V. 1970. Diseases of forage grass crops. First Workshop on Forage Production and Utilisation, Hisar.

Tewari, A.N. 2000. Identification of plant diseases and their control: Field diagnosis manual. G.B. Pant University of Agriculture and Technology, Pantnagar. 88p.

Thind, B.S. and Soni, P.S. 1983. Persistance of chlorine in maize plants and soil in relation to control to bacterial stalk rot of maize. *Indian Phytopath*. 36: 687-690.

Chapter 24

Seed Production Technology of Major Forage and Fodder Crops

Birendra Prasad[1] *and Anil Kumar Singh*[2]

[1]*Assistant Professor, Department of Seed Science and Technology, GBPUA&T, Hill Campus, Ranichauri, Tehri Garhwal, Uttarakhand*
E-mail: bprasadsst@yahoo.co.in

[2]*Senior Scientist, Agronomy, Division of Land and Water Management, ICAR Research Complex for Eastern Region, ICAR Parisar (P.O. - B.V.College), Patna – 800 014, Bihar*
E-mail: aksingh_14k@yahoo.co.in

Introduction

The production, productivity and availability of quality seed are the most important factors for forage crops. It is more so important because the crops have been bred substantially for vegetative purpose and as such they are shy seeders with very low seed productivity and the forage crops are not allowed to come to maturity and cut at the vegetative stage and as such the opportunity of producing seed is limited. Therefore, efforts are required to augment the seed production of cultivated fodder crops and also the range grasses and pastures.

Seed Production of Forage Sorghum

Sorghum, being the most important fodder crop of both summer and kharif season and with proven potential for yields, may realm as the guiding element in the present scenario of fodder deficit. Sorghum is an often cross pollinated crop. Out crossing occurs to an extent of 15.0 per cent depending upon the variety, location and season (Rao and Rachie, 1965).

Land Requirement

Field that to be used for seed production should be free from volunteer plants and there must be absence of Johnson grass in the seed field or within the isolation distance. Crop thrives well in heavier soil with good fertility status and tilth. However, sorghum may be grown in all type of soil.

Isolation Distance

Since sorghum is an often cross pollinated crops. Different varieties adjacent to one another may easily pollinate to each other and also with other sorghum and broom corn. Therefore, seed field must be minimum isolated by 200 and 100 meters for foundation and certified seed respectively from fields of other varieties of sorghum and also for same variety not conforming to varietal purity requirements for certification. In case of Johnson grass (*Sorghum halepense*), the seed field should be at least 400 metre isolated for both types of seed field *i.e.* foundation and certified seed. Forage sorghum seed producer take care in mind that differential blooming dates for modifying isolation distances are not permitted.

Field Preparation

With the onset of summer rains, ploughing four times followed by levelling is must to get the land into proper tilth. Requisite soil moisture has to be ensured by pre-sowing irrigation if the rains are not sufficient to moist the soil. The field must be free from weeds at the time of sowing.

Seed and Sowing

Procurement of Seed

If a farmer or seed producer want to produce foundation seed, it should have procure breeder seed and to produce certified seed must have obtain foundation seed from a reliable source approved by seed certification agency.

Time of Sowing

In northern India seed should be sown in February as summer crop and for kharif seed should be sown from last week of June to second week of July. While, in southern India sorghum taken as Rabi crop and sown in September.

Spacing and Seed Rate

Seed crop must invariably be sowing in rows with a distance of 45-50 cm and plant to plant 15 cm, while depth of seeding should be kept 3-4 cm. In any case seed should not be sown more than 5 cm hence, sorghum seed do not have enough energy to lift the soil below 5 cm depth and emerged in the field. To ensure good stand 12-15 kg seed per hectare is considered to be sufficient.

Fertilization

Sorghum is heavy fertilizer feeder crop. Fertilizer should be applied as per the fertility status of the field. However, when soil test results are not available, apply

100-120 kg nitrogen, 50 kg P_2O_5 and 40 kg K_2O per hectare. Half amount of nitrogen along with total dose of phosphorous and potash should be applied at the time of sowing. The remaining half amount of nitrogen should be applied as top dressing at 30-40 days of sowing.

Irrigation

Irrigate the crop as and when needed. It must be ensure that there is sufficient moisture in the soil at the time of flowering. Proper drainage facilities should have otherwise stagnation of water will damage the seed crop. Precaution should be taken that no irrigation should be given after the seed has reached the dough stage.

Interculture

To facilitate the field free from weeds a shallow interculture with the hand hoe when crop is about 20-25 days. Do not intercultivate too near the plants in sorghum. For controlling the weeds pre-emergence spraying of atrazine @ 0.5 -1.0 kg a.i. per hectare in 1000 litres of water can also be used.

Plant Protection

Plant protection measures should be adopted as per requirement as scheduled for commercial crops.

Roguing

Seed field must be rogued out of off- type plants, volunteer plants, Johnson grass, Sudan grass, grain Sorghum plants and diseased plants, before the shedding of pollen. Kernel smut or grain smut diseased affected plant and head smut plant must be rogued out from time to time as required.

Harvesting and Threshing

It is ensure that seed crop must be fully ripe before harvesting. The harvested heads should be dried on the threshing floor separately for 5-7 days before threshing. Dried head should be threshed by appropriate threshers and before storage seed must be again dried to 10 per cent moisture content for safe storage.

Seed Yield

30-40 quintals per hectare.

Forage Sorghum Hybrid Seed Production

In forage sorghum hybrid seed production involving cytoplasmic genetic male sterile lines, effective pollen spread from the male parents and pollination are important in determining the seed yield. Therefore, optimum planting ratio is necessary for better seed setting. Perfect synchronized flowering of the parental lines is most important in hybrid seed production. The differential behavior of parental lines in flowering habit results in non-synchronization of parents giving rise to a poor seed set. Therefore, knowledge of the behavior of parental lines for the flowering habit is not only essential but also very useful for careful planting of suitable staggering to ensure nicking and thereby ensuring maximum seed set. The various steps involved in hybrid seed production are as follows:

Maintenance of Parental Lines

- *A-line*: Carrying cytoplasmic genetic male sterility
- *B-line*: Maintainer line, sister strain of A-line having male fertile, non pollen restoring
- *R-line*: Restorer line, used as male parent for the purpose of producing hybrid seed male fertile

Maintenance of A and B Line

As earlier said the male sterile line 'A' carried male sterility due to cytoplasmic genetic factors and it is maintained by crossing with B-line plants having male fertile, non-pollen restoring capacity, which is the sister line of A in separate an isolated field. In an isolated crossing field, the usual planting ratio of A-line and B-line is 4:2 or 6:2 depending on the pollen producing nature of B-line along with 4-6 border rows of B-line seed are planted in around the seed plot. The seed produced from A-line is again male sterile and this seed is used for further seed production of hybrid seed and further increase of A-line in subsequent year.

A-Line x B-Line = A-Line

B-Line x B- Line = B-Line

Maintenance of Restorer Line (R-Line)

The seed of restorer line (R-Line) is also required to produce in a separate an isolated plot in the manner described earlier for open pollinated varieties.

Isolation Requirement for Sorghum Inbreds

For maintaining the A, B and R line fields must be isolated at least 300 metre from other varieties and the same variety not conforming to varietal purity requirement for certification, while incase of Johnson grass and other forage sorghum the minimum isolation requirement is 400 metre. Here seed producer also attented that the differential blooming dates for modifying isolation distance are not permitted.

Production of Hybrid Seed

In this stage involving the crossing of male sterile line (A-Line) with pollen restorer line (R-Line). In hybrid seed production through A, B, R Lines, the maintenance of these parental lines in an isolated field is referred to as foundation seed stage, and further production of hybrid (A X R) seed is referred to as certified seed stage.

A-line x R-Line -- Hybrid seed

The seed harvested from the A-Line is hybrid seed.

Land Requirement

Same as mentioned earlier as for open pollinated varieties of sorghum.

Isolation Distance

The hybrid seed production plot must be minimum isolated 200 metre from field of other varieties and same hybrid not conforming to varietal purity requirement of certification. However, in case of Johnson grass the isolation to be extent at least 400 metres and this stage isolation distances are not modified for differential blooming dates.

Agronomic Practices

For good crop stand the agronomic practices are the same as described earlier for the seed production of OPV (open pollinated varieties).

Planting Geometry

1. *Planting ratio*: The planting ratio of female and male row vary from 4:2 to 8:2. It totally depends on nature of pollen production of R-line (male parent). Some male parent shy in pollen production, while others produce abundant pollen.
2. *Spacing*: It also depends on the growth habit of parental lines. Higher the plant height greater the spacing. However, 50-75 cm row spacing and 10-20 cm plant to plant distance kept to facilitate rouging and inspection. Seed should not be placed at more than 4 cm below the soil surface.
3. *Border rows*: To ensure an abundance pollen on the edges to the seed plots field at least 4-6 boarder rows of R-line should be planted on all sides, especially at the end of rows.

Roguing

Roguing in an essential operation in hybrid seed production to maintain the identity of quality seed. Before flowering, remove all the off-type plants from both seed and pollinator parent. If seed production is being carried out at large scale it is necessary to commerce roguing at each alternate day. Roguing must be done at both stages *i.e.* maintaining the parental lines and hybrid seed production stage.

The process of rouging should be done at proper manner as follows:

1. Strictly rogued out the off-types, volunteers and pollen shedders in female rows before shedding pollen.
2. All pollen shedders, rogued and volunteer plants should be pulled out prevents re-growth and subsequent contamination.
3. Plant that grow out of place *i.e.* between the two lines or male parent in female row or female plant in male row also to be removed as quickly as possible. Seed producer take a special attention at the end of the border rows and seed row as male seed may fall in female rows or female in male rows at the time of sowing.

Pollen Shedder

Refers to the presence of male fertile pollen shedding plant in male sterile row *i.e.* presence of plant of B-line in A-line called pollen shedder.

Harvesting

Firstly harvest male parent (R-line) at any time after the flowering period than seed parent and harvested produce are removed of the male parent row from the seed production field that avoid admixture of seed at the time of harvesting.

Yield

Six to ten quintals per hectare depending upon the receptivity of female line, pollen producing nature of R-line and synchronization of female and male flowers.

Seed Production of Pearl Millet (Bajra) (Open Pollinated Varieties)

Bajra named as 'Pearl millet', 'Cattail millet' or 'Bulrush millet' is a short day warm weather annual kharif grass crop and considered to be a poor man's food. In India it is one of the important millet as well as forage crop which flourish well even under adverse conditions of weather. Besides coarse grain it is also used as feed for poultry and green fodder or dry karbi for cattle.

Land Requirement

The field selected for seed production should be free from volunteer plants, it means previous year pearl millet was not grown in that field. The field should have drainage facility and well aerated.

Isolation Distance

Pearl millet is highly cross pollinated crop and the extent of cross pollination is about 80 per cent. For pearl millet seed production field should be isolated by 400 meters and 200 metres for foundation and certified seed class respectively from other varieties along with same variety not conforming to varietal purity requirements for certification.

Soil and Field Preparation

It grown a wide range of soil, but water logging is not suitable for this crop. It grows well in drained sandy loam. Pearl millet requires fine soil tilth free from clods as the seeds are quite small. A deep ploughing followed by three harrowing along with levelling makes good soil for sowing.

Seed and Sowing

Seed Source

If the foundation seed is being produced breeder seed must be sown, while certified seed is being to produce, then foundation seed used for sowing through obtain a reliable source approved by the certification agency.

Time of Sowing

Sowing time in most of the states is June-July, while pearl millet grown round the year of assured water supply in South India.

Seed Rate and Spacing

4-5 kg seed per hectare for direct seeding and 1.5 kg seed per hectare for transplanting. Seed should be sown at 50 cm row apart and plant to plant distance adjusted at 10 cm apart in rows after a fortnight by thinning or gaps filling.

Method of Sowing

- ☆ *Direct sowing*: Seed crop should be sown in row with the help of bajra seed drill or desi plough below about 3 cm depth.
- ☆ *Transplanting*: Where sowing in delayed and in areas of adequate moisture transplanting method has found to give better seed yield as compared to direct sowing. For transplanting seed sown in raised bed of 500 square metre area nurseries to get seedling of a hectare. Healthy, robust seedlings are transplanted after 15-21 days. In delay transplanting difficulties in establishing the seedlings and if the growth is excessive leaves should be clipped to ensure better seedling establishment.

Fertilization

Amount of fertilizer should be applied on the basis of soil test result. But the absence of soil test general recommendations is 100-120 kg nitrogen, 40-60 kg phosphorus and 30-40 kg potash per hectare. Half dose of nitrogen and full doses of phosphorus and potassium should be applied at the time of sowing. The rest nitrogen should be top dressed in two doses, one 3-4 weeks after sowing and at ear formation stage.

Irrigation and Drainage

If rains are not adequate, generally 1 or 2 irrigations during growing period are enough. Pearl millet does not tolerate water stagnation, so proper arrangement for drain out the excess water should be made.

Interculture and Weed Control

For aeration of roots interculture the crop 3-4 weeks after sowing through wheel hoe, triphali or hand hoe. Pre-emergence application of Atrazine or Propazine @ 0.5 kg per hectare in 800 litres of water controls most of the monocot and dicot weeds.

Plant Protection

Plant protection measures should be adopted as and when required as commercial crops.

Roguing

Strict roguing must be done by removing of off-type plants and volunteers before shedding the pollen. The rouged plant should be pulled out to check the regrowth. Seed producer distinguish the off-type plant on the basis of plant phenotypic character *i.e.* stem shape, pith, juiciness, hairiness, anthocyanin colour of nodes and leaves, plant height peduncle and other panicle characteristics. At ear head stage rogued out the plants affected with green ear, ergot or grain smut disease.

Harvesting and Threshing

Harvest the crop when grains are hard enough having about 20-22 per cent moisture. The harvested ear head should be sorted out to diseased or other undesirable heads then it is dried in sun before threshing for about 5-7 days. Threshing should be done with thresher. The threshed seed should be dried up to 10 per cent moisture before safe storage.

Yield

Seed yield vary on the improved varieties, cultural practices and irrigated conditions. However the seed yield varies from 15-25 qs per hectare.

Seed Production of Oat

Oat is an erect annual grass classified in the genus Avena and adapted to cool subtropical places. It is one of the most important Rabi fodder crops in India. Oat is used as green fodder, straw, hay or silage.

Land Requirement

Seed plot must be free from volunteer plants.

Isolation

It is a self -pollinated crops, therefore 3 metre isolation distance is sufficient for foundation as well as certified seed crop. An isolation of 150 metre should be kept if infection of loose smut in excess of 0.1 and 0.5 per cent for foundation and certified seed class respectively, from infected fields of oat.

Soil and Land Preparation

Oats make their best growth generally on loam soil. However it can be grown on all types of soil except alkaline and water logged soil. The seed plot must be prepared to secure a fine tilth and firm seed bed by deep ploughing followed by 3-4 harrowing and planking for levelling.

Seed and Sowing

Procure appropriate seed from reliable source approved by the seed certification agency. The accurate time for sowing in north India is from middle of October to middle of November. The seed should be sown in rows 20-25 cm apart with a seed drill or behind plough. 80-100 kg seed is sufficient to plant one hectare seed field.

Fertilization

Oat is not heavy feeder of nutrient. Oat demands 80 kg nitrogen and 40 kg phosphorus, out of which half nitrogen and full amount of phosphorus should be incorporated at sowing time and rest half of the nitrogen should be applied in two splits after 20-25 days of sowing and at the first cutting respectively.

Irrigation

3-4 irrigations are sufficient for seed crop growth including the pre-sowing irrigation. Irrigate the crop at an interval of 20-25 days as and when required.

Interculture and Weed Control

Seed crop require one weeding after 20-30 days of sowing for controlling the weeds.

Plant Protection

Oat crop is not affected and do not cause much damage by insect pests. Only termite and ants have created problems under unirrigated and sandy loam soil. Application of 5 per cent Aldrin dust @ 25 kg per hectare in the field before sowing may check the damage by termites and white ants.

Cutting

Only one cut is allowed to get the good seed and quality for early sown crops, while there is no cutting is permitted for late sown crops. First cut should be taken 50–55 days after sowing and cutting ensures 8-10 cm above the soil surface for good regrowth.

Roguing

An intensive roguing is required at each and every growth phase of oat crop. Just before the flowering or at flowering stage first roguing should be done. All the off-type, smutted plants and early heading plants must be rogued out. Second roguing should be done after flowering is over and panicle begins to change in colour. At this stage all off-types, smutted and tall varieties which are late in growth should be rogued out and third roguing completed after the panicle turn colour and begin to ripe.

Harvesting and Threshing

Harvest the crop when it matures. Delay in harvesting leads to poor seed quality. After harvesting crops should be dried at threshing floor before threshing. The crop threshed with stationary thresher on well cleaned threshing floor. Threshed seed must be dried up to 10 per cent moisture for safe seed storage.

Seed Yield

25-35 quintals seed per hectare

Seed Production of Forage Legume

Seed Production of Berseem

Trifolium alexandrinum L. commonly known as Berseem or Egyptian clover. It is most important winter legume forage and highly nutritive and succulent liked by the all animal even pet rat and rabbit also. It covers large area in India and increasing day by day. It also improves the physical properties of soil and add total nitrogen, organic carbon and available phosphorus. In Bihar and U.P. farmer's grown the paddy nursery in berseem field as previous crop and get benefited. It's fodder fed to the animal as a sole after chaff cutting or mixed with hay or dry fodder. Berseem fodder is very useful for animal especially for milch cattle.

Soil and Land requirement

It can easily grow on every soil except too much sandy. It can grow on alkaline soil that have enough water holding capacity, while the crop does not perform well in acidic soils. Fields that are used for seed production must be free from volunteer plants and in berseem care should be taken, there is no organic manures or other contaminating material be applied during the period in which seed is produced. The crop respond well on well drained, medium loam soil that contain phosphorus and calcium.

Isolation Distance

In berseem natural crossing is totally depends on insect activities. 400 metre and 100 metre isolated berseem seed field is minimum requirement for foundation and certified seed respectively, from field of other varieties and fields of the same variety not conforming to varietal purity requirements for certification.

Field Preparation

Hence berseem seed is quite small, a well pulverized and levelled land is necessary to get the more and uniform germination. For this purpose first ploughing must be done with mould board plough or disc plough and 3-4 harrowing then planking is necessary to obtain the fine tilth for sowing.

Seed and Sowing

Source of Seed

Desired class of seed must be obtained from reliable source approved by the seed certification agency and seed producer handled the tag and container/bag at the end of the crop period.

Sowing Time

The appropriate time to get maximum benefit is first fort night of October.

Seed Rate

25-30 kg seed is sufficient for sowing in one hectare land.

Method of Sowing

Berseem crop sown in the following two methods.

1. *Wet Bed Method*: In this method seeds are sown in flooding beds with 4-6 cm standing water through broad casting. Water of the field must be stirred properly before seed sowing that make soil suspension through puddler that helps in setting of seeds and proper germination.
2. *Dry Bed Method*: In this case berseem seed is broadcasted and covered with one centimeter friable soil. Condition is that field should have sufficient moisture that would help the germination of seed. After germination satisfactorily irrigation should be provided to the crop.

Seed Inoculation

It is necessary if berseem is going to be sown for first time in the field, seed should be inoculated with bacterial culture *Rhizobium trifolii* that supports in nodulation in the crop. For bacterial inoculation 1000 gram culture added with one litre of water and 100 gram sugar is used for 20 kg seed. Mix the mixture properly with the seed and subjected the seed for drying in the shade. The inoculated seed should be broadcasted within 24 hours after inoculation.

In case of non-availability of berseem culture 50 kg of upper surface soil of a field where berseem was grown regularly, taken and spread properly along with the seed in prepared field.

Fertilization

A light dose of nitrogen (25-30 kg) along with 50-60 kg phosphorous per hectare are required for berseem, since it is a leguminous forage crop that accumulates the free atmospheric nitrogen by symbiotic nitrogen fixing bacteria. Full amount of nitrogen and phosphorus should be given as basal at the time of sowing.

Irrigation

To get proper germination, early establishment and growth, first irrigation should be given immediately after seedling emergence. Further 2-3 irrigations at one week intervals may be done. Subsequent irrigations may be applied as per requirement of the crop and soil conditions.

Interculture

Producing maximum pure quality seed, field must be free from weeds. Chicory and Wild Palak are predominantly seemed in the berseem field and chicory is the objectionable weed for this crop. Since chicory seed one mix very difficult to separate them. For control of this objectionable weed, apply pre-emergence of butachlor at 3 kg/ha or post emergence spray of Dinoseb @ 4 kg/ha in 600-800 litre of water.

Plant Protection

Appearance of insect infestation, green crop should be cut-off as soon as to check the spread of infection. To check the attack of gram caterpillar, spray the crop with 1 litre of endosulfan 35 EC in 800 litre of water per hectare. If there is attack of black ant, that take away the germinating seed, it can be destroyed by mixing 25 kg aldrin 5 per cent dust in the soil.

Berseem crop suffering with major disease is stem rot caused by a fungus, *Sclerotinia sclerotiorum* present in the field. For management of this disease drench the soil with 0.4 per cent solution of Brasicol @ 4g/litre of water. 10 litre solution required for 1 sq. metre area of the soil surface.

Cuttings

The crop grown for seed production, no cutting should be allowed from the first week of March. Generally 3-4 cuttings may be taken from the seed crop then left for seed production. More cuttings lead to not only poor seed setting and a reduction in a signification seed yield due to poor pollination in the ensuring dry and hot weather, but also produce seeds of poor viability and vigour. A greater gap taking by the last cut earlier, may lead to lodging the crop to excessive growth that leads to poor pollination and seed setting.

Roguing

Strict and frequent roguing is required for berseem crop. Chicory plant create greater problem in this crop, so it must be rogued very sincerely at vegetative, pre flowering, flowering and at maturity stage. All types of weed plants, off-types, chicory plants as well as other crop plants must be rogued out regularly. This process must be ensured before the crop harvested for seed.

Harvesting

Seed crop matures in the end of May, when pods turn yellow to brown in colour. To prevent seed shattering harvesting may be done preferably in the morning of the day, when some dew are still on the plants. Threshed the harvested crop when crop ensure to dry in a well cleaned or separate threshing floor. Threshed and cleaned seeds require further drying to 8-10 per cent moisture content for safe storage.

Seed Yield

5-8 quintals of seed may be obtained from a hectare of land.

Seed Production of Lucerne

Lucerne (*Medicago sativa* L.) or alfalfa is one of the most important legume fodder crop next to berseem of India. It provides very much nutritious and digestible fodder. It is a good substitute of grains and concentrates rich in calcium, protein and vitamins.

Soil and Land Requirements

It is grown all kinds of soil but better performance reflected on well drained loam soil. It exhibited poor results on heavy and water logged soils. Also land must be free from volunteer plants.

Isolation Distance

Same as berseem *i.e.* 400 metre and 100 metres for foundation and certified seed respectively.

Field Preparation

It requires fine levelled field with sufficient moisture. One ploughing with a mould board plough and further 3-4 harrowing followed by planking ensures fine and firm for an ideal seed bed.

Seed and Sowing

Seed Source

Obtain appropriate kind of seed from a reputed source approved by seed certification agency.

Sowing Time

The best time for sowing of Lucerne as seed crop is October.

Spacing

Lucerne grown for seed should be sown in rows from 50-60 centimetre apart with in 1-1.5 centimetre layer of soil.

Seed Rate

12-15 kg quality seed is sufficient to grown the one hectare seed crop.

Seed Inoculation

If the Lucerne crop sown in the field first time, seed should be inoculated with bacterial culture *Rhizobium melilotii* that promote in nodulation and nitrogen fixation. The Rhizobiuim culture inoculation may be accomplished as the manner similar to the berseem.

Fertilization

It is being a leguminous crop, therefore, requires only 20-25 kg nitrogen alongwith 60-80 kg phosphorus and 20-25 kg potash as a basal dose are sufficient for growing a good crop. Since it is a heavy feeder of nutrient, so care should be taken to supply the micro-nutrients *i.e.* boron, zinc, manganese and iron if deficiency occurs.

Irrigation

Lucerne requires greater time to establish in early stage, therefore a irrigation should be make available at short interval *i.e.* 7-8 days. Further irrigation interval extended up to 15-20 days as and when required. Water at blooming and at pod filling stage are essential for getting better and quality seed yield.

Interculture

It provide ample opportunity for weed growth since takes a long time to establish itself in the field. To obtain quality seed it is necessary to keep the field free of weeds, mainly dodder. After the last cutting of crop do not allow dodder plants to set the seeds, because seeds of dodder (Cuscuta) can viable in the soil for several years. It can be destroyed by spraying crude oil at the time of flowering.

Plant Protection

Plant protection for Lucerne is similar to those of berseem mentioned earlier.

Honeybees for Pollination Supplement

Since it is a cross pollinated crops, the prospects of better seed setting if bee hives may set up for tripping flowers and pollinating them in or near the field.

Cutting

In seed production plot, the crop is allowed to flower after taking cutting in the February.

Roguing

Roguing must be done strictly and carefully at pre-flowering, flowering and at maturity stages. All the abnormal, dissimilar, off-types, other crop plants and cuscuta (dodder) plants must be removed before harvest.

Harvesting

Harvesting the seed crop in the end of May and all the procedures for harvesting and threshing should be done as that of bereseem, which is earlier discussed.

Seed Yield

A well managed seed crop produces 2.5-3.0 quintals of seed per hectare.

References

Agrawal, R.L. (1996). Seed Technology (2nd Edn.), Oxford and IBH Publishing Co. Pvt,. Ltd., New Delhi, pp. 829.

Balasubraman iyan,P. and Palaniappan, S.P. (2007). Principle and practices of agronomy (2nd Edn.), Agrobios (India), Jodhpur, pp. 576.

Boka I., Daci F., Tahiraj K., Karaj S. (1999). Results of the experimentation and cultivation of Lucerne in Albania. Proc. of XIII EUCARPI Medicago spp. Group "Lucerne and Medics for the XXI Century" Perugia (Italy) 13-16 September 1999, (in press).

Dharamalingam, C., Sivasubramaniam, K. and Yadav, S.K. (2001). A dictionary of Seed Technology Terms. Kalyani Publisher, pp. 109.

Dabadghao, P.M. and Shankarnary, K.A. (1975). The grass cover of India, ICAR, New Delhi, p. 713.

Fairey, D.T., and Hampton, J.G. (1997). Forage Seed Production (Vol.I) Temperate Species, Cab International, New York. pp.420.

Falcinelli, M., Martiniello, P. (1998). Forage seed production in Itlay. J.Appl. Seed Prod., 16, 61-65.

Hazra and Sinha (1996). Forage Seed Production (A technological development). South Asian Publisher Pvt. Ltd. New Delhi, pp 253.

Krishna, K. (1996). The Kerala experience with forage seed production and supply system. In: P.M.Horne, C.Phaikaew, and W.W. Stur, (eds.). Forage Seed Supply Systems: Proc. Of an International Workshop held in Khon Kaen, Thailand, 31 October to 1 November 1996. Los Banos, Philippines CIAT Working Document Number 175.

Loch, D.S. and Ferguson, J.E. (1999). Forage Seed Production (Vol.II) Tropical and Subtropical Species. Cabi Publishing, New York. pp.479.

Prasad, B.(2000). Effect of row ratio and staggered planting of parental liens on seed yield and quality of forage sorghum hybrid, PCH-106. Ph.D. Thesis, GBPUAT, Pantnagar, Uttarakhand.

Prasad, B., Shukla, P.S. and Singh, R. (2005). Effect of staggered planting of male parent on seed yield and quality of forage sorghum hybrid, PCH-106., Seed Research, Vol.33(2), pp. 213-214.

Rao, D.V.N. and Rachie, K.O. (1965). Natural crossing in sorghum as affected by locality and season. Indian. J.Agric. Sci. 35:8-13.

Rivoira, G., Bulliata, S., Porqueddu, C., Roggero, P.P.(1989). Attivita di ricerca 1984-1989 del Centro di Studio sul miglioramento della produttivita dei Pascoli, CNR-Sassari, pp. 77.

Singh Punjab and Pathak, P.S. (1988). Rangelands Resource and Management, Proceeding of the National Rangeland Symposium, IGFRI, Jhansi, November 9-12, 1987. pp. 473.

Singh, R.P. (1986). Handbook of Temperate Forage Production. IBD, Dehradun, p. 109.

Singh, C. (1991). Modern techniques of raising field crops. Seventh (Edn), Oxford and IBH Pub. Co. Pvt. Ltd. New Delhi, pp. 523.

Turton, C. and P. Baumann (1996). Beyond the formal sector: fodder seed networks in India. In: Horne, P.M., C.Phaikaew, and W.W.Stur, (eds.). Forage Seed Supply Systems: Proc. International Workshop held in Khon Kaen, Thailand, 31October to 1 November, 1996. Los Banos, Philippines, CIAT Working Document Number 175.

Chapter 25

Constraints and Safeguards of Forage and Fodder Crops Seed Production

Birendra Prasad[1] and Anil Kumar Singh[2]

[1]*Assistant Professor, Department of Seed Science and Technology, GBPUA&T, Hill Campus, Ranichauri, Tehri Garhwal, Uttarakhand*
E-mail: bprasadsst@yahoo.co.in

[2]*Senior Scientist, Agronomy, Division of Land and Water Management, ICAR Research Complex for Eastern Region, ICAR Parisar (P.O. - B.V.College), Patna – 800 014, Bihar*
E-mail: aksingh_14k@yahoo.co.in

Introduction

In India, the importance of livestock is well recognized for agriculture and dairy industry. However, the low productivity of livestock is a matter of concern, which is mainly due to poor fodder and feed resources. The lower milk production per cattle in India is mainly attributed to production and feeding of poor quality forage and that too inadequate amount (46.6 per cent) of requirement. To increase the milk production in the country, emphasis has to be given on bridging the gap between supply and demand of fodder. The area under forage crops in India is also very meager *i.e.* 4.4 per cent of the total cultivated land. It is, therefore, essential to maximize forage production per unit area. Looking to the need of bridging the huge gap between demand and supply, it is imperative to lay adequate emphasis on forages.

The plants that we refer to as forage or herbage grasses and legumes are actually utilized and exploited in a multiplicity of ways, and these horizons are likely to continue to expand as we address the challenges of future. The most critical factor which is coming in the way of popularization of the forage crops in the non availability of good quality seeds. Therefore, an effort is being required in order to increase the area, production, productivity and creating awareness among the farmers regarding importance of forage to their milch as well as draft animal and also necessary to do this direction so that available technology could be passed on to the forage and seed producing agencies and also to farmers for their eventual use in augmenting forage resources. Majority of our animals depends largely on agricultural wastes *i.e.* dry leaves, straw etc.

The scientific input for forage research has been strengthened by the Indian Council of Agricultural Research by establishing of All India Coordinated Projects on Forage crops in 1970 and also with the establishment of the Indian Grassland and Fodder Research Institute, Jhansi in 1962. The network on forage research has come out with very sound and valuable results with respect to enhancing the production and productivity of forage crops by utilizing the improved varieties and taller hybrids as well as by suitable management agronomical practices. However, the output of these research have not been translated strongly and effectively at the farmer's level and major constraints being faced in this regard is non-availability of good quality seed. Due to lack of awareness and interest towards forage production of Indian farmers and meager area covered by forage crops, private seed industry could not take keen interest to supply the forage seed as like vegetable, hybrids or other cereals crops. Therefore, it is regarded as orphan as well as neglected crops in India. Hence at present time efforts are needed to produce good quality seed at all tier of seed production, *i.e.* nucleus, breeder, foundation and certified seeds. It is also realize the need to have proper coordination of all seed producing agencies (public and private sector), so that we will in position to achieve the maximum coverage under high yielding varieties through improved quality seed in few coming years.

In this chapter strengths and weaknesses of seed supply, challenges for the future and seed production technology of major forage crops are discussed with particular reference of cultivated forage cereals, legumes and range grasses.

Strengths and Weaknesses of Seed Supply

Strengths

The present seed supply system has facilitated the large-scale planting of pastures of government stations and large farms and in backyard forage programmes in the villages. It has also enabled many thousand of kilometers of roadsides to be over sown with suitable grasses.

A government subsidy for seed is helping to break down old prejudices. Many farmers believe that grass is a free commodity supplied by nature and that to buy grass seed is not money well spent (Phaikaew and Hare, 1996). This attitude is changing as more and more farmers are encouraged to use grass seeds and come to realize the economic benefits of growing good quality forage rather than buying expensive concentrates for their cattle.

Village seed production has brought economic benefits to many smallholders through higher returns per hectare. It has also enabled them to grow crops that do not deplete soil fertility. Probably the most important underlying factors contributing to the development of the present system has been continuity and commitment by the government. In India, there has been continuous research and development for more than 50 years and in particular continuity of key personnel and experienced support staff to facilitate progress. There has been strong commitment shown at all stages, firstly to the success of the various pilot projects and then to the development of production on a larger scale including government financing of the seed enterprise. The extensive network of AICRP on forage crops and IGFRI scientist and its forage stations in nearly all the tropical, subtropical and temperate zones has enabled village seed production and pasture development to expand rapidly.

Weaknesses

Limited Range of Species

The present seed supply system has been restricted to a limited range of species. A wider range of species is required to maximize the benefits from pasture development in India. In case of dairying, where high quality forage is needed, none of the current major species really meets this criterion. Also, research on evaluation of different forage species for seed production and dry season tolerance is being conducted with promising varieties emerging.

Government Support Needed

Government of India in the region may play a crucial role in the forage seed supply system. Farmer seed production would not be possible without initial government support (Phaikaew, 1997). In many cases, government agencies produce seed on government stations and utilize the harvested seed in government projects. The demand for seed fluctuates from year to year, making seed production a risky enterprise for farmers. In India, the government has limited the risk for farmers by guaranteeing farmers a contract price for a certain amount of seed. Purchase of excess product is not guaranteed. But many farmers produce seed above quota and sell this seed directly to other farmers.

Low Price of Seed

In the past, a lot of seeds was given free of charge of farmers, or very cheaply at subsidized prices. In such circumstances, many farmers have not looked after their pasture properly. For example, they were not concerned if their pasture was damaged by overgrazing during the dry season because they could replant cheaply again at the start of the wet season. For many such farmers, pasture establishment is an annual event. Nowadays, farmers generally have to pay for seed although seed prices are low.

Need More Research, Seed Production and Extension

The present seed supply system involves many government personnel who could be utilized more fully in research as well as in seed production and extension. They should be involved in breeding, evaluation and initial seed multiplication. Many producers in India now have considerable experience, so the need for strong government supervision of production and its heavy involvement in marketing is decreasing.

Many of these problems can be overcome through greater private sector involvement in forage seed marketing.

Future Development

Not only in India but also in Southeast Asia, there are considerable demands for forage seeds which can not be met within the country. Thus, there is a good prospect in increase forage seed production within countries and to develop international trade links between countries in the region by importing species which we find difficult to produce and exporting others that we can produce well.

There is a need to expand the range of species grown to service a wider range of markets, *e.g.* high quality forage for dairy production, salt tolerant forages, amenity roadside planting for recreation use, rehabilitation of degraded land, for turf and even ornamental use.

For this to be realized, countries in the region must decide on the most useful and widely adapted forage species. In addition, common seed quality standards, seed certification, storage and shipping guidelines must be developed.

Constraints for Forage Seed Production

1. The major constraints for the assessment of the highly valuable native genetic resources and the development of the seed industry that hinder the production of seed of competitive costs appears to be due to several intrinsic or structural links.
2. The presence of restricted and diversified demand (small pockets of markets and many species or varieties) in relationship to the different environmental agro-ecological areas to edaphic covering and climatic variations *i.e.* arid, semiarid, sub-humid, humid and irrigated areas and the different farming systems.
3. Lack of importance given to the choice of variety. Farmers when buying large seeds often do not exercise the same attention to the choice of the variety as they do for cereals and seed price is the chief driving force in the seed market.
4. Underdeveloped organization of forage seed production. In India the seed is locally produced and re-utilized by the farmers themselves.
5. In the irrigated areas there are horticultural fruit, flower and vegetable crops along with cereals.

6. No seed production technology as well as minimum seed certification standard was standardized till date with regards to many grasses.
7. Government agricultural policies favors mainly to cereals, pulses, oil seed as well as cash crops not to forage crops.
8. A low amount of research funds were addressed to this area.

In agriculture, as well as in other sector, the research was carried out mainly in the more developed area with the aim of solving problems related to the main and more profitable crops and little interest in the problems of less favored areas (Rivoira *et al.*, 1989).

Safeguard for Maintaining Genetic Purity during Seed Production

Before beginning the seed production programme of any kind or variety, researchers, seed producers, students or farmers must have to know about the mechanism of maintaining the genetic purity which is the prime object of quality seed production. The important safeguard for maintaining genetic purity during seed production is:

Land Requirement

The seed field must be free from volunteer plants, weed plant and other crop plant and in the preceding season the same crop should not have grown on the same land. Hence, volunteer plants are difficult to remove during the roguing of the seed field. Soil texture and fertility status of the land should be as the requirement of seed crop that help to reflect the full genetic potential by the seed crop. The soil of the seed field must be free from soil borne diseases and insect pests. The seed field must be levelled and isolated as per requirement of the crop concerns.

Volunteer Plants

A plant belonging to the same crop germinated from previous crop seed growing along with the seed sown in the current season in the same field.

Isolation

"It is a prescribed minimum distance which separates seed crop from the commercial crop of the same variety or another variety or other species of the same genera and family or diseases etc. required for maintaining the desired purity and health of the crop".

Isolation distances are affected by following means:

1 *Category of seed is to be produced*: Breeder and foundation seed crop require greater isolation compared to certified seed crop.

2 *Nature of pollination of the crop:* Cross pollinated seed crop are raised with greater isolation distances compared to self pollinated crop.

3 *Mechanical admixture*: Keep the crop to avoid mechanical admixture at the time of sowing and harvesting.

4. *Seed borne disease:* To keep the crop away from seed borne disease *e.g.* in barley and wheat for loose smut affected field isolation distances goes to 150 metre, while normal isolation is 3 metre only.
5. *Boarder rows:* In hybrid seed production one addition border rows minimize the isolation distances of 12.5 metre.

Types of Isolation

1. *Distance isolation*: It is required in aerial distance (metre) for avoiding contamination of natural crossing with other contaminating crop or diseased field.
2. *Time isolation*: The distance isolation may be reduced, modified or avoided by time isolation. In these conditions the receptivity of stigma of the flower of seed crop do not synchronized with contaminating pollen.
3. *Physical barrier*: With the use of physical barrier isolation distance may be modified for seed production. It is of two types.
   (*a*) Natural: Natural topography, wind breaks etc.
   (*b*) Artificial: Hedge, net, poly house, grow dense and taller plant around the seed field.

Controlling Seed Source

The use of an appropriate class from an approved source is compulsory for raising of any seed crop.

Classes of Seed

Now a day's three tier system is adopted in seed multiplication:

Breeder Seed

Breeder seed is seed or vegetative propagating material directly control by originating or sponsoring plant breeder of the breeding programme or institution and/or seed whose production is personally supervised by a qualified plant breeder and which provide the source for the initial and recurring increase of foundation seed. Breeder seed has 100 per cent genetically pureness.Normally Breeder seed is produced by the agricultural university and research institutes. In some cases State Farms Corporation of India(SFCI) and National Seed Corporation (NSC) also produced breeder seed.

Breeder seed is not certified by certification agency but it is monitored by a joint inspection team consisted of crop breeder and representative of each crop coordinators, State Seed Certification Agency (SSCA) and NSC to check the quality of breeder seed during production. The breeder seed is supplied to NSC, SFCI, SSC and reputed private seed companies for production of foundation seed. Yellow/golden tag is used for breeder seed.

Foundation Seed

Foundation seed is the progeny of breeder seed or it may be progeny of foundation seed. Foundation seed produced directly from breeder seed shall be designated as foundation seed stage I, while foundation seed produced from foundation seed stage I shall be designated as foundation seed stage II. Foundation seed is produced by agricultural universities, NSC, SFCI, SSC and few reputed private companies of seed. The state seed corporation (SSC) produces foundation seed to meet their local requirements while foundation seed of national varieties are produced by National Seed Corporation (NSC) and State Farms Corporation of India (SFCI). Preference for the production of foundation seed in given of agricultural universities. White tag is used for foundation seed.

Certified Seed

Certified seed is the progeny of foundation seed or it may be progeny of certified seed. Certified seed produced directly from foundation seed shall be designated as certified seed stage I.While certified seed produced from certified seed stage I shall be designated as certified seeds stage II. Certified seed production is done by the NSC, SFCI, SSC and private seed companies. The SFCI produces certified seed at its own farms where as other agencies do so through contract growers. Blue tag is used for certified seed. Certified seed may be the progeny of certified seed provided reproduction does not exceed 3 generation beyond foundation seed. stage I.

Class of Seed

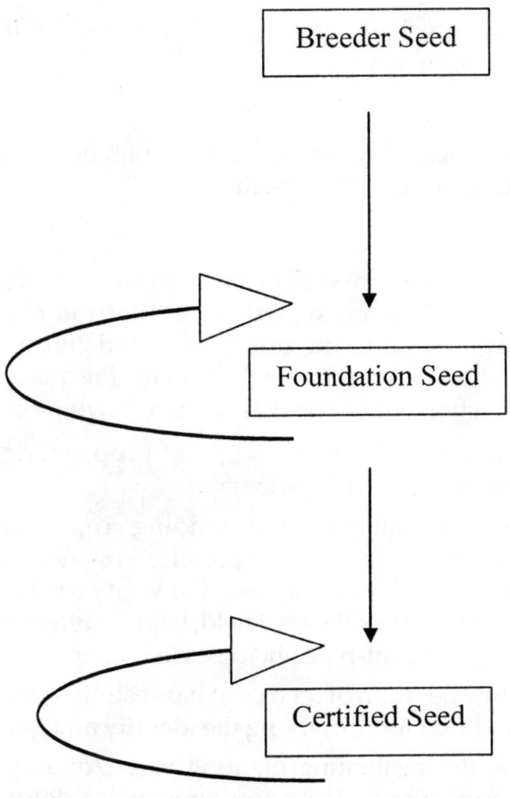

Figure 25.1: 3-Tier System in Seed Multiplication

Roguing of Seed Crop

"Removal of off types, rogues, volunteer plants, diseased plant, designated weeds etc. from a seed crop in called roguing". For maintains the genetic purity of the seed adequate and timely roguing is necessary in seed production, otherwise may cause quick deterioration in seed stock by the opportunities offered for cross pollination and transmission of diseases. Therefore, be removed at early as possible date before flowering. Roguing in seed crops can be done at least following three stages.

1 Vegetative/before flowering

2 During flowering

3 At maturity

Off Type

Plants differing in their characteristics from those of seed crop phenotypically or genotypically.

Rogue

An off- type, undesirable and plants with designated disease that deviates from seed crop genetically and phenotypically are called rogue.

Designated Disease

Refers to the diseases specified for certification of seeds and in whose regard certification standards must be met.

Designated Weed

Refers to weed species specified for certification of seed of a given kind and in whose regard certification standard must be met.

Seed Certification

The genetic purity in commercial seed production is often maintained through a system of seed certification. The main goal of certification of seed is to ensure genuineness and quality of seed to the purchaser. Seed certification is a legally sanctioned system under seed Act. 1966 to Section (8) for quality control of seed multiplication and production which consists of the following control measure.

1 An administrative check on the origin of propagating material for the purpose of determining trueness to type.

2 *Field inspection*: An examination of standing crop is an essential step in verifying conformity of seed crop to prescribed standards for certification. Field inspections are done to achieve the verify seed source, identity of variety,cropping history of the seed field, land requirement, isolation, other crop and weed plants, off-type and diseased plants.

3 Supervising the operation of seed crop harvesting, storage, transport and processing condition for preserving the identity and quality of seed.

4 *Seed inspection*: For evaluating planting value/quality a representative sample drawn by the certification agency for determining the seed

germination, moisture content, weed seed, admixture and purity. Drawn seed samples are sending to notified seed testing laboratory. After qualifying the prescribed minimum field and seed standards seed certificating agency issued a certification tag of qualified seed lot.

Genetic Purity

Those varieties being grown for seed production should periodically time to time be tested for genetic purity by grow out tests or by molecular/genetic marker to make confirm that they are being maintained in their true form.

References

Agrawal, R.L. (1996). Seed Technology (2nd Edn.), Oxford and IBH Publishing Co. Pvt,. Ltd., New Delhi, pp. 829.

Balasubramaniyan, P. and Palaniappan, S. P. (2007). Principle and practices of agronomy (2nd Edn.), Agrobios (India), Jodhpur, pp. 576.

Boka I., Daci F., Tahiraj K., Karaj S. (1999). Results of the experimentation and cultivation of Lucerne in Albania. Proc. of XIII EUCARPI Medicago spp. Group "Lucerne and Medics for the XXI Century" Perugia (Italy) 13-16 September 1999, (in press).

Dharamalingam, C., Sivasubramaniam, K. and Yadav, S.K. (2001). A dictionary of Seed Technology Terms. Kalyani Publisher, pp. 109.

Dabadghao, P.M. and Shankarnary, K.A. (1975). The grass cover of India, ICAR, New Delhi, p. 713.

Fairey, D.T., and Hampton, J.G. (1997). Forage Seed Production (Vol.I) Temperate Species, Cab International, New York. pp.420.

Falcinelli, M., Martiniello, P. (1998). Forage seed production in Itlay. J.Appl. Seed Prod., 16, 61-65.

Hazra and Sinha (1996). Forage Seed Production (A technological development). South Asian Publisher Pvt. Ltd. New Delhi, pp 253.

Krishna, K. (1996). The Kerala experience with forage seed production and supply system. In: P.M.Horne, C.Phaikaew, and W.W. Stur, (eds.). Forage Seed Supply Systems: Proc. Of an International Workshop held in Khon Kaen, Thailand, 31 October to 1 November 1996. Los Banos, Philippines CIAT Working Document Number 175.

Loch, D.S. and Ferguson, J.E. (1999). Forage Seed Production (Vol.II) Tropical and Subtropical Species. Cabi Publishing, New York. pp.479.

Prasad, B.(2000). Effect of row ratio and staggered planting of parental lines on seed yield and quality of forage sorghum hybrid, PCH-106. Ph.D. Thesis, GBPUAT, Pantnagar, Uttarakhand.

Prasad, B., Shukla, P.S. and Singh, R. (2005). Effect of staggered planting of male parent on seed yield and quality of forage sorghum hybrid, PCH-106., Seed Research, Vol.33(2), pp. 213-214.

Rao, D.V.N. and Rachie, K.O. (1965). Natural crossing in sorghum as affected by locality and season. Indian. J.Agric. Sci. 35:8-13.

Rivoira, G., Bulliata, S., Porqueddu, C., Roggero, P.P.(1989). Attivita di ricerca 1984-1989 del Centro di Studio sul miglioramento della produttivita dei Pascoli, CNR-Sassari, pp. 77.

Singh Punjab and Pathak, P.S. (1988). Rangelands Resource and Management, Proceeding of the National Rangeland Symposium, IGFRI, Jhansi, November 9-12, 1987. pp. 473.

Singh, R.P. (1986). Handbook of Temperate Forage Production. IBD, Dehradun, p. 109.

Singh, C. (1991). Modern techniques of raising field crops. Seventh (Edn), Oxford and IBH Pub. Co. Pvt. Ltd. New Delhi, pp. 523.

Turton, C. and P. Baumann (1996). Beyond the formal sector: fodder seed networks in India. In: Horne, P.M., C.Phaikaew, and W.W.Stur, (eds.). Forage Seed Supply Systems: Proc. International Workshop held in Khon Kaen, Thailand, 31October to 1 November, 1996. Los Banos, Philippines, CIAT Working Document Number 175.

Chapter 26

Cultivated Green Fodder for Dairy Animals

Amitava Dey[1*] *and P.K. Singh*[2]

[1]*ICAR Research Complex for Eastern Region, Patna*

[*]*E-mail: amitavdey_icar@yahoo.co.in*

[2]*Department of Animal Nutrition, Bihar Veterinary College, Patna – 14*

Introduction

The sustainability of dairy industry in India largely depends upon the quality of herbage based feed and fodder resources. Green fodder is the essential component of feeding high yielding milch animals to obtain desired level of milk production. Only 8.3 million hectare area are cultivated under fodder crops in our country and further there is little scope for its expansions Thus, to produce targeted quantity of green fodder, the best option is to maximize the fodder production per unit area per unit time. The technology of year round fodder production has been tested in All India Coordinated Research Project on Forage Crops and different crop sequences have been identified on regional basis. In northern region the green fodder yield varies from123-201t/ha/year. Similarly, the green fodder yield varies as 168-253, 85-131 and 75-225 t/ha/year in western, eastern and southern regions respectively. Fodder scarcity is one of the reasons for insufficient spread of crossbreeding programme in many states. In eastern region of India, only 4 per cent cropped area is under fodder crop while it is more than 10 per cent in Punjab and Haryana. However, for economic milk production and sound health of animals fodder is necessary. On one acre of irrigated land, 3 milch animals can be maintained through year round fodder production; however, in mixed crop-livestock farming system 5 animals can be reared (Tomer and Wali, 1995). Practically, the number is much higher in majority of the states. On the other hand, there is continuous shrinkage of pasture and grazing land with poor quality biomass yield. In the states of Punjab, Haryana and western parts of UP, there is hardly any land available for grazing except by the sides of rivers, canals, roads and railway tracks. At the same time, it is also difficult to increase the area under fodder cultivation. So, cultivation of high yielding varieties and intercropping of fodder with food/cash crops are the only alternatives to increase the green fodder availability. Lack of precise information on area under fodder, improved varities used, nutritive value of different cultivars are lacking in different agro-climatic zones which are hampering the effective feeding policies for livestock sector.

Factors Affecting Forage Quality

Though total biomass yield of forage crop is important for the farmers, at the same time nutritional quality is equally important. Animals are the best judge of any feed which can be examined through palatability test, growth and milk yield. The quality of forage crops is determined by genetic and non-genetic factors. The genetic factors include type of forage crops (leguminous or non leguminous or mixed grasses), cultivars and genetic modification (like QPM). The non genetic factors affecting the forage quality are environment, crop management practices, soil types etc.

Low light intensity decreases the forage digestibility of grasses due to decrease in photosynthesis and lower soluble carbohydrate content. High environmental temperature results in higher content of NDFand lower fibre digestion due to rapid growth of stem. High soil moisture may cause flux of soluble nutrients from leaves to stems and grains with maturity resulting in lower straw/stover quality. In stress conditions, soluble carbohydrate content is increased resulting improvement of forage quality. Stage of maturity is also an important factor determining the forage quality. Method of processing of forage is equally important when animal feeding is concerned. Application of fertilizer and optimum and timely irrigation may increase forage biomass yield. However, excess nitrogen application may also increase lignin content of forage due to increase in aromatic amino acids. Excess irrigation may decrease forage quality.

Fodder in the Ration of Dairy Cows and Buffalo

Ramchandran *et al.* (2005) have showed that increased feed resources have been one of the major contributing factors for the increased milk and meat production. The states of Punjab, Haryana, UP and HP are self sufficient in feed resources and produce 54 per cent of the total milk in India. Dairy animals giving upto 10 kg of milk may be maintained on good quality forage based diet; however, for high yielders concentrate supplementation is required. Minimum amount of forage to be required of milk production and sound health varies in different experimentations depending on the quality. Large breeds of cattle can produce 6-8 kg milk per day on good quality forage alone while smaller breeds can sustain 4-6 kg milk per day (Pachauri, 1990). It has been also reported that cows fed ad lib Lucerne could produce 10 kg of milk per day, however, ad lib berseem feeding with 1 kg concentrate could sustain only 7.5 kg milk per day. In case of non-leguminous fodder, feeding of ad lib maize fodder with concentrate supplementation gave significantly higher energetic efficiency in buffaloes than feeding of straw and concentrate. Lactating buffaloes yielding 8 kg milk per day can give better economic return on green maize based diet with roughage to concentrate ratio of 3:1. In high environmental temperature, green forage based diet is helpful due to lower heat increment. Gupta *et al.* (1983) did not observe any adverse effect of feeding lucern as sole diet on milk productionin buffalo. However, buffaloes on sole forage diet had less DM intake and milk yield though not statistically significant (Table 26.1). Danial *et al.* (1967) reported that cows fed ad lib lucern could produce 10 kg milk per day while Jackson and Gupta (1971) observed that buffalo can sustain 7.5 kg milk yield per day on ad lib barseem feeding. Lal *et al.* (2005) reported that supplementation of barseem improved the DMI and digestibilities of oat straw in crossbred heifers. A combination of oat straw and barseem at the ratio of 1:1 gave the best results. Singh and Chaudhary (2007) have reported that daily feeding of green maize at the rate of 10 and 15 kg improved growth rate and nutrient digestibility of crossbred heifers. Singh *et al.* (2007) have also reported that inclusion of 40 kg green maize improved nutrient utilization, milk yield and composition in crossbred cows.Amount of fodder given to dairy animals varies according to season, household fodder availability in different regions. In Haryana, it is observed that 20-23 per cent of total cost of rearing is spent for green fodder feeding for milch buffalo (Bharadwaj *et al.*, 2006). In Tarai region of Uttarakhand, Tiwari *et al.* (2007) have reported that cattle and buffaloes were offered 1.42 kg DM as green fodder per LSU. Bardhan *et al.* (2005) have reported that 52 per cent farmers in Tarai region of Uttarakhand face the problem of green fodder feeding to livestock. Patange *et al.* (2002) have reported that milch buffalo takes 5.01 kg DM through green fodder. The proportion of green fodder on DM basis rose from summer to winter (11.4 to 20 per cent). Green fodder production in the summer months is constraint for buffalo production in Etah district (Singh and Singh, 2005).

Tabe 26.1: Feeding of Berseem or Lucern Fodder on Milk Production in Buffaloes

Feeding Schedule	*Straw+ Concentrate*	*Sraw+Conc.+ Berseem*	*Sole Berseem*	*Sraw+Conc.+ Lucern*	*Sole Lucerne*
DMI (kg/100 kg body weight)	2.39	2.51	2.74	2.53	2.62
Milk yield (kg/d)	7.96	7.91	6.74	7.14	6.99
Net energy efficieny (per cent)	86.59	62.27	59.84	58.52	59.80
Net protein efficiency (per cent)	44.12	29.70	28.11	23.64	23.46

Source: Gupta *et al.*, 1983.

With the feeding of high quality leguminous fodder, concentrate can be spared. Pachauri *et al.* (1990) observed that a concentrate mixture having 9-10 per cent protein and 68 per cent TDN was sufficient with high quality leguminous fodder in buffalo

Mudgal *et al.* (1985) observed that feeding of a mixture of berseem and mustard fodder (35-40 kg/d) along with 2-3 kg wheat straw during winter months can meet the nutritional requirement of a cow for its maintenance and for producing 6-8 kg milk per day. Over and above the forage diet for every 2.5 kg of milk 1 kg concentrate should be supplied. In case of non leguminous fodder Singh *et al.* (1988) have reported that ad lib feeding of maize fodder along with concentrate gave higher energetic efficiency in buffalo for milk production. It is an fact that high roughage diet leads to higher fat content in milk by producing more acetate. A mixture of maize and cowpea fodder during summer months can support nutrional requirement of a cow for maintenance and producing 4-6 kg of milk per day (Mudgal, 1985). During winter a mixture of mustard and oat fodder can support 6-8 kg milk production per day (Mudgal and Mallikarjunappa, 1985).

Table 26.2: Dry Matter Intake, Milk Yield and Feed Conversion Efficiency in Buffalo Fed Green Forage Based Diets

Parameters	*Green Maize+ Concentrate*	*Green Maize+ Straw+Concentrate*	*Straw+ Concentrate*
Dry matter intake (Kg/d)			
Through straw	-	3.33	9.0
Through green maize fodder	9.0	3.33	-
Through concentrate	0.7	3.33	1.6
Total	9.7	10.0	10.6
FCM milk yield (kg/d/animal)	10.0	8.91	8.02
Energy efficiency (per cent)	40.22	33.34	20.29
Gross protein efficiency (per cent)	38.71	39.90	45.66

Source: Singh *et al.*, 1988.

Chemical Composition of Fodder Crops

Chemical compositions of fodder depend on variety, soil, environment, management practices and stage of maturity. The chemical compositions of different fodder are presented in the Table 26.3.

Mineral Content of Fodder Crop

The need of minerals for growth, reproduction and production as well as normal physiological functions of animal body is well recognized. In most of the tropical and subtropical countries mineral deficiency/imbalances is frequently encountered in livestock feeds and forages (McDowell *et al.*, 1993) which often limits the production performance (Corah, 1996). Furthur, the mineral contents of soil keeps on changing due to pressure on land for maximum crop production, fertilizer application and natuiral calamities, which may alter their contents in feeds and fodders thereby affecting the mineral status of animals. The mineral contents of different fodder from different zones are presented in Table 26.4. Mineral contents of fodder in 5 different states has been presented below:

Table 26.3: Chemical Composition of Fodder (Per cent DM basis)

Fodder	*Region*	*DM*	*CP*	*OM*	*NDF*	*ADF*	*Lignin*	*EE*	*CF*	*Variety*	*Reference*
Oat straw	Pantnagar	83.4	7.2	90.7	65.9	42.5	9.9				Lal *et al.*, 2005
Oat	Uttarakhand	29.7	10.8	89.3	58.9	43.0		3.02	28.5	Kent	Raghuvanshi *et al.*, 2002
Berseem	Pantnagar	25.2	14.9	89.3	50.0	35.9	6.9				Lal *et al.*, 2005
Berseem	Maharashtra	86.5	17.8	91.3				1.8	33.4		Kumbhar *et al.*, 2003
Berseem	Maharashtra	12.28	18.17	86.74	37.06	28.23	6.86	3.18	25.91	Combinations of varieties	Fernandes and Waditake, 2006
Lucerne	Maharashtra	88.9	18.2	90.3				1.8	39.5		Kumbhar *et al.*, 2003
Lucerne	Maharashtra	28.46	21.65	89.78	47.55	38.02	8.33	2.68			Fernandes *et al.*, 2005
Cowpea	Maharashtra	91.1	18.1	92.5				0-.9	31.0		Kumbhar *et al.*, 2003
Cowpea	Tarai, Uttrakhand	18.22	23.41	83.09				2.10	22.54		Tiwari *et al.*, 2007
Stylo	Maharashtra	89.4	18.7	91.6				1.0	39.0		Kumbhar *et al.*, 2003
Anjan	Maharashtra	91.5	12.3	91.9				1.1	42.1		Kumbhar *et al.*, 2003
Hybrid Napier	Maharashtra	92.9	10.1	89.8				1.3	40.2	Yashwant	Kumbhar *et al.*, 2003
Maize	Maharashtra	92.4	7.3	88.9				1.0	41.1	African Tall	Kumbhar *et al.*, 2003
Maize	Tripura		8.65	89.47				1.20	27.01	Vijay	Datt *et al.*, 2006
Maize	Tarai, Uttrakhand	15.07	8.86	88.16				1.56	27.01		Tiwari *et al.*, 2007
Sorghum	Tarai, Uttrakhand	18.25	9.49	88.02				2.31	26.98		Tiwari *et al.*, 2007
Sorghum	Pantnagar	25.5	7.6	95.0	74.2	52.8					Khan *et al.*, 2002
Jowar	Maharashtra	91.7	7.6	90.7				0.9	40.1		Kumbhar *et al.*, 2003

Table 26.4: Mineral Content of Green Fodder (Ca and P g/100 g. Trace minerals microgram per gram)

Fodder	Region	Variety	Ca	P	Cu	Fe	Mn	Zn	Reference
Berseem	Faridabad	NA	1.41	0.21	41.9	386	45.2	30.7	Mandal *et al.*, 2004.
Oat	Faridabad	NA	1.56	0.19	28.5	342	38	24.1	Mandal *et al.*, 2004.
Lucern	Maharashtra		0.8	0.2	39	47	16	18	Kumbhar *et al.*, 2003.
Berseem	Maharashtra		0.7	0.1	40	38	14	13	Kumbhar *et al.*, 2003
Cowpea	Maharashtra		1.0	0.0	11	63	17	28	Kumbhar *et al.*, 2003
Stylo	Maharashtra		0.9	0.1	11	74	11	26	Kumbhar *et al.*, 2003
Maize	Maharashtra		0.2	0.0	10	21	13	25	Kumbhar *et al.*, 2003
Jowar	Maharashtra		0.1	0.1	62	37	19	10	Kumbhar *et al.*, 2003
Hybrid Napier	Maharashtra		0.2	0.1	10	35	9	29	Kumbhar *et al.*, 2003
Napier	Assam		0.42	0.18	9.42	177.6	116.25	41	Kalita *et al.*, 2003
Guinea	Assam		0.36	0.24	18.20	131.66	122.25	33.7	Kalita *et al.*, 2003
Green fodder	Uttarakhand, Tarai		0.32	0.26	6.69	317.50	60.45	64.01	Tiwari *et al.*, 2007
Bajra	Gujrat		0.52	0.23					Garg *et al.*, 2004
Sorghum	Gujrat		0.54	0.22					Garg *et al.*, 2004
Hybrid Napier	Gujrat		0.69	0.24					Garg *et al.*, 2004
Bajra	Gujrat				11.86	205	61.9	44.45	Garg *et al.*, 2004
Cowpea	Gujrat				11.58	1063	79.77	43.17	Garg *et al.*, 2004
Guar	Gujrat				14.49	471.33	54.03	42.7	Garg *et al.*, 2004
Hybrid Napier	Gujrat				13.53	304	78.35	42	Garg *et al.*, 2004
Lucern	Gujrat				11.38	603.1	59.9	27.39	Garg *et al.*, 2004
Maize	Gujrat				4.64	602.5	48.60	34.25	Garg *et al.*, 2004

Contd...

Table 26.4–Contd...

Fodder	*Region*	*Variety*	*Ca*	*P*	*Cu*	*Fe*	*Mn*	*Zn*	*Reference*
Sorghum	Gujrat				6.72	374.69	59.31	33.95	Garg *et al.*, 2004
Lucern	Rajasthan				13.20	541.5	49.65	26.70	Garg *et al.*, 2003
Maize	Rajasthan				11.50	422	92.30	30.40	Garg *et al.*, 2003
Sorghum	Rajasthan				11.20	384.75	57.72	36.17	Garg *et al.*, 2003
Napier	Assam		0.89		7.02	165.27	158.24	28.49	Buragohain *et al.*, 2006
Guinea grass	Assam		0.79		5.58	564.75	113.04	25.97	Buragohain *et al.*, 2006
Berseem	Maharashtra		1.83	0.12					Fernandes and Waditake, 2006
Guinea grass	U.P.		0.48	0.3	5.93	110	39.8	27.5	Singh and Singh, 2005
Oat	U.P.		0.54	0.22	12.6	370	64.9	32.1	Singh and Singh, 2005
Lucern	U.P.		1.08	0.35	8.93	228	40.5	22.7	Singh and Singh, 2005
Stylo	U.P.		1.58	0.28	8.1	213	57	19.2	Singh and Singh, 2005
Jowar	Karnataka		0.30	0.41	16	361		14	Ramana *et al.*, 2000
Hybrid Napier	Karnataka		0.51	0.22	8	740		14	Ramana *et al.*, 2000
Maize	Karnataka		0.31	0.34	20	206		9	Ramana *et al.*, 2000
Hybrid Napier	West Bengal		0.46	0.32	14.28	362.68	46.66	69.25	Nasker *et al.*, 2003

West Bengal

Pasture grasses and green fodder in West Bengal were reported to be deficient in Ca (14.3 to 33.3 per cent), Phosphorus (42.9 to 72.2 per cent), Zn (57.1 to 100 per cent)and Cu (72.2 to 100 per cent), however not deficient in Fe and Mn (Das *et al.*, 2003a). Das *et al.* (2003) have reported that grazing for 6-8 hours on pasture containing Doob grass which happens to be the main bulk of the feed to the animals in red lateririte zone of West Bengal could not support the mineral requirement.

Assam

On the basis of the requirements of livestock for normal growth and production, 33 per cent of the fodder contained low amount of Ca, 67 per cent of fodder contained low in phosphorus, 56 per cent low in magnesium, 61 per cent low in zinc, 30 per cent low in copper and 22 per cent low in manganese. All fodder showed higher content of iron (Kalita, *et al.*, 2003).

Haryana

Mandal *et al.* (2004) have reported that the concentration of Ca, P, Cu and Mn in barseem and oat was sufficient to meet the requirement for maintenance and low production in Faridabad district of Haryana. However, animals yielding more than 10 liters of milk need to be supplemented with concentrate to meet out phosphorus requirement. Also animals receiving wheat straw and concentrate in any ratio were unable to meet Ca requirement of buffaloes producing more than 10 liters of milk daily unless supplemented from extraneous sources. About 75 per cent samples of berseem and all the samples of oat were deficient in Zn due to low Zn in soil of Haryana.

Uttrakhanad

In Garwahal region Cu deficiency in fodder was observed by Sharma and Joshi (2004). The variation in the mineral content of different fodder may vary with the herbage plant species, cultivar differences and soil and climatic conditions in which plants are grown (Turner *et al.*, 1978). Bardhan *et al.* (2005) have reported that 52 per cent farmers in Tarai region of Uttarakhand face the problem of green fodder feeding to livestock.

Gujrat

The average copper and zinc content was found low in green fodder whereas manganese and iron content were found adequate in green fodder (Garg *et al.*, 2004).

Conclusion and Scope of Research

It may be concluded that for economic milk production and good health of dairy animals, fodder has paramount importance. By growing different crop rotation, crop mixture and intercropping, it is possible to provide year round fodder supply to dairy animals which can greatly reduce the production cost of milk. In dry areas, watershed approach, bringing wasteland under silvipastoral system are some of the measures to increase fodder availability which can be implemented through farmers' awareness programme. Successful integration of forage crops in the existing cropping systems (rice-wheat or jowar/bajra/maize-wheat) may be profitable as well as beneficial from the viewpoint of soil health. There are reports of more profitable forage crop sequences *viz.*, rice-berseem and cowpea-wheat when compared to rice-wheat and jowar-wheat systems, respectively. Similarly, integration of *Sesbania* on bunds (accounting 5-7 per cent of field area) may provide nutritious forage during summers. Food-feed crop cultivation also has importance to ensure fodder supply for dairy animals.

References

Bardhan, D.,Srivastava, R.S.L. and Dabas, Y.P.S. 2005. Study of constraints perceived by farmers in rearing dairy animals. Indian J. Dairy Sci 58(3): 214-218.

Bharadwaj, A., Dixit, V.B. and Sethi, R.K.2006. Economics of buffalo milk production in Hisar district of Haryana state. Indian J. Dairy Sci 59(5): 322-327.

Burqgohain, R., Ghosh, M.K., Baruah, K.K., Pathak, P.K.and Bhattacharya, M. 2006. Mineral status of feeds, fodder and tree leaves for dairy animals in Jorhat district of Assam, Anim. Nutr. And Feed Technology 6: 289-293.

Conrah, L. 1996. Trace mineral requirement of grazing cattle. Animal Feed Science and Technology 59: 61-70.

Dannial, S.J., Bhoserekar, M.R. and Mullick, D.N. 1967. Studies on the utilization of homogeneous roughage by dairy cows. Indian J. Dairy Sci., 20: 191-195.

Das, A. and Singh, G.P. 1999. Effect of different proportions of berseem and wheat straw on in vitro dry matter digestibility and total gas production. Indian J. Anim. Nutr. 16: 60-64.

Das, A., Halder, S. and Ghosh, T.K. 2003. Mineral status of soil, feeds and grazing buffaloes in two agro-climatic zones of West Bengal. Indian J. Dairy Sci 56(4): 223-229.

Das, A., Haldar, S., Biswas, P. and Ghosh, T.K. 2003a. Distribution of some major and trace elements in soil, feed, fodder and livestock in red laterite zone of West Bengal. Indian J Anim. Nutr. 20(2): 136-142.

Datt, C., Niranjan, M., Chhabra, A., Chattopadhyay and Dhiman, K.R. 2006. Forage yield, chemical composition and in vitro digestibilities of different cultivars of maize. Indian J. Dairy Sci 59(3): 177-180.

Fernandes, A.P. and Waditake, S.K. 2006. Comparative evaluation of berseem varieties for yield and fodder quality. Animal Nutrition and Feed Technology 6: 301-306.

Garg, M.R., Bhandari, B.M. and Sherasia, P.L. 2004. The status of certain trace minerals in feeds and fodders in Kutch district of Gujrat. Indian J,. Anim. Nutr. 21(1): 8-12.

Garg, M.R., Bhandari, B.M. and Sherasia, P.L. 2003. Macro and micromineral status of feeds and fodders in Kota district of Rajasthan. Indian J,. Anim. Nutr. 20(3): 252-261.

Gupta, P.C., Singh, K., Lodhi, G.P., Gupta, L.R.and Sharma, D.P. 1983. Effect of feeding lucern and berseem on the milk yield, efficiency of utilization of nutrients and cost of production in Murrah buffaloes. Indian J. Anim. Sci. 53(11): 1181-1185.

Jackson, M.G. and Gupta, D.C. 1971. The value of concentrate supplementation of berseem forage in milk production in buffaloes. Indian J. Anim. Sci., 41: 86-91.

Kalita, D.J., Sarmah, B.C., Sarmah, D.N. and Mili, D.C. 2003. Mineral status and their retentionin lactating cows in relation to soil, fodder and feed in Kamrup district of Assam. Indian J Anim. Nutri 20 (4): 421-429

Khan, M.I., Kumar, A., Singh, M. and Sohane, R.K. 2002. Rumen fermentation pattern as affected by feeding oat hay and green sorghum based diets. Indian J. Anim. Nutr. 19(1): 36-40.

Kumbhar, B.R., Fulpagare, Y.G. and Anarase, S.A. 2003. Nutritional evaluation of commonly utilized forages in cattle. Indian J. Anim. Nutr. 20(4): 457-460.

Lal, B., Rastogi, S.K., Korde, J.P., Madan, A.K., Kumar, R. and Kumar, A. 2005. Nutrient utilization from oat straw and green barseem combinations. Indian J. Animal Nutrition 22 (1): 48-50.

Mandal, A.B., Yadav, P.S., Sunaria, K.R., Kapoor, V. and Mann, N.S. 1996. Mineral status of buffaloes in Mahindergarh district of Haryana. Indian J. Anim. Sci 66: 849-851.

Mann, N.S., Mandal, A.B., Yadav, P.S., Lal, d. and Gupta, P.C. 2003. Mineral status of feed and fodder in Bhiwani district of Haryana. Indian J. Anim. Sci 73: 1150-1154.

Mcdowell, L.R., Conrad, J.H. and Glenhembry, F. 1993. Minerals for grazing ruminants in Tropical region. 2nd Edition, Bulletin, Department of animal science, University of Florida, Gainesville, USA.

Mudgal, V.D. and Mallikarjunappa, S.1985. Influence of various roughage and concentrate ratios on lactation performance and feed conversion efficiency of Murrah buffaloes. Indian J. Dairy Sci., 39: 378-382.

Mudgal, V.D. 1985. Feeding dairy cattle economically. NDRI, Karnal, Publication No. 101.

Nasker, M., Mondal, P., Rajendran, D. and Biswas, P. 2003. Assessment of micro-nutrients status in soil, plants and small ruminants in coastal saline zone of West Bengal. Animal Nutrition and Feed Technology 3: 107-115.

Pachauri, V.C. 1990. Herbage based ration schedule for economic milk production. Lecture note in summer institute on "Forage production technology, IGFRI, Jhansi.

Patange, D.D., Kulkarni, A.N., Gujar, B.V. and Kalyanrkar, S.D. 2002. Nutrient availability to milch Marathwadi buffaloes in their home tract. Indian J. Anim. Nutr. 19(1): 41-46.

Raghuvanshi, N., Verma, M.L. and Jaiswal, R.S. 2002. Nutritional evaluation of oat varieties in crossbred heifers. Indian J. Animal Science 19(3):232-235.

Ramana, J.V., Prasad, C.S. and Gowda, S.K. 2000. Mineral profile of soil, feeds, fodders and blood plasma in southern transition zone of Karnataka. Indian J. Anim. Nutr. 17(3): 179-183.

Reddy, G.V.N., Reddy, M.R. and Das, T.C. 1996. Effect of feeding fodder based complete diet on the production performance of crossbred cow. Indian J. Anim. Sci. 6: 63

Sharma, M.C. and Joshi, C. 2004. Soil, fodder and serum micro-minerals status and haematobiochemical profile in cattle of Garhwal region of Uttarakhand. Indian J. Anim. Sci 74: 775-779.

Singh, B. and Chaudhary, J.L. 2007. Effect of different levels of maize fodder on t he performance of crossbred heifers. Indian J. Anim.Nutr. 24(4): 256-257.

Singh, B., Chaudhary, J.L. and Gupta, L.2007. Effect of feeding different levels of green maize on performance of lactating crossbred cows. Indian J. Anim.Nutr. 24(3): 164-166.

Singh, S.P. and Singh. M. 2005. Problems of buffalo keepers in Etah district of Uttar Pradesh. Indian Dairyman 57(10): 25-28.

Singh, N., Walli, T.K. and Mudgal, V.D.1988. Level of green fodder in the ration of buffaloes for optimum milk production. Indian J. Dairy Sci., 40: 2-6.

Singh, S., Katiyar, P.K. and Singh, S. 1999. Variation in nutritive value of fodder maize genotypes. Indian J. Anim. Sci 69: 827-829.

Tiwary, M.K., Tiwari, D.P., Mondal, B.C. and Kumar, A. 2007. Macro and micro-mineral profile in soil, feeds and animals in Haridwar district of Uttarakhand. Anim. Nutr. Feed Technol. 7: 187-195.

Tomar, O.S. and Wali, T.K. 1995. Forages and milk production. In: National Symposium on Forage Production System for Sustainable Agricultural Development. IGFRI, Jhansi, 21-23 December, 1995.

Turner, M.A., Neall, V.E. and Wilson, G.F. 1978. Survey of magnesium content of soil and pastures and incidence of grass tetany in three selected areas of Taranaki, New Zealand. J.Agric. Res.21: 583-592.

Chapter 27

Modern Techniques of Conservation of Forages: Silage

P.K. Singh[1] *and Amitava Dey*[2]*

[1]*Department of Animal Nutrition, Bihar Veterinary College, Patna – 14*

[2] *ICAR Research Complex for Eastern Region, Patna*

**E-mail: amitavdey_icar@yahoo.co.in*

Feed shortages, both quantitative and qualitative, limit livestock productivity. It is recognized in developed countries that the preparation of silage of high quality cultivated forage can be a valuable component for the development of a high-performing and low-cost system of animal production, using a relatively low level of purchased concentrates. Quality of feed is essential for optimum animal productivity. When forage growth exceeds herd requirements, one of several strategies available to keep pasture quality high is fodder conservation. Silage or hay should only ever be made from grass that is a genuine surplus to animal requirements.

Silage

Silage is the material produced by controlled fermentation of forages or crop residues with high moisture content. The purpose is to preserve forages by natural fermentation by achieving anaerobic conditions and discouraging clostridial growth. Fresh forage crops, such as maize, grasses, legumes, wheat and lucerne, can be preserved by ensiling. It is essential to have a good microbial fermentation process to produce high quality silage. A good fermentation process is not only dependent on the type and quality of the forage crop, but also on the harvesting and ensiling technique.

Steps of Silage Formation

Although the ensiling process appears quite simple, many factors affect type and products of fermentation that takes place in a silo. Acid production is essential in the keeping qualities of silage. The rate, amount and kind of acid produced are influenced by the moisture content of the chopped forage and the readily available carbohydrate content of the forage.

Harvest Forage at Proper Stage of Maturity

The crop for silage is harvested at proper stage of maturity, which ensures good quality silage. The crop is generally harvested at the flowering stage when it has the maximum amount of nutrients. Maize is harvested at the early dent stage whereas sorghum is cut at an early dent stage.

Chop to Proper Length

Prior to ensiling, plant materials should be chopped. Chopping facilitates facilitate good compaction while filling. A good compaction results in a good anaerobic fermentation and prevents fungal growth and rotting of the silage. When most oxygen is removed, clostridial growth is discouraged and lactic acid fermentation encouraged. Secondly, chopping releases plant juices, stimulating the growth of lactic acid bacteria. Also, chopping may increase silage intake by improving quality of fermentation and by accelerating rate of passage of feed particles through the rumen. Chopping forages too long makes compaction difficult and air will remain trapped in the silage resulting in heating and spoilage, whereas, chopping the forage to cut any finer can have a negative impact upon milk fat production and the incidence of displaced abomasums in dairy cattle due to inadequate scratch factor. Thus, 10-15 per cent of the silage material should be above 25 mm in length in order to maintain an effective fibre function.

Usually forage is chopped into short lengths (1-3 cm) before ensiling. But, the fineness of chopping varies with moisture content and nature of the material. Rough and hard materials should be finely chopped, while delicate and soft materials can be roughly chopped. The following guidelines can be used:

High moisture forage (Moisture >75 per cent)	chop to 6.5-25 mm
Wilted materials (Moisture 60-70 per cent)	chop to 6.5 mm
Whole maize plant	chop to 6.5-13 mm

Control of Moisture Content in Raw Materials

Adequate moisture is essential for bacterial fermentation and aids in packing to help exclude oxygen from the silage. Although silage may be made within a large range of moisture contents, DM should be over 35-40 per cent to assure good quality silage. Forages with a DM content above 50 per cent are considered difficult to ensile. The dry matter content of the forage can also have major effects on the ensiling process via a number of different mechanisms. First, drier silages do not pack well and thus it is difficult to exclude all of the air from the forage mass. Second, as the dry matter content increases, growth of lactic acid bacteria is curtailed and the rate and extent of fermentation is reduced. (For example, acidification occurs at a slower rate and the amount of total acid produced is less).

There are many disadvantages to ensiling crops with high moisture content. First, ensiling of wet materials results in the generation of a large volume of effluent, which not only poses disposal problems, but also carries off valuable, highly digestible nutrients in solution. Effluent is produced when moisture is above 75 percent. Secondly, ensiling wet crops will encourage clostridial fermentation, resulting in excessive protein degradation, high DM loss, reduced nutritive value and production of toxins. Even if the water-soluble carbohydrate levels are adequate to ensure lactic fermentation, very wet silages may still be nutritionally undesirable because voluntary dry matter intake of these is frequently low. Where weather permits, wilting forage above 30-35 per cent DM prior to ensiling can reduce the incidence of clostridia because these organisms are not very osmo-tolerant (they do not thrive well in dry conditions).

Control of Water Soluble Carbohydrates

Another factor that can affect the ensiling process is the amount of water-soluble carbohydrates present for good fermentation to take place. The amount of water-soluble carbohydrates necessary to obtain sufficient fermentation depends on the dry matter content and the buffer capacity of the crop. The buffering content of the forage mass can have an effect on silage fermentation. Alfalfa has a high buffering capacity in comparison to corn. Thus, it takes more acid production to lower the pH in alfalfa than in corn silage, resulting in the former being more difficult to make. Corn silage harvested at the proper growth stage or at physiological maturity has a high level of readily available carbohydrates for lactic acid production. If legumes and grass crops are not wilted in the field to an average of 65 per cent moisture or less depending on the type of storage, then the addition of a carbohydrate-rich feedstuff will enhance fermentation.

Forages with insufficient fermentable substrate or too low a DM content, addition of sugars directly (*e.g.* molasses) or by adding enzymes that release extra sugars from the crop ensures sufficient substrate for lactic acid bacteria thus causing the desired reduction in pH.

Filling, Packing and Sealing

The silo should be filled with the crop harvested as rapidly as possible. Filling delays will result in excessive respiration and increased silage losses. Remove all air from the silo by compacting the fodder by jumping and stamping on the fodder after addition of each load. Any air left in the silo or any air entering the silo will render the microenvironment within the silo conducive for undesirable fungal and bacterial growth that will decompose the silage. Packing should begin immediately when storing silage in bunker silos. A wheeled tractor is preferred as a packing vehicle, as it will supply greater weight per surface unit than a tracked vehicle. The silo should be sealed with an air-tight cover once it is filled to prevent penetration of air and rainfall into the silage. A good grade of plastic weighted down with discarded tires will generally provide an adequate seal. Cover the silo and make it airtight by adding heavy materials like stones on the top and waterproof by providing a roof. The silage is generally ready after 4-6 weeks. Maintain airtight seal until feeding out. It has been observed that water soluble carbohydrate dramatically decreased and dry matter losses increased when corn forage was not immediately packed into silos after chopping. Losses increased with prolonged times of delay.

Additives

Most silage additives are designed to aid fermentation by providing fermentation bacteria, enzymes or fermentable substrate. Although not a replacement for good management, they are tools to help ensure that the ensiling process stays within acceptable boundaries. The types and numbers of bacteria on the plant also have profound effects on silage fermentation. Natural populations of lactic acid bacteria on plant material are often low in number. In addition, if air is not removed from the silage mass, other types of fermentation can occur. Proper maturity assures adequate fermentable sugars for silage bacteria and maximum nutritional value for livestock. In high DM silages with reduced water availability, the presence of suitable, osmo-tolerant lactic acid bacteria could become a limiting factor in the ensiling process. It has been shown that these bacteria represent only a small percentage of the indigenous microflora on forage crops. Inoculants that increase lactic acid fermentation might be useful to inhibit clostridial activity. The minimum number of lactic acid bacteria required to inhibit clostridial activity was found to be at least 100 000 colony-forming units per gram of fresh crop (Kaiser and Weiss, 1997).

Feed to Cattle

If the cattle refrain from eating it while introduced for the first time, let the silage rest in open air for a while to get ride of some of the heavy odour gasses, and then feed it to the cattle.

The Ensiling Process

Ensiling is the term given for all physical and chemical changes that take place when forage with sufficient moisture content (60-65 per cent) is stored in the absence of air for silage preservation. Silage fermentation is an exceptionally complex process involving biochemical interactions among the forage, microbial populations and the ensiling environment. Successful silage production depends upon the promotion of the fermentation brought by beneficial bacteria. The process is based on spontaneous lactic acid fermentation under anaerobic conditions. The lactic acid bacteria ferment the water-soluble carbohydrates in the crop to lactic acid, and to a lesser extent to acetic acid. Due to the production of these acids, the pH of the ensiled material decreases and spoilage micro-organisms are inhibited. The ideal characteristics of material for silage preservation are: DM content above 20 per cent and an adequate level of fermentable substrate (8-10 per cent of DM) in the form of water soluble carbohydrate. The ensiling material should also ideally have a physical form that allows easy compaction in the silo. Materials such as maize stover, sorghum stover and grass can be ensiled successfully, while crop residues such as rice and wheat straw, with low water soluble content, do not fulfill these requirements, and therefore pre-treatments, such as fine chopping or use of additives, or both, may be necessary. From a practical view, the three most important things that must occur in order to make good silage are 1) the rapid removal of air, 2) the rapid production of lactic acid that results in a rapid drop in pH, and 3) continued exclusion of air from the silage mass during storage and feed out.

Once the fresh material has been harvested, chopped, compacted and sealed well to exclude air, the ensiling process then begins and undergoes, which can be divided into 4 stages.

Phase 1–Aerobic Phase

Until plant cells are dead or have no oxygen available, they respire to generate the energy they need for normal activity. As the forage is harvested, aerobic organisms predominate on the forage surface. During the initial ensiling process, the freshly cut plant material, and the aerobic and facultative aerobic micro-organisms such as yeasts and enterobacteria continue to respire within the silo structure. Plant respiration continues for several hours (and perhaps days if silage is poorly packed) until air is used up. Thus, any oxygen trapped between the forage particles is eliminated as a result of the respiration of plant material and the aerobic and facultative aerobic micro-organisms (yeasts and bacteria).

If air is not removed quickly, high temperatures and prolonged heating are commonly observed. Air can be eliminated by wilting plant material to recommended dry matters for the specific crop and storage structure, chopping forage to a correct length, quick packing, good compacting, even distribution of forage in the storage structure, and immediately sealing the silo. Air must be removed before optimal fermentation can take place. Once air is removed, fermentation can begin. Lactic acid bacteria utilize water-soluble carbohydrates to produce lactic acid, the primary acid, responsible for decreasing the pH in silage. The acidity of silage can be determined by measuring its pH. Depending on the crop, plant material in the field can range from a pH of about 5 to 6 and decrease to a pH of 3.6 to 4.5 after acid is produced. In addition, a rapid decrease in pH will inhibit the growth of undesirable anaerobic micro-organisms such as *enterobacteria* and *clostridia.* Another important chemical change that occurs during this early phase is the breakdown of plant proteins. Plant enzymes such as proteases and carbohydrases are active during this phase until air is used up, provided the pH is within the normal range for fresh forage juice (pH 6.5-6.0). Proteins are first reduced to amino acids and then to ammonia and amines. Up to 50 per cent of the total plant protein may be broken down during this process. The extent of protein breakdown (proteolysis) is dependent on the rate of pH decline in the silage. A quick reduction in silage pH will help to limit the breakdown of protein in the silo by inactivating plant proteases. Phase I ends once the oxygen has been eliminated from the silage mass.

This phase is undesirable since the aerobic bacteria consume soluble carbohydrates that might otherwise be available for the beneficial lactic acid bacteria or the animal consuming the forage. Although, this phase reduces the oxygen to create the desired anaerobic conditions for silage fermentation but the respiration process produces water and heat in the silage mass. Excessive heat build-up resulting from an extended Phase I period can greatly reduce the digestibility of nutrients such as proteins. Under ideal crop and storage conditions, this phase will last only a few hours. With improper management, this phase could continue for several weeks. Delayed filling that result in excessive amounts of air trapped in the forage mass can have detrimental effects on the ensiling process. Rapid removal of air is important because it prevents the growth of unwanted aerobic bacteria, yeasts, and molds that compete with beneficial bacteria for substrate. Key management practices are proper maturity, moisture, chop length and rapid filling with adequate packing and proper sealing of the storage structure.

Practical Aspect of the Aerobic Phase

- ✫ Chop the material as short as possible (1-3 cm)
- ✫ Fill the silo quickly
- ✫ Compact the silo as well as possible, as fingers should not be able to insert into the compacted forage.
- ✫ Seal the silo air tight
- ✫ Weight the top of the stack to maintain air tight seal between the cover and compacted forage
- ✫ Seal as soon as possible.

Phase 2–Fermentation Phase

After the oxygen in the ensiled forage has been utilized by the aerobic bacteria and the silage becomes anaerobic, Phase II begins. This is an anaerobic fermentation where the growth and development of acetic acid-producing bacteria occurs. These bacteria ferment soluble carbohydrates and produce acetic acid as an end product. Acetic acid production initiative the pH drop necessary to set up the following fermentation phases. As the pH of the ensiled mass falls below 5.0, the acetic bacteria decline in numbers as this pH level inhibits their growth. The lower pH enhances the growth and development of lactic acid producing bacteria. If the fermentation proceeds successfully, lactic acid bacteria develop and become the predominant population. The lactic-acid bacteria ferment soluble carbohydrates and produce lactic acid. Lactic acid is the most desirable fermentation acids and for efficient preservation, the acid should comprise greater than 60 per cent of the total silage organic acids produced. When silage is consumed, lactic acid will also be utilized by cattle as an energy source. Due to the production of lactic and other acids, the pH decreases to 3.8-5.0. This low pH inhibits the growth of all bacteria. When this pH is reached, the forage is in a preserved state. No further destructive processes will occur as long as oxygen is kept away from the silage. Instead of lactic acid producing bacteria developing, large populations of clostridia bacteria may grow in the silage. These anaerobic bacteria produce butyric acid rather than lactic acid, which results in sour silage. With this type of fermentation, the pH may be 5.0 or above. The final pH of the ensiled forage depends largely on the type of forage being ensiled and the condition at the time of ensiling.

This is the longest phase in the ensiling process and continues for between several days to several weeks, depending on the properties of the ensiled forage crop.

Practical Aspect of the Fermentation Phase

- ☆ Mix molasses (@3-5 per cent on wet basis), a substrate source for the bacteria, to encourage lactic acid fermentation.
- ☆ Maintain the dry matter about 30 per cent.

Phase 3–Stable Phase

In general, good silage will remain stable once air is eliminated and it has achieved a low pH. For as long as air is prevented from entering the silo or container, relatively little changes occurs. Most micro-organisms of phase 2 slowly decrease in numbers. Some acid-tolerant micro-organisms survive this period in an almost inactive state; others, such as clostridia and bacilli, survive as spores. Only some acid-tolerant proteases and carbohydrases and some specialized micro-organisms, such as *Lactobacillus buchneri*, continue to be active at a low level.

Practical Aspect of the Stable Phase

- ☆ Maintain the airtight seal around the silage.
- ☆ Repair holes as soon as they are noticed.

Phase 4–Feed-out Phase or Aerobic Spoilage Phase

This phase refers to the silage as it is being fed out from the storage structure in which surface of the silage is exposed to air. This phase is important because research shows that nearly 50 per cent of the silage dry matter losses occur from secondary aerobic decomposition. High populations of yeast and mold or the mishandling of stressed crops can lead to significant losses due to aerobic deterioration of the silage. During feed-out this is unavoidable, but it can start earlier due to damage to the silage covering (*e.g.* by rodents or birds). The process of spoilage can be divided into two stages. The primary spoilage stage is the onset of deterioration due to the degradation of preserving organic acids by yeasts and, occasionally, by acetic acid bacteria. This will cause a rise in pH, and thus the second spoilage stage is started, which is associated with increasing temperature, and activity of spoilage micro-organisms such as moulds and enterobacteria. When exposed to air, spoilage micro-organisms metabolize lactic acid that causes the pH of the silage to increase, thus allowing bacteria that where inhibited by low pH to grow and further spoil the mass. However, the rate of spoilage is highly dependent on the numbers and activity of the spoilage organisms in the silage. Spoilage losses of 1.5-4.5 per cent DM loss per day can be observed in affected areas. These losses are in the same range as losses that can occur in airtight silos during several months of storage (Honig and Woolford, 1980). Aerobic spoilage occurs in almost all silages that are opened and exposed to air. Airtight silos and removal of sufficient silage during feed-out can help to prevent aerobic spoilage.

To avoid failures, it is important to control and optimize each phase of the ensiling process. In phase I, good silo filling techniques will help to minimize the amount of oxygen present between the plant particles in the silo. Good harvesting techniques combined with good silo filling techniques will thus minimize water soluble carbohydrates losses through aerobic respiration in the field and in the silo, and in turn will leave more water soluble cabohydrates available for lactic acid fermentation in phase 2. During phases 2 and 3, the farmer cannot actively control the ensiling process. Methods to optimize phases 2 and 3 are therefore based on the use of silage additives applied at the time of ensiling, as will be discussed in the section on additives, below. Phase 4 will start as soon as oxygen is available. To minimize spoilage losses during storage, an airtight silo is required, and any damage to the silo covering should be repaired as soon as possible. During feed-out, spoilage by air ingress can be minimized by a sufficiently high feed-out rate. In addition, silage additives capable of decreasing spoilage losses can be applied at the time of ensiling. Proper management is vital to reduce these losses.

Practical Aspect of the Feed-out Phase

- ☆ Maintain air tight seal
- ☆ Feed out to ensure about 20-30 cm removal from the entire silage face each day.
- ☆ If the silage gets hot, feed it out at faster rate.

Silage Microbiology

Successful preservation of high-moisture forage and other crops depends upon the controlling the activities of microbes, particularly bacteria. The silage microflora plays a key role in the successful outcome of the conservation process. The flora can basically be divided into two group, namely the desirable and the undesirable micro-organisms. The desirable micro-organisms are lactic acid bacteria. The undesirable ones are the organisms that can cause anaerobic spoilage (*e.g.* clostridia and enterobacteria) or aerobic spoilage (*e.g.* yeasts, bacilli, listeria and moulds).

Desirable Micro-organisms

Lactic Acid Bacteria

Lactic acid bacteria are normally present on growing crops in small numbers but usually multiply rapidly after harvesting, particularly if the crop is chopped or lacerated. Based on their sugar metabolism lactic acid bacteria can be classified as obligate homofermenters (*Pediococcus damnosus* and *Lactobacillus ruminis*), facultative heterofermenters (*Lactobacillus plantarum, L. pentosus, Pediococcus acidilactici, P. pentosaceus* and *Enterococcus faecium*) or obligate heterofermenters (*Leuconostoc, Lactobacillus brevis* and *Lactobacillus buchneri*. Obligate homofermenters produce more than 85 per cent lactic acid from hexoses (C_6 sugars) such as glucose, but cannot degrade pentoses (C_5 sugars) such as xylose. Facultative heterofermenters also produce mainly lactic acid from hexoses, but in addition they also at least degrade some pentoses to lactic acid, and acetic acid and/or ethanol. Obligate heterofermenters degrade both hexoses and pentoses, but unlike homofermenters they degrade hexoses to equimolar mounts of lactic acid, CO_2 and acetic acid and/or ethanol (Balows *et al.*, 1992; Schleifer and Ludwig 1995). The homofermentative groups are more efficient in converting forage sugars into lactic acid. Of these species *L. plantarum* is generally considered to be the most competitive, rapidly producing large amounts of lactic acid in freshly ensiled forage. As a result of lactic acid production, pH of the silage reduced to 4 and 5, depending on the species and the type of forage crop.

Undesirable Microorganisms

Clostridia

Clostridia are endospore forming anaerobic bacteria. They can be divided into two major groups, saccharolytic clostridia and proteolytic clostridia. The saccharolytic clostridia (*Clostridium butyricum* and *C. tyrobutyricum* can degrade lactic acid and residual water soluble carbohydrates to butyric acid, H_2 and CO_2, causing a rise in pH (Gibson, 1965; McDonald *et al.*, 1991).

$$2 \text{ Lactic acid} \longrightarrow 1 \text{ Butyric acid} + 2\, H_2 + 2\, CO_2$$

The proteolytic group includes *C. bifermentans* and *C. sporogenes*. These bacteria ferment amino acids to a number of different products. Many clostridia ferment carbohydrates as well as proteins, thus causing problems such as reduction in feeding value and the production of biogenic amines.

Enterobacteria

Enterobacteria (*Escherichia coli and Erwinia herbicola*) associated with silage are facultative anaerobes and usually are active in the early stage of fermentation when the pH (7.0) is favourable for their growth. The enterobacteria, sometimes described as 'acetic acid bacteria' or 'coliform bacteria', are usually present in very low numbers on crop. They compete with the lactic acid bacteria for the water soluble carbohydrates and ferment them to mixture of products including acetic acid, ethanol and hydrogen. In addition they can degrade protein, decarboxylate and deaminate amino acids leading to the production of large concentration of ammonia ultimately reduction in feeding value of silage.

Bacilli

Bacilli are like endospore-forming, rod-shaped facultative aerobes bacteria, Facultative aerobic bacilli ferment a wide range of carbohydrates to compounds such as organic acids (*e.g.* acetate, lactate and butyrate) or ethanol, 2,3-butanediol and glycerol (Shlegel,1987). Proliferation of bacilli in silage is generally considered undesirable. Not only are bacilli less efficient lactic and acetic acid producers than lactic acid bacteria (McDonald *et al.*, 1991), they can also enhance (later stages of) aerobic deterioration (Lindgren *et al.*, 1985). To decrease bacillus growth in silage, storage temperatures should not be too high (Gibson *et al.*, 1958).

Yeasts

Yeasts are eukaryotic, facultative anaerobic, heterotrophic micro-organisms. Under anaerobic silage conditions, yeasts ferment sugars to ethanol and CO_2 (Schlegel, 1987; McDonald *et al.*, 1991). This ethanol production in silage not only decreases the amount of sugar available for lactic acid fermentation, but it can also have a negative effect on milk taste (Randby *et al.*, 1999). Under aerobic conditions, many yeast species degrade the lactic acid to CO_2 and H_2O.

Table 27.1: Summary of Microbiology and Fermentation Pathways of Silage

Organisms	*Conditions Required*	*Major Products/Effects*
Lactic acid bacteria	☆ Anaerobic condition	*Homofermentative pathway*:
	☆ Wilting of crop	*Glucose* ⟶ 2 Lactic acid
	☆ Chopping of crop for rapid establishment of lactic acid bacteria.	*Fructose* ⟶ 2 Lactic acid
		Pentose ⟶ Lactic acid + Acetic acid
		Heterofermentative pathway:
		Glucose ⟶ Lactic acid + Ethanol +CO_2
		3 Fructose ⟶ Lactic acid + Methanol +Acetic acid + CO_2
		Pentose ⟶ Lactic acid +Acetic acid
Clostridia	☆ Anaerobic condition	*Saccharolytic*
	☆ Wet forage	2 Lactic acid ⟶ Butyric acid + $2CO_2$ + $2H_2$
		Proteolytic
		Glutamic acid ⟶ Acetic acid + Pyruvic acid +NH_3
		Lysine ⟶ Acetic acid + Butyric acid + $2NH_3$
Enterobacteria	☆ Anaerobic condition	Glucose ⟶ Acetic acid + Ethanol + $2CO_2$ + $2H_2$
	☆ Optimum pH 7.0	
	☆ Active in early stages of fermentation	

Silo

A silo is an airtight to semi-airtight structure designed for the purpose of preservation and storage of high moisture feeds as silage. Silos are of following different types:

Upright or Tower Silo

Tower silos are circular in shape and constructed from brick. The advantages of this type of silo include: long life, small space required, minimum top and side spoilage losses, and possibility for mechanization. However, tower silos are expensive due to its cost of construction. The size varies from about 12-20ft in diameter and 40-80ft. in length depending upon the requirement.

Cellar Silo

The cellar type is the most common silo on individual farms. Round or square concrete silos are usually built inside houses for protection from the weather. Advantages are lower cost and easy management. Size can be adjusted according to scale of production. Cellar silos are suitable for rural conditions. A disadvantage is high effluent loss, especially with clay walls.

Trench Silo

This type is generally built underground or semi-underground, with two solid walls of 1.5-2 m in height. It is most popular in areas where weather is not too severe and where there is good drainage. Low initial cost, ease of construction and more suitability for mechanization are advantages of trench silo. The tractor can be driven on top from one side to the other for compaction purposes. After compaction, it is sealed with a plastic sheet pressed down with soil, sandbags or straw bales to maintain anaerobic conditions.

Stack Silo

This type of silo implies a pile of material on the ground surface. On flat and dry ground, plastic sheet is placed underneath and the material is laid in a stack. The top is covered with plastic and sealed all round with soil. Sandbags or old types, or any other suitable objects, are placed on top to prevent the top cover from being blown away by the wind. The advantages of the stack silo are low cost and flexibility of placement.

Plastic Silo

The plastic silo is similar to the stack silo but it is covered with plastic sheets of polyvinyl chloride (PVC) or polyethylene. Alternately, the silo can be made in bags with sealed tops. The stack silo is also inexpensive and can be placed anywhere. However, labour requirements are high due to manual filling and handling.

Losses during Silag Making

Objective o the conservation of forage is to preserve as much as of the crop nutrient as possible. However, during ensilage loss of nutrients occurs. There are five sources of loss. Theses are: (*a*) Field losses, (*b*) Oxidation or respiration losses, (*c*) Fermentation losses, (*d*) Effluent losses, (*f*) Aerobic deterioration.

Field Losses

Filed loss is a result of respiration activity of the harvested crop and physical factors such as cutting, shattering, leaching by rain, and the efficiency of pick-up by the forage harvester. The main nutrients affected are the water-soluble carbohydrate, and protein which are hydrolyzed to amino acids. Prolonged period of wilting can result in substantial dry matter losses. With crops harvested and ensiled the same day, nutrient losses are negligible and even over 24-hours wilting period, losses of dry matter of 1-2 per cent occur. Dry matter losses as high as 6 per cent after five days, and 10 per cent after eight days of wilting in the filed have been reported.

Oxidation Losses

Oxidation losses result form oxidation of sugar by the plant and microbial enzymes leading to the formation of carbon dioxide and water. As long as aerobic condition prevails in silo, and the pH of the forage has not been reduced, respiration by both plant cells and the aerobic microflora will continue. Therefore, rapid filling of silo can reduce oxidation losses.

Fermentation Losses

Fermentation loss is the result of the activities of anaerobic or facultative anaerobic microorganisms under aerobic conditions. In well preserved silage in which dominant microorganism is lactic acid bacteria, the DM recovery would be 100 per cent. If one consider heterolactic fermentation of glucose, there is a 24 per cent loss of DM through evolution of carbon dioxide. In case of clostridial and entrobacterial fermentation, losses occur due to the evolution of the gaseous carbon dioxide, hydrogen and ammonia.

Effluent Losses

Effluent contains sugar, soluble nitrogenous compounds, minerals and fermentable acids, which are of high nutritional value. In most silos, free drainage occurs and the liquid or effluent carries with it soluble nutrients. The quantity of effluent produced depends largely on the moisture content of the forage at ensiling. Crops ensiled with a dry matter content of 15 per cent may result in dry matter effluent loss of as high as 10 per cent, whereas crops wilted to about 30 per cent dry matter produce little effluent loss, if any. Effluent loss increases if the silo is left uncovered so that rain enters.

Aerobic Deterioration

Aerobic deterioration occurs when the surface of silage is exposed to air. Such losses may consists of surface loss and feed- out loss. Surface loss occurs at the surface and sides of silo and is more characteristics of bunker silo than tower silo. Feed-out loss occur as soon silo is opened for feeding of silage to livestock.

Source	*Comment*	*Loss of Net Energy*
Respiration loss	Unavoidable	1-2
Fermentation (Lactic acid) loss	Unavoidable	4
Fermentation (Butyric acid) loss	Avoidable	0-5
Effluent loss	Avoidable	5-7
Aerobic deterioration	Avoidable	0-10

Evaluation of Silage Quality

Although silage quality can be estimated by visually examining the silage, it can only be estimated accurately by chemical analysis. The most important analyses obtained from feed testing laboratories include dry matter, pH, crude protein, fibre, calcium and phosphorus.

Silage pH

The best single indicator of the effect of ensiling on the nutritive value of high moisture silage is pH. The pH of an ensiled sample is a measure of its acidity. The relationship between pH and silage quality can be determined. In general the lower the pH, better the quality of silage, since it indicates that a lactic acid type of fermentation has occurred. pH is, however, not a good indicator of quality for silages which contain less than 65 per cent water. In general, legume silages have a higher pH than non-leguminous silages and take longer to ensile because of their higher buffering capacity.

Some common reasons for a high silage pH are as follows:

- ☆ Dry silage (> 50 per cent DM)
- ☆ Silage not fully fermented due to early sampling time relative to harvest, cold weather during harvest, and slow or poor packing
- ☆ Legume silages with extremely high ash contents (> 15 per cent of DM) and (or) high protein content (> 23-24 per cent CP)
- ☆ Silage with excess ammonia or urea
- ☆ Clostridial silages
- ☆ Spoiled or moldy silages
- ☆ Silages containing manure.

Lactic Acid

Lactic acid should be the primary acid in good silages. Higher the lactic acid, better the quality of silage. Lactic acid should be at least 65 to 70 per cent of the total silage acids in good silage. This acid is stronger than other acids in silage (acetic, propionic and butyric) and thus usually responsible for most of the drop in silage pH. Fermentations that produce lactic acid result in the fewest losses of dry matter and energy from the crop during storage.

Some common reasons for low lactic acid content are as follows:

- ☆ Restricted fermentation due to high DM content (especially legumes and grasses with > 50 per cent DM).
- ☆ Restricted fermentation due to cold weather.
- ☆ Silages high in butyric acid (Clostridial silages) are usually low in lactic acid.

Acetic Acid

Extremely wet silages (< 25 per cent DM), prolonged fermentations (due to high buffering capacity), loose packing, or slow silo filling can result in silages with high concentrations of acetic acid (>3 to 4 per cent of DM). In such silages, energy and DM recovery are probably less than ideal. Silages treated with ammonia also tend to have higher concentrations of acetic acid than untreated silage, because the fermentation is prolonged by the addition of the ammonia that raises pH.

A new microbial inoculant (*Lactobacillus buchneri*) designed for improving the aerobic stability of silages causes higher than normal concentrations of acetic acid in silages. However, production of acetic acid from this organism should not be mistaken for a poor fermentation and feeding treated silages with a high concentration of acetic acid does not appear to cause negative effects on animal intake.

Propionic Acid

Most silage contains very low concentrations of propionic acid (< 0.2 to 0.3 per cent) unless the silage is very wet (< 25 per cent DM). In silages with more typical concentrations of DM (35 to 45 per cent DM), concentrations of propionic acid may be undetectable. Biological additives that theoretically increase the propionic acid concentration of silage usually contain bacteria from the Propionibacteria family. However, these organisms are usually unable to compete in normal silage environments and are thus, usually ineffective. Chemical additives containing propionic acid are more effective for increasing the concentration of this acid in silages. These additives can range markedly in their percentage of active ingredients but most mainstream products will increase the concentration of propionic acid at ensiling from 0 to about 0.15 to 0.30 per cent (DM basis) if added at 2 to 4 lb per ton of wet (~35 per cent DM) silage.

Butyric Acid

A high concentration of butyric acid (>0.5 per cent of DM) indicates that the silage poorly preserved and has undergone clostridial fermentation, which is one of the poorest fermentations. Silages high in butyric acid are usually low in nutritive value and have higher ADF and NDF levels because many of the soluble nutrients have been degraded. Such silages may also be high in concentrations of soluble proteins and may contain small protein compounds called amines that have sometimes shown to adversely affect animal performance. High concentrations of butyric acid have sometimes induced ketosis in lactating cows and because the energy value of silage is low, intake and production can suffer. As with other poor quality silages, total removal or dilution of the poor silage is advised.

Ammonia

Low values for ammonia-N indicate minimal breakdown of protein in the silage, usually because pH has fallen quickly to a low level in the silage. High concentrations of ammonia (>10 to 15 per cent of CP) are a result of excessive protein breakdown in the silo caused by a slow drop in pH or clostridial action. In general, wetter silages have higher concentrations of ammonia. Extremely wet silages (< 30 per cent DM) have even higher ammonia concentrations because of the potential for clostridial fermentation. Silages packed too loosely and filled too slowly also tend to have high ammonia concentrations. Often times, silage with high concentrations of ammonia coupled with butyric acid may also have significant concentrations of other undesirable end products, such as amines, that may reduce animal performance

Ethanol

High concentrations of ethanol are usually an indicator of excessive metabolism by yeasts. Dry matter recovery is usually worse in silages with large numbers of yeasts. These silages are also usually very prone to spoilage when the silage is exposed to air. Usual amounts of ethanol in silages are low (< 1 to 2 per cent of DM). Extremely high amounts of ethanol (> 3 to 4 per cent of DM) in silages may cause off flavors in milk.

Silage Additives

Silage fermentation is a dynamic process that is affected by variety of factors. The primary goal of making silage is to maximize the preservation of original nutrients in the forage crop for feeding at a later date. Unfortunately, fermentation in the silo is a very uncontrolled process usually leading to less than optimal preservation of nutrients. In order to assist in the fermentation process, various silage additives have been used to improve the nutrient and energy recovery in silage, often with subsequent improvements in animal performance (Bolsen *et al.*, 1996). Their main functions are to either increase nutritional value of silage or improve fermentation so that storage losses are reduced. The benefits obtained from silage additives depend upon their influence on the silage fermentation process. These benefits are usually measured by the reduction in fermentation losses and/or improvement in silage quality and feeding value.

Silage additives or preservatives function in the following ways:

- Alter the rate, amount and kind of acid production
- Acidify the silage
- Inhibit bacterial and mold growth
- Culture silage (inoculants) to stimulate acid production
- Increase nutrients content of the silage

Silage additives have been classified into various categories that generally include (1) stimulants of fermentation (microbial inoculants, enzymes, fermentable substrates), (2) inhibitors of fermentation (acids, other preservatives), and (3) nutrient additives (ammonia, urea, minerals).

Fermentation Stimulants

This group of additives operates by encouraging the fermentation. Silage fermentation can be sub-optimal due to a lack of sufficient numbers of suitable lactic acid bacteria or a lack of sufficient amounts of suitable water soluble carbohydrates, or both. Stimulation of fermentation can be achieved at least in three ways: by (a). The addition of efficient lactic acid producing micro-organism (b). The addition of a substrate which will promote a lactic acid fermentation or (c). The enzymatic breakdown of forage plant components rich in fermentable substrate,

Microbial Inoculants

Since silage is a product resulting from the action of bacterial enzymes on the material stored, attempts have been made to alter or regulate silage fermentation through the addition of bacteria, yeasts and molds. It is known that growing crops are often poor source of lactic acid bacteria and that some strains of theses organisms are not ideally suited for ensiling purposes. Therefore, adding suitable lactic acid bacteria can accelerate and improve the ensiling process. The primary purpose for adding bacterial inoculants is to increase the number of lactic acid-producing bacteria, thus encouraging more lactic acid production and well-preserved forage mass. Lactic acid bacteria, propionic acid bacteria and yeast are major microbial inoculants which used as fermentation stimulant.

Efficacy of microbial inoculant depends upon the inoculation rate as well as the presence of adequate concentrations of sugar. The organism(s) from microbial inoculants must be present in sufficient numbers to effectively dominate the fermentation. The minimum number of lactic acid bacteria should be at least 100 000 colony-forming units per gram of fresh crop (Kaiser and Weiss, 1997). Consequently, a number of commercial inoculants containing freeze-dried cultures of homofermentative lactic acid bacteria have become available for application to harvested forage.

Propionibacteria are also used as silage inoculant which are able to convert lactic acid and glucose to acetic and propionic acids that are more antifungal than lactic acid. Addition of *propionibacteria* improves aerobic stability and prevents the growth of molds and yeast in high moisture silage (Florez-Galaraza *et al.*, 1985; Bolsen *et al.*, 1996).

Fermentable Substrates

Molasses, which is a by-product of the sugar cane and sugar beet industries, can be used as a source of water soluble carbohydrates in silage. Molasses is an effective silage additive in terms of promoting lactic fermentation, reducing silage pH, discouraging a clostridial fermentation and proteolysis, and generally decreasing organic matter losses. It is of particular benefit when applied to forage crops low in fermentable carbohydrates for lactobacilli.

Enzymes

A variety of enzymes, particularly those that digest plant fiber and starch have used as silages additives. Forage plants have a large reserve of carbohydrate as polysaccharides. Enzymes are added to silage to degrade polysaccharides in cell wall of plants to fermentable water-soluble carbohydrates for use by lactic acid bacteria because these organisms cannot use polysaccharides as an energy source to make lactic acid. Cellulases and hemicellulases are the most widely used enzyme additives.

Nutrient Additives

Certain crops are deficient in dietary components essential for ruminants. The nutritional quality of these crops can be improved by supplementation with specific additives at the time of ensiling. Materials such as cereal grains, molasses, dry forages, limestone, urea and anhydrous ammonia are examples of nutrient additions to silage.

Nitrogen Additives

Urea and anhydrous ammonia, non-protein nitrogen (NPN) sources, are sometimes used as nitrogen addivies for silage. Ammonia additions can result in addition of an economical source of crude protein (Huber *et al.*, 1979; McDonald *et al.*, 1991). The recommended application rates for anhydrous ammonia is 7-9 lbs./ton of wet silage. A benefit of adding NPN to corn silage includes increasing the CP content by 2-6 percentage units. Ammoniated or urea-treated corn silage usually has CP concentrations of 12-13 per cent (DM basis). This can result in substantial savings by reducing the need for protein supplementation. Another benefit of adding NPN to silage crops is that it reduces plant protein destruction during fermentation.

Energy Additives

Cereal grains are another source of carbohydrates. Feeds such as corn and small grains can be added to forage at the time of ensiling. Addition of grain to corn silage is not useful, but adding it to hay crop silage has two benefits. First, adding grain to hay crop silage increases the energy content of the silage. This will reduce the amount of supplemental grain that has to be fed. If silage will be the main or only feed offered, then adding some grain to the forage at ensiling will make it a more complete feed. However, grain mixed with silage prior to ensiling or at feeding is nutritionally equal; therefore, if supplemental grain must be fed anyway, no true benefit is realized. Secondly, adding grain to forage will increase the dry matter content of the silage. Hay crops that are not wilted sufficiently prior to ensiling can cause seepage and result in an undesirable fermentation.

Minerals

Minerals such as calcium, phosphorous, sulfur and magnesium have been added to forage at the time of ensiling. Usually these either have no effect on fermentation or act as buffers resulting in higher pH silage. The only reason for adding minerals at the time of ensiling is if the silage will be the only feed offered to the animals. Addition of minerals will make the silage more nutritionally complete. If concentrates are going to be supplemented, it is better to add the minerals to the concentrate mix. Limestone (calcium carbonate) and $MgSO_4$ to increase the calcium and magnesium contents is sometimes added @ 10 to 20 pounds per ton of corn silage to increase the calcium content and extend the fermentation process.

Fermentation Inhibitors

Inorganic Acids and the AIV Process

A number of chemical compounds have been tested as potential fermentation inhibitors. Acids are added to forage at ensiling to cause an immediate drop in pH, increase bunk life and limit fermentation losses of protein and carbohydrates. One of the earliest was mixture of mineral acids proposed by A.I.Virtanen, the technique being referred to as A.I.V process. Usually hydrochloric acid and sulphuric acid were added to the herbage during ensiling in sufficient quantity to lower the pH value below 4.0. These acids are extremely caustic and hazardous to use.

Organic Acid

To inhibit aerobic spoilage, spoilage organisms (*i.e.* yeasts, acetic acid bacteria) have to be inhibited in their activity and growth. Some additives that have proven to be effective in this respect include chemical additives such as propionic acid, acetic acid, formic acid, sorbic acid and benzoic acid (McDonald *et al.*, 1991; Phillip and Fellner, 1992; Weinberg and Muck, 1996). The major benefit of adding weak acids to silage appears to be in reducing spoilage in open storage structures.

Propionic Acid

Of the short-chain fatty acids, propionic acid has the greatest antimycotic activity. It is effective in reducing yeast and molds which are responsible for aerobic deterioration in silages. The antimycotic effect of propionic acid is enhanced as pH declines, making it an ideal candidate for improving the aerobic stability of corn silage where pH is low. In the past, aerobic stability was improved when large amounts of propionic acid (1 to 2 per cent of the DM) were added to silage, but the high percentage of acid restricted fermentation in these cases. The application rate of propionic acid additives depends on moisture content of the forage, length of storage and formulation with other preservatives. For corn silage, propionic acid at usage rates of 0.2 to 0.5 per cent is effective. Propionic acid is also difficult to handle because it is corrosive. Thus, salts of acids, *e.g.*, calcium, sodium and ammonium propionate is the most popular fermentation inhibitors.

Formic Acid

Formic acid is widely used to restrict fermentation by artificially reducing pH to below 4.0. Formic acid is added to silages at 0.45 per cent of the wet weight or 2.25 percent of the dry matter weight (Woolford, 1975; McDonald *et al.*, 1991).

Acetic Acid

Acetic acid also inhibits aerobic spoilage. Acetic acid-type preservative is sodium diacetate. This is mixture of acetic acid and its sodium salt. Recommended rate of application is 1-2 lbs. active ingredient/ton of wet silage. This compound is quite effective in reducing top spoilage. It was established that at pH 5, acetic acid @ 20 per cent of dry matter content of crop would inhibit endospore forming bacteria.

Lactic Acid

Lactic acid is fairly effective inhibitor of the endospore forming bacteria, particularly at pH5.

Advantages of Silage

1. It can be efficiently used for strategic off-season feeding.
2. It is a means of increasing feed resource availability and a form of insurance, especially for calving dairy cows.
3. It can be fed to reduce pressure on pasture when required.
4. It can be an efficient supplement to grazing cattle during the dry season.
5. It is an inexpensive home-made feed, resulting in the animal production at lower cost.
6. It improves palatability, reduces significantly toxic substances present in some fresh vegetables to safe level concentrations (such as cyanogenic glucosides in fresh cassava leaves) and destroys harmful micro-organisms possibly present in poultry litter or fish wastes.
7. It can provide a major diet source, as basal ration as well as a feed supplement for grazing animals.

References

Bolsen, K. K., D. R. Bonilla., G. L. Huck, M. A. Young, R. A. Thakur, and A. Joyeaux. 1996. Effect of a propionic acid bacterial inoculant on fermentation and aerobic stability of whole-plant corn silage. *J. Anim. Sci.* 74(Suppl. 1):274.

Flores-Galaraza, R. O., B. A. Glatz, C. J. Bern, and L. D. Van Fossen. 1985. Preservation of high moisture corn by microbial fermentation. *J. Food Protection*. 48: 407-411.

Gibson, T., Stirling, A.C., Keddie, R.M., and Rosenberger, R.F. 1958. Bacteriological changes in silage made at controlled temperatures. *J. Gen. Microbiol.*, 19: 112-129.

Gibson, J. 1965. Clostridia in silage. *J. Appl. Bacteriol.*, 28: 56-62.

Huber, J. T., J. Foldager, and N. E. Smith. 1979. Nitrogen distribution in corn silage treated with varying levels of ammonia. *J. Anim. Sci.*, 48:1509.

Kaiser, E., and Weiss, K. 1997. Fermentation process during the ensiling of green forage low in nitrate. 2. Fermentation process after supplementation of nitrate, nitrite, lactic acid bacteria and formic acid. *Arch. Anim. Nutr.*, 50: 187-200.

Lindgren, S., Petterson, K., Kaspersson, A., Jonsson, A., and Lingvall, P. 1985. Microbial dynamics during aerobic deterioration of silages. *J. Sci. Food Agr.*, 36: 765-774.

McDonald, P., Henderson, A.R., and Heron, S.J.E. 1991. *The Biochemistry of Silage.* 2nd ed. Chalcombe Publications, Marlow, UK.

Phillip, L.E., and Fellner, V. 1992. Effects of bacterial inoculation of high-moisture ear corn on its aerobic stability, digestion, and utilization for growth by beef steers. *J. Anim. Sci.*, 70: 3178-3187.

Randby, A.T., Selmer-Olsen, I., and Baevre, L. 1999. Effect of ethanol in feed on milk flavour and chemical composition. *J. Dairy. Sci.*, 82: 420-428.

Schlegel, H.G. 1987. *General Microbiology*. 6th ed. Cambridge University Press, Cambridge, UK.

Weinberg, Z.G., and Muck, R.E. 1996. New trends and opportunities in the development and use of inoculants for silage. *FEMS Microbiol. Rev.*, 19: 53-68.

Woolford, M.K. 1975a. Microbiological screening of the straight chain fatty acids (C_1-C_{12}) as potential silage additives. *J. Sci. Food Agr.*, 26: 219-228.

Chapter 28

Modern Techniques of Conservation of Forages: Hay

*P.K. Singh[1] and Amitava Dey[2]**

[1]*Department of Animal Nutrition, Bihar Veterinary College, Patna – 14*
[2] *ICAR Research Complex for Eastern Region, Patna*
**E-mail: amitavdey_icar@yahoo.co.in*

Hay refers to forage that are harvested and dried and stored as 85-90 per cent dry matter. Hay is a leafy dry fodder, green in colour, and free from moulds. It should contain less than 15 per cent moisture.

Principles of Hay Making

The aim of hay making is to reduce the moisture level of the green crop to a level low enough so that it can be safely be stored in mass without undergoing fermentation or becoming mouldy. Drying of forage inhibits the action of plant and microbial enzyme. The process of drying the green crop without significant change in aroma, flavour and nutritive quality of forage is called "curing". This involves reducing the moisture content of green forages, so that they can be stored without spoilage or further nutrient loss. Green forage with 80-85 per cent dry matter preserves most of the nutrients.

Suitable Crops for Making Hay

Crops with thin stems and more leaves are better suited for hay making as they dry faster than those with thick stem and small leaves.

Steps of Hay Making

Harvest at Proper Stage of Plant Maturity

A good forage crop is only the first requisite in producing a quality hay crop. The way forage is harvested and stored determines how well the quality of the standing crop is preserved. Forage needs to be cut before it is fully mature (long before it has seeded) to maximize its nutritive value. As plants mature, their lignin content (a component of fiber) increases and traps the nutrients within indigestible cell walls. Leguminous fodder crops (*e.g.* Cow pea, Berseem, Lucerne, etc) should be harvested at the flower initiation stage or when crown buds start to grow, while grasses (*e.g.* oats) and similar fodder crops should be harvested at the pre-flowering stage. At this stage, the crop has maximum nutrients and green matter. After flowering and seeding, grasses contain fewer nutrients.

In order to make the process of curing easier, the fodder should preferably be harvested when air humidity is low. Cut the fodder in the evening (but do not spread) or in the early morning. Natural physiological processes in plants cause the concentrations of soluble carbohydrates and other highly digestible nutrients to peak in the evening. Recent research suggests that hay cut at or near sundown is higher in energy than hay harvested at sunup.

Dry the Forage at 15 to 20 Per cent Moisture Content

When the forage is cut, it needs to be laid out in the sun in as thin a layer as possible, and it should be raked a few times and turned regularly to hasten drying. If there is sufficient labour, then chopping the forage after it has been cut will hasten drying. Leaves are more nutritious than the stems, and so when cutting forage, it is important that it is cut with as much leaf and as little stem as possible. However, during drying, the leaf (being more brittle) will tend to shatter. Hay should therefore be handled with care, to try and minimize the amount of leaf that is lost in this way. The dried forage is then collected and baled when the moisture content is low, ideally less than 15 per cent. This helps storage and requires less space. Crops with thick and juicy stems can be dried after chaffing to speed up the drying process and to prevent loss of nutrients.

The rate of drying is accelerated by low relative humidity, high air temperature and good air movement around the cut forage. Since leaves of cut herbages lose water more rapidly than stems, mechanical conditioning (crushing or crimping) also can reduce the time required for curing. However, this effect generally is greater for legumes than grasses. Reducing the curing time is of critical importance in haymaking. Field curing is conducted during bright sunny weather. Spread the cut fodder right after the dew has dried up and the sun begins to shine. However, this may result in bleaching of the forage and loss of leaves due to shattering. One can also dry fodder on a fence, tripod or any similar structure, but this method is rather laborious. One must heap up the hay at night to protect it from dew, and spread it twice on the first day of drying and at least once during the succeeding days. To avoid this, drying can be done in barns by passing hot air through the forage. Although artificial drying produces hay of good quality, it is expensive and beyond the reach of small and marginal farmer but can be attempted on a community basis in areas where there is a need, and the necessary facilities.

Store on a Well-Drained Site

Well dried hay should be stored properly for feeding to livestock. Hay should not be stored until it has dried completely, since wet or moist hay (in addition to encouraging the growth of moulds) may also ferment. The heat produced during fermentation can be a fire risk. Hay must be stored in a dry environment. Hay can also be stored by creating hay stacks. These may be created in a field near the source, or close to where the hay will be required later in the year. Stacks may be covered by plastic sheets to keep out rain. The surface layer of a stack may also be "thatched", in the same manner as a thatched roof to a house. Hay of similar quality should be stored together. High-quality hay should be stored inside or should be protected from the weather and raised off the ground on old tires, poles, or crushed rock. Break direct contact between damp soil and hay using rock, tires or poles.

Hay Quality

Hay quality really means feed value. The ultimate test of hay quality is animal performance. Quality can be considered satisfactory when animals consuming the hay give the desired performance.

Factors Affecting Hay Quality

Two principal methods can be used to determine forage quality. The first is to have a chemical analysis. Where a chemical analysis is not available, the second method of hay quality evaluation involves a visual inspection of hay and a judgement based on certain physical characteristics of forage. Hay is judged visually for relative feeding value and overall desirability as a feed. The factors known to influence hay quality and animal performance include the following: (1) stage of maturity at harvest, (2) leafiness, (3) color, (4) foreign material, and (5) odor and condition.

Stage of Maturity

Stage of maturity refers to a plant's stage of development at the time it is harvested. Of all the factors affecting hay quality, stage of maturity at the time of harvest is the important factor influencing quality. Younger the plant, better the quality;. As legumes and grasses advance from the vegetative to reproductive (seed) stage, they become higher in fiber and lignin content and lower in protein content, digestibility, and acceptability to livestock. Determining the maturity of legumes and grasses is easy before harvesting but becomes more difficult after cutting and baling. Weathering or sun bleaching after cutting or the delay of normal development of legume flowers due to cool, cloudy weather, especially with first cutting, further complicate the determination of maturity. For top quality hay, legumes should be cut at the 20 per cent bloom stage. Grasses should be cut in the fully headed to early bloom stage. Most forages will have a 20 per cent loss in TDN (total digestible nutrients) and a 40 per cent loss in protein by a delay of only 10 days past the most desirable stage of harvest.

Leafiness

Leafiness is an excellent indicator of hay quality. This refers to the ratio of leaves to stems present and is also related to the stage of maturity, especially in grass hays. Younger the plant, the higher the proportion of leaves. As a grass matures, stems increase, thus decreasing the quality of the forage. Leafiness in legumes is particularly critical because legumes lose their leaves during curing and handling more readily than grasses. A high percentage of leaves also indicate good harvest and handling methods. In general, hay with high leaf content has a higher percentage of minerals and vitamins and a greater energy value than hay with few leaves. As legumes and non-leguminous forages advance in maturity, the stems mass increases, lower leaves fall from the plant, and the leaf-to-stem mass ratio decreases. The method of curing of hay, the method of handling it from field to storage, and the weather conditions during curing and baling also influence leafiness. Leaves shattered from the stems and loose in the bales may be wasted when the hay is fed. To preserve leafiness, hay must be cut early and carefully cured and handled.

Colour

Color can be a definitive characteristic of hay, which tells more about the curing process of the hay than its quality. The most desirable hay color is the bright green of the immature crop in the field. Hays having this bright green color were most likely cut at an early, desirable stage of maturity and were well and rapidly cured, with no damage from rain, molds, or overheating during storage. Green hay is rich in carotene, whereas, straw-colored or brown hay is very poor in carotene. Yellow color is often a result of sun bleaching and does not seriously reduce quality. Dark brown or black is often an indicator that the hay was exposed to rain or high humidity and is usually accompanied by a distinctive musty odor. Overall, slight discolorations from sun bleaching, dew or moderate fermentation are not as serious as the loss of green color from maturity, rain damage or excessive heating or fermentation. Slight discolorations from sun bleaching, dew, or moderate fermentation are not as serious as the loss of green color from maturity, rain damage, or excessive fermentation or heating.

Golden Yellow Coloured Hay/Sun-Bleached Hay

Sun-bleached hay is light golden yellow in colour. Sun bleaching reduces palatability and carotene content. In general, the amount of carotene in hay is directly proportional to its degree of greenness.

Yellow Coloured Hay

Yellowing, especially in grass hay, usually indicates that the plants were over-matured when cut. Grasses with fully ripe seeds usually have yellowish-brown stems and heads and many brown leaves. Yellowing due to maturing can be distinguished from sun bleaching because all of the plants, rather than just those on the outside, have the same yellowish color.

Dark Brown Coloured or Rain Damaged Hay

Hay that has been exposed to rain or to heavy dews or fog has a characteristic dark brown or black appearance. Rain falling on hay between the time of cutting and baling can leach out soluble nutrients and reactivate plant respiration. Hard rain can also shatter leaves, especially if the rain occurs when the hay is nearly dry. The stems of hay that has been sun-bleached or discolored by rain are usually harsh and brittle. Avoiding rain damage is a goal in haymaking but delayed cutting must be balanced against the lower feeding value of late cut hay.

Brown Coloured Hay/Heat Damaged Hay

Brown hay indicates heating from microbial (mold) growth and fermentation. This results when the hay is stored at too high a moisture content. This hay has a distinctive musty, moldy odor and when the bale is opened the flakes are often caked and show visible mold. Molds consume nutrients in the hay, particularly sugars and starches, producing carbon dioxide and water. Extensive heating, as with brown hay, results in considerable loss of dry matter, digestible protein, and energy and destroys much of the carotene and other vitamins. Slight discoloration from sun bleaching, dew, or moderate fermentation is not as serious as the loss of green from maturity, rain damage, or excessive fermentation.

Foreign Materials

Foreign materials in hay can be divided into injurious and non injurious categories. Injurious foreign material is material that will harm the animal if eaten. This includes poisonous plants and matter such as wire or nails. Non-injurious foreign material is matter that is commonly wasted in feeding operations but is not harmful to livestock if eaten. This includes weeds, stubble, chaff, and sticks. Weeds are the most common non-injurious foreign material found in hay. Livestock do not relish most weeds and if they are eaten they usually have little or no feed value. Weed seeds usually pass undigested through livestock and when the manure is spread on land it becomes a source of weed infestation.

Smell

The smell of new mown hay is the standard by which hay odor is judged. A fresh aroma, freedom from mold and relatively high carotene content add to its palatability and feed value. Mildew, mustiness, or a putrefied (rotten) odor result from weather damage or insufficient drying before baling and indicate lower quality hay. Odour problems usually result in lower acceptability by livestock. Hay should be free from dust, mold, insect and disease damage.

Characteristics of Good Quality Hay

1. Hay should be nutritious therefore prepared form plants cut at an appropriate stage of maturity when it has the maximum nutrients.
2. Good hay should be leafy. The leaves are generally richer in proteins, minerals and vitamins then other parts of the plant.
3. Hay should be green in colour. The green colour indicates the amount of carotene which is precursor of vitamin A.
4. Hay should be soft and pliable
5. Hay should be free from weeds and stubbles.
6. The moisture content in hay shouldn't exceed 15 per cent.

7. It should have the smell of and aroma of characteristics of the crop from which it is made.

Quality Characteristics of Hay

Excellent Quality

Legumes cut in the late bud to early bloom stage for alfalfa or 20 per cent bloom stage for clovers; grasses cut in the boot stage. Hay is bright green, leafy, and free from mold or mustiness. This feed is high in carotene, proteins, minerals, and energy and low in fiber.

Good Quality

Legumes cut by half bloom; grasses cut in the early heading stage. Good quality hay is leafy, has good green color and is free from mold or mustiness. Hay that was rained on after partial curing can fall into this category if it was cut early. The nutritional value is not as high as excellent quality hay but good results can be obtained if it is properly supplemented.

Fair Quality

Legumes or grasses cut at full bloom. Fair quality hay lacks greenness and is stemmy and low in carotene, minerals, proteins, and energy and high in fiber.

Poor Quality

Any legume or grass cut after full bloom. Poor quality hay is stemmy and lacks leaves. It may be severely weather damaged, bleached, musty or moldy.

Benefits of Hay Making

Hay is one of the most versatile of stored forages in that

1. It can be kept for long periods of time with little loss of nutrients if protected from weather;
2. A large number of crops can be successfully used for hay production;
3. It can be produced and fed in small or large amounts;
4. It can be harvested, stored and fed by hand or the production and feeding can be completely mechanized;
5. Hay can supply most nutrients needed by many classes of livestock.

Hay is, therefore, the most commonly used stored feed on most farms.

Hay Preservatives

Storage of hay at higher moisture contents may lead to some heating due to the activity of aerobic micro-organisms and possibly some plant respiration. The warm, moist conditions in hay will provide the ideal environment for growth of spoilage bacteria, *e.g.* Bacilli and yeasts, moulds and fungi. These organisms utilize the energy and protein of the hay and can lead to a substantial increase in their respective populations. Their action leads to the following reaction:

Hay (plant sugars) + oxygen ⟶ Carbon dioxide + water + heat

The resultant heating causes a reaction between the carbohydrates and proteins rendering both less digestible as temperatures continue to increase.

Hay preservatives inhibit or reduce the growth and activity of these aerobic micro-organisms (bacteria, yeast and moulds) in the hay so stops production of the water and heat. Moulding and heating is avoided, so negligible nutritive value is lost. Most hay preservatives do not improve the nutritional quality of the forage, but do prevent the decline in quality caused by the micro-organisms. Hay preservatives allow the conservation of hay at higher moisture (of up to 35 per cent) thereby reducing drying times and leaf shattering losses associated with handling of dry forage. The use of hay preservatives permits greater flexibility in hay making operations.

Types of Hay Preservatives

Hay preservatives have been classed into several types. These are: organic acids and their salts, bacterial inoculants, sulphur based preservatives, ammonia-based additives and common salt. Some products may also include enzymes, antioxidants and nutrients.

Organic Acids and their Salts

Organic acids have been the most widely accepted hay preservatives. Organic acids act as fungicides by producing an acid environment which is not conducive to mould, yeast or bacterial growth. Propionic and acetic and being naturally occurring acids in the rumen, are safe for all types of livestock, including horses. The propionic acid is highly effective against mould growth whilst acetic acid is more effective against bacteria and yeast. Walgenbach (1989) indicates that the expected results of propionic acid on hay depends upon the complex relationships between: (a) the level of inhibitory, free propionic acid, (b) the level of the non-inhibitory, ionized, propionate form of the acid, (c) the buffering capacity of the crop and (d) the hay microbial populations, some of which are capable of metabolizing the protective, free acid form.

These products are corrosive on machinery and can be dangerous for operators to use in their pure form. To overcome these problems, "buffered" acids, some times referred to as "neutralised or pH balanced" acids, have been developed and commonly include salts of propionic, acetic and formic acids. Certain precautions should be observed in using organic acid-based preservatives. Protective clothing should be used in mixing or transferring the material, and water should be available at all times to flush affected areas if an accident occurs. Equipment surfaces and applicator systems should be thoroughly washed and flushed immediately after use to prevent excessive corrosion. Treated hay should be stored under protective cover because the preservatives can be leached or diluted easily on outer surfaces by rainfall. Preserved hay should not be stored in contact with normally field-cured hay, since the drier hay can absorb moisture, making mold development possible. Although hays containing moisture levels higher than 35 per cent moisture can be effectively preserved with these materials, the practice is not recommended because of preservative costs and difficulty of handling wet bales.

Bacterial Inoculants

Some silage inoculants developed for silage use are also effective for hay preservation. There are three basic groups of bacterial or microbial inoculants products used as hay preservatives:

1. Fermentation enhancers
2. Aerobic spoilage inhibitors
3. Antibiotic-producing, bacterial inoculants

Fermentation Enhancers

Some silage inoculants developed as fermentation enhancers are also effective for hay preservation. Most contain lactic acid producing bacteria that compete with yeasts and mould forming organisms aiming to maintain forage quality. Commonly used lactic acid bacteria are *Lactobacillus, Pediococcus* and *Streptococcus.* When using this class of inoculants on hay, the plant sugars must be high (substrate for the bacteria) to provide the best protection against moulds and yeasts. For pasture plants this usually occurs when they are vegetative, in mid to late spring.

Aerobic Spoilage Inhibitors

These recently developed inoculants use *Lactobacillus buchneri,* a group of bacteria which restrict the growth of spoilage type organisms such as yeasts and moulds. Their mode of action is thought to be due to secondary metabolites called "buchnericides" and may reduce the need for paddock curing unless extremely wet. Recent research indicates that they are usually much more effective than the fermentation enhancer group.

Antibiotic-producing Bacterial Inoculants

Another group of bacteria is *Bacillus amyloliquefaciens.* These bacteria produce antimicrobial compounds which inhibit, sometimes stopping, the growth and reproduction of moulds and other spoilage micro-organisms. These bacteria also increase the rate of water loss which makes hay less conducive to spoilage organisms. The increased drying rate reduces the rate and amount of plant respiration, resulting in less nutrient loss and reduces the period for population growth of spoilage organisms.

Sulphur Based Preservatives

Sulphur based preservatives are another product that control of microbial proliferation. Sulphur compounds scavenge oxygen. In fodder, they create an environment within the hay bale that is unconducive to microbial growth stemming mould and yeast development. The application rate at 0.8-1.0 kg per tonne is much lower then with acids. Sulphur compounds are not acidic, and as such are generally fairly user friendly.

Ammonia Based Preservatives

Anhydrous Ammonia

Anhydrous ammonia has fungicidal properties and has been used successfully in the preservation of high-moisture hays. Use of 1 per cent anhydrous ammonia (dry matter basis) to hay containing up to 30 per cent moisture has been shown to reduce storage dry matter losses and prevent heating and mold development. Increased crude protein content is an additional benefit of ammonia preservation. However, this method of chemical preservation has not received wide acceptance because application of anhydrous ammonia to large volumes of hay is difficult and it is a hazardous chemical.

Urea

Dry urea could be used as an alternative to anhydrous ammonia in preserving high-moisture hays and increasing the crude protein content of poor-quality hays. Applying urea, which is then converted to ammonia by the bacteria, is simpler than applying ammonia gas. Relatively large amounts (5 to 7 per cent) of urea applied during baling can be effective up to 30 per cent moisture. However, the treated hay must be covered tightly with plastic sheeting as soon as possible after baling.

Common Salt

The use of salt (NaCl) on wet hay does have a biological basis in that sufficient concentrations will absorb free water on the surface of the hay and thereby inhibit microbial growth. Common salt has often been used between layers of moist hay to both preserve the hay and improve its palatability. However, it is ineffective as a preservative unless applied in such amounts as to be physiologically harmful to animals. The salt itself would be detrimental to microbial and mould growth, and being hygroscopic in nature, absorbs moisture in proximity to the salt granules. The placing of salt between layers of large rectangular bales will only be effective near the salt itself, possibly having an effect of 1 to 3 cm into the bale itself. The only way salt may be effective would be if the salt was spread throughout forage in the bale (Lacefield, 1987).

Hay Desiccants/Drying Agents

Hay desiccants are used to reduce the length of time needed for hay drying. The chemical drying agent (potassium or sodium carbonate, sodium silicate and citric acid) are applied to the stem of the alfalfa plant at the time of mowing. These naturally-occurring salts reduce drying time by acting on the moisture-conserving, waxy-cutin layer of forage (Rohweder *et al.*, 1983). These chemicals are most effective when applied at 5-7 pounds of active ingredient in 30 gallons of water per acre. This allows the stem to dry faster and can reduce the interval between cutting and baling. Hay desiccants may decrease time needed for drying by one third. Hay desiccants work the best in good drying conditions and are less effective in poor drying conditions. Leaching losses may be greater in desiccant-treated alfalfa if rained on before baling.

Losses during Hay Making

Chemical changes resulting in losses of nutrients arise during the drying process.

Respiration Losses

Even after harvesting, plant cells alive and functioning until the moisture content falls below 35 to 40 per cent. Sugars are the primary plant carbohydrates lost during storage respiration. Hay that dries quickly will lose 2 to 6 per cent dry matter due to respiration. Hay that dries very slowly may lose 15 per cent dry matter due to respiration. During extended storage periods, microbial and fungal respiration reduces oxygen levels and produce carbon dioxide, water, and heat, causing the reduction in dry matter. Dry matter losses could exceed 10 percent if moisture levels are between 20 and 30 per cent (Klinner and Shepperson, 1975). Harvesting hay when good drying weather is expected will reduce respiration losses considerably.

Management practices that shorten drying time resulting in reduced respiration and harvest losses include: (a) cutting early in the day to maximize solar drying, (b) cutting when anticipated weather will allow for relative humidity of the air to be below the equilibrium humidity of the forage, (c) the use of mechanical or chemical conditioning to reduce the cuticular resistance to water escape and (d) maximizing hay exposure to wind and sunlight by creating wide and thin windrows.

Losses by Leaching

The uncertainty of weather conditions always makes haymaking difficult. If hay is almost cured and is exposed to heavy and prolonged rains, especially when it is in the field, loss of soluble carbohydrate, minerals, water soluble vitamins and nitrogenous constituents through leaching. Leaching by rain can cause up to 20 per cent nutrient loss. Rain may prolong the enzyme action within the cell cells, thus causing greater losses of soluble nutrients, and may also encourage the growth of molds.

Losses by Shattering

During the drying process the leaves loose moisture more rapidly than the stems, so becoming brittle and easily shattered by handling. The loss due to shattering of leaves is in hay making is of importance, especially in the case of legumes. The leaves at the hay stage are richer in digestible nutrients then the stems and hence losses by shattering of leaves decrease the nutritive value of hay. Hay moisture content is the largest single factor contributing to leaf loss. When the hay becomes too dry and brittle and losses become excessive. To avoid these losses, hay should never be over dried or handled during warm periods of the hay. Hay stored at moisture content above 15 per cent has much less leaf loss than hay stored below 15 per cent moisture. If hay becomes too dry, stop baling and resume in the evening or morning when the leaf moisture level increases. This dew-moistened hay can be stored at a slightly higher moisture level than when it was drying down because dew moisture in the hay is more easily released during curing than internal moisture.

The upper moisture limit to prevent mold growth and other hay deterioration for large round alfalfa bales is typically 18 to 20 per cent. Hay stored above 25 per cent moisture will usually spoil unless chemical preservatives are added to the hay. Hay that is put up too wet also increases the chance of fire, especially when stored indoors.

Losses by Bleaching

During the process of drying, much of the carotene, a precursor of vitamin A is lost with bleaching. In general, the carotene content of freshly cured hay is proportional to the greenness. With severe bleaching, more than 90 per cent carotene may be lost.

Storage Losses

If hay is baled at higher moistures and not protected by a preservative or inoculant, heating may occur. The first temperature peak will generally occur within a few days and can be the result of aerobic bacterial growth, fungal growth and/or plant respiration. If oxygen and a favorable moisture level are available, microorganisms begin to multiply, generating heat up to 130 to 140° F. The hay may, however, undergo several heating cycles during the next few weeks as various populations of microorganisms increase and decrease. Eventually the temperature will stabilize near ambient temperature (Prather, 1988). Aerobic fungi are the primary microbes responsible for the breakdown of complex carbohydrate and subsequent generation of heat (Martin, 1980; Tomes, 1989).

The primary nutritional losses that occur during storage are due to microbial growth and the subsequent heating. Excessive heat damage can reduce protein and energy digestibility of the hay. Heat damaged protein is measured by determining the nitrogen content of the fiber in tests such as the Acid Detergent Insoluble Nitrogen (ADIN) analysis. It is generally felt that excessive heat damage has occurred when ADIN approaches 10 per cent or more of the total nitrogen (Ricketts *et al.*, 1982; Shelford, 1983)).

The amount of storage losses are directly related to several factors *viz.* moisture content at baling and the time of storage, storage conditions, environmental conditions (relative humidity, air temperature, and air movement), forage species and the epiphytic microbial populations present on the hay (Tomes, 1989). Hay that is stored at moisture contents greater than 20 per cent can develop mold and lose dry matter and quality to bacterial degradation. Storage losses are directly related to microbial growth and to subsequent heating. All hay stored at moisture contents between 15 and 20 per cent will undergo some elevation in temperature in the first 2 to 3 weeks after baling. This heat buildup is referred to as "sweating" which leads to measurable losses of 4 to 5 per cent in hay DM.

Benefits of Using Hay Preservatives

Benefits of using hay preservatives are:

1. They prevent dry matter and quality loss in storage due to bacterial, yeast and mould activity.
2. They allow the safe storage of hay from slightly above target moisture levels up to 25 per cent (or 30 per cent) moisture depending on preservative type.
3. They allow baling after a shorter curing period which reduces risk of rain damage and sun bleaching. It may also allow baling earlier in a season in certain areas, when fodder is less mature and nutritive value higher.
4. They reduce dry matter and nutrient loss caused by leaf loss and shatter, microbial activity and moulds.
5. They maintain hay colour (due to increased leaf retention) and often smells better.
6. They reduce risk of spontaneous combustion.

7. They may increase animal intake.
8. Animal and human health is not affected due to lack of mould spores.

References

Klinner, W.E. and G. Shepperson. 1975. The State of Haymaking Technology: A Review. *J. Br. Grassland Society*. p. 303.

Lacefield, G.D. 1987. Why Salt Hay? *Hoard's Dairyman*. May 25, 1987. p. 463.

Martin, N.P. 1980. Harvesting and Storage of Quality Hay. Am. Forage and Grassland Council Proceedings. p. 177.

Prather, T.G. 1988. Hay Fires - Prevention and Control. University of Tennessee Agricultural Extension Publication PB 1306.

Ricketts, R., R. Belyea, H. Sewell and G. Garner. 1982. Understanding and Interpreting Feed Analysis Reports. University of Missouri Extension Publication.

Rohweder, D.A., M. Collins. M. Finner and R. Walgenbach. 1983. A Hay making System To Help Make "Hay in a Day". University of Wisconsin Agricultural Extension Publication.

Shelford, J.A. 1983. Personal Communication.Department of Animal Science, University of British Columbia.

Tomes, N.J. 1989. Personal Conversation.Research Microbiologist, Microbial Genetics, West Des Moines, IA.

Chapter 29

Forage and Fodder Production: An Economic View

K.M. Singh[1] and M.S. Meena[2]

[1]*Principal Scientist and Head,*
Division of Socio-Economic, Extension and Training,
ICAR Research Complex for Eastern Region,
ICAR Parisar (P.O.- B.V.College), Patna – 800 014, Bihar
E-mail: m.krishna.singh@gmail.com
[2]*Senior Scientist (Agricultural Extension)*
ICAR-Research Complex for Eastern Region, Patna (Bihar)
E-mail: ms101@sify.com

Introduction

More than 70 per cent Indian population live in rural areas and livestock are the imperative for their subsistence farming and sustainable livelihood. India alone has more than 30 per cent of the world's bovine population which has resulted not only egalitarian ownership of cattle but also an inseparable cultural and symbiotic relationship between rural families and their farm animals. The mixed crop-livestock systems are underpinned by the crop residues which contribute on an average 40-60 per cent of the total dry matter intake per livestock unit. Crop residues and pastures grazing are the major feed resource for this activity. Climatic, topographic, physiographic, altitude and related factors have influenced distribution of various crop and grass species, which determine the fodder and forages production both qualitatively and quantitatively. Although, livestock rearing is an important occupation of farmers since immemorial but the forage cultivation has remained almost neglected. Grazing in forest areas and pastures is the mainstay for livestock but fodder trees and shrubs also contribute significantly. Now-a-days, reckless cutting of trees, indiscriminate use of grazing areas and absence of rehabilitation programmes has lead to denudation of hill slopes, which has resulted in critically low biomass availability and adverse effects on livestock rearing. Consequently the livestock productivity is very low. There is however considerable regional variation in dominant type of crop residue; rice and wheat straws in irrigated regions compared to coarse cereal straws and hay from leguminous crops in the arid and semi-arid regions. While the nutritive value of fodder from dual-purpose crops can be determined.

Why Should we Consider Forages?

Benefits of Forages/fodders in Crop Rotation : Forages can be a simple answer to soil erosion and decline in organic matter and fertility, a problem caused by modern cultivation and fallowing practices on much of the farmland world wide. Forages can also help to reduce nitrogen fertilizer costs and energy costs associated with applying nutrients. Many farmers are using forages for positive results on any land, but particularly on marginal crop land. Forage also requires fewer cash inputs than most grain crops. Hence, the numerous benefits in any situation include:

- Increased soil fertility when legumes are used
- Increased soil quality
- Better water filtration and internal drainage
- Less disease in subsequent cereal crops
- Reduced weed populations
- Increased yields in subsequent crops
- Better economics in subsequent crops
- Greater and deeper carbon sequestering for greenhouse gas reduction.

Systems of Fodder Productior

The systems of fodder production vary from region to region, place to place and farmer to farmer. It depends on availability of inputs (*viz.*, fertilizers, irrigation, insecticides, pesticides, etc.) and topography. An ideal fodder system is that which gives the maximum outturn of digestible nutrients/hectare, or maximum livestock products from a unit area. It should also ensure the availability of succulent, palatable and nutritive fodder throughout the year. Some of the important intensive fodder-crop rotations and the expected yields are given in Table 29.1 for different regions.

Fodder Production for Intensive Dairy Farming

The requisites for intensive dairy-farming are with the reasons that (*i*) fodder is required in uniform quantity throughout the year (*ii*) fodder crops in the rotation should be high-yielding (*iii*) area for fodder production should be fully irrigated, and (*iv*) other inputs like fertilizers and pesticides, should be made able in optimum quantity. The different systems of fodder production fall into two categories, *viz.* the overlapping cropping and the relay-cropping. In the overlapping system, a fodder crop is introduced in the field before the other crop completes its life-cycle. In relay-cropping, the fodder crops are grown in successions, *i.e.* one after another, the gap between the two crops being very small.

Overlapping System

The overlapping cropping system evolved by taking advantage of the growtn periods of different species ensures a uniform supply of green fodder throughout the year. One such system continues for three years. The best rotation in this system is *berseem* + Japan *sarson*-Hybrid Napier+cowpea-Hybrid Napier; (October-April)-(April-June)-(June-October).

Intensive Fodder Production under Relay Cropping

There is ample scope for increasing fodder production from the high-input areas, either by growing high-yielding fodder crops singly or in mixture. The growing of three or four successive fodder crops helps to boost fodder production per unit area. Some of the important intensive fodder-crops, crop-rotations and the expected yields from each are summarized in Table 29.1.

Table 29.1: Different Cropping Sequence for Fodder Crops Production

1.	Maize + cowpea - maize + cowpea + *berseem* + mustard (300 q/ha) - (450 q/ha) - (1,000 q/ha)
2.	Sweet sudan + cowpea - *berseem* + oats (1,000 q/ha) - (1,000 q/ha)
3.	Hybrid Napier + Lucerne (1,250 q/ha) - (850 q/ha)
4.	Maize + cowpea - *jowar* + cowpea - *berseem* + mustard (300 q/ha) - (400 q/ha) - (1,000 q/ha)
5.	Teosinte + *bajra* + cowpea - *berseem* + oats (1,000 q/ha) - (1,000 q/ha)
6.	Sweet sudan + cowpea - mustard - oats + peas (1,000 q/ha) - (250 q/ha) - (500 q/ha)
7.	*Jowar* - turnips - oats - 1800 q/ha

Fodder Production in Arable Farming

There is ample scope for fitting in the short-duration fodder crops, either single or in mixture, with the other crops during the gap period between two main cash crops. Two distinct fallow periods are available for raising short-duration fodder crops, provided adequate resources are available. In the case of the wheat-jowar rotation, gap periods between April and June and between October and November are available for each crop as fodders. Thus, in first rotation, M. P. Chari + Cowpea, Maize + Cowpea, Bajra + Cowpea is successfully grown and an additional green-fodder yield to the tune of 300-350 q per ha is obtained. Similarly, in the second gap period (October-November), which is rather short, the growing of fodder turnips and short-duration mustard varieties helps to get 250-300 q per ha of fodder without disturbing the normal cropping systems.

Fodder Production Under Dryland Farming

A large proportion of the area of our country is located in the dryland regions. In these areas, the farmers usually grow at least one crop in the *rabi* season after conserving the soil moisture. Thus there is a great scope for raising two crops under such situations. First, the growing of a fodder crop which gets ready in 45-50 days after sowing (cowpea, *jowar*, *guar*, *sanwa*, *moth*, etc.), yield 150-250 q per ha of green fodder. After harvesting the fodder crops, crops such as gram, linseed, barley, wheat and safflower are raised on the conserved moisture.

Economic Aspects of Forage and Fodder Production*

Forages are an essential part of ruminant's and other grazing animal's diets and are an important part of a profitable livestock enterprise. Growing forages represents a significant cost. This cost is affected by the choice of forage crop and how it is produced, harvested, stored, and fed. Forage availability and quality affects livestock performance, including growth rates, milk production, and body condition. Variable weather conditions can cause low yields and risk management strategies create added costs.

When making decisions about forage, consider the following points;

- ☆ Cost of production, measured at the point where the animal consumes the forage
- ☆ Impact of forage choices on total feed cost
- ☆ Impact on animal performance, and
- ☆ Impact of year-to-year variations in yield and quality.

Cost and quality considerations are important when choosing among alternatives. Yields and moisture content at harvest have a big impact on dry matter production and costs. However, in addition to these production costs, there are hidden costs in the form of crop losses through chemical changes, spoilage, and waste. Losses will vary among different crops and different harvesting, storage, and feeding systems. Field to mouth losses can range from 15 to 50 per cent and have a significant impact on costs. Farm equipments and labor cost estimates can be helpful when evaluating the cost and profitability of custom work alternatives. Also consider the reliability of the custom operator and the timeliness and quality of his or her work. The most effective way to compare alternative forage crop production or procurement options is to develop balanced rations capable of achieving animal performance targets, using the various forage alternatives and other available feedstuffs. Evaluate

---

* Geoffrey A. Benson and James T. Green, Jr., 2006. Forage Economics. Department of Agricultural and Resource Economics, College of Agriculture and Life Sciences. **North Carolina Cooperative Extension Service,** NC State University P-8.
http://www.ag-econ.ncsu.edu/faculty/benson/tb305.pdf

the total feed costs, including all the ration components. Forage type and forage quality can affect animal performance. Consider the impact of different levels of animal performance on profitability as well as any differences in forage production costs and total feed costs. Strategies to cope with short crops include buying additional forages or stretching the forage supplies already on hand by purchasing commodity and by-product feeds with significant effective fiber content. Farmers can plant additional acreages of forage crops. In normal or above-normal years, the surplus can be used to build buffer stocks for future use or can be sold. Additionally, they can also diversify the types of crops grown. Each of these risk-management strategies has associated costs that must analyze to identify the most cost-effective strategy. Clearly, there are no simple answers to questions on the economics of alternative forage crops and different production and procurement systems. Each alternative has several aspects that should be considered. However, every decision must start with a clear understanding of the costs involved and the impact on animal performance and income.

Costs

Production Costs

Production costs are important considerations when choosing among alternative forage crops. These costs include operating expenses for items that are used up within one cropping season and fixed costs associated with investments in machinery and equipment. The cost structure is different when comparing annual crops, such as corn silage or winter rye for grazing, to perennial forage crops, such as fescue and Bermuda grass. For annual forage crops, all the production costs are incurred during the production cycle for a single crop. For perennial crops, costs can be separated into the start-up or establishment costs and the annual costs incurred thereafter. Enterprise budgets are only guidelines and should be carefully evaluated and modified according to specific farm situation. These establishment costs can be thought of as an up-front investment that must be allocated over the life of the crop. The costs are defined as follows;

- ☆ Materials and services include seed, suckers and other planting materials.
- ☆ Inoculants, herbicides, lime and fertilizers and custom application.
- ☆ Labor costs-if the work is done by own self, then the opportunity cost of alternative uses for own labor.
- ☆ Machinery operating costs consist of fuel, lubricants, and maintenance and repairs.
- ☆ Machinery ownership costs, also called fixed costs, include depreciation, property taxes, insurance premiums, and an interest charge on the investment.

For perennials, these include the amortized cost of establishing the crop. Many forage crops are grown for grazing and hay production. Because moisture content varies widely and because nutrients are contained in the dry matter of the forage, these cost estimates are calculated on a dry matter basis for ease of comparison.

However, the nutrient composition of the dry matter varies among forages, so cost/ton of dry matter should not be the only criterion for selecting among forage types.

Harvesting, Storage, and Feeding Costs

The methods used for storage and feeding also affect the cost of feeding livestock. Each alternative has different operating, labor and investment costs. This includes the cost of labor and the full cost of the equipment used to move hay from storage to livestock on pasture fields or a sacrifice area. Considerable variation will occur from farm to farm, however. Similarly, the budgeted cost of pasture for grazing does not include any charges for managing cattle on pastures. This includes the cost or value of time spent traveling to the pasture fields and moving cattle. It also includes the full cost of equipment used for transportation. These costs do not include the ownership or fixed costs associated with investments in fencing, lanes, and watering systems.

In addition to the costs described above, there are "hidden" costs in the form of crop losses through spoilage, and waste. The losses will depend on many factors including the specific crop; the particular harvesting, storage, and feeding systems in place on the farm; and the level of management. Total losses for forage crops range from 15 to 50 per cent of the standing crop at the time harvesting begins. Storage and feeding losses for concentrate feeds are likely to be around 5 per cent of the purchased amount in a well-managed storage and feeding system. However, treat these loss estimates as guidelines; there is likely to be wide variation from farm to farm. Clearly, these losses can have a major impact on costs, if storage and feeding losses increase, then the cost of the hay actually consumed by the livestock also increases significantly.

Impact on Animal Performance, Total Feed Costs, and Profitability

Forage type, quantity, and quality determine the amounts and balance of specific nutrients available to the animal. Sample and analyze each of the major forages used on the farm every year in order to develop balanced and economically formulated rations needed for animal performance. Use the analysis to evaluate the impact of different forages on animal performance, including growth rates, milk production, reproduction, and body condition. The most effective way to compare alternative, forage crops and procurement options is to develop nutritionally balanced rations capable of achieving the desired level of animal performance. Consider the various forage alternatives, and other feedstuffs and their costs. The value of different forages and feeds can change over time. Many ration balancing programs generate a "shadow price," which is the break-even price for any one of the available feedstuffs or ingredients. Use this shadow price to evaluate the maximum economic value of individual forages. If the price or cost of particular forage is greater than its shadow price, there is a more economical way of feeding the animals to achieve a target level of performance. Repeat this analysis periodically because the shadow price (value) of one feed is affected by the prices of other feeds and ingredients. Rethink about forage production strategy if the costs of production exceed the value of the forage. Budgets can be developed to compare the profitability of alternative forage production and feeding systems. These budgets should incorporate any animal performance differences and the resulting effects on income or costs.

Thumb Rule for Economic Forage Production

The following production suggestions can help to get an edge on establishment, and when appropriate, to terminate the stand in the most efficient manner. Stands should be terminated sooner rather than later for maximum nitrogen benefits; two to three years is usually optimum.

Consider No-Till Seeding

Forage establishment in a no-till situation is usually better than in a conventional system, especially in drier years, because forage seeds are small and are vulnerable to dry seedbeds and erosion that often occur with conventional techniques. Some residue on the soil surface can provide some of the same benefits (shading, lower soil temperatures and reduced blowing soil) as companion crops, although excessive residue from the previous crops should be removed for better establishment. The relative firmness of no-till soils also provides firm seed beds for excellent soil-to-seed contact.

Choose Less-Competitive Companion Crops

Although companion crops can often reduce forage yields in the second year by hindering stand establishment. They can also provide much needed shade and moisture conservation for new forage seedlings. There are situations where you may find a cover crop more economical than none at all, especially if you harvest the cover crop early for silage. In these cases, it is important to reduce the seeding rate of the companion crop to minimize the amount of competition for the forage stand being established.

Consider No-Till Stand Termination

One can often get more-efficient stand termination by substituting herbicides for tillage. Tilling is expensive, uses fossil fuel energy, dries the soil, and in wet years it may not kill the stand completely. However, depending on the forage species, herbicides use may be less costly and more effective. As well, because nitrogen release is slower, herbicides can improve the availability of nitrogen for uptake into subsequent crops.

Use an Effective Herbicide Combination

Glyphosate/2, 4-D Amine, Lontrel/2, 4-D or glyphosate/Banvel are all highly effective combinations for stand termination, although higher rates of glyphosate are required for mixtures with higher grass content. Apply to at least 8 inches of growth for greatest kill efficiency. Most glyphosate products can be used as a pre-harvest treatment, but allow 3 to 4 days after spraying before grazing or cutting the treated crop for silage or hay. All glyphosate products are more effective when grasses have 3 to 4 or more leaves per stem, and when legumes are in the bud or later stage of maturity.

Evaluate when to Terminate

Although maximum agronomic benefits from forages can be obtain after two or three years of production, the cost of establishment may dictate that a stand be left longer. Costs of production should be considered, so that both agronomic and economic benefits are balanced.

Cost-effective Considerations

Because of reduced inputs and fuel costs, the cost of production for rotations that include forages have proven to be lower than those for rotations based on continuous grain crops. Furthermore, net returns tend to be more stable across a range of crop prices for rotations that include forages. Studies continually show that including 2 to 3 years of forage crops in 6 year rotations significantly reduce income variability, even more than crop insurance.

Marketing the Forage

In recent years, forage markets have opened up. Although in hotter regions forage digestibility is lesser than the cooler places. There is need to strengthen the

Future Thrust

- ✰ Forage production must be taken up as a first management goal and some forest area should be put under trees with regulated accessibility to the farmers.
- ✰ Growing forage grasses and fodder trees along village roads and panchayat lands.
- ✰ Growing forage grasses and fodder trees on terrace risers/bunds-a non competitive land use system.
- ✰ Conservation of native biodiversity for future improvement.
- ✰ Breeding biotic, abiotic, stress tolerant cultivars of forage species suitable for area not used under arable agriculture.
- ✰ Participatory techniques to be adopted to identify the problems and to carry out the improvement programme.
- ✰ In-depth studies on migratory graziers and forage based agro-forestry systems must be undertaken.
- ✰ Controlled grazing to maintain the productivity of pasture land.

References

Department of Agricultural and Resource Economics at NC State University,

http://www.ag-econ.ncsu.edu/extension/Ag_budgets/html

Feed and Fodder Requirements for Milk Production in India.

http://www.love4cow.com/feedandfodder.htm. *Accessed on 14.04.2010*

Geoffrey A. Benson and James T. Green, Jr., 2006. Forage Economics. Department of Agricultural and Resource Economics, College of Agriculture and Life Sciences. North Carolina Cooperative Extension Service, NC State University, P-8. http://www.ag-econ.ncsu.edu/faculty/benson/tb305.pdf

Krishi World, Forage Crops and Grasses, available at http://www.krishiworld.com/html/for_crop_grass1.html, accesed on 04.06.2010

Misra, A. K, Rama Rao, C.A, Subrahmanyam, K.V. Ramakrishna, Y.S. (2009). Improving dairy production in India's rain-fed agro ecosystem: constraints and strategies, *Outlook on Agriculture*, 38 (3): 284-292.

Sharma, J. R. and Jindal, K. K. (1989). Introduction of superior varieties of grasses in orchards. Paper presented in the workshop on pasture and grassland improvement at HPKV, Palampur on 12 to 13 October, 1989.

Singh, V. (1995). Technology for forage production in hills of Kumaon. In : New Vistas in Forage Production (ed. Harzra, C.R and Misri Bimal). AICRPF (IGFRI). Publication Information Directorate, New Delhi. pp.197-202.

Singh, K.A., Prasad, R.N., Stapathy, K.K. and Sharma, U.C (1993). Need for forage resource development in the hills of eastern Himalayas. Indian Farming. 43 (8):21-26

Chapter 30

Scenario and Strategies for Revitalizing Fodder Production Technologies

M.S. Meena[1] and K.M. Singh[2]

[1]*Senior Scientist (Agricultural Extension)*
ICAR-Research Complex for Eastern Region, Patna (Bihar)
E-mail: ms101@sify.com
[2]*Principal Scientist and Head,*
Division of Socio-economic, Extension and Training,
ICAR Research Complex for Eastern Region,
ICAR Parisar (P.O.- B.V.College), Patna – 800 014, Bihar
E-mail: m.krishna.singh@gmail.com

Introduction

Livestock production is the backbone of Indian agriculture and also plays a key role in providing employment especially in rural areas. This sector has been the primary source of energy for agriculture operation and major source of animal protein for masses. Therefore, India has been the home of major draught, milch and dual-purpose breeds of cattle. Indian dairy production system is complex and generally based on traditional and socioeconomic considerations. However, there has been a rapid change in way of agriculture (*i.e.* cropping system, water resources, diversification of crops, intensification of agriculture), increasing use of mechanical power, transformation from sustenance farming to market oriented farming, changing food habits etc., All these factors have their impact on animal husbandry practices. India has 15 per cent of world cattle population and due to ever increasing population pressure of human, arable land is mainly used for food and cash crops, thus there is little chance of having good quality arable land available for fodder production, until milk production becomes remunerative to the farmers as compared to other field crops. In India, there is no practice of fodder production in rural areas and animals generally consume naturally grown grasses and shrubs which are of low quality in terms of protein and available energy, they are thus heavily dependent on seasonal variations and this results in fluctuation in fodder supply round the year affecting supply of milk round the year.

Significance of Feed and Fodder

Livestock rearing in India is changing fast and there has been a rise in demand of milch cattle as compared to dual or draught breeds. Population of indigenous breeds like Haryana, Nagori, Khilar *i.e.* dual and draught purpose breeds has declined more than milch breeds. In this age of market economy, the agri-economy and milk production has to compete for growing fodder on good quality land, required for high productivity and reproductive efficiency of dairy animals. Hence, its significance can be understood from the following points.

Economy in Production

Feed and fodder cost constitute about 60-70 per cent of cost of milk production thus cultivated fodder has an important role in meeting requirement of various nutrients and roughage in our country to produce milk most economically as compared to concentrates. Feeding not only meets nutrient requirement but also fills rumen to satisfy the animals. Feed has to meet requirement of cattle maintenance, production and requirement of microbes to promote digestion.

Better Feeding for Ruminants

In view of the peculiar digestive system, provided by nature, ruminants need feeds, which not only meet their nutritional requirements but also fill the rumen and satisfy the animal. In view of microbial digestion system the feeds have to meet requirements of the animal, its production as well as the needs of microbes for promoting digestion. The fodder crops meet these requirements very effectively and hence are important for ruminant production system. As evident from reports that mixed with coarse roughages, like wheat straw, its intake and digestion are improved.

Good Source of Critical Elements

Fodder from common cereal crops like Maize, Sorghum and Oats are rich in energy and the leguminous crops like Lucerne, Berseem and Cowpea are rich in proteins. These leguminous crops are good source of macro- and micro-minerals, which are critical for rumen microbes as well as animal system. The green fodder crops are known to be cheaper source of nutrients as compared to concentrates and hence useful in bringing down the cost of feeding and reduce the need for purchase of feeds/concentrates from the market. In case surplus fodder is available in some season it can be stored in form of silage or hay for lean season.

Scenario of Feed and Fodder Availability and Future Requirement

There is tremendous pressure of livestock on available feed and fodder, as land available for fodder production has been decreasing. Scenario of feed and fodder availability till 2025 is as below.

Table 30.1: Scenario of Feed and Fodder Availability and Future Requirements (in million tones)

Year	Supply		Demand		Deficit as % of Demand (Actual Demand)	
	Green	Dry	Green	Dry	Green	Dry
1995	379.3	421	947	526	59.95 (568)	19.95 (105)
2000	384.5	428	988	549	61.10 (604)	21.93 (121)
2005	389.9	443	1025	569	61.96 (635)	22.08 (126)
2010	395.2	451	1061	589	62.76 (666)	23.46 (138)
2015	400.6	466	1097	609	63.50 (696)	23.56 (143)
2020	405.9	473	1134	630	64.21 (728)	24.81 (157)
2025	411.3	488	1170	650	64.87 (759)	24.92 (162)

Source: Draft Report of Working Group on Animal Husbandry and Dairying for Five-Year Plan (2002-2007, Govt. of India, Planning Commission, August-2001).

Table 30.2: Feed Production (in million tones)

Particulars	2002-03	2003-04	2004-05	2005-06	2006-07
Concentrates available	41.96	43.14	44.35	45.63	48.27
Concentrates required	117.44	120.52	123.59	127.09	130.55
Concentrate Deficit	64.27	64.21	64.12	64.10	63.03

Source: Draft Report of Working Group on Animal Husbandry and Dairying for Five-Year Plan (2002-2007, Govt. of India, Planning Commission, August-2001).

It is obvious from Table 30.1 that deficit in green and dry fodder is increasing every year, while for concentrates, the gap is almost static. However, this gap is critical and is going to determine the type of animals and husbandry practices to be followed. Scarcity of feed and fodder resources (both quantity and quality), low production potential of animals, non-availability of critical inputs or services in time along with access to capital and markets, are primary reasons for low productivity of dairy animals (Mishra *et al.*, 2009).

Constraints in Achieving Higher Fodder Productivity

India is presently under heavy stress on account of a large-scale exploitation for fuel wood, timber and fodder, mismanagement of forest resources and frequent fires. There is acute shortage of fodder especially green nutritious fodder, which is major cause of low productivity of livestock, especially in hilly areas (Roy *et al.*, 1989). The main reasons for low productivity is insufficient and low quality fodder and feed including grazing facilities (Roy, 1993). The main constraints can be described as;

Reduced Area Under Fodder Crops

The division of the families has fragmented the land. At present land holdings are very small and farmers are always biased in choice of the crops. Due to these reasons agricultural land ratio does not permit diversion of land from food production to cultivated fodder. Thus, area under fodder crops is meagre.

Uncontrolled Grazing of Dairy Animals

Uncontrolled grazing has led to a decline in biomass availability. The grazing pattern has created manifold problems in these pastures. Obnoxious weeds have invaded the pastures. Excessive and continuous grazing has severely damaged these lands.

Poor Management Practices

Management practices play an important role in determining productivity of grasslands. Presence of inferior and unproductive grass species, lack of fertilizer application, and absence of legume component, improper cutting and indiscriminate grazing are some of the important factors responsible for poor productivity of grasslands. There exists a wealth of indigenous knowledge for its proper utilization and management of natural resource base but farmers because of increasing population pressure and declining land productivity are not using it.

Intense Livestock Population

Livestock is the integral component of Indian agriculture since time immemorial. Its contribution to national economy through milk, meat, wool as well as farmyard manure is enormous. We have approximately 20 per cent of world's cattle, 50 per cent of buffaloes, more than 120 million goats and 60 million sheep (Roy, 1993). Due to religious beliefs, population of unproductive cattle is increasing. This huge population and poor fodder availability has widened the gap between demand and supply of forage crops. It is a fact that considerable fodder resources are wasted on maintenance of an excessive number of poorly fed and low yielding animals, which contributed to process of pasture destruction.

Fodder Tree Use

Indian sub-continent is one of the richest in biodiversities on the globe. For instance, Himalaya supports about 84 trees and 40 shrubs of fodder value, yet not more than 20 trees are extensively used by farmers (Misri, 1997). Tree leaf fodder is the major feed resource during lean periods. Over exploitation and unscientific management of fodder trees has depleted this resource at huge environmental cost.

Future Guidelines for Revitalizing Fodder Production Technology

Changing agriculture production practices, globalization of economy, market oriented production system, decentralized form of governance etc, have its impact on livestock production system in India. The production may be milk, dung, urine etc., has to compete in economic terms for allocation of resources of production with other competing options of crops. In India, for many families especially landless, small and marginal farmers, agriculture and livestock have been a livelihood issues and economic issues. Therefore, our policies must ensure both these aspects. Following policies may be focused in future.

1. Look for good indigenous milch breeds of cattle particularly for semi-arid/arid climate; upgrade these breeds through modern techniques as animal husbandry is main activity in such areas. There is need to adopt intensive and well-defined mile stones to achieve growth in productivity of indigenous breeds in such areas.
2. Adopt breeds/cross breeds in areas commensurate with productivity of land *i.e.* in case of high agriculture production area, cross breeds with high milk potential will be able to compete with agriculture crops.
3. In upcoming days, when need for fat is going to drop and cheaper fats are likely to come from different countries it is certain that cow milk production will be preferable to buffalo milk.
4. There should be focused programme on regeneration, promotion of Silvi-pasture, revenues and wastelands, which will not only meet shortage of feed and fodders but will give equal access to poor and improve environment also.
5. There is need to promote scientific fodder crop production through improved agronomic practices and improved seed. Extension in this sector is totally neglected because it is part of animal husbandry department for which it has never been a priority. Thus we should look into the possibility of attaching it to agriculture department.
6. Promotion of techniques of treatment of straws and feed supplements as entrepreneurial activity than treatment at farmer level.
7. Though a number of fodder varieties have been developed but seeds are not available because it is trapped in vicious cycle of lack of demand due to lack of extension, which inhibits production of seed etc. Thus this cycle need to be broken through proper extension.
8. Presently research has been mainly conducted on cultivation of green fodder in irrigated areas but it is high time to emphasize to dry land fodder or partially irrigated fodder crops.
9. Extension to promote balance feed, feeding chaffed feed and proper storage of fodder to avoid losses need also to be emphasized.

xtension Strategies for Revitalizing Fodder Production

Extension strategies can bring the desirable changes in behavior of the fodder rowers. The components of extension strategy can be described below:

1. *Awareness creation about fodder production technology*: There is utmost need to organize method/result demonstrations and organizing field days showing the monetary gain and benefits of cultivation of high yielding varieties fodder crops.
2. *Strengthening the extension and development activities*: Farmers can be motivated through campaigning for growing perennial fodder crops (*e.g.* Napier) in pond bank, farmhouse, road side, embankment etc., Extension personnel should also help in identification of effective technologies and their transfer to fields, hence, it can be easily adopted by the stakeholders.
3. *Capacity building of farmers and extension functionaries*: The skilled extension staffs are heavily loaded with veterinary and artificial insemination activities alone. There is need to strengthen the manpower of animal husbandry departments across the country who should be trained in latest technologies to support the livestock owners both in terms of animal health as well as management aspects. Trainings must also be conducted to train the fodder growers to keep them abreast with latest technical know how.
4. *On-farm evaluation of fodder technologies*: On farm evaluation and demonstration of existing technologies may be attempted to narrow the gap between yields realized on farmers' fields and those on research stations. Providing the basic advice to the farmers is very essential which enable them to withstand in competitive market. Adaptive research on fodder production technology must be encouraged through providing necessary feed back from the farmers' field.
5. *Conservation of forages to meet the demand in crisis*: Fodder scarcity is mostly observed in dry season and during floods. Conserved forages which are enriched with nutrients like energy, protein and vitamins and low cost methods of silage making are to be promoted among the farmers.
6. *Motivate farmers for Indigenous Technical Knowledge (ITKs)*: Farmers are using ITKs since immemorial and it is the part of their culture. Hence, there is need for the evaluation, screening and utilization of indigenous potential of forages crops in hilly, coastal and other areas as animal feeds. Shrubs and small trees (like *Gliricidia, Desmanthus, Leucaena, Sesbania* spp.) are very good and cheap source of proteins and minerals and can be introduced between farm plots and have multipurpose utility.

Scientific Interventions for Revitalising Fodder Production

There is a need to understand the existing resource utilization pattern in totality. Fodder production is a component of farming system, hence; efforts are needed for increasing forage production in a farming system approach. The holistic approach of integrated resource management will be based on maintaining the fragile balance between productivity functions and conservation practices for ecological sustainability. The strategies for improvement and conservation of forage resources will have to be dictated by actual users *i.e.* farmers who are the native inhabitants of that region. Some of scientific interventions, which could help in improving productivity of forages, are described here.

Agronomic Management of Forage Crops

The herbage production from grasslands and meadows can be enhanced with the adoption of improved technology. Important components of this technology are:

- ✰ Control of bushes and weeds
- ✰ Pasture establishment
- ✰ Introduction of legumes/grasses
- ✰ Fertilizers application
- ✰ Cutting and grazing management

Scientific Cultivation of Fodder Crops

For augmenting the fodder availability, emphasis needs to be given to cultivate fodder crops on large area. Important fodder crops of temperate region are; *Avena sativa, Brassica sp., Medicago sativa, Pisum sativum* etc. (Singh, 1987). Foliage of fodder trees could be fed to livestock in mixture with crop residues and hay. Mixing of tree foliage with dry roughage improves their palatability and nutritive value.

Adoption of Silvi-pastoral System

Silvi-pasture implies sustained and combined management of the same land for herbaceous fodder, top feeds and fuel wood, thereby leading to optimization of production. Himalayan rangelands exhibited enormous gain in forage production over existing situation due to multi-tier silvi-pasture techniques amalgamated with an adaptable complementary plant species. Silvi-pastoral systems are most important for increasing fodder production from marginal, sub-marginal and other wastelands. It comprises about 50 per cent of total land area. It involves planting of multipurpose trees in existing pastures/grazing lands or planting such trees on wasteland/ denuded lands followed by sowing/planting of grasses and or legumes in between the inter-spaces of trees. Atul (1996) obtained 5-7 t/ha green fodder under silvipastoral system, where as it was only 3-4 t/ha with out a tree component. Sharma and Koranne (1988) found that maximum production of 300 g/m^2/annum under existing grasslands, while under modified network of silvipastoral system of *Digitaria decumbens+Bauhinia pupurea/Quercus incana/Grewia optiva/Celtis australis* production varied from 1800-2450 g/m^2/annum.

Adoption of Agri-silvipastoral System

Under agri-silvicultural system multipurpose trees including fodder cum fuel trees can be grown in association with crops. Trees are pruned annually, yielding fodder as well as fuel wood. In addition to annual pruning, few trees are also cut down in order to allow light penetration and minimization of competition with the crops. Under alley cropping system multipurpose trees like *Leucaena leucocephala* and even perennial pigeon pea etc. are pruned frequently to provide leaf fodder to get better crop production.

Agri-horti-silvicultural System

Under this system besides growing fruit trees and fodder crops, fast growing NFTs like *Leucaena leucocephala* can be lopped two to three times in a year to provide fodder (2.5-3.0 t/ha) and fuel wood (1.8-2.5 t/ha). These fodder trees also provide some protection to fruit trees during summer and cold winters.

Horti-pastoral System

In this system forage are grown in wide inter-row spaces of fruit trees for economic utilization of orchard lands. Horti-pasture up to an elevation of 2000 m is catching up with the orchadist. Forage from horti-pasture is consumed fresh and is also conserved as hay for winters. Sharma and Jindal (1989) found that introduction of Fescue in apple orchard gave 83.50 per cent higher fodder yield over local grasses in Shimla hills of Himachal Pradesh. There is considerable area under orchards in temperate regions. Inter spaces between fruit trees could be utilized for the production of fodder by growing perennial grasses and legumes. In U.P hills (Singh, 1995) reported that Rye grass and orchard grass are the best perennial grasses for introduction in apple orchards. Soil nitrogen build up was maximum with white clover introduction.

Forage Production on Terrace Risers or Bunds

A non-competitive land use systems for forage production in the hills is to grow forage on terrace bunds and risers (Singh *et al.*, 1993_a). Forage grasses/legumes/ fodder trees grown on terrace risers and bunds arrest the nutrient loss in runoff water under high rainfall conditions of this region. This gives an added advantage to produce forage with out any fertilizer or manure.

Future Opportunities and Lessons Learnt Replication

The process of documentation, validation, dissemination and practical applications are vital in effective re-integration and revitalization of traditional knowledge systems pertaining to fodder. Community participation in their traditional knowledge and experience has great significance especially regarding the nutritive value of tree fodder.

1. Female farmers have sound traditional knowledge regarding nutritive value of different fodders and grasses. Community-led action research coupled with scientific validation methodologies is important in blending people's knowledge. This enabled a positive shift in the attitude towards traditional fodder varieties by the various stakeholders.
2. Traditional feed and fodder species are more suitable for rain fed areas compared to new fodder varieties being introduced that need irrigation facilities and other requirements. Government policies as well as top down interventions can be counter productive if not based on the needs and requirements of livestock keepers as shift towards non-food cash crops, planting timber varieties in forests etc resulted in reduction in fodder resources for the livestock.
3. Documentation of traditional knowledge concerning plants, feed and fodder species has to be undertaken so that this knowledge is not lost and future

generations can bank upon documented resource base. This needs concerted efforts, financial and human for a considerable period of time and can lead to revitalization of traditional knowledge systems both by individuals and as a collective, which in turn, can result in positive changes in their livelihood systems.

4. Lopping of branches of trees undertaken in traditional ways facilitates enhanced growth of branches leading to more fodder production demonstrating that the traditional practice of lopping of trees in forest areas does not harm the forest in any way.
5. There is great scope for forest department to plant fodder yielding traditional trees with active community participation instead of growing timber yielding trees.
6. This experience shows that the tribal community has been empowered to take back control over their own knowledge and related genetic resources, and utilize it in ways that are making a positive impact on their livelihoods. Therefore, community based institutions are best suited to conserve and propagate traditional species of feed and fodder.

Conclusions

Importance of forage production in maintaining food security as well as nutritional security has been felt since long. The overall scene of forage production is very alarming and corrective measures have to be taken to improve this problem. A comprehensive grazing policy needs to be formulated and both grazing and forage cultivation has to be considered complementary to each other and simultaneous efforts are required to improve both. Fodder tree improvement programmes for higher leaf fodder have to be initiated. For the improvement of grasslands, its management needs to be considered holistically promoting interaction between grassland, livestock and grazing communities. Therefore, the vast natural resource can serve human society substantially, more particularly grazing communities. A favorable policy environment in terms of access to micro-credit and assured market will have to be provided and simultaneously there is need to address the socio-economic and technical constraints.

References

Atul (1996). Silvipastoral system a tool for reclaiming wastelands. In: Agro forestry manual (*ed.* Atul and Punam) HPKV, pp.73-97.

Feed and Fodder Requirements for Milk Production in India. http://www.love4cow.com/feedandfodder.htm. *Accessed on 14.04.2010.*

Misri, B. K. (1997). Important fodder trees/shrubs of temperate Himalaya: Distribution pattern and habitat affinities. In Training Course on Management and Utilization of fodder trees and shrubs in sub-tropical and temperate Himalaya. Sponsored by FAO Rome, organised by IGFRI, Jhansi (September 22-30, 1997). pp. 18-23.

Misra, A. K, Rama Rao, C.A, Subrahmanyam, K.V. Ramakrishna, Y.S. (2009). Improving dairy production in India's rain-fed agro-ecosystem: constraints and strategies, *Outlook on Agriculture*, 38 (3): 284-292.

Roy, D. (1993). Reap more biomass through diversity in forestry, Intensive Agriculture. XXXI (5-8): 23-26.

Roy, D, Shankaranarayan, K. A. and Pathak, R. S. (1989). The Fodder Trees of India and their importance. Indian Forester, 106: 306-311.

Sharma, B. R and Koranne, K. D (1988). Present status and management strategies for increasing biomass production in North-Western Himalayan rangelands. In: Rangelands-resources and management (ed. Punjab Singh and P.S. Pathak). Range Management Society of India. Indian Grassland and Fodder Research Institute, Jhansi. pp. 138-147.

Sharma, J. R. and Jindal, K. K. (1989). Introduction of superior varieties of grasses in orchards. Paper presented in the workshop on pasture and grassland improvement at HPKV, Palampur on 12 to 13 October, 1989.

Singh, L. N. (1987). Fodder production strategies for temperate and sub-temperate regions of India. In: Forage Production in India (ed. Punjab Singh). Range Management Society of India, IGFRI, Jhansi, pp. 21-27.

Singh, V. (1995). Technology for forage production in Hills of Kumaon. In : New Vistas in Forage Production (ed. Harzra, C.R and Misri Bimal). AICRPF (IGFRI). Publication Information Directorate, New Delhi. pp.197-202.

Singh, K.A., Prasad, R.N., Stapathy, K.K. and Sharma, U.C (1993). Need for forage resource development in the hills of eastern Himalayas. Indian Farming. 43 (8):21-26

Chapter 31

Role of Women in Fodder and Livestock Management

Ujjwal Kumar

ICAR Research Complex for Eastern Region, Patna – 800 014

Women make a substantial contribution towards rural economy of India. About 70 percent of the total working population of women is extensively involved in agricultural activities. Nature and extent of their involvement differ with the variations in agro-production systems. Further, their mode of participation in agricultural activities varies with the ownership of land of farm households. Their role ranges from managers/decision makers to landless laborers. In the highly diversified Indian context, no simple gender division of labour exists with regard to crop production. In certain areas in India, women play key roles as seed selectors and in seedling production. Their knowledge on seeds and seed storage contribute to viability of agricultural diversity and production. Women prepare and apply green and farmyard manure. As integrated pest management practices are introduced, it could be expected that women's work would increase due to more labour-intensive activities. In addition to their role in crop production, women are gainfully employed in agri-based allied activities like dairying, animal husbandry, poultry, goatery, rabbitry, apiary, floriculture, horticulture, fruit preservation etc. In case of livestock more than 90 per cent of the work related to animal care is done by women. In Livestock management their role vary widely ranging from care of animals, grazing, fodder collection, cleaning of animal sheds, processing milk and livestock products. In livestock sector, indoor jobs like milking, feeding, cleaning etc. are done by women in 90 per cent of families while management of animals and fodder production are done by men. One can easily find women with the sickle in hand harvesting green fodder for their own animals or for the sale which also contributes to family income and savings. Mostly, this operation is done in the non owning farms which sometimes happen to be cause of conflicts. Though women play a significant role in livestock management and production, women's control over livestock and its products is negligible.

Despite their considerable involvement and contribution, women's role in livestock production has been underestimated, undervalued and widely ignored. This is due to a paternalistic bias from men but also from women, who are often conditioned by their culture and society to undervalue their own worth.

Nature of Participation of Farm Women in Fodder and Livestock Activities

Women's participation in fodder and livestock management was studied in few villages of Karnataka. It was observed that majority of women were involved in almost all the livestock activities (Table 31.1) *viz.* feeding animals (96 per cent), fodder collection (96 per cent), maintenance of cattle shed (94 per cent), cow dung making for fuel (94 per cent) milking (88 per cent), marketing of milk and milk products(84 per cent), cleaning and health care of animals (80 per cent).Percentage of farm women under "doing" were found to be fairly high in the activities like cow dung making (88 per cent), fodder collection (84 per cent), maintenance of cattle shed (80 per cent) and feeding of animals (78 per cent).

Table 31.1: Nature of Participation of Farm Women in Fodder and Livestock Activities

Sl.No.	*Activities*	*Women's Participation (per cent)*		
		Doing	*Supervision*	*No Participation*
1.	Milking	70	18	12
2.	Marketing of milk and milk products	76	08	16
3.	Feeding animals	78	18	04
4.	Maintenance of cattle shed	80	14	06
5.	Fodder collection	84	12	14
6.	Cleaning of cattle	56	24	20
7.	Cow dung making for fuel	88	06	06
8.	Health care	64	36	20

Source: Nataraju, M.S *et al.*, 2009.

The time Use Survey of 2000 shows that on the average, women spend more time than men on activities related to animal husbandry in states in which animal husbandry contributes substantially to the agricultural GDP, as in Gujarat, and Haryana. In activities such as collection of fodder, fuel, wild food, water etc., women spend more time than men do in all the States (Tables 31.2 and 31.3).

Table 31.2: Average Weekly Time Spent on Activities Related to Collection of Food, Fodder, Fuel, Water etc. (Hrs/week)

Sl.No.	States	Women	Men
1.	Haryana	7.78	1.57
2.	Madhya Pradesh	1.52	0.91
3.	Gujarat	3.0	1.4
4.	Orissa	5.95	1.92
5.	Tamil Nadu	2.78	0.55
6.	Meghalya	5.29	2.95

Table 31.3: Average Weekly Time Spent on Activities Related to Animal Husbandry (Hrs/week)

Sl.No.	States	Women	Men
1.	Haryana	9.8	4.78
2.	Madhya Pradesh	3.75	7.72
3.	Gujarat	5.93	4.31
4.	Orissa	3.15	3.35
5.	Tamil Nadu	2.42	3.69
6.	Meghalya	1.22	3.92

It has been reported in several studies that majority of the farm women actively participate in almost all the house keeping activities, farm operations and livestock production activities. Participation of rural women is more in terms of "doing" than that of supervision in case of agricultural and livestock production activities.

References

Nataraju, M.S *et al.* (2007) "Proceedings of National Symposium on Women in Agriculture, UAS, Bangalore". April 10-12,2007.

Sridhara, Shakunthala *et al.* (2009) Women in Agriculture and Rural Development. New India Publishing Agency, New Delhi.

Chapter 32
Management of Rodents in Pastures and Range Ecosystem

Mohd. Idris
ICAR Research Complex for Eastern Region, Patna – 800 014, Bihar

The Indian desert "Thar desert" is unique among world's desert and most populated in terms of arid land standard having an average density of 90 person/km^2 as against 3 person/km^2 in other desert. It also maintains a livestock population of over 23 million (135 cattle heads/km^2). The Thar harbours 18 species of rodents belonging to three families. Out of 18, Twelve species are often encountered in pasture ecosystem. Some of these rodents' species are habitat specifics and act as bio-indicators. The rodent fauna has direct relationship with vegetation and rainfall pattern. Rodents of Thar inflict 4.4 to 20 per cent damages to flora of the region. Rodents mainly cause damage to vegetation by debarking or slicing activities. It is estimated that rodents caused damage in *Acasia* sps, *Prosopis juliflora* and *Albizzia lebbek* was ranged from 10 to 40 per cent which main source of forage in the Thar Desert. Rodents that maintain perpetual pressure on the range land, difficult groups of control. There are two methods to control their population *i.e.* lethal and non lethal. Various methods have been suggested for their control. However, an integrated approach is also important for management of these pests, and it has been suggested that such operations should be undertaken twice a year during summer and winter months.

Introduction

The Thar desert is eastern limit of vast arid tracts of Sahara and Arabian desert, bounded roughly by the latitudes 24°30′N and 30° N and longitudes of 69° 30′E and76° E (Singhvi and Kar, 1992) Like other deserts of the world, Thar is characterized by low and erratic rainfall, extreme variations in diurnal and annual temperatures and high evaporation rate. In spite of these climatic vagaries, the Thar is the most populated desert of the world. In recent decades, the Thar has witnessed tremendous increase in human and livestock populations with density being respectively more than 90 person/km^2 and 135 cattle heads/km^2. during last decades, human population has icreased by 50 per cent and livestock population by 25 per cent. Such an increase in population is exerting greater pressure on its meager natural resources. Soils of arid region show morphological and physiochemical variation, depending upon the parent material, age and evolutionary history of landscape. The soil of Thar desert is pale brown in colour, structure less, single grain, friable and calcareous. Because of harsh climatic conditions, the vegetation in the Thar is scanty. True forest cover is found only in vegetation can be seen in 0.69 per cent of total area (Saxena, 1977). Xerophytic annuals form main flora of the region. A clear stratification in vegetation can be seen in relation to various rainfall zones. *Calligonum polygonoides, Haloxylon salicornicum, Leptadenia pyrotechnica,Clerodendrum phlomoides*, and *Acacia jacquemontii* are major flora of lowest rainfall zone (100-150mm).In 250-350 mm rain fall zone, *Prosopis cineraria, A. senegal, Calotropis* spp., *Citrullus colocynthis, Aerva persica* and *Indigophora cordifolia* are predominant species.*Lasiurus sindicus, Cenchrus ciliaris, Eleusine compressa* and *Aristida* spp are major grasses of the region, on which herbivores of the region thrive. Some exotic species, like *Prosopis juliflora* have been introduced in the arid land to increase potential of the region. These desert adapted species are spreading very fast mainly at the cost of endemic flora of the region.

In spite of erratic and low rain fall, great temperature fluctuations and other harsh climatic conditions, the Thar is teeming with rich diversity. In addition to large number of reptiles, birds and amphibians, about 65 species of mammals are found in Thar. Rodents constitute about 30 per cent of this mammalian diversity.These small mammals play a significant role in deforestation of arid and semi arid regions. They are devasting as they cut apical buds and other growing vegetative points of trees ages 1 to 3 years. They cause damage to mature trees by debarking them. By consuming seeds and small sapling, they inflict severe damage to nurseries. Being fussorial habit, rodents cause severe damage to fibrous root system of the grasses by tunneling activities. While digging the rodents gnaw at the roots and kill them by exposing roots to dry air. As the most of the rodents are seedivorous, they consume seeds of the grasses trees and shrubs, hampering the further regeneration of the vegetation. They also cause severe damage to barseem, cowpea, Lucerne and great napier.

Pasture ecosystem in arid tracts provide ideal and safe abode to rodents, where these species can easily fulfill their food requirement. Thus, continuously available food in the pasture ecosystem enhances the breeding potential of rodent's manifolds and population grows in geometric ratio.

Rodent Diversity in the Thar

In the world 29 families of order rodentia are found. Out of that 4 families *viz.*, Sciuridae, Muridae, Diplodidae and Histricidae, occur in India. In the Thar desert, 18 species of rodents belonging to 11 genera and three families are quite commonly found. Porcupine, gerbils, rats; mice, mole and squirrels are the major groups of rodents found in Thar desert. Prakash *et al.* (1971) and Prakash (1974, 1975) have discussed the distribution of rodents and their relationship with flora and other fauna in details. Table 32.1 depicts the rodent diversity in arid region.

Table 32.1: Association of Major Rodent Species with their Ecosystem

Family	*Species*	*Creating Problems in*
Hystricidae	*Hystrix indica*	Forest trees, shrubs and tuber crops
Sciuridae	*Funambulus pennanti*	Forest trees nurseries, kitchen gardens and orchards
Muridae		
Sub family		
(1) Gerbillinae	*Tateva indica* Hardwicke	Agriculture, pastures and forestry
	Meriones hurrianae Jerdon	Agriculture, pastures and forestry
	Gerbillus gleadowi Murray	Sand dune and Plantation
	G. nanus Thomas	Sand dune and Plantation
(2) Murinae	*Vandeleuria oleracea* Bennet	Forest trees
	Rattus rattus Linneus	Nursery, seed godowns and hutments
	Cremnomys cutchicus Wroughton	Hilly out crops and foot hills
	Millardia meltada Gray	Irrigated agriculture
	Rattus gleadowi Gray	Arid agriculture, pasture and forestry
	Mus musculus Linneus	Commensal
	M. cervicolor Hodgson	Agriculture, Pasture and forestry
	M.platythrix Bennet	Agriculture, Pasture and forestry
	M. booduga Gray	Agriculture, Pasture and forestry
	Golunda ellioti Gray	Agriculture, Pasture and forestry
	Nesokia indica Gray and Hardwicke	Forest plantation
	Bandicota bengalensis Gray	Agriculture and forestry

Indian rodents are either Palaeotropical or Oriental in origin (Blanford,1877; Prakash,1995; Prakash and Singh,2005). Rodents inhabiting Thar desert have Saharan and Oriental affinities. It is conjectured that with the onset of aridity in this region. Some rodent species of Saharan and Oriental origin migrated to the Thar desert and certain Oriental species advanced from east to this region. Recently, the Thar is witnessing a great change in rodent diversity because of gigantic Indira Gandhi Canal (Singh and Prakash, 2004).

Association of Major Rodent Species with Arid Pasture

Out of 18 species that are found in the Thar desert, 12 species are often encountered in the pasture land of this region. Besides naturally occurring trees, several indigenous as well as exotic species of trees have been planted in rocky, sandy and gravel habitats of desert under Desert Development Programme. Main objective of this programme is soil conservation and social forestry. The rodents cause considerable damage to trees, grasses by cutting, debarking and by exposing roots of during excavation of burrows. Major rodent species associated with various tree types are given in Table 32.2.

Table 32.2: Association of Various Rodents with Tree Types and in Arid Land Forest

Tree Species	*Rodent Pest Species*
Albizzia lebbek	*T. indica* and *M.hurrianae*
Anogeissus pendula	*H. indica, F. pennanti, C. cutchicus, M. platythrix,* and *M.cervicolor*
Prosopis cineraria	*G. gleadowi* and *M. hurrianae*
P. juliflora	*T. indica, M. hurrianae* and *N. indica*
Acacia tortilis	*T. indica, M. hurrianae G. gleadowi* and *N. indica*
Parkinsonia aculeata	*T. indica* and *M. hurrianae*
Acacia senegal	*H. indica, C. cutchicus* and *M.cervicolor*
Azadirachta indica	*H. indica, T. indica* and *R. rattus*

Source: Rana, 1983.

Habitat Preferences

Rodents show a great degree of habitat specificity in this region. Prakash (1962, 1963) classified the desert biome into sandy, gravel and ruderal habitats. Prakash *et al.* (1971), Prakash and Rana (1970, 1972,1973), Prakash and Jain (1971) and Rogovin *et al.* (1994) have carried out extensive studies in all these four habitats. These studies reveal that some rodents of the region are very habitat specific. For example, little Wagner's gerbil, *Gerbillus nanus* prefers the sandy habitat while Cutch rock rat, *Cremnomys cutchicus* and *Mus cervicolor* prefer rocky outcrops of the desert. *Mus musculus* and *M.booduga* prefer ruderal habitat to others. Other species of rodents are not much habitat specific and have been collected from more than one habitat (Table 32.3).

Table 32.3: Preference of Habitats of Desert Dwelling Rodent Species

Sandy Habitat		*Gravel Habitat*	*Rocky Habitat*	*Ruderal Habitat*	
Sand Dunes	*Sandy Plains*			*Residential Areas*	*Crop Fields*
G. gleadowi	*G. nanus* *M. hurrianae* *R. gleadowi*	*M. hurrianae* *T. indica* *M. meltada* *R. gleadowi* *M. platythrix* *G. ellioti*	*H. indica,* *C. cutchicus,* *M. cervicolor*	*F. pennanti,* *R. rattus* *M. musculus*	*T. indica,* *M. meltada,* *M. booduga,* *G. ellioti* *N. indica* *B. bengalensis*

Gravel habitat was least preferred habitat among other habitats. Though *Rattus gleadowi, Meriones hurrianae, Mus platythrix, Golunda ellioti* and *Tatera indica* were collected from this habitat by Prakash *et al.* (1971), their relative abundance was quite low.

Distributional Pattern in Relation to Vegetation and Rainfall Zone

The amount of precipitation determines the distribution pattern of flora, which influence the distribution of fauna. Distribution of various rodent species in relation to various rainfall zones and vegetation is as follows:

100-250 mm Rainfall Zone

This lowest rainfall zone of the Thar is predominated by shrubs and under shrubs due to the climatic constraints. *Colligonum polygonoides* is the major shrubs of the zone with which associated *Haloxylon salicornis. Leptadenia pyrotechnica, Clerodendrum* spp. and *Acacia* spp. are other associates of shrubs. Due to less rainfall and scanty vegetation only desert adapted gerbils are found in the lowest rainfall zone.

250-350 mm Rainfall Zone

Shrubs like, *Aerva persica, Calotropis procera* and *Citrullus colocynthism* dominate Churu-Bikaner-Naguar tracts of this rainfall. *Indigofera* spp., *Sericostemma* spp., *Acacia senegal, Tecomella undulata, Prosopis cineraria, P.juliflora* and *Colophospermus mopane* are predominant tree species of the zone. Unlike 100-150mm rainfall zone, this zone supports gerbils and some other murid rodent. *Tatera indica, Nesokia indica* and *Meroines hurrianae* are the predominant rodent species of this rainfall zone.

350 mm and Above Rainfall Zone

The northern frindge of the Thar has more than 350mm rainfall and supports *Prosopis-Acacia* type of vegetation. The southern part of this rainfall zone support *Salvadora-Prosopis-Capparis* type vegetation.Flat alluvial plains with heavy textured soil in this zone support good vegetation cover and better regeneration capacity. *P. cineraria, Alianthus excela, Capparis decidua Ficus bengalensis, Azardirachta indica, Moringa oleifera* and *Codia myxa* are major plant species of the zone. Many mesic species of rodents make their appearance in this rainfall zone. In addition to two gerbils (*T.indica and M.hurrianae*), *Millardia meltada, Nesokia indica* and *Bandicota bengalensis* are the major rodent species of the zone.

Major rodent species in relation to various rainfall zones are given in Table 32.4.

Nature and Extent of Damage

Indian arid region of the western Rajasthan harbours considerable population of rodents, which exert greater pressure on its natural resources. Rodents are found to exploit natural grassland, crop fields and meager forest plantation of the region. Most of the rodents are opportunistic feeder and their feeding habits vary greatly according to availability of food (Singh, 1995). During monsoon, when their reproductive activities are at the peak, they consume leaves and shoots of the plants. In winter, they feed on seeds. In summer when no plant material is available for consumption they even start feeding on insects (Prakash, 1962).This rotation in feeding habits poses a great threat to vegetation of desert as they consume almost all parts of living plants, greatly affecting their generation process.

Table 32.4: Association of Rodent Species with Vegetation in Relation to Rainfall Zones

Rainfall Zone	*Major Vegetation*	*Associated Rodent*
100-250mm	*Calligonum, Haloxylon, Leptadenia, Clerodendrum*	*G. gleadowi, M. hurrianae, T.indica*
250-350mm	*Aerva, Calotropis, Citrullus, Colophosphermus*	*M. hurrianae, T.indica* and *N. indica*
350mm and above	*Prosopis,Tecomella, Capparis, Salvadora*	*M. hurrianae, T.indica, M.meltada, B. bengalensis*

These rodents cause severe damage in grasses and forest plantations during summer. In inimical condition of summer when gerbils faces scarcity of water and food, therefore, they start feeding on rhizomes of grasses and debarked *P. aculeata, A. tortalis, Albizzia lebbek* and *P. juliflora* plantations. The debarking activity is restricted to about 0.5 m above ground surface and sometimes even extends to lateral branches. Usually the cortical cells of the stem are debarked with a detrimental effect on tree's growth. At times even the xylem vessels of the tree are injured, causing its eventual death. This activity of debarking has been observed in trees of 3 to 4 years of age (Prakash, 1976). *M. hurrianae, T. indica* and *M. meltada* are major species associated with debarking activity. According to one estimate of Jain and Tripathi (1988), *Nesokia indica* inflict 44 per cent damage to *A.tortilis,* 10 per cent to *P. juliflora* and 10 per cent to *A. nilotica* due to slicing activities. Similar type damages in range of 10-12 per cent to *A. tortilis* have been observed by the author in the Talchhapar Blackbuck Sanctuary, situated in Churu district of Rajasthan. Barnett and Prakash (1975) have reported as much as 20 per cent loss to *P. juliflora, Albizzia lebbek* and *A.tortilis* due to girdling activity of merion gerbil, *M.hurrianae.*

As the population of rodents in Indian arid zone is dense as compared to that in the adjoining areas. The dietary demand of the rodent population, which fluctuates from 7.4 to 523 per hectare, is so insatiable that it maintains an appreciable pressure on the already sparse vegetation and rangelands. The total food requirement of the rodent population (477/ha.) in the Rajasthan desert was found to be 1044 kg/ha. The annual production of edible (for sheep) grass species was estimated to be, on an average 865 kg/ha and total forage production to be of 1100 kg/ha. Rodents can consume over a year the entire production of edible fodder, leaving virtually nothing for the livestock.

The forest nursery in the arid ecosystem is another habitat where these rodents inflict maximum damage. Due to perpetual availability of food, water and shelter, these rodents increase unabated in these microhabitats. *Funambulus pennanti, Hystrix indica*, *T. indica* and *M.hurrianae* are major rodent species associated with forest nurseries. These rodents cause immense damage to pre-plantation stage by consuming sown seeds and at post plantation stage by gnawing samplings and young trees.

Strategies for Rodent Pest Management

Pasture ecosystem as such provides congenial environment to arboreal as well as burrowing rodents. For sustainable and successful implementation of forestation programmes, the management of rodent pests is prerequisite. Rodents are highly evolved mammals and can readily adapt to new habitats created by man. Certain behaviour of rodents, like trap avoidance, bait shyness, neophobia, resistance to antocoagulants and shyness to acute poisons, make them one of the most difficult group of pest to control.

Several methods have been tried and tested but no single method or strategy is good enough to keep their population below thresholds level. It is thus necessary that more than one management strategy should be applied in an integrated manner to control their population. Generally two methods, lethal and nonlethal, are being used to manage rodent pests.

Nonlethal Methods for Rodent Control

Nonlethal methods for rodent pest management include habitat manipulation, biological control and mechanical methods. The idea of this method is to remove basic needs of rodents such as food, water and shelter. Chemosterilants and repellants are two other methods of nonlethal control of rodent pests. Following nonlethal methods are used to control rodent population:

Habital Manipulation

Food, Water and Shelter are main requirements for any species to thrive well. By manipulation these factor, rodents pests and their numbers can be reduced (Prakash and Mathur, 1987). Reduction of these factors will increase inter and intra-specific competition and thus cause decrease in the population through mortality or migration. Rodents prefer to build their burrows on bunds of the nurseries and plantation site. Such bunds should be of minimum possible height to reduce rodent infestation (Rana *et al.*, 1994, Kumar *et al.*, 1995). If necessary to have bunds then it is suggested that *Opuntia* and other cacti should be grown on them to deter rodents from establishing colonies on them. Habitat manipulation by physical barriers to protect structure from rodent invasion is known as "Rodent Proofing" and is another method of rodent control (Fitzwater, 1988). Metal bands of 1.5 to 2.0 widths have been used around the palm and other trees to keep away squirrels and rats. These bands should be placed so high that rodent's can not jump over them.

Polythene bags, straw and other material are used in preparation of nursery beds and such leftovers are dumped in the nurseries as garbage. Such garbage provides shelter to rodents and helps in establishing colonies there. Likewise, weeds in the area also provide food as well as hiding places to rodents, therefore, should not be allowed to grow in the nurseries and their vicinity.

Biological Control

Biological controls used to suppress rodent populations are parasites, predators and antifertility agents. Various developmental activities such as over exploitation of land and forest resources, developments of roads, industrialization, urbanization and changed agriculture patterns have produced imbalance in natural control of rodents.

Predators

Rodents have important role in ecosystem, as they are food of many carnivorous animals like, desert cat, jungle cat, monitor lizard, mongoose, hedgehogs, jackals, foxes, snakes and many raptors (Prakash and Mathur, 1987). About 60.2 percent of the winter food of spotted owlet, *Athene brama indica* was composed of rodents, especially *Mus* (Jain *et al.*, 1982). However, due to habitat destruction and other anthropogenic activities the population of these carnivorous animals is declining very fast. As a result, the prey population (rodents) is increasing unabated.

Parasites and Diseases

Various microorganisms (viruses, bacteria and protozoa) and macroparasites (Helminthes and arthropods) have been successfully used as bio-control agents against rodents in other countries (Singleton and Redhead, 1990). Very little research on this aspect of rodent control has been carried out in India. Deoras (1964) tried *Salmonella typhimurium* and *S. enteritidis* as parasites on *Bandicota bengalensis* and *Rattus rattus*, but recorded poor results. Bindra and Mann (1975) recorded 20 per cent mortality in *Mus musculus bacterianus* and 40 per cent mortality in *Tatera indica* in laboratory trials with typhus fever bacterium strain – 5170.

Antifertility Agents

What makes rodents most successful animals on the earth is their ability to produce large population in relatively short period. Using chemicals to inhibit reproduction in rodents is known as Chemosterilants. A number of Chemosterilants like clomiphene (Kalra and Prasad, 1966), tetradifon (Sagar and Batth, 1968), cadmium chloride (Kar and Das, 1963), furadantin and colchicine (Srivastava, 1966; Srivastava and Nigam, 1975), alpha chlorohydrin (Saini and Parshad, 1991, 1993) and ethyl methane sulphaonate (Kaur and Parshad,, 1997) have been evaluated against several Indian rodents. The use of reproductive inhabitants is, of course, good approach for integrated rodent pest management but requires further laboratory and field trials on a large scale to establish its potential rodent control.

Mechanical Methods

This technique is applicable in cases where the use of rodenticides may cause secondary poisoning. Some tribes of Tamil Nadu, Andhra Pradesh and northeast states consume rodents as food. In such areas, mechanical removal of rodents is the best way of controlling their population.

Deep ploughing, flooding of burrows with rain or irrigated water are some of the methods used by farmers to destroy rodents before sowing. Irulas and Kuruva tribes of Tamil Nadu, Musahar of Bihar and Erukulas of Madhya Pradesh are professional rat catchers. Larger rodents like *Tatera indica and B. bengalensis* are caught physically using nets and catapults. They will dig out the rodents from burrows or even fumigate the burrows to flush out them. However, this physical killing is done often around the time of ripening of crop when maximum damage to crops has already been incurred. The catchers do not catch pregnant females because of that the rodent population revives very fast.

Trapping is one of the oldest techniques to minimize the rodent population (Fitzwater and Prakash, 1978). Mainly two types of traps *i.e.* kill traps and live traps are used for capturing rodents. Snap traps or break – back trap, Tanjore bow trap, Sherman live trap, Wonder trap or multi-catch trap and glue trap are a few types of traps used for trapping various rodent species. Trapping can be and effective means of rodent control for small population of *Funambulus pennanti, T. indica, R. rattus* and *M. musculus*. The desert gerbil, *Meriones hurrianae* tend to avoid trap because of neophobia and is not an easy to trap rodent. For such rodent traps should be laid in "off" condition for 3-4 days for acclimatization. Traps should be kept on runways or in the areas where fresh sign of rodent activities (dropping etc) are visible. Peanut butter should be used as bait material and traps should be checked regularly to avoid predation of captured rodents. Traps should be washed after each catch to avoid adverse effect caused by odour of previously captured rodent. Trapping can be used as follow up action after a chemical control operation.

Lethal Methods of Rodent Control

The use of toxic chemicals is the most effective, cheap and humane method of rodent control. In India, Zinc phosphide (acute poison), Aluminium phosphide (fumigant), Warfarin, Racumin (Ist generation anticoagulants) and Bromadiolone (2nd generation anticoagulant) are being commonly used rodenticides. The effectiveness of chemical control depends on the selection of an appropriate compound method and timing of application.

Number of acute rodenticides *viz.* Zinc phosphide, Aluminium phosphide, Arsenic trioxide, Alpha naphthyl thio urea (ANTU), Barium carbonate, Vacour (RH-787) and many other have been tested against Indian rodents for evaluation their efficacy and toxicity. Among all these, zinc phosphide is most widely used and effective rodenticide. With acute rodenticides, pre-baiting is essential. Pre-baiting (one kg bajra grain mixed with 20g vegetable/edible oil) should be carried out for 2 to 3 days. On fourth day, 2 percent zinc phosphide should be added to the bait (one kg bajra grains +20 g vegetable oil +20g zinc phosphide) and rolled deep into the active burrow with the help of paper or pipe at the rate of 8-10gm/burrow. The death may occur within 2 hours of bait intake but in some cases may take place after 2-3 days. Major problem with acute rodenticides is poison shyness among sub-lethally poisoned rodents. It is observed that rodents avoid zinc phosphide for 6 to 170 days after sub – lethal administration of the dose. It is, therefore, suggested that repeated use of zinc phosphide in same habitat should be avoided.

The anticoagulants have characteristic of an ideal rodenticide slow action, high palatability, low risk and easy application. Because of slow action of these poisons, pre-baiting is not essential. The remaining population of rodents, after zinc phosphide operation, can be eliminated using second-generation anticoagulants or by fumigation. Bromadiolone, Brodifacoum, Flocoumafen and Difethiolone are a few second-generation anticoagulants tested against Indian rodents (Mathur and Prakash, 1981a, 1982b). After 8-10 days of application of zinc phosphide, Bromadiolone (0.005 per cent) ready to use loose bait or wax block should be rolled deep into freshly opened burrow for managing residual population. The toxic effect of second-generation anticoagulants begins after 2-3 days of ingestion and most deaths occur between 4 to 10 days. Poison-shy rodents also eat sufficient poison bait for a complete kill, as they do not develop aversion to it (Parshad and Kochar, 1995)

Fumigation technique is applied where baiting and trapping are not very effective. Hydrogen cyanide, aluminum phosphide, methyl bromide, ethylene dibromide, carbon monoxide, carbon dioxide are a few fumigants tried against Indian rodents (Prakash and Mathur, 1987). Barnett and Prakash (1975) stated that aluminium phosphide pellets are more effective than other fumigants. Krishnamurthy and Singh (1967) advocated that fumigation is better choice for rodent control because danger of poisoning of other animals is nil, lethal concentration is built up quickly in the burrow and persist for considerable duration and involves low cost.

Many chemicals have been tried and tested for there repellent action Malathion and cyclohexamides were effectively used against Norway rats. Majumdar *et, al.* (1964) reported repellents effect of malathion based formulation of *Rattus rattus*. Two fungicides – copper oxychloride and thiram (Tetra methyl thiuram disulphide) have been used against *R. rattus* and *B. bengalensis* to see their behavioural response. Parshad *et al.* (1993) observed that the surface application of 1.5 per cent and 4.5 per cent solution of copper oxychloride and thirom in water and peanut butter respectively to the cartons and their contents protected them from *R. rattus* for 30 days. Application of these chemicals on tree trunks may provide protection form debarking activities.

Timing of Control Operation

Best period of rodent control operation is the lean period. Studies on population dynamics indicate that lowest number of rodents occur during May and June. During this period, food and shelter are scarce and the rodents easily accept poison bats. It is thus evident that summer months are most appropriate months for rodent control operation. Studies under social Engineering Programme on rodent control at CAZRI, Jodhpur indicate that if control operations are taken up twice a year (1st time in May –June and 2nd time in November –December, regularly for four years, 95 per cent of the rodent population can be eliminated.

References

Barnett, S. A. and Prakash, I. 1975. *Rodents of Economic Importance in India.* Arnold Heinemann, New Delhi and London, pp 175.

Bindra, O.S. and Mann, G.S. 1975. Ineffectiveness of murinae typhus fever bacterium against field rats and mice at Ludhiana. *Indian J. Plant Prot.*, 3: 98:99.

Blanford, W. T. 1877. *The Fauna of British India, Mammalia.* Taylor and Francis, London.

Deoras, P. J. 1964. Rats and their control: A review of an account of work done on rats in India. *Indian J. Ent.*, 26:407-408.

Fitzwater, W. D. 1988. Nonlethal methods in rodent control. In: *Rodent Pest Management* (Ed. I. Prakash). CRC Press,

Fitzwater, W. D. and Prakash, I 1969. Burrows behavior and home range of the Indian desert gerbil, *Meriones hurrianae* (Jerdon). *Mammalia*. 33:598-606.

Fitzwater, W. D. and Prakash, I. 1978. *Hand book of vertebrate Pest Control*. ICAR, New Delhi: 1-92.

Jain, A. P. Advani, R. and Prakash I. 1982. Winter food of spotted owlet, *Athene brama indica*. *J. Bombay nat Hist*. Soc. 80-:515-416.

Jain, A. P. and Tripathi, R. S. 1988. Incidence and extent of damage by *Nesokia indica* in the afforestation ecosystem of Rajasthan. Paper presented in seminar on *Advances in Economic Zoology*, Jodhpur univ. Jodhpur.

Kalra, S. P. and Prasad, M.R.M. 1966. Effect of clomiphene on fertility in female rats. *Indian J. Exptl. Bio*. 5:5-8.

Kar, A. B. and Das, R. P. 1963. The nature of protective action of selenium on cadmium-induced degeneration of the rat testis. *Proc. Natl. Inst. Sci.*, India. 92:297-305.

Kaur, R. and Parshad, V.R. 1997. Ethyl methane sulphonate induced changes in the differentiation, structure and function of spermatozoa of the house rat, *Rattus rattus*. *J. Biosci*. 22:357-365.

Krishnamurthy, K. and Singh, P. 1967. Studies on rats and their control,. II Control of field rats with aluminium phosphide. *Bull. Grain Technol*. 5: 173-175.

Kumar, P., Pasahan, S.C., Sabhlok, V.P. and Singhal, R. K. 1995. Impact of bund dimension on rodent infestation. *Rodnt Newsl*. 19:5-6.

Majumdar, S.K., Krishna Kumari, M. K. and Krishna Rao, J. K. 1964. Malathion as a repellent for rats. *Curr. Sci*. 33: 212.

Mathur, R. P. and Prakash, I. 1981a. Comparative efficacy of three anticoagulant rodenticides for the control of desert rodents. *Prot. Ecol*. 3:327-331.

Mathur, R. P. and Prakash I. 1981b. Evaluation of Brodifacoum against *T. indica*, *M. hurrianae* and *R. Rattus*. *J. Hyg. Camb*. 87: 463-468.

Parshad, V. R. and Kochar, J. K. 1995. Potential of three rodenticides to induce conditional aversion to their baits in the Indian mole rat, *Bandicota bengalensis*. *Appl. Anim. Behav. Sci*. 45:267-276.

Parshad, V. R., Saini, M.S. and Jindal, S. 1993. Repellent action of two fungicides against the house rat, *Rattus rattus* and the Indian mole rat, *Bandicota bengalensis*. *Int. Bidet. Biodge*. 31:77-82.

Praksah, I. 1962. Taxonomical and ecological account to the mammals of Rajasthan Desert. *Ann. Arid Zone*. (1 and 2): 142-163.

Prakash, I. 1963. Zoogeography and evolution of the mammalian fauna of Rajasthan desert. *Mammalia*, 27(3): 342-351

Prakash, I. 1974. The ecology of vertebrates of the Indian desert. In : *Biogeography and Ecology in India* (Ed. M. Mani) Dr. Junk. Ve. Verlag, The Hague. 369-420.

Prakash, I. 1975. Population and ecology of rodents in the Rajasthan desert, India. In: *Rodents in Desert Environment*. (Eds. I. Prakash and P. K. Ghosh). Dr. W. Junk. Publishers, The Hague. 75-116.

Prakash, I. 1976. *Rodent Pest Management, Principles and Practices*. Manograph no. 4, Centrals Arid Zone Research Institute, Jodhpur. Pp28

Prakash, I. 1995. Invasion of peninsular small mammals towards the Aravalli ranges and the Thar Desert. *Intl.J. Ecol. and Envtl. Sci.*, 21(1) : 17-24.

Prakash, I. and Jain, A. P. 1971. Some observation on the Wagner's Gerbil, *Gerbillus nanus indus* Thomas, in the Indian Desert. *Mammalia*, 35 (4): 614-628.

Prakash, I. and Mathur, R. P. 1987. *Management of Rodent Pest*. Indian Council of Agricultural Research, New Delhi. Pp 133

Prakash, I. and Rana, B.D. 1972. A study of field population of rodent in the Indian desert. II. Rocky and piedmont zones. *Z. angew. Zool.* 59:129-139.

Prakash, I. and Rana, B. D. 1973. A study of the field population of rodents in the Thar desert. III. Sand dunes in 100mm rainfall zone. *Z angew. Zool.* 60:31-41.

Prakash, I. and Rana. B. D. 1970. A study of field population of rodents in the Indian Desert. *Z. angew Zool.* 62: 339-348.

Prakash, I. and Singh, P. 2005. *Ecology of Small Mammals of Desert and Montane Ecosystems*. Scientific Publishers, Jodhpur.

Prakash, I. Gupta, R. K., Jain, A. P., Rana, B. D. and Dutta, B.K. 1971. Ecological evaluation of rodent population in the desert biome of Rajasthan. *Mammalian*, 35 (3): 384-423.

Ragovin, K. A., Shenbrot, G.I., Surov, A. V. and Idris, M. 1994. Spatial organization of a rodent community in the western Rajasthan desert (India). *Mammalia*, 58(2):243-260.

Rana, B. D. 1983. The role of rodents in silvi pastoral system in Rajasthan. *Intl. Tree Crop Journal*. 2: 267-272.

Rana, B. D., Jain, A.P. and Tripathi, R.S. 1994. *Fifteen years of coordinated research on rodent control*. AICRP on Rodent control, CAZRI, Jodhpur pp 148.

Sagar, P. and Batth, S.S. 1968. Preliminary studies on Tetradifon (Tedion) as Chemosterilants for albino rat, *Mus norvegicus*. *Proc. Intl. Symp on Bionomics and control of Rodents*. 121-123.

Saini, M.S. and Prashad, V. R. 1991. Control of Indian mole rat with alpha chlorohydrin: Laboratory studies on bait acceptance and antifertility effects. *Ann. Biol.* 188:239-437.

Saini, M. S. and Prashad, V.R. 1993. Field evaluation of alpha chlorohydrin again the Indian mole rat : Studies on toxic and antifertility effects. *Ann. Appl. Biol.* 122:153-160.

Saxena, S. K. 1977. Vegetation and its succession in the Indian desert. In: *Desertification and Its Control* (Ed. P.L. Jaiswal) ICAR, New Delhi. 177-192.

Singh, P. 1995. Ecology, Population Structure and Behaviour of Cutch rock-rat, *Cremnomys cutchicus* in the Aravallis. Ph.D. Thesis (unpublished), JNV University, Jodhpur.

Singh, P. and Prakash, I. 2004. Ecological Impact of Indira Gandhi Canal on the Thar Desert. In: *Biodivrsity and Environment* (Ed. A. Kumar). APH Publishing Corp., New Delhi.

Singhvi, A. K. and Kar, A. 1992. (Eds.). *Thar Desert in Rajasthan: Land, Man and Environment Geological Society of India*, Bangalore: pp 191.

Singleton, G. and Redhead, T. 1990. Future prospects for biological control of rodents using micro and macroparasites. In: *Rodents and Rice* (Ed. G.R. Quick). Manila. 75-82.

Srivastava, A. S. 1966. Mixture of furadantin and colchicines acts as effective Chemosterilants against male female rats. *Labdev. J. Sci. Technol.* 4:178-180.

Srivastava, A. S. and Nigam, P. M. 1975. Mechanism of action of Chemosterilants in field rats. I bengales Role of furadanitin and colchicines in causing sterility in lesser bandicoot, *Bandicota bengalensis bengalensis* Gray. *Proc. All India Rod. Semi.* Ahmedabad. 239-246.

Annexures

Table 1: Demand, Supply and Deficits of Fodder in the Country (million tonnes): Past, Present and Future

Year	*Supply*		*Demand*		*Deficit as % Demand*	
	Green	*Dry*	*Green*	*Dry*	*Green*	*Dry*
2003	387.7	437.3	1006	560.1	61.51	21.81
2005	389.8	441.6	1021	568.0	61.83	22.12
2010	395.2	452.7	1057	588.2	62.63	22.91
2020	406.6	475.7	1134	630.9	64.26	24.57

Table 2: List of Genera and some of their Wild and Weedy Relatives that have been/ can be Used as Potential Forage

Genus	*Species including Wild and Weedy Relatives*
Acacia sp.	*A. difficilis, A. mangiu, A. torulosa* and *A. tumida var. tumida*
Agropyron sp.	*Agropyron cristateum, A. cimmericum*
Astragalus sp.	*Astragalus alpinus, Astragalus adsurgens*
Avena sp.	*A. abyssinica kochisi, A. barleota, A. byzantina, A. canariensis, A. claida, A. faba, A. hirtula, A. insalavis, A. longiglumis, A. lucovicana, A. macrostachya, A. pilosa, A. saviloviana, A. strigosa* and *A. weistii*
Brachiaria sp.	*B. ruziziensis*
Brassica sp.	*Brassica abyssinica B. orientalis*
Centrosema	*Centrosema pubescens*
Casuarina sp.	*Casuarina cunninghamiana* ssp. *cunninghamiana*
Chenopodium sp.	*C. capitatum, C. gigantenum, C. pallidicaule* and *C. quinoa*
Clitoria sp.	*Clitoria ternetea*
Centrosema sp.	*Chloris gayana, Chloris dactylis*
Cyamopsis sp.	*C. senegalensis* and *C. serrata*
Dactylis sp.	*Dactylis glomerata,*
Desmodium sp.	*Desmodium barbatum, D. intortum* and *D. uncinotum*
Eleusine sp.	*Eleusine indica, E. africana E. fleceifolia, E. jaegeri, E. multiflora* and *E. tristachya*
Eragrostis sp.	*Eragrostis curvul*
Elymus sp.	*Elymus haffmannii*
Eucalyptus sp.	*E. argophloia, E. camaldulens var. obtusa E. tereticornis* ssp. *tereticornis*
Festuca sp.	*Festuca arundinacea*
Glycine sp.	*G. argyrea, G. canescens, G. centennial, G. clandestina, G. cyrtoloba, G. falcata, G. javanica, G. latifolia, G. latrobeana, G. microphylla, G. soja, G. tabacina and G. tomentella*
Hevea sp.	*H. camarguana, H. colina, H. fousiflora, H. guyanensis, H. nitida* and *H. pauciflora*

Table 3: Minimum Seed Standards for Forage Crops

Crop	Class of Seed	Minimum		Maximum Permissible Limit							
		Germination (%)	Pure Seeds (%)	Inert Matter (%)	Other Crop Seeds (No.kg)	Weed Seeds (No./kg)	Objection-able Weeds (No./kg)	Diseased Seeds (%/by no.)	Other Distinguish Variety Seeds (No./kg)	Moisture Ordinary Pack	Moisture Vapour Proof
Sorghum, Sudan grass	FS	75	97	3	05	5	–	0.02	10	12	8
	CS	75	97	3	10	10	–	0.04	20	12	8
Pearlmillet	FS	75	98	2	10	10	–	0.02	–	12	8
	CS	75	98	2	20	20	–	0.04	–	12	8
Berseem, Lucerne	FS	80	98	2	10	10	5	–	–	10	7
	CS	80	98	2	20	20	10	–	–	10	7
Buffle grass,	FS	30	80	20	20	20	–	–	–	10	8
Birdwood grass	CS	30	80	20	40	40	–	–	–	10	8
Dharaf grass	FS	15	80	20	20	20	–	–	–	10	8
	CS	15	80	20	40	40	–	–	–	10	8
Dinanath grass	FS	50	95	5	20	20	–	–	–	10	8
	CS	50	95	5	40	40	–	–	–	10	8
Gwar	FS	70	98	2	10	None	–	–	10	9	8
	CS	70	98	2	20	None	–	–	20	9	8
Gunia grass	FS	20	80	20	20	20	–	–	–	10	8
	CS	20	80	20	40	40	–	–	–	10	8
Senji	FS	65	98	2	10	10	–	–	10	10	7
	CS	65	98	2	20	20	–	–	20	10	7

Contd...

Table 3–Contd...

Crop	Class of Seed	Minimum		Maximum Permissible Limit							
		Germination (%)	Pure Seeds (%)	Inert Matter (%)	Other Crop Seeds (No.kg)	Weed Seeds (No./kg)	Objection-able Weeds (No./kg)	Diseased Seeds (%/by no.)	Other Distinguish Variety Seeds (No./kg)	Moisture Ordinary Pack	Moisture Vapour Proof
Marvel grass	FS	40	90	10	10	–	–	–	10	10	8
	CS	40	90	10	20	–	–	–	20	10	8
Oats	FS	85	98	2	10	10	2	–	10	12	8
	CS	85	98	2	20	20	5	–	20	12	8
Ricebean	FS	70	98	2	None	5	–	–	10	9	8
	CS	70	98	2	5	10	–	–	20	9	8
Setaria grass	FS	50	95	5	20	–	–	–	–	10	8
	CS	50	95	5	40	–	–	–	–	10	8
Stylo	FS	40	90	10	10	10	–	–	10	10	8
	CS	40	90	10	20	20	–	–	20	10	8
Teosinte	FS	80	98	2	5	None	–	–	–	12	8
	CS	80	98	2	10	None	–	–	–	12	8

FS: Foundation Seed; CS: Certified Seed.

Table 4: Minimum Field Standards for Forage Crops

Crop	Class of Seed	Minimum		Maximum Permissible Level (Per cent)			
		Isolation (Metre)	No. of Field Inspections	Off Types	Inseparable Other Crop Plants	Objectionable Weed Plants	Plants/ Heads Affected by Designated Diseases
Sorghum	FS	200	3	0.10	–	–	0.05
	CS	100	3	0.20	–	–	0.10
Pearlmillet	FS	400	3	0.05	–	–	0.05
	CS	200	3	0.10	–	–	0.10
	CS		4				
Berseem, Lucerne	FS	400	2	0.2	–	None	–
	CS	100	2	1.0	–	0.05	–
Buffle grass, Birdwood grass,	FS	20	3	0.1	–	–	–
Guinea grass, Dharaf grass	CS	10	3	1.0	–	–	–
Marvel grass,	FS	50	3	0.1	–	–	–
Dinanath grass, Stylo	CS	25	3	1.0	–	–	–
Setaria grass	FS	400	3	0.20	–	–	–
	CS	200	3	0.10	–	–	–
Indian clover (Senji)	FS	50	2	1.0	–	–	–
	CS	25	2	0.20	–	–	–
Oats	FS	3	2	0.05	0.01	0.01	0.10
	CS	3	2	0.20	0.05	0.02	0.50
Ricebean	FS	50	3	0.10	–	–	–
	CS	20	3	0.20	–	–	–
Teosinte	FS	200	3	0.10	–	–	–
	CS	100	3	0.50	–	–	–

Table 5: Testing of Forage Seeds (Germiantion Methods)

Crop	Substrata	Temperature °C	First Count (Days)	Final Count (Days)
Sorghum	BP, TP	20–30, 25	4	10
Pearl millet	BP, TP	20–30, 25	3	7
Berseem	BP, TP	20	3	7
Lucerne	TP, BP	20	4	10
Gunia grass	TP	15–35	10	28
Napier grass	BP	20–30	3	10
Buffel grass	TP, S	20–35, 20–30, 30	7	28

Contd...

Table 5–Contd...

Crop	*Substrata*	*Temperature °C*	*First Count (Days)*	*Final Count (Days)*
Birdwood grass	TP	20–35	3	14
Doob grass	TP	20–35	7	21
Dharaf grass	TP	20–35	7	28
Dinanath grass	TP	35, 20–35	7	28
Senji	TP, BP	20	4	7
Marvel grass	TP	20–30	7	21
Para grass	TP	20–30	–	21
Rice bean	BP	20–30	5	8
Setaria grass	TP	20–35	7	21
Sudan grass	TP, BP	20–30	4	10
Stylo	TP	20–35	4	10
Teosinte	BP,S	20–30, 25	–	7
Clovers	BP, TP	20	3–5	7–14
Lupins	BP, S	20	4–10	10–21
Vetches	BP, S	20	4–5	8–14

BP: Between paper; TP: Top of the paper; S: Sand.

Table 6: Lot Size and Sample Weights of Forage Crops

Crop	*Maximum Weight of Seed Lots (kg)*	*Submitted Sample (g)*	*Working Sample for Purity (g)*	*Working Sample for Count of Other Species (g)*
Sorghum	10000	900	90	900
Pearl millet	1000	150	15	150
Oats	20000	1000	120	1000
Berseem	10000	60	6	60
Lucerne	10000	50	5	50
Indian clover	10000	100	10	100
Teosinte	20000	1000	900	1000
Gunia grass	10000	25	2	20
Napier grass	10000	150	15	150
Buffel grass, Birdwood grass	10000	25	3	25
Doob grass	10000	25	1	10
Marvel grass, Para grass	10000	30	3	30
Setaria grass	10000	25	2	20
Stylo	10000	70	7	70
Sudan grass	10000	250	25	250
Lupins	20000	1000	450	1000

Table 7: Conversion Factors for English and Metric Units

To Convert Column 1 Into Column 2, Multiply by	*Column 1*	*Column 2*	*To Convert Column 2 Into Column 1 Multiply by*
Length			
0.621	kilometer, km	mile, mi	1.609
1.094	meter, m	yard, yd	0.914
0.394	centimeter, cm	inch, in.	2.54
Area			
0.386	kilometer², km²	mile², mi²	2.590
247.1	kilometer², km²	acre, a	0.00405
2.471	hectare, ha	acre, a	0.405
Volume			
0.00973	meter³, m³	acre-inch, a-in.	102.8
3.532	hectoliter, hi	cubic foot, ft³	0.2832
2.838	hectoliter, hi	bushel, bu	0.352
0.0284	liter, l	bushel, bu	35.24
1.057	liter, l	quart (liquid), qt	0.946
Mass			
1.l02	ton (metric)	ton (English)	0.9072
2.205	quintal, q	hundredweight, cwt (short)	0.454
2.205	kilogram, kg	pound,lb	0.454
0.035	gram, g	ounce (avdp.), oz	28.35
Pressure			
14.50	bar	lb/in.², psi	0.06895
0.9869	bar	atmosphere, atm	1.013
0.9678	kg weight)/cm²	atmosphere, atm	1.033
14.22	kg weight)/cm²	lb/in. ², psi	0.07031
14.70	atmosphere, atm	lb/in. ², psi	0.06805
0.1450	kilopascal, kPa	lb/in.², psi	6.895
0.009869	kilopascal, kPa	atmosphere, atm	101.30
Yield or Rate			
0.446	ton(metric)/hectare, ton/a	ton (English)/acre, ton/a	2.240
0.891	kg/ha	Lb/a	1.12
0.891	quintal/hectare, q/ha	hundredweight/acre, cwt/a	1.12
1.15	hectoliter/hectare, hl/ha	bushel/acre, bula	0.87

Table 8: Conversion Factors

Multiply	*By*	*To get*
acres	0.4048	hectare
acres	43,560	square feet
acres	160	square rods
acres	4,840	square yards
bushels	4	pecks
bushels	64	pints
bushds	32	quarts
centimeters	0.3937	inches
centimeters	0.01	meters
cubic feet	0.03382	ounces (liquid)
cubic feet	1,728	cubic inches
cubic feet	0.03704	cubic yards
cubic feet	7.4805	gallons
	29.92	quarts (liquid)
cubic yards	27	cubic feet
cubic yards	46,656	cubic inches
cubic yards	202	gallons
feet	30.48	centimeters
feet	12	inches
feet	0.3048	meters
feet	0.060606	rods
feet	Y3 or 0.33333	yards
feet	0.01136	miles per hour
gallons	0.1337	cubic feet
gallons	4	quarts (liquid)
gallons of water	8.3453	pounds of water
grams	15.43	grams
grams	0.001	kilograms
grams	1,000	milligrams
grams	0.0353	ounces
grams per liter	1,000	parts per million
hectares	2.471	acres
inches	2.54	centimeters
inches	0.08333	feet
kilograms	1,000	grams
kilograms	2.205	pounds
kilograms per hectare	0.892	pounds per acre
kilometers	3,281	feet

Table 9: Forage Crop-wise Status of Germplasm in India

Sl.No.	Forage Crop	Germplasm Status (Location-wise)
(A)	**Cereak Forages**	
1.	Sorghum (Jowar)	Hisar (290), Pantnagar (150), Jhansi (410), Dharwad IGFRI-RRS (49), Anand (752), Jabalpur (153), Hyderabad (260), Rahuri (9), Ludhiana (12)
2.	Pearl millet (Bajra)	Faizabad (51), Avikanagar IGFRI-RRS (70), Dharwad IGFRI-RRS (89), Anand (8), Hyderabad (27), Vellayani (85), Rahuri (4), Ludhiana (6), Coimbatore (102)
3.	Maize	Bhubneswar (25), Anand (102), Hyderabad (160), Rahuri (43), Pusa (15), Jhansi (156)
4.	Teosinte	Jhansi (9), Jorhat (1), Ludhiana (51)
5.	Oats	Hisar (520),Jhansi (1200), Rajouri (8)
6.	Coix	Jorhat (1), Kalyani (26)
7.	Sudan grass	Ludhiana (34)
(B)	**Leguminous Forages**	
8.	Cowpea	Hisar (34), Pantnagar (615), Dharwad IGFRI-RRS (6), Anand (67), Jabalpur (58), Vellayani (59), Rahuri (58), Bikaner (100), Ludhiana (292), Jhansi (232), Coimbatore (58), Rajouri (10), Hyderabad (1)
9.	Guar	Hisar (390), Ludhiana (126), Avikanagar IGFRI-RRS (231), Vellayani (2)
10.	Ricebean	Bhubneswar (97), Jorhat (97), Vellayani (100), Kalyani (20)
11.	Berseem	Hisar (252), Jhansi (894), Palampur CSK HPKV (5), Rajouri (10), Palampur IGFRI-RRS (5)
12.	Lucerne	Jhansi (450), Dharwad IGFRI-RRS (28), Anand (88), Hyderabad (13)
13.	Lablab purpureus	Jhansi (220), Dharwad IGFRI-RRS (5), Jorhat (12), Rahuri (16), Jodhpur (46), Vellayani (3)
14.	Soybean	Jabalpur (27)
(C)	**Grasses**	
15.	Hybrid Napier	Jorhat (7), Hyderabad (21), Vellayani (2), Coimbatore (54), Ludhiana (36), Faizabad (36), Anand (49), Rahuri (63), Palampur IGFRI-RRS (1)
16.	Guinea grass	Jhansi (198), Hyderabad (8), Vellayani (90), Rahuri (29), Ludhiana (32), Faizabad (43), Dharwad IGFRI-RRS (2), Coimbatore (114)
17.	Cenchrus ciliaris	Jhansi (352), Hyderabad (12), Rahuri (16), Jodhpur (188)
18.	Cenchrus setigerus	Jhansi (56), Jodhpur (49)
19.	Sehima nervosum	Jhansi (25), Hyderabad (12), Rahuri (7)

Contd...

Table 9–Contd...

Sl.No.	*Forage Crop*	*Germplasm Status (Location-wise)*
20.	*Lasiurus sindicus*	Jodhpur (172)
21.	Cocksffot	Palampur IGFRI-RRS (2), Palampur CSK HPKV (25)
22.	Rye grass	Palampur IGFRI-RRS (2)
23.	Tall fescue	Palampur CSK HPKV (30), Palampur IGFRI-RRS (2)
24.	Setaria	Palampur CSK HPKV (40), Almora (5), Dharwad IGFRI –RRS (1), Palampur IGFRI-RRS (2), Rajouri (1)
25.	*Dichanthium annulatum*	Jhansi (187), Coimbatore (150)
26.	*Chrysopogon fulvus*	Jhansi (18), Palampur CSK HPKV (6), Hyderabad (6), Rahuri (7)
27.	Heteropogon	Jhansi (75)
28.	Napier grass	Dharwad IGFRI-RRS (62), Jajouri (2), Vellayani (40), Ludhiana (23)
29.	Dinanath grass	Ranchi (3)
30.	*Pennitum* spp.	Jhansi (15)
31.	Other grasses	Faizabad (31), Jhansi (95), Dharwad IGFRI-RRS (18), Palampur IGFRI-RRS (12), Jorhut (28), Anand (20), Hyderabad (13), Udaipur MLSU (269), Palampur CSK HPKV (115), Jodhpur (91)
(D)	**Range Legumes**	
32.	Stylosanthes	Jhansi (416), Dharwad IGFRI-RRS (57), Anand (19), Hyderabad (36), Vellayani (22), Ranchi (14), Rahuri (30), Urulikanchan (42), Kalyani (20), Coimbatore (23)
33.	Siratro	Jhansi (54), Rahuri (2)
34.	Clitoria	Jhansi (98), Dharwad IGFRI-RRS (3), Kalyani (6), Jorhat (3), Jodhpur (30), Rahuri (16)
35.	Clovers	Palampur IGFRI-RRS (184), Palampur CSK HPKV (70)
36.	Other *Trifolium* spp.	Jhansi (15), Palampur IGFRI-RRS (18)
37.	*Lotus comiculatus*	Palampur IGFRI-RRS (23), Palampur CSSK HPKV (10)
38.	*Desmanthus virgatus*	Jhansi (57), Dharwad IGFRI-RRS (4)
39.	*Centrosema pubescens*	Jhansi (88)
40.	Other legumes	Faizabad (14), Jhansi (90), Dharwad IGFRI-RRS (27), Palampur IGFRI-RRS (2), Anand (10), Hyderabad (13), Palampur CSK HPKV (25), Ranchi (2), Jodhpur (4)
(E)	**Other minor forages**	
		Coimbatore (150), Kalyani (16), Hyderabad (6)

Table 10: Forage Varieties Released/Identified by IGFRI during 2008

Crops	*Varieties*	*Green Fodder Yield (t/ha)*	*Area*
Oat	Bundel Jai 822	44–50	Central zone
	Bundel Jai 851	40–50	Whole country
	Bundel sheet Jai 829	55–60	Hill temperate
Anjan grass	Bundel Anjan 1	30–35	Whole country
Dinanath grass	Bundel Dinanath 1	55–60	Whole country
	Bundel Dinanath 2	60–65	Whole country
Napier-	IGFRI–6	90–160	Central & NE zone
bajra hybrid	IGFRI–10	150–180	Whole country
Berseem	Wardan	70–80	Whole country
	Bundel Berseem–1	65–70	South zone
	Bundel Berseem–2	65–85	Central & NW zone
Cowpea	Bundel Lobia 1	25–30	Whole country
	Bundel Lobia 2	25–30	NW zone
Guar	Bundel Guar 1	25–35	Whole country
	Bundel Guar 2	30–40	Whole country
	Bundel Guar 3	30–40	Whole country
Field bean	Bundel Sem 1	25–35	Whole country
Lucerne	Chetak	45–50	Whole country

Table 11: Forage Varieties Released/Identified by AICRPF during 2008

Sl.No.	*Name of Crop*	*Varities Recommended*
1	Cowpea	UPC-9202, UPC-607, UPC- 618
2	Pearl millet	AVKB-19
3	Sorghum	HJ 523
4	Guar	Bundel Guar-3
5	Ricebean	Bidhan-1, KRB-4
6	Coix	KCA-3, KCA-4
7	Guinea grass	PGG-616, JHGG 96-5
8	Anjan grass	Bundel Anjan-3
9	Sehima grass	IGS-9901
10	Chrysopogon grass	IGC-9903
11	Berseem	Bundel Berseem-3, BL-180
12	Oat	JHO 99-1JHO 99-2,JHO 2001-3, RO-19 (multicut)

Table 12: Important High-Yielding Cultivars/Varieties of Fodder Grasses, Cereal Fodders and Legumes Recommende for Cultivation in Different Parts of India

1.	**Jowar (*Sorghum vulgare*)**	
(A)	*Northern region*	
(*i*)	'J.S.-20', 'J.S.-29/1', 'J.S.-263', 'J3/53', 'Swarna', M.P. chari', 'S.L.-44', 'Pusa chari', 'Haryana chari'	Uttar Pradesh
(*ii*)	'T-3', 'T-4', '8B', 'M.P. chari', 'S-700', 'H_1', 'H_2', 'Rio'	Punjab, Haryana, Delhi
(B)	*Western region*	
(*iii*)	Sundhia 1049', 'Chhastia 10-2', 'Dudhia'	Gujarat
(*iv*)	'Red Khaki', 'Nilwa', 'Nandyal', 'M-35-1'	Maharashtra
(C)	*Central region*	
(*v*)	'Gwalior-82', 'Gwalior-304', 'Vidisha 60-1', 'Ujjain-6', 'Ujjain-8', 'J-195'	Madhya Pradesh
(D)	*Southern region*	
(*i*)	'K 3', 'Co 11', 'Co 18', 'Co 19', 'Rungu 1', 'M.P. chari'	Andhra Pradesh and Tamil Nadu
2.	**Maize (*Zea mays*)**	
(*i*)	*Hybrids*–'Ganga-Safed-2', 'Ganga-3', 'Ganga-5', 'Ganga-7'	All India
	Composites–'Jawahar', 'Amber', 'Kissan', 'Vijay', 'Sona', 'Vikram'	
	Open-pollinated–'N.P. Yellow-2', 'K-41', 'Bassi', 'Jaunpur'	
(*ii*)	'Emenillo de cuba', 'Kalimpong	West Bengal
3.	**Bajra (*Pennisetum typhoides*)**	
(*i*)	'A-1/3', 'H.B.3', 'S-530', 'T-55' 'D 1941', '2291'	All India
4.	**Oat (*Avena sativa*)**	
(*i*)	*Early varieties*–'Western-11'	All India
	Mid-season varieties–'Kent', 'Craig', 'Afterlee', 'Green Mountain', 'A-17', 'Flaminagolds', 'Fulgham', 'Bamboo-966', 'IGFRI-Soil-3021', 'IGFRI-Soil-2688'	
	Late varieties–'Algerian', '37/14', 'FOS-1/29', 'Kharsai'	
5.	**Cowpea (*Vigna sinensis*)**	
(*i*)	'FOS-1', 'FOS-10', 'K-395', 'K-585', 'EC 4216', 'IGFRI-S-450', 'IGFRI-S-457	Haryana, Punjab and Delhi
(*ii*)	'Russian Giant', 'IGFRI-S-978', 'IGFRI-S-985', 'Russian Giant'	Uttar Pradesh and Karnataka
(*iii*)	'E.C. 4216', 'Russian Giant', 'Chhrodi 14-20', 'Chhrodi 26-28'	Gujarat
(*iv*)	'Co 1', 'Russian Giant', 'E.C. 4216'	Southern states
(*v*)	'Co 1', 'E.C. 4216'	West Bengal

Contd...

Table 12–Contd...

6.	**Guar (*Cyamopsis tetragonoloba*)**	
(*i*)	'FOS 1', 'FOS 2', 'E.C. 4216F.S. 277', 'IGFRI-S 212', 'No. 2'	Punjab, Haryana, Delhi, Uttar Pradesh
(*ii*)	'Durgapura safed'	Rajasthan
7.	**Velvet bean (*Stizolobium niveum*)**	
(*i*)	'IGFRI-S 2276-5'	Uttar Pradesh, Rajasthan, Andhra Pradesh, Madhya Pradesh and Haryana
8	**Field bean (*Dolichos lablab* var. *lignosus*)**	
(*i*)	'IGFRI-S 2214-II',–Broad leaf, erect	
(*ii*)	'IGFRI-S 2218-1',–Medium leaf, decumbent	
9.	**Berseem (*Trifolium alexandrinum*)**	
(*i*)	'Meskawi-Diploid', 'Pusa giant-Tetraploid', 'IGFRI-S-29-1', 'Chhindwara'	
10.	**Lucerne (*Medicago sativa*)**	
(*i*)	'Type-8', 'Type-9','Anand II', 'IGFRI-S-244', 'Moopa', 'IGFRI-S-54', N.D.R.I.-1	
11.	**Senji (*Melilotus parviflora*)**	
(*i*)	'FOS-1', 'F.S. 14', 'F.S. 18'	
12.	**Methi (*Trigonella foenumgraecum*)**	
(*i*)	'FOS-8'	
13.	**Hybrid Napier**	
(*i*)	'Pusa Giant Napier', 'NB 21', 'EB 4', or 'Gajraj', 'N.B. 5', 'Coimbatore'	
14.	**Sudan grass (*Sorghum sudanese*)**	
(*i*)	'SS-59-3', 'G 287', 'Piper', 'J-69'	
15.	**Dinanath grass (*Pennisetum pedicellatum*)**	
(*i*)	'Type-3', '10', '15', or 'IGFRI-S-3808', 'G-73-1', 'T-12', 'IGFRI-S-866-1'	
16.	**Blue panicum (*Panicum antidotale'*)**	
(*i*)	'S-297'	
17.	***Anjan* or Buffel (*Cenchrus ciliaris'*)**	
(*i*)	'Pusa Giant Anjan', 'IGFRI-S-3108', 'IGFRI-S-3133', 'C-357', 'C-358', *Cenchrus glaucus*	
18.	**Bird wood (*Cenchrus setigerus'*)**	
(*i*)	'Pusa yellow Anjan'	
19.	**Marvel (*Dichanthium annulatum*)**	
(*i*)	'M-8', 'IGFRI-S-495-1', 'IGFRI-S-495-5'*Cenchrus glaucus*	
20.	**Mustard (*Brassica* spp.)**	
(*i*)	'Japan *sarson*', 'IM-98', 'IM-100', 'Laha 101', 'Chinese cabbage'	
21.	**Butterfly pea (*Clitoria ternatea'*)**	
(*i*)	'IGFRI-S-23-1', 'IGFRI-S-12'	
22.	**Turnips**	
(*i*)	'Green Top', 'Purple Top', 'Kenshin—Kaba'	

Table 13: Important Legume Used as Forage and Fodder

- ☆ *Trifolium africanum*
- ☆ *Trifolium albopurpureum*
- ☆ *Trifolium alexandrinum*
- ☆ *Trifolium amabile*
- ☆ *Trifolium ambiguum*
- ☆ *Trifolium amoenum* Greene - Showy Indian Clover (California in the United States)
- ☆ *Trifolium andersonii*
- ☆ *Trifolium andinum*
- ☆ *Trifolium angustifolium*
- ☆ *Trifolium arvense* L. - Hare's-foot clover (Europe, Western Asia)
- ☆ *Trifolium attenuatum*
- ☆ *Trifolium aureum* Pollich - Large Hop Trefoil (central and southern Europe)
- ☆ *Trifolium barbigerum*
- ☆ *Trifolium beckwithii*
- ☆ *Trifolium bejariense*
- ☆ *Trifolium bifidum*
- ☆ *Trifolium bolanderi*
- ☆ *Trifolium brandegeei*
- ☆ *Trifolium breweri*
- ☆ *Trifolium buckwestiorum*
- ☆ *Trifolium calcaricum*
- ☆ *Trifolium campestre* Schreb. - Hop Trefoil (Europe, Western Asia)
- ☆ *Trifolium carolinianum*
- ☆ *Trifolium cernuum*
- ☆ *Trifolium ciliolatum*
- ☆ *Trifolium cyathiferum* Cup clover (Western United States)
- ☆ *Trifolium dalmaticum*
- ☆ *Trifolium dasyphyllum*
- ☆ *Trifolium dedeckerae*
- ☆ *Trifolium depauperatum*
- ☆ *Trifolium dichotomum*
- ☆ *Trifolium douglasii*
- ☆ *Trifolium dubium* Sibth. - Lesser Hop Trefoil
- ☆ *Trifolium echinatum*
- ☆ *Trifolium eriocephalum*

Contd..

Table 13–Contd...

- ☆ *Trifolium fragiferum*
- ☆ *Trifolium friscanum*
- ☆ *Trifolium fucatum*
- ☆ *Trifolium glomeratum*
- ☆ *Trifolium gracilentum*
- ☆ *Trifolium gymnocarpon*
- ☆ *Trifolium haydenii*
- ☆ *Trifolium hirtum*
- ☆ *Trifolium howellii*
- ☆ *Trifolium hybridum* L. - Alsike Clover
- ☆ *Trifolium incarnatum* L. - Crimson Clover (Europe)
- ☆ *Trifolium jokerstii*
- ☆ *Trifolium kingii*
- ☆ *Trifolium lappaceum*
- ☆ *Trifolium latifolium*
- ☆ *Trifolium leibergii*
- ☆ *Trifolium lemmonii*
- ☆ *Trifolium longipes*
- ☆ *Trifolium lupinaster*
- ☆ *Trifolium macraei*
- ☆ *Trifolium macrocephalum*
- ☆ *Trifolium medium* L.
- ☆ *Trifolium michelianum*
- ☆ *Trifolium microcephalum*
- ☆ *Trifolium microdon*
- ☆ *Trifolium minutissimum*
- ☆ *Trifolium monanthum*
- ☆ *Trifolium mucronatum*
- ☆ *Trifolium nanum*
- ☆ *Trifolium neurophyllum*
- ☆ *Trifolium nigrescens* Viv. (Mediterranean Basin)
- ☆ *Trifolium obtusiflorum*
- ☆ *Trifolium oliganthum*
- ☆ *Trifolium olivaceum*
- ☆ *Trifolium ornithopodioides*

Contd..

Table 13–Contd...

- ☆ *Trifolium owyheense*
- ☆ *Trifolium parryi*
- ☆ *Trifolium patens* Schreb.
- ☆ *Trifolium pinetorum*
- ☆ *Trifolium plumosum*
- ☆ *Trifolium polymorphum*
- ☆ *Trifolium pratense* L. - Red clover (Europe, Western Asia, Northwestern Africa)
- ☆ *Trifolium productum*
- ☆ *Trifolium purpureum*
- ☆ *Trifolium pygmaeum*
- ☆ *Trifolium reflexum*
- ☆ *Trifolium repens* L. - Shamrock (white clover) (Europe, Northern Africa, Western Asia)
- ☆ *Trifolium resupinatum*
- ☆ *Trifolium rollinsii*
- ☆ *Trifolium rueppellianum*
- ☆ *Trifolium scabrum*
- ☆ *Trifolium semipilosum*
- ☆ *Trifolium siskiyouense*
- ☆ *Trifolium spumosum*
- ☆ *Trifolium squamosum*
- ☆ *Trifolium stoloniferum* Muhl. ex A. Eaton - Running Buffalo Clover (Eastern and Midwestern United States)
- ☆ *Trifolium striatum*
- ☆ *Trifolium subterraneum* L. - Subterranean clover (Northwestern Europe)
- ☆ *Trifolium suffocatum*
- ☆ *Trifolium thompsonii*
- ☆ *Trifolium tomentosum*
- ☆ *Trifolium trichocalyx*
- ☆ *Trifolium uniflorum*
- ☆ *Trifolium variegatum*
- ☆ *Trifolium vesiculosum*
- ☆ *Trifolium virginicum*
- ☆ *Trifolium willdenovii* Spreng. - Tomcat clover (Western United States, British Columbia in Canada)
- ☆ *Trifolium wormskioldii* Lehm. - Cow clover (Western United States, British Columbia in Canada, northern Mexico)

Table 14: Acronyms

ACIAR	–	Australian Centre for International Agricultural Research
AIT	–	Asian Institute of Technology
BOES	–	Bureau of Economics and Statistics
CADA	–	Command Area Development Authority
CBIP	–	Central Board of Irrigation and Power
CG Centres	–	Consultative Group Centres
CGIAR	–	Consultative Group on International Agricultural Research
CIAT	–	International Centre for Tropical Agriculture
CIFA	–	Central Institute of Fisheries and Aquaculture
CIFRI	–	Central Inland Fisheries Research Institute
CIMMYT	–	Centro Internacional de Mejoramiento de Maiz y Trigo (The International Maize and Wheat Improvement Center)
CRIDA	–	Central Research Institute for Dryland Agriculture
CRRI	–	Central Rice Research Institute
CSIR	–	Council of Scientific and Industrial Research
CSIRO	–	Commonwealth Scientific and Industrial Research Organization, Australia
CSU	–	Colorado State University
CSWCRTI	–	Central Soil and Water Conservation Research and Training Institute,
CTRI	–	Central Tobacco Research Institute, Rajahmundry
CWC	–	Central Water Commission
DSE	–	Delhi School of Economics
DST	–	Department of Science and Technology, Govt. of India
DWR	–	Directorate of Wheat Research
DWMR	–	Directorate of Water Management Research, Patna
ER	–	Eastern Region
FAO	–	Food and Agriculture Organization of the United Nations
GATT	–	General Agreement on Trade and Tariffs
GFCC	–	Ganga Flood Control Commission
IARI	–	Indian Agricultural Research Institute
IASRI	–	Indian Agricultural Statistics Research Institute
IBPGR	–	International Board for Plant Genetic Resources
ICARDA	–	International Center for Agricultural Research in the Dry Areas
ICFRE	–	Indian Council of Forestry Research and Education
ICRAF	–	International Council for Research on Agro–forestry
ICRISAT	–	International Crops Research Institute for the Semi-Arid Tropics
IDMI	–	International Data Management Institute

Contd..

Table 14–Contd...

IFAD	–	International Fund for Agricultural Development
IFPRI	–	International Food Policy Research Institute, Washington
IGFRI	–	Indian Grassland and Fodder Research Institute
IIHR	–	Indian Institute of Horticultural Research, Bangalore
IIP and DM	–	Indian Institute of Pests and Disease Management
IIPR	–	Indian Institute of Pulses Research
IIVR	–	Indian Institute of Vegetable Research
IIT	–	Indian Institute of Technology
IITM	–	Indian Institute of Tropical Meteorology
ILRI	–	International Livestock Research Institute, Nairobi, Kenya
IMD	–	Indian Meteorological Department, Pune
INCID	–	Indian National Committee on Irrigation and Drainage
IPGRI	–	International Plant Genetic Resources Institute
IRRI	–	International Rice Research Institute, Los Banos,
IVRI	–	Indian Veterinary Research Institute
IWMI	–	International Water Management Institute
MOA	–	Union Ministry of Agriculture
MORD	–	Union Ministry of Rural Development
MOWR	–	Union Ministry of Water Resources
MRC	–	Mekong River Commission
NAARM	–	National Academy of Agricultural Research and Management
NATP	–	National Agricultural Technology Project of ICAR
NBFGR	–	National Bureau of Fish Genetic Resources
NBPGR	–	National Bureau of Plant Genetic Resources
NBSS & LUP	–	National Bureau of Soil Survey and Land Use Planning
NCAP	–	National Centre for Agricultural Economics and Policy Research, New Delhi
NEH	–	ICAR Complex for North–east Hill Region
NGOs	–	Non–Government Organizations
NIH	–	National Institute of Hydrology, Roorkee
NHB	–	National Horticultural Board
NRCW	–	National Research Centre for Women
NRSA	–	National Remote Sensing Agency, Hyderabad
PDCSR	–	Project Directorate for Cropping System Research
RAU	–	Rajendra Agricultural University
SAD	–	State Agricultural Department
SAUs	–	State Agricultural Universities
SIWI	–	Stockholm International Water Institute

Contd..

Table 14–Contd...

UA	–	University of Arizona, Tueson, USA
UC	–	University of California, Davis Campus, USA
UEA	–	University of East Anglia
UR	–	University of Reading
USAID	–	United States Agency for International Development
USDA	–	United States Department of Agriculture
WALMI	–	Water and Land Management Institute
WFC	–	World Financial Centre
WMO	–	World Meteorological Organisation
WRI	–	World Resources Institute, Washington
WTC	–	Water Technology Centre
WTCER	–	Water Technology Centre for Eastern Region
ZT	–	Zero Tillage

Glossary

A horizon: The surface horizon of a mineral soil having maximum organic matter accumulation, maximum biological activity and/or eluviation of materials such as iron and aluminum oxides and silicate clays.

A line: The male sterile parent used in crossing to evolve/produce hybrid seed. It is the male sterile line.

ABC soil: A soil with a distinctly developed profile, including A, B, and C horizons.

Abscidc acid (ABA): An important growth-inhibitor of plants, initially called dormin, has wide range of physiological effects including promotion of senescence and abscission, causing dormancy in buds and seeds, closure of stomata, retardation and inhibition of growth known as *'antigibberrelin'* since it inhibits gibberrelin-stimulated growth.

Abscission layer: A zone of cells in the petiole or other plant structure whose cells separate and thereby bring about leaf fall, fruit drop etc.

Abscission: Detachment of fruit, leaf or other parts from a plant.

Absolute advantage: The ability to produce a greater physical output with a given set of resources than a similar set of resources can produce elsewhere.

Absolute growth rate (AGR): The rate of increase in size of a growing plant (or part of it) in a given time under specific condition.

Absorption: The process by which a substance is taken into or includes within another substance, like intake of water by soil or intake of nutrients by plants.

AC soil: A soil having a profile containing only A and C horizons with no clearly developed B horizon.

Acclimatization: An adjustment in the morphology or physiology of an organism in response to a change in the environment.

Accumulation: The uptake of substances against a concentration gradient and hence an energy-requiring process.

Achene: A dry indehiscent, one-seeded fruit in which the ovary wall remains free from the seed-coat.

Acid equivalent: Refers to the portion of a formulation (herbicide) which can be theoretically converted into acid. It is written as a.e.

Acid rain: Atmospheric precipitation with ph values less than about 5.6, the acidity being due to inorganic acids such as nitric and sulfuric that are formed when oxides of nitrogen and sulfur are emitted into the atmosphere.

Acid soil: A soil with a pH reaction of less than 7.0 (usually less than 6.6). An acid soil has a preponderance of hydrogen ions over hydroxyl ions, and blue litmus paper turns red in contact with moist acid soil.

Acid-sulphate soil is the one which intrinsically postulates a pH value lower than 3.5 and simultaneously represents a peaty saline soil. The accumulation of soluble salts as well as free-acid substances chiefly comprising ferric sulphate and aluminum sulphate form these soils. To illustrate, *Kari* soil in Kerala represents a monumental acid sulphate soil. This soil is highly acidic, deep black in colour, contains high organic matter, is poorly drained and above all remains under water during the monsoon season.

Actinomycetes: A group of organisms intermediate between the bacteria and the true fungi that usually produce a characteristic branched mycelium. Includes many, but not all, organisms belonging to the order of Actinomycetales.

Active absorption: Movement of ions and water into the plant root as a result of metabolic processes by the root, frequently against an activity gradient.

Active absorption: Refers to the movement of ions and water into the plant root due to metabolic processes of the root, frequently against an activity gradient.

Active ingredient is the portion of a formulation (herbicide) which is directly responsible for the pesticidal (herbicidal) influence. It is written as a.i.

Acute toxicity is generally used with respect to herbicides and refers to the intensive and rapid killing of plants by herbicide.

Adaptation is the ability of an organism to alter the organs quantitatively as well as qualitatively to meet the requirement of life. The ability to bring the inner forces into adjustment with the exterior forces is also called adaptation.

Adenosine triphosphate (ATP): ADP with an additional phosphate group attached by a high-energy bond, its decomposition into ADP and phosphate results in the release of energy needed for cellular metabolism.

Administrative regulation: Rules and regulations established by administrative decision.

Adsorption: The attraction of ions or compounds to the surface of a solid. Soil colloids adsorb large amounts of ions and water.

Adventitious bud: Bud formed at an uncommon location on a plant or it develops at an unusual time during plant growth and development.

Adventitious roots: The roots which appear at an unusual location or are produced at an unusual time during plant growth, like the nodal roots in grasses.

Adventitious: Organs arising in unusual positions, as buds from roots.

Aerobic condition: The presence of more oxygen as a part of the environment or growth only in the presence of molecular oxygen (referred to as aerobic organisms) or the occurrence of a process in the presence of molecular oxygen (aerobic decomposition).

Agent middlemen Negotiate the transfer of goods from seller to buyer without themselves taking title to those goods.

Aggregate demand curve: A curve showing the negative relationship between the price level and the quantities of real goods and services demanded during a given period of time.

Aggregate supplies curve: A curve showing the positive relationship between the quantities of real goods and services supplied by businesses and the price level during a given period of time.

Agribusiness: Involves the manufacture and distribution of farm supplies; production operations on the farm; and the storage, processing, and distribution of farm commodities and items made for them.

Agric horizon: A mineral soil horizon in which clay, silt, and humus derived from an overlying cultivated and fertilized layer have accumulated. The wormholes and clay, silt and humus occupy at least 5 per cent of the horizon by volume.

Agricultural bargaining: A group of producers organized to gain greater power in the market for its members.

Agricultural economics: An applied social science dealing with how mankind uses technical knowledge and scarce productive resources to produce food and fiber and to distribute them to society for consumption over time.

Agricultural fundamentalism: The school of thought holding that there is something special and unique about the farm way of life.

Agricultural Lime: In strict chemical terms, calcium oxide. In practical terms, a material containing the carbonates, oxides and/or hydroxides of calcium and/or magnesium used to neutralize soil and acidity.

Agricultural options: A type of financial instrument that gives the holder the right to buy or sell futures contracts.

Agricultural policy: Public policies developed to achieve specific objectives desired for agriculture.

Agrology: The study of applied phases of soil science and soil management.

Air porosity: The proportion of the bulk volume of soil that is filled with air at any given time or under a given condition, such as a specified moisture potential; usually the large pores.

Aleurone (layer): Ttype of protein generally present as a thin layer of the endosperm in maize and the small grained seeds.

Alfisols: Soils with grey to brown surface horizons, medium to high supply of bases, and B horizons of alluvial clay accumulation. These soils form mostly under forest or savanna vegetation in climates with slight to pronounced seasonal moisture deficit.

Alkaline soil: Any soil that has pH > 7. Usually applied to surface layer or root zone but may be used to characterize any horizon or a sample thereof. *See also* reaction, soil.

Allelopathy: Direct or indirect harmful effect that one plant has on another mutually on each other through the production of harmful compounds that escape into the environment.

Alluvial soil: A soil developing from recently deposited alluvium and exhibiting essentially no horizon development or modification of the recently deposited materials.

Amino acids: Nitrogen containing organic acids that couple together to form proteins. Each acid molecule has one or more amino group (NH_2) and at least one carboxyl group (-COOH). Some amino acids contain sulphur.

Amphi-photoperiodic plants: Plants which flower either under shorter or longer photoperiods, but not in the intermediate range, for example, *Media elegans* and *Setaria verticillata*.

Amylopectin: Type of starch in which glucose molecules are connected so as to form a branched chain.

Amylose: Amylose is the type of starch in which glucose molecules are connected so as to form a straight chain.

Annidation: The complementary effects of resources by exploiting the environmental supplies in differing ways by the components of a community

Anthesis:The process of dehiscence of anthers and spells out the period of the distribution of pollen.

Anti-auxin: A chemical or substance which inhibits the effect of auxin in the stimulation of cell enlargement.

Antitranspirants: Substances that retard or inhibit transpiration, e.g. PMA.

Apical dominance: Inhibiting effect of a tenninal bud upon the development of lateral buds.

Apical meristem: Tissue at the tip of a root or shoot where active cell division occurs.

Apomixis: The development of an embryo from cell other than a fertilized egg. The setting of seed without fertilization as nuclear embryo in citrus.

Aquatic weed: Denotes a foreign aquatic (water) plant making an inroad to spend at least a part of its life cycle in water.

Arable land: land currently being cropped or capable of being cropped without requiring additional costs of development.

Arbitrage: The simultaneous purchase and sale of a commodity in two different markets to take advantage of differences in the prices of that commodity in the markets.

Arid climate: Climate in regions that lack sufficient moisture for crop production without irrigation. In cool regions annual precipitation is usually less than 25 cm. It may be as high as 50 cm in tropical regions. Natural vegetation is desert shrubs. The evapotranspiration far exceeds the rainfall in these areas.

Aridisols: Soils of dry climates. They have pedogenic horizons, low in organic matters, which are never moist as long as 3 consecutive months. They have an ochric epipedon and one or more of the following diagnostic horizons: argillic, natric, cambic, calcic, petrocalcic, gypsic or a duripan.

Assets: Items of money value owned by a business, including land, buildings, tractors, combines, and so on.

Auxins: A group of growth-regulators that may stimulate cell growth, root development and other growth process including seed germination.

Available nutrient: Refers to that portion of any element or compound in the soil that can be readily absorbed and assimilated by the growing plants. It is different from an exchangeable nutrient.

Average fixed cost (AFC): Fixed cost per unit of output, determined by dividing total fixed cost by output at each level of output.

Average physical product (APP): Units of output produced per unit of input for each level of variable input use.

Average total cost (ATC): Total cost per unit of output, determined by dividing total cost by output at each level of output.

Average value product (AVP): The value of output per unit of input for each level of variable input use.

Average variable cost (AVC): Variable cost per unit of output, determined by dividing total variable cost by output at each level of output.

B horizon: A soil horizon usually beneath the A that is characterized by one or more of the following: (a) a concentration of silicate clays, iron and aluminum oxides, and humus, alone or in combination; (b) a blocky or prismatic structure, and (c) coatings of iron and aluminum oxides that give darker, stronger, or redder color.

Back cross: In plant breeding a cross of hybrid with one of its parents and the purpose is to transfer a specific gene from an undesirable variety to another commercially desirable one lacking in that particular character.

Balance of trade: The value of merchandise exports minus the value of merchandise imports.

Band treatment is the application of herbicide directly either in the crop row or on the specified strip.

Bank reserves: The amount that a bank has on deposit with the Federal Reserve Bank or held as cash in its vault.

Basis: The difference between a futures price and the cash price of a commodity at a particular time and place.

BC soil: A soil profile with B and C horizons but with little or no a horizon. Most BC soils have lost their A horizon by erosion.

Biological resources: Natural resources, such as forests and fisheries, which produce a harvestable yield from the resource stock.

Bio-type means a group of plants possessing an identical genetic constitution which may be homozygous or heterozygous.

Biroype: A population within a species that has a distinct genetic variation.

Blind cultivation is the cultivation done before the emergence of the plants from the soil.

'B'line is the fertile counter parent of the' A' line. It is also called the maintainer line.

Border-strip irrigation methods: The water is applied at the upper end of a strip with earth borders to confine the water to the strip.

Boron toxicity: Symptom are the appearance of a yellow discolouration of the leaf tips that spreads along the margins. Large brown elliptical spots appear along the leaf margins. Affected parts turn brown and wither. Boron toxicity occurs on coastal soil, arid region soils, soils irrigated with high–boron water and geothermal areas.

Breeder seed: It is the seed directly controlled by the originating or sponsoring plant institution or individual, and the source of the production of seed for the certified classes. It is genetically purest seed stock.

Broom rape Orobanche: Tobacco crop is very often infested by Orobanche (*Orobanche cernua*) or Broom rape a phanerogamic, total root parasite. It is reddish brown or yellowish brown unbranched herb growing to a height 60 cm with leaves scale like ovate to lanceolate, 6-20 mm long, acute, sessile and devoid of chlorophyll. The stem is pubescent.

Bud: A very young or unelongated stem.

Budget line: Shows all possible combinations of goods that a consumer can buy with a given money income.

Bulk density: The ratio of the weight of oven dried soil to its bulk maximum volume. It is also called the volume weight.

Bulk density of soil: The mass of dry soil per unit of bulk volume, including the air space. The bulk volume is determined before drying to constant weight at 105°C.

Buoyancy: The upward force excreted on the volume of fluid by virtue of the density difference between the volume of fluid and that of surrounding fluid.

C horizon: A mineral horizon generally beneath the solum that is relatively unaffected by biological activity and pedogenesis and is lacking properties diagnostic of an A or B horizon. It mayor may not be like the material from which the A and B have formed.

C4 group of plants: Plants like maize, soybean, finger millet, grasses etc. *Cylodon dactylon*. sudan grass have capacity to fix most of the C4into the form of organic acid like malic acid, aspartic acid etc. and convert it into carbohydrate durng the day-time, these called C_{-4} pathway. The carboxylating enzyme: Which has a high potential activity and a high affinity for CO_2, helps in a high rate of CO_2 fixation and maintenance at relatively restricted stomatal opening? These physiological changes use to inhibit the photorespiration and thereby the net photosynthesis is enhanced.

Calcareous soil: Soil containing sufficient calcium carbonate (often with magnesium carbonate) to effervesce visibly when treated with cold 0.1 N hydrochloric acid.

Calcic horizon: A mineral soil horizon of secondary carbonate enrichment that is more than 15 cm thick, has a calcium carbonate equivalent of more than 15 per cent, and has at least 5 per cent more calcium carbonate equivalent than the underlying C horizon.

Cambic horizon: lacks cementation or induration and has too few evidences of illuviation to meet the requirements of the argillic or spodic horizon.

Cambric horizon: A mineral soil horizon that has a texture of loamy very fine sand or finer, contains some weatherable minerals, and is characterized by the alteration or removal of mineral material.

Capillary water (Obsolete): The water held in the "capillary" or *small* pores of a soil, usually with a tension >60 cm of water.

Capital: Productive resources (goods) that are available, as a result of past human decisions, to produce other want-satisfying goods.

Capitalism: A type of economic organization in which private individuals or groups own and manage all resources.

Capitalized value: The present value of a resource obtained by discounting the value of its future net income stream.

Carbon Credits: One Carbon Credit is equal to one ton of Carbon. Carbon trading is application of emissions trading approach.

Carbon cycle: The sequence of transformations whereby carbon dioxide is fixed in living organisms by photosynthesis or by chemosynthesis, liberated by respiration and by the death and decomposition of the fixing organism, used by heterotrophic species, and ultimately returned to its original state.

Carbon Offset Credits: It consist of clean forms of energy production, wind, solar, hydro and biofuels.

Carbon offsets: Its enable individuals and businesses to reduce the CO_2 emissions they are responsible for by offsetting, reducing or displacing the CO_2 in another place, typically where it is more economical to do so. Carbon offsets typically include renewable energy, energy efficiency and reforestation projects.

Carbon Reduction Credits: It consists of the collection and storage of Carbon from our atmosphere through reforestation, forestation, ocean and soil collection and storage efforts. Both approaches are recognized as effective ways to reduce the Global Carbon Emissions crises.

Carbon Sequestration: It is process of capture and safe storage of carbon dioxide in living organism specifically in soil and vegetation by carbon assimilation (photosynthesis). This is one of the best practices to mitigate carbon dioxide emissions, include reducing emissions from present sources.

Carbon/nitrogen ratio: The ratio of the weight of organic carbon (C) to the weight of total nitrogen (N) in a soil or in organic material.

Carotene: The yellow colouring material found in a number of fruits and vegetables, also in leaves and green plant parts. It is the precursor of Vitamin A.

Caryopsis: The one seeded fruit of the plants of the grass family.

Cat clays: Wet clay soils high in reduced forms of sulfur that upon being drained, become extremely acid because of the oxidation of the sulfur compounds and the formation of sulfuric acid.

Catena: A sequence of soils of about the same age, derived from similar parent material, and occurring under similar climatic conditions, but having different characteristics because of variation in relief and in drainage.

Cation exchange capacity: The sum total of exchangeable cations that a soil can adsorb. Expressed in centimoles per kilogram (cmol/kg) of soil (or of other adsorbing material such as clay).

Cation exchange: The interchange between a cation in solution and another cation on the surface of any surface-active material s.uch as clay or organic matter.

Cation: A positively charged ion; during electrolysis it is attracted to the negatively charged cathode.

Centre of origin: An area where the given plant species shows the greatest genetic diversity.

Cereal: Grass which is primarily grown for its seeds used as feed or food. *Chlorophyll* is the green pigment contained in the chloroplast and consists of large and complex molecules which can absorb and capture radiant energy.

Certified seed: It is the progeny of breeder, foundation or registered seed, which is handled to maintain satisfactory genetic purity and identity in a manner acceptable to certifying agency.

Ceteris paribus: Holding some variables constant, while letting specific variables change.

Change in demand: A shift in the entire demand schedule.

Change in quantity demanded: A movement along a given demand curve in response to a price change.

Change in quantity supplied: A movement along a given supply curve in response to a price change.

Change in supply: A shift in the entire supply schedule.

Check-basin irrigation methods: The water is applied rapidly to relatively level plots surrounded by levees. The basin is a small check.

Chlorosis: A condition in plants relating to the failure of chlorophyll (the green coloring matter) to develop. Chlorotic leaves range from light green through yellow to almost white.

Choice: Deciding on one action over another or one good in preference to another, according to certain criteria.

Chronic toxicity: Refers to the effect of an herbicide. The toxicity occurs at slow rate over a period of time due to herbicides applied to a plant.

Clay mineral: Naturally occurring inorganic material (usually crystalline) found in soils and other earthy deposits, the particles being of clay size, that is, <0.002 mm in diameter.

Clone: Plant produced vegetatively by rooting a certain part of a single plant. The plant, thus produced is identical to the parent plant in all the characteristics.

Coarse texture: The texture exhibited by sands, loamy sands, and sandy loams except very fine sandy loam. Cobblestone rounded or partially rounded rock or mineral fragments 7.5-25 cm (3-10 in.) in diameter. Cohesion holding together: force holding a solid or liquid together, owing to attraction between like molecules.

Coleoptile: It is indicative of the first leaf immediately above the cotyledon. It encloses the tip of the stem and other leaves, that is, the sheath.

Collective farm: Typically, a farm in the USSR, owned by the government, but operated by a number of families, which share in the farm-derived revenue.

Colluvium: A deposit of rock fragments and soil material. Accumulated at the base of steep slopes as a result of gravitational action to the order Coniferae, usually evergreen, with cones and needle-shaped or scale-like leaves and producing wood known commercially as "softwood."

Common market: A customs union that also provides for the free mobility of factors of production between the member nations.

Common property resources: Resources that are owned by the government in the name of the public.

Comparative advantage: A situation in which an individual, region or nation is relatively superior at producing some goods because of its lower opportunity costs, and gains by trading for goods that another is relatively more proficient at producing.

Complements Using two goods in combination because the presence of one good enhances the benefit obtained from the other.

Conglomerate merger: The merging of two or more firms that operate in unrelated industries.

Conjunctive use of water: Represents the planned use of surface and sub-surface water to augment canal supply, combat water logging and also to facilitate irrigation with poor quality ground water. The conjunctive use of surface and ground waters has been practiced in India to a limited extent.

Conservation tillage: Means any tillage sequence which reduces the loss of soil and water relative to conventional tillage including minimum tillage, mulch tillage, no-tillage, plough-planting, sod-planting, sub-surface tillage and *wheat* track planting.

Conservation tillage: Any tillage sequence that reduces loss of sailor water relative to conventional tillage, including the following systems.

Conservation: Preserving or extending the productive life of a resource.

Constant returns: A constant input-output ratio.

Consumer Price Index: The average price of a market basket of goods and services in index form to permit comparisons of the prices of goods and services over a period of years.

Consumption component: The utility or satisfaction an individual derives from an education.

Consumptive use: The water used by plants in transpiration and growth, plus water vapor loss from adjacent soil or snow, or from intercepted precipitation in any specified time. Usually expressed as equivalent depth of free water per unit of time.

Consumptive use of water is the total amount of water used by the vegetation of a given area in transpiration, building of plant tissues and that which evaporates from the adjacent soil, snow or intercepted precipitation on the area, in a specified time. It is generally expressed in inches or centimeters with respect to time. Evapotranspiration has been often used synonymously with CUW, although strictly speaking CUW also includes the water used for building the plant tissues. Quantitatively, there is only a slight difference and hence both are used interchangeably.

Contact herbicide: A herbicide which causes the death of a plant or its part only after coming in contact with the plant or the plant parts covered by the applied herbicide.

Contour strip cropping: Layout of crops in comparatively narrow strips in which the farming operations are performed approximately on the contour. Usually strips of grass, close-growing crops, or fallow are alternated with those of cultivated crops.

Contour: An imaginary line connecting points of equal elevation on the surface of the soil. A contour terrace is laid out on a sloping soil at right angles to the direction of the slope and nearly level throughout its course.

Contract production: Contractual agreements between producers and their input suppliers or product marketing firms.

Conventional tillage: The combined primary and secondary tillage operations normally performed in preparing a seedbed for a given crop grown in a given geographic area.

Cooperative: An association of member-owners operating a business that provides services at cost to its patrons.

Corrugation Irrigation methods: The water is applied to small, closely spaced furrows, frequently in grain and forage crops, to confine the flow of irrigation water to one direction.

Cost-benefit ratio (CBR): Equivalent to the per rupee investment which is derived by dividing the gross returns by the total cost of cultivation.

Cotyledon: First leaves arising from the embryo. Their number varies from one in monocotyledonous plants to two or more in dicotyledonous plants.

Cover crop: A close-growing crop grown primarily for the purpose of protecting and improving soil between periods of regular crop production or between trees and vines in orchards and vineyards.

Creep: Slow mass movement of soil and soil material down relatively steep slopes primarily under the influence of gravity, but facilitated by saturation with water and by alternate freezing and thawing.

Critical period (stage): The period during which a crop is affected severely due to soil moisture or nutrient(s) stress and the loss cannot be made good by adequate supply in other periods or stages.

Critical stages of tobacco for irrigation: Grand growth initiation stage i.e. 25-30 days after planting and development stage i.e. 55-60 days after planting (after topping and piercing) were found to be most critical stages for irrigation.

Crop coefficient: Ratio between crop evapotranspiration to pan evaporation when crop is raised in large field under optimum growing conditions.

Crop geometry: Crop geometry defines the pattern of distribution of plants over the ground or the shape of the area available to the individual plant.

Crop rotation: A planned sequence of crops growing in a regularly recurring succession on the same area of land, as contrasted to continuous culture of one crop or growing different crops in haphazard order.

Crop water-use efficiency: The ratio of crop yield to the amount of water depleted by the crop plants through evapotranspiration.

Cropping system management: Cropping system management is basically the design of spatial and temporal combinations of crops in an area.

Cropping system: Cropping system is the cropping pattern used on a farm and their interactions with farm resources, other farm enterprises and available technology that determine their makeup.

Cross fertilization: The fertilization of an ovule or a number of ovules of one flower with the pollen from another flower, or it may be defined as the fertilization of the flower of one plant by the pollen from another plant.

Cross-pollinate: To apply pollen of one flower to the stigma of another, commonly refers as pollinating the flowers of one plant by pollen from another plant.

Crushing strength: The force required to crush a mass of dry soil or, conversely, the resistance of the dry soil mass to crushing. Expressed in units of force per unit area (pressure).

Crust: A surface layer on soils, ranging in thickness from a few millimeters to perhaps as much as 3 cm, that is much more compact, hard, and brittle, when dry, than the material immediately beneath it.

Crystal: A homogeneous inorganic substance of definite chemical composition bounded by plane surfaces that form definite angles with each other, thus giving the substance a regular geometrical form crystal structure.

Crystalline rock: A rock consisting of various minerals that have crystallized in place from magma.

Cultivar: It is equivalent to 'variety' and is defined as an assemblage of cultivated plants which is clearly distinguishable by any character (morphological, physiological, cytological, chemical or others) and when reproduced (sexually or asexually) retains its distinguishing characters. The term is derived from cultivated variety.

Cultivation: A tillage operation used in preparing land for seeding or transplanting or later for weed control and for loosening the soil.

Cusec: The quantity of water flowing at the rate of 1 cubic foot per second. One cubic foot of water weights 62.4 lbs or 6.24 gallons.

Cytokinins: A term used for substances which promote cell-division and exert other growth regulatory functions, e.g. zeatin (naturally occurring) and 6-benzyl amino-purine (synthetic).

Cytoplasmic male sterility: A type of male sterility conditioned by the cytoplasm rather than by nuclear genes and transmitted only through the female parent.

Day cusec: The quantity of water in cusecs flowing for 24 hours.

Delta (water): The total depth of water required by a crop.

Delta: An alluvial deposit formed where a stream or river drops its sediment load upon entering a quieter body of water.

Denitrification: The biochemical reduction of nitrate or nitrite to gaseous nitrogen, either as molecular nitrogen or as an oxide of nitrogen.

Deoxyribose nucleic acid (DNA): The nucleic acid of the chromosomes, the main carrier of genetic information.

Deoxyribose: A 5-carbon sugar which is present in deoxyribose nucleic acid.

Depth of irrigation: Water depth required to bring soil water content of root zone to field capacity.

Desert crust: A hard layer, containing calcium carbonate, gypsum, or other binding material, exposed at the surface in desert regions.

Diagnostic horizons: Combinations of specific soil characteristics that are indicative of certain classes of soils.

Dihybrid: Hybrid that is result of crossing of homozygous parents that differ with respect to 2 loci.

Dioctahedral: An octahedral sheet, or a mineral containing such a sheet, that has two thirds of the octahedral sites filled with trivalent ions such as aluminum or ferric iron.

Diploid: A plant with 2 sets of chromosomes, one set coming from each parent. Most organisms are diploid.

Disintegration: The breakdown of rock and mineral particles into smaller particles by physical forces such as frost action.

Disposable personal income The after-tax income of individuals available for spending on goods and services.

Distribution efficiency: Ratio of water made directly available to the crop and that released at the inlet of a block of fields.

Diversion canal: A canal to divert water from one point to another. In irrigation practice, it extends from the point of diversion at the main canal to the beginning of the distribution system.

Double cross: A hybrid between 2 single crosses involving 4 different inbred lines.

Drainage requirement: The amount of water that drain must convey in a given time for satisfactory crop growth.

Drainage water: Also called gravitational water, free water and excess water. The water that the soil is unable to hold against the force of gravity, Also water from surface, ground or storm water flowing into a drain.

Drainage soil: The frequency and duration of periods when the soil is free from saturation with water.

Drainage: The removal of excess surface or ground water from land by means of surface or subsurface drains.

Drip irrigation method: A planned irrigation system where all necessary facilities have been installed for the efficient application of water directly to the root zone of plants by means of applicators (orifices, emitters, Porous tubing, perforated pipe, etc.) Operated under low pressure. The applicators may be placed on or below the surface of the ground.

Drought avoidance mechanism: Drought avoidance is the ability of plants to maintain relatively high tissue water potential despite a shortage of soil-moisture, whereas drought tolerance is the ability to withstand water-deficit with low tissue water potential. Mechanisms for improving water uptake, storing in plant cell and reducing water loss confer drought avoidance. The responses of plants to tissue water-deficit determine their level of drought tolerance. Drought avoidance is performed by maintenance of turgor through increased rooting depth, efficient root system and increased hydraulic conductance and by reduction of water loss through reduced epidermal (stomatal and lenticular) conductance, reduced absorption of radiation by leaf rolling or foldingand reduced evaporation surface (leaf area). Plants under drought condition survive by doing a balancing act between maintenance of turgor and reduction of water loss.

Drought enduring plants: Neither the ephemerals nor the succulents can be regarded truly drought resistant in the sense that their cells can ensure a severe reduction in water content for extended period of time without injury but the plants of drought enduring group can only resist the injuries caused by drought. The example of such among higher plants is creosote bush (*Lerrea tridentata*) which is dominant plant throughout large areas in the semi-arid regions.

Drought escape mechanism: Drought escape is defined as the ability of a plant to complete its life cycle before serious soil and plant water deficits develop. This mechanism involves rapid phenological development (early flowering and early maturity), developmental plasticity (variation in duration of growth period depending on the extent of water-deficit) and remobilization of preanthesis assimilates to grain.

Drought index: A drought index which provides a measure of drought based on loss of yield under drought-condition in comparison to moist condition has been used for screening drought-resistant genotype. An artificially created water-stress environment is used to provide the opportunity in selecting superior genotype out of a large population. Visual scoring or measurement for maturity, leaf rolling, leaf length, angle, root morphology and other morphological characters of direct relevance to drought resistance are also taken into consideration.

Drought tolerance mechanism: The mechanisms of drought tolerance are maintenance of turgor through osmotic adjustment (a process which induces solute accumulation in cell), increase in elasticity in cell and decrease in cell size and desiccation tolerance by protoplasmic resistance.

Dry land farming: The practice of crop production in low rainfall areas without irrigation.

Duty (water): It is the area irrigated by 1 cusec discharge of water during the crop period; it is equal to twice the base divided by delta.

Earthworms: Animals of the Lumbricidae family that burrow into and live in the soil. They mix plant residues into the soil and improve soil aeration.

Economic profit: The amount by which a firm's revenue exceeds its total (explicit plus implicit) costs. Also referred to as *pure profit*.

Ecotype: Plant type or strain within a species, resulting from exposure to a particular environment.

Effective irrigation: A controlled and uniform application of water to crop land in required amount at required time, with minimum cost to produce optimum yields without waste of water and adverse effect on soil.

Entisols: Soils have no diagnostic pedogenic horizons. They may be found in virtually any climate on very recent geomorphic surfaces.

Entrepreneurship: Organizing resources to produce and market goods and services.

Ephemeral plants: In real sense these pants are not true drought resistant types. Under semi-arid regions the seed of such plants germinate and the plants complete their life-cycle within a few weeks time. The new crop of seeds survives the entire dry period until rainy season comes. Such plants are termed drought escaping or ephemerals.

Equilibrium price: The price in a market at which quantity supplied and quantity demanded are equal.

Equilibrium: A condition in which opposing forces within a system just offset one another.

European Community (EC-IO): A group of countries which have reduced or abolished tariffs among themselves and established a common and uniform tariff to outsiders. The EC includes 10 countries: West Germany, France, Italy, Belgium, the Netherlands, Luxembourg, Great Britain, Denmark, Ireland, and Greece.

European Free Trade Association (EFTA): A group of countries which have reduced or abolished tariffs among themselves, but have not established a common tariff to outsiders. The EFTA includes: Sweden, Norway, Switzerland, Austria, Portugal, Iceland, and Finland.

Evolution: The development of a race, species or other group. Exotic (plant): A newly introduced plant not native to a place. Ft: The first filial generation of a cross between 2 individuals.

Exchange capacity: The total ionic charge of the adsorption complex active in the adsorption of ions.

Exchange: Function includes all the activities throughout the market channels as buyers and sellers make their transactions.

Exchangeable sodium percentage(ESP): The extent to which the adsorption complex of a soil is occupied by sodium.

Expansion path: A line connecting the least cost combination points along the production surface. Also, a line connecting the most profitable combination points for each of a number of production possibilities curves.

Expansionary fiscal policy: Increasing government expenditures or reducing taxes to cause economic activity to expand and increase the nation's GNP.

Expansionary monetary policy: Government action to increase the money supply and lower the interest rate so as to expand economic activity and increase the nation's GNP.

Expenditure method: Measuring GNP by totaling the value of all final goods produced in the economy during an accounting period.

Explicit cost: An expenditure made for the use of a resource.

Exports: Products and services sold to foreign countries.

External growth: Expanding the size of a firm by adding another firm through its purchase or other means of merger.

Externalities: Costs or benefits that are external to the decision maker and are imposed on others.

Facilitating functions: Those activities throughout the marketing system that improve its operational and pricing efficiency.

Facilitative organizations: Institutions that provide information or facilities for marketing commodities.

Factor markets: Markets in which resources are bought and sold. One loop in a circular flow diagram that measures the earnings of all resources used in producing a nation's goods and services.

Factor-factor relationship: The relationship between two factors or inputs used in production.

Factor-product relationship: The functional relationship between a factor of production and its product.

Family farm: A farm in which the family provides most of the labor required in its operation.

Fertilizer grade: Refers to the guaranteed minimum analysis expressed in percentage of the major plant nutrient elements contained by a fertilizer material or by a mixed fertilizer generally in terms of N.

Fertilizer requirement: Represents the quantity of certain plant nutrient elements required, over and above the amount supplied by the soil to augment plant growth to a predetermined optimum level.

Fertilizer requirement: The quantity of certain plant nutrient elements needed, in addition to the amount supplied by the soil, to increase plant growth to a designated optimum.

Field capacity (field moisture capacity): The percentage of water remaining in a soil two or three days after its having been saturated and after free drainage has practically ceased.

Field irrigation efficiency: It is the ratio expressed in percentage of irrigation water available for use of a crop to that delivered to a field. It excludes ditch losses of water.

Final product value: The market value of a final product.

Final product: The finished product ready for sale to the ultimate user or consumer.

Fine texture: Consisting of or containing large quantities of the fine fractions, particularly of silt and clay.

Fine-grained mica: Silicate clay having a 2: 1-type lattice structure with much of the silicon in the tetrahedral sheet having been replaced by aluminum and with considerable interlayer potassium, which binds the layers together and prevents interlayer expansion and swelling, and limits interlayer cation exchange capacity.

Firm: A decision-making business entity that uses resources hired from households to produce goods and services for sale to households or other consuming units.

Fiscal policy: The use of government policy to achieve specific economic goals by manipulating expenditures or the tax rate.

Fixed costs: Those costs incurred for resources which do not change as output is increased or decreased.

Flocculate: To aggregate or clump together individual, tiny soil particles, especially fine clay, into small clumps or floccules.

Flooding Irrigation methods: The water is released from field ditches and allowed to flood over the land.

Floodplain: The land bordering a stream, built up of sediments from overflow of the stream and subject to inundation when the stream is at flood stage.

Flora: The sum total of the kinds of plants in an area at one time.

Flow resources: Resources whose available quantities are constantly being replenished. For example, using the wind's power or the heat energy from the sun does not reduce the supply of these resources to others.

Fluorapatite: A member of the apatite group of minerals containing fluorine.

Foliar diagnosis: An estimation of mineral nutrients deficiencies (excesses) of plants based on examination of the chemical composition of selected plant parts, and the color and growth characteristics of the foliage of the plants.

Form utility: The utility or satisfaction generated by the processing function in marketing as it delivers products in the desired forms.

Free market: Amarket system in which buyers and sellers decide and act at their own initiative and in their own economic interests.

Free trade area: Group of nations that abolishes trade restrictions among themselves without also imposing common tariffs on other nations.

Full employment: A rate of unemployment no greater than that due to the normal functioning of the labor market.

Fulvic acid: A term of varied usage but usually referring to the mixture of organic substances remaining in solution upon acidification of a dilute alkali extract from the soil.

Fund or stock resources: Resources whose quantities are fixed in their natural state. Coal, oil, and iron ore are examples of these non-renewable resources.

Fungi: Simple plants that lack a photosynthetic pigment. The individual cells have a nucleus surrounded by a membrane, and they may be linked together in long filaments called *hyphae*, which may grow together to form a visible body.

Furrow Irrigation methods: The water is applied to row crops in ditches made by tillage implements.

Futures contracts: Standardized contracts for future delivery of commodities that are traded on organized exchanges.

Gamete: A sex cell which, after union with another develops into a new plant.

Gap filling of tobacco seedling: Gap filling operation is done to maintain optimum plant population by replacing dead plants those are unable to bear transplanting shocked. Gap filling should be done after 8-10 days of transplanting.

Genetics: The science of heredity including the study of its chemical foundation, its development, expression and its bearing on variation, selection, adaptation, evolution, breeding and the activities of man.

Genotype: It indicates the hereditary constitution of an individual (plant/animal) which dictates the characteristics of that individual. Its expression is influenced by the environmental conditions. The characteristics may be the plant type, shape as well as arrangement of leaves, colour of the flower and seed in plants, and skin colour, body make up, colour of eye balls in animals.

Genus: A group of closely related species. In a scientific name the genus is the first of the 2 names given for an organism.

Germination: The resumption of growth by the embryo and the development of the seedlings from the seed placed in conditions necessary for the resumption of growth.

Germplasm: The hereditary material or the genetic traits which can be incorporated into new cultivars.

Glabrous: It means having a smooth surface *i.e.*, without hairs. *Glumes* are the pairs of the bracts located at the base of a spikelet in the members of the Gramineae or grass family.

Global warming and climate change: The green house gases are the main culprits of the global warming and climate change. The green house gases like carbon dioxide, methane, and nitrous oxide are playing hazards in the environment.

Global warming potential (GWP): GWP is the ratio of the warming that would result from the emission of one kilogram of a greenhouse gas to that from the emission of one kilogram of carbon dioxide over a fixed period of time such as 100 years. To standardize all gases for comparison in this paper, we use the global warming potentials (N_2O = 310 and that of CH_4 = 21).To calculate the effect of emissions on atmospheric forcing in terms of carbon dioxide equivalents (CO_2e), values expressed as carbon dioxide equivalents (CO_2e) are calculated based on their global warming potential (GWP). Thus, for example, one tonne of N_2 emitted will be equivalent to 310 tonnes of CO_2.

GNP Price Deflator Index: An index used to remove the effects of inflation or deflation to determine changes in real GNP over time.

Grassed waterway: A natural or constructed waterway covered with erosion-resistant grasses that permits removal of runoff water without excessive erosion.

Gravitational potential: That portion of the total soil water potential due to differences in elevation of the reference pool of pure water and that of the soil water. Since the soil water elevation is usually chosen to be higher than that of the reference pool, the gravitational potential is usually positive.

Green manure: Plant material incorporated with the soil while green, or soon after maturity, for improving the soil.

Greenhouse effect: The entrapment of heat by upper atmosphere gases such as carbon dioxide, water vapor, and methane just as glass traps heat for a greenhouse. Increases in the quantities of these gases:in the atmosphere will likely result in global warming that may have serious consequences for humankind.

Gross national product (GNP):The total market value of all final goods and services produced in an economy during a given period of time.

Growth curve: A graph in which the size of some plant characteristics (height, weight, leaf area etc.) is plotted against time or age. The rates of growth of any single cell or mass of cells are sigmoid (S-shaped); slow at first, increase with time to a maximum and then all off at about the same rate to zero. The period of maximum growth is called' grand period' of growth.

Growth-regulator (plant): Organic substances which in minute amounts may participate in the control of growth processes. In plant 4 types have been recognized to date: auxins, gibberellins, kinins and inhibitors.

Growth-retardant: A chemical (such as CCC) that selectively interferes with normal hormonal promotion of growth but without appreciable toxic effects.

Guard cropping: Means growing main crop in the centre surrounded by hardy and thorny crop, such as safflower around pea, mesta around sugarcane and sorghum around maize.

Gully erosion: The erosion process whereby water accumulates in narrow channels and, over short periods, removes the soil from this.narrow area to considerable depths, ranging from 1-2 feet to as much as 23-30 m (75-100 ft) natural erosion Wearing away of the Earth's surface by water, ice, or other natural agents under natural environmental conditions of climate, vegetation, and so on, undisturbed by man.

Habitat: It refers to the sum total of the conditions of environment, apart from the organism in consideration or to the sum total of all the external conditions effecting the development, special response and growth of plants.

Haploid: Having 1 set of chromosomes, as in gametes.

Hard pan: It is a hard and impermeable layer formed in the soil profile due to the accumulation of clay or salts like calcium carbonate which move down into the profile in a dissolved form in rain or irrigation water.

Hectare: A metric measure of land area (one hectare equals 2.4710 acres).

Hedging: Establishing opposite positions in the futures and cash markets as a protection against adverse price changes.

Herbicide: A chemical that kills plants or inhibits their growth; intended for weed control.

Heterogeneous: Units within a group have dissimilar characteristics.

Heterotroph: An organism capable of deriving energy for life processes only from the decomposition of organic compounds and incapable of using inorganic compounds as sole sources of energy or for organic synthesis. *Contrast with* autotroph.

Heterozygous: Not true-breeding for a specific hereditary character. Homogeneity test: A test to measure variation within a seed lot.

Histosols: Soils formed from materials high in organic matter. Histosols with essentially no clay must have at least 20 per cent organic matter by weight (about 78 per cent by volume). This minimum organic matter content rises with increasing clay content to 30 per cent (85 per cent by volume) in soils with at least 60 per cent clay.

Homogeneous: All units within a group are alike in their characteristics.

Homozygous: Having identical genes at the same locus in each member of a pair of homologous chromosomes.

Hookah tobacco (*Nicotiana rustica*): Bihar is one of the major producers of hookah tobacco and grown mainly in Purnia Katihar and Saharsa districts. Around 10 per cent of total area of tobacco is under this category. The average yield is very poor and is being only 900-1100 kg/ha.

Horizons: That occur at the soil surface are called epipedons; those below the surface, diagnostic subsurface horizons.

Horizontal merger: A combination of two or more firms operating in the same industry.

Hormone: An organic substance produced naturally in higher plants, controlling growth or other physiological functions at a site remote from its place of production and active in minute amounts.

Horticulture: The art and science of growing fruits, vegetables, and ornamental plants.

Household responsibility system (HRS): China's restructured system for production and marketing decision making by individual households rather than at the commune (production team) level.

Household: The basic economic entity in society that consumes the final goods and services produced by the system.

Human capital: The educational investment that improves the knowledge and productivity of people.

Humic acid: It is a complex organic acids, precipitated upon acidification of a dilute alkali extract from soil.

Humid climate: Climate in regions where moisture, when distributed normally throughout the year, should not limit crop production. In cool climate annual precipitation may be as little as 25 cm; in hot climates, 150 cm or even more. Natural vegetation in uncultivated areas is forests.

Humification: The processes involved in the decomposition of organic matter and leading to the formation of humus.

Humin: The fraction of the soil organic matter that is not dissolved upon extraction of the soil with dilute alkali.

Humus: That more or less stable fraction of the soil organic matter remaining after the major portions of added plant and animal residues have decomposed. Usually it is dark in color.

Hybrid vigour: The enhanced vigor of progeny resulting from the crossing of 2 different inbred plants.

Hybrid: The fist generation progeny or offspring from a cross of different varieties or strains or inbred lines.

Hydathode: A pore generally at the tip of a vein, capable of exuding liquid water.

Hydration: The association of one or more molecules of water with an ion, molecule or micelle. The water being of atmospheric or magnetic origin.

Hydraulic conductivity: An expression of the readiness with which a liquid such as water flows through a solid such as soil in response to a given potential gradient.

Hydrogen bond: The chemical bond between a hydrogen atom in one molecule and a highly electronegative atom such as oxygen or nitrogen in another polar molecule.

Hydrologic cycle: The circuit of water movement from the atmosphere to the Earth and back to the atmosphere through various stages or processes, as precipitation, interception, runoff, infiltration, percolation, storage, evaporation, and transpiration.

Hydrophytic plants: Subclasses include freshwater and saltwater marshes.

Hydroponics: The growing of plants in water solutions of essential nutrients.

Hydrotropism: The growth movement in response to unequal distribution of water, as when roots bend towards moist soil.

Hydroxyapatite: A member of the apatite group of minerals rich in hydroxyl groups. A nearly insoluble calcium phosphate.

Hygroscopic coefficient: The amount of moisture in a dry soil when it is in equilibrium with some standard relative humidity near a saturated atmosphere (about 98 per cent), expressed in terms of percentage on the basis of oven-dry soil.

Hypocotyl: The part of the embryo or seedling below the cotyledon node and above the root or the transition region connecting the stem and root.

Ideotype: Refers to plant type in which morphological and physiological characteristics are ideally suited to achieve high production potential and yield reliability.

Illuvial horizon: A soil layer or horizon in which material carried from an overlying layer has been precipitated from solution or deposited from suspension. The layer of accumulation.

Immature soil: A soil with indistinct or only slightly developed horizons because of the relatively short time it has been subjected to the various soil-forming processes. A soil that has not reached equilibrium with its environment.

Imperfect competition: Any market structure in which firms do not exhibit the characteristics of perfect competition.

Imperfect substitutes: Resources whose marginal rate of substitution changes as their proportions are changed, ceteris paribus.

Implicit cost: A cost for which there is no cash outlay at the time a resource is being used, or for which no cash payment is required.

Import quota: A maximum limit on the quantity of a commodity that can be imported.

Imports: Goods and services purchased from foreign countries.

Improved seed: The genetically and physically pure seed of an improved crop variety. Its different categories are nucleus seed, breeder's stock seed, foundation seed, registered seed and certified seed.

Inbred: The progeny of either a single cross-pollination plant obtained by selfing or two closely related plants obtained by inbreeding.

Inceptisols: Soils that is usually moist with pedogenic horizons of alteration of parent materials but not of illuviation. Generally, the direction of soil development is not yet evident from the marks left by various soil forming processes or the marks are too weak to classify in another order.

Income effect: The change in a consumer's real income that occurs as the price of good changes.

Income elasticity: The responsiveness of quantity purchased to a one percent change in income, ceteris paribus.

Income method: Measuring GNP by totaling all labor and other resource earnings throughout the economy during an accounting period.

Independent variable: A variable whose changes cause the quantity of another variable to be changed.

Indifference curve: Shows all the combinations of goods that yield an individual the same amount of satisfaction or utility.

Indifference map: The utility surface shown by a family of indifference curves.

Induced mutations: The mutations artificially produced with the help of mutagens.

Infiltration rate: A soil characteristic determining or describing the maximum rate at which water can enter the soil under specified conditions, including the presence of an excess of water.

Infiltration: The downward entry of water into the soil.

Inheritance: The reception or acquisitions of characters or qualities by transmission from parent to offspring.

Inoculation: The process of introducing pure or mixed cultures of microorganisms into natural or artificial culture media.

Inorganic compounds: All chemical compounds in nature except compounds of carbon other than carbon monoxide, carbon dioxide, and carbonates.

Insecticide: It is a chemical material applied as a spray or dust to kill harmful insects.

Insecticide: A chemical that kills insects.

Insurance hedging (or hedging): Establishing an equal but opposite position in the futures market from the one taken in the cash market.

Intensive cropping: Intensive cropping is the process of growing a number of crops on the same piece of land during the given period of time. In other words, when the area is limited and the number of crops to be grown is increased within a definite period of time, this cropping method is termed as intensive cropping.

Inter grade: A soil that possesses moderately well developed distinguishing characteristics of two or more genetically related great soil groups.

Intercropping: This is a process of growing subsidiary crops between two widely spaced rows of main crop. The main object of this type of cropping is to utilize the space left between two rows of main crop and to produce more grain per unit area. Parallel cropping, companion cropping, multistoried cropping and synergistic cropping are major kind of intercropping system.

Interdependent demand curves: A market in which each firm's demand curve is influenced by the pricing decisions of other firms in that market.

Interest rate: The price of borrowed money.

Interest: A charge made for borrowed money. The rate at which we discount future economic goods.

Internal growth: Expanding a firm's capital and facilities as the market for that firm's product grows.

International trade: Tansactions with foreign nations in which goods and services are bought or sold.

Internodes: The portion of the stem which is located between two nodes.

Interstratification: Mixing of silicate layers within the structural framework of a given silicate clay.

Irrigation efficiency: Represents the ratio, expressed in percentage, of water stored in the root zone depth of the soil to the water delivered in the field from the farm supply source.

Irrigation efficiency: The ratio of the water actually consumed by crops on an irrigated area to the amount of water diverted from the source onto the area.

Irrigation methods: Methods by which water is artificially applied to an area. The methods and the manner of applying the water are as follows.

Irrigation requirement of tobacco: In, general 2-4 irrigations at 30 days interval starting from grand growth initiation stage were found to be sufficient. At pre flowering stage, little deep irrigation may be sufficient to obtain optimum yields of high quality.

Irrigation requirement: Refers to the amount of water exclusive of precipitation and the ground water contribution required for crop production. It represents the net irrigation requirement plus other economically unavoidable water losses. It is usually expressed as the depth of water for a given time.

Isolation distance: It is the recommended distance to be maintained between the seed crop and the likely source of contamination.

Isomorphous substitution: The replacement of one atom by another of similar size in a crystal lattice without disrupting or changing the' crystal structure of the' mineral.

Isotopes: One of a set of chemically identical species of atom which have the same atomic number but different mass numbers.

Kolkhozes: Collective farms in the USSR.

Kyoto Protocol: Kyoto Protocol is an agreement made under the United Nations Framework Convention on Climate Change (UNFCCC). The treaty was negotiated in Kyoto, Japan in December 1997 and the protocol came into force on February 16, 2005. The aim is to lower overall emissions of six greenhouse gases–carbon dioxide, methane, nitrous oxide, sulfur hexafluoride, HFCs (Hydrofluro Carbon), and PFCs. It has been established that per capita GHG emission is strongly correlated with economic prosperity. Further, it is recognized that without increase in GHG emissions or access to appropriate alternative technology options, developing countries would not be able to pursue their socio-economic goals.

Labor: The physical effort of humans that is combined with other resources to produce valued goods and services.

Lacustrine deposit: Material deposited in lake water and later exposed either by lowering of the water level or by the elevation of the land.

Land: All the productive attributes of the earth's surface, including space and the natural environment.

Land capability classification: A grouping of kinds of soil into special units, subclasses, and classes according to their capability for intensive use and the treatments required for sustained use. One such system has been prepared by the USDA Soil Conservation Service.

Land classification: The arrangement of land units into various categories based upon the properties of the land or its suitability for some particular purpose.

Land: A broad term embodying the total natural environmental of the areas of the Earth not covered by water. In addition to soil, its attributes include other physical conditions such as mineral deposits and water supply; location in relation to centers of commerce, populations, and other land; the size of the individual tracts or holdings; and existing plant cover, works of improvement, and the like.

Land-use planning: The development of plans for the uses of land that, over long periods, will best serve the general welfare, together with the formulation of ways and means for achieving such uses.

Law of comparative advantage: The principle that each person, area, or nation will gain economically if each specializes in producing those goods and services where they have the greatest relative advantage, then trades with other similarly specialized individuals, areas, or nations.

Law of diminishing marginal utility: As an individual consumes additional units of a specific good, holding everything else constant, and the amount of satisfaction from each additional unit of those good decreases.

Law of diminishing returns: As successive amounts of a variable input are combined with a fixed input, the total product will increase, reach a maximum, and eventually decline.

Leaching: The removal of materials in solution from the soil by percolating waters

Leaf area index (LAI): Area of leaves per unit of land on which the plants are growing.

Least cost combination: A combination of two or more resources in such a way that the resource cost of producing a given level of output is a minimum.

Legume: A pod-bearing member of the Leguminosae family, one of the most important and widely distributed plant families. Includes many valuable food and forage species, such as peas, beans, peanuts, clovers, alfalfas, sweet clovers, lespedezas, vetches, and kudzu. Nearly all legumes are associated with nitrogen-fixing organisms.

Length-of-run: A planning concept relating to the proportion of a firm's inputs that may be fixed or varied in a given time period.

Liabilities: The total claims of creditors against the value of assets owned by a business.

Lichchivi: New chewing tobacco variety Lichchivi has been released in 2001for cultivation in all chewing tobacco growing areas of Bihar and U.P. Its specific feature is its thick bodied cured leaf. Its total average yield is 28 q/ha with 15 q/ha first grade leaf outturn.

Liebig's law of minimum: The growth and reproduction of an organism are determined by the nutrient substance (oxygen, carbon dioxide, calcium, etc.) That is available in minimum quantity, the limiting factor.

Light soil: A coarse-textured soil; a soil with a low drawbar pull and hence easy to cultivate. *See also* coarse texture; soil texture.

Lignin: The complex organic constituent of woody fibers in plant tissue that, along with cellulose, cements the cells together and provides strength. Lignins resist microbial attack and after some modification may become part of the soil organic matter.

Lime requirement: The mass of agricultural limestone, or the equivalent of other specified liming material, required to raise the ph of the soil to a desired value under field conditions.

Limestone: A sedimentary rock composed primarily of calcite ($CaCO_3$). If dolomite ($CaCO_3MgCO_3$) is present in appreciable quantities, it is called a dolomitic limestone.

Loam: The textural class name for soil having a moderate amount of sand, silt, and clay. Loam soils contain 7-27 per cent clay, 28-50 per cent silt, and 23-52 per cent sand.

Local variety: A mixture of different types and is well adapted to local environments. It is endemic to an area with its origin dating back to several hundred years.

Long position: Purchases of futures contracts not offset by sales.

Long run: A time period long enough that all factors of production can be varied.

Luxury consumption: The intake by a plant of an essential nutrient in amounts exceeding what it needs. For example, if potassium is abundant in the soil, alfalfa may take in more than it requires.

Macro-economic equilibrium: A situation where aggregate supply equals aggregate demand and there are no forces in the economy acting either to increase or decrease the nation's real GNP.

Macro-economics The area of economics that deals with output, employment, incomes, or other activities in the aggregate.

Macro-nutrient: A chemical element necessary in large amounts (usually 50 mg/kg in the plant) for the growth of plants. Includes C, H, O, N, P, K, Ca, Mg, and S. ("Macro" refers to quantity and not to the essentiality of the element.) *See also* micronutrient.

Management: The process of controlling or directing a situation. Also one of the four factors of production, with responsibility for decision making.

Marginal cost (MC): The change in total cost when output is changed by one unit.

Marginal factor cost (MFC): The amount added to total cost when one more unit of the variable input is used in production.

Marginal physical product (MPP): The amount added to total physical product when another unit of the variable input is used in production.

Marginal rate of product substitution (MRPS:) The rate at which one product substitutes for another, measured along the production possibilities curve.

Marginal rate of substitution (MRS): The rate at which one good (or resource) can be substituted for another at the margin without changing satisfaction (or output).

Marginal revenue: The amount added to total revenue when an additional unit of output is produced and sold.

Marginal value product (MVP): The amount added to total value product when another unit of the variable input is used.

Market demand curve: A horizontal summation of individual demand curves for a product.

Market equilibrium: That point where the market demand curve intersects the market supply curve. The market clearing price where quantity demanded equals quantity supplied.

Market orders and agreements: Government authorized programs that producers can use to determine the terms and conditions under which a commodity can be marketed.

Market power: A firm's ability to influence product prices either on the buying or selling side of the market.

Market supply curve:A horizontal summation of individual supply curves for a product.

Market: Consists of buyers and sellers with facilities to communicate with each other.

Marketing agreement: A voluntary agreement among handlers of a commodity with respect to the quality, quantity, and price of that good.

Marketing efficiency: A comparison of the value of output to the value of inputs used in the marketing process.

Marketing margin: The difference between the price that consumers pay for the final product and the price received by producers for the raw product.

Marketing order: A price agreement that, once issued, is binding on all handlers of the commodity covered by that order.

Marketing: Business activities that direct the flow of goods and services from producer to consumers.

Marl: Soft and unconsolidated calcium carbonate, usually mixed with varying amounts of clay or other impurities.

Marsh: Periodically wet or continually flooded area with the surface not deeply submerged. Covered dominantly with sedges, cattails, rushes, or other.

Matric potential: That portion of the total soil water potential due to the attractive forces between water and soil solids as represented through adsorption and capillarity. It will always be negative.

Mature soil: A soil with well-developed soil horizons produced by the natural processes of soil formation and essentially in equilibrium with its present environment.

Maximum water-holding capacity: The average moisture content of a disturbed sample of soil, 1 cm high, which is at equilibrium with a water table at its lower surface.

Mellow soil: A very soft, very friable, porous soil without any tendency toward hardness or harshness. *See also* consistence.

Merchant middlemen: Purchasers of goods from wholesalers for later sale to buyers.

Merger: Combining two or more formerly independent firms under a single management entity.

Metamorphic rock: A rock that has been greatly altered from its previous condition through the combined action of heat and pressure. For example, marble is a metamorphic rock produced from limestone; gneiss is produced from granite, and slate from shale.

Methane (CH_4): An odorless, colorless gas commonly produced under anaerobic conditions. When released to the upper atmosphere, methane contributes to global warming.

Metric ton: A metric measure of weight (one metric ton equals 2204.622 pounds).

Micas: Primary aluminosilicate minerals in which two silica tetrahedral sheets alternate with one alumina/ magnesia octahedral sheet with entrapped potassium atoms fitting between sheets. They separate readily into thin sheets or flakes.

Micro relief: Small-scale local differences in topography, including mounds, swales, or pits that are only a meter or so in diameter and with elevation differences of up to 2 m.

Microeconomics: The area of economics that deals with individual decision units people, firms, or markets-within the economy.

Microfauna: That part of the animal population which consists of individuals too small to be clearly distinguished without the use of a microscope. Includes protozoans and nematodes.

Microflora: That part of the plant population which consists of individuals too small to be clearly distinguished without the use of 'a microscope. Includes actinomycetes, algae, bacteria, and fungi.

Micro-nutrient: A chemical element necessary in only extremely small amounts «50 mg/kg in the plant) for the growth of plants. Examples are B, Cl, Cu, Fe, Mn, and Zn. ("Micro" refers to the amount used rather than to its essentiality.)

Middleman: Functions performed in moving goods and services from producer to consumer.

Mineral soil: A soil consisting predominantly of, and having its properties determined predominantly by, mineral matter. Usually contains <20 per cent organic matter, but may contain an organic surface layer up to 30 cm thick.

Mineralization: The conversion of an element from an organic form to an inorganic state as a result of microbial decomposition.

Minimum tillage: The minimum soil manipulation necessary for crop production or meeting tillage requirements under the existing soil and climatic conditions.

Mixed cropping: Mixed cropping is the process of growing two or more crops together in the same piece of land. This system of cropping is generally practiced in areas where climatic hazards such as flood, drought, frost etc., are frequent and common. *e.g.*, wheat + gram or wheat+barley or wheat+ mustard or they may mature at different times, *e.g.*, arhar + jowar + mung and til or groundnut+ bajra etc. Mixed cropping may be classified into the mixed crops, companion crops, guard crops and augmenting crops based on their method of sowing.

Moisture equivalent: The weight percentage of water retained by a previously saturated sample of soil 1 cm in thickness after it has been subjected to a centrifugal force of 1000 times gravity for 30 minute.

Mollic epipedon: A surface horizon of mineral soil that is dark colored and relatively thick, contains at least 0.6 per cent organic carbon, is not massive and hard when dry, has a base saturation of more than 50 per cent, has less than 250 mg/kg P_2O_S soluble in 1 per cent citric acid, and is dominantly saturated with bivalent cations.

Mollisols: Soils with nearly black, organic-rich surface horizons and high supply of bases. hey have mollic epipedons and base saturation greater than 50 per cent in any cambic or argillic horizon. They lack the characteristics of vertisols and must not have oxic or spodic horizons.

Monetary policy: Manipulation of the money supply and the interest rate to achieve specific economic goals.

Monopoly: Single seller in an industry.

Monopsony: Single buyer in an industry.

Montmorillonite: An aluminosilicate clay mineral in the smectite group with a 2: 1 expanding crystal lattice, with two silicon tetrahedral sheets enclosing an aluminum octahedral sheet. Isomorphous substitution of magnesium for some of the aluminum has occurred in the octahedral sheet. Considerable expansion may be caused by water moving between silica sheets of contiguous layers.

MOR: Raw humus type of forest humus layer of incorporated organic material, usually matted or compacted or both; distinct from the mineral soil, unless the latter has been blackened by washing in organic matter.

Moraine: An accumulation of drift, with an initial topographic expression of its own, built within a glaciated region chiefly by the direct action of glacial ice. Examples are ground, lateral, recessional, and terminal moraines.

Mucigel: The gelatinous material at the surface of roots grown in unsterilized soil.

Muck soil: A soil containing 20-50 per cent organic matter. An organic soil in which the organic matter is well decomposed.

Mulch: Any material such as straw, sawdust, leaves, plastic film, and loose soil that is spread upon the surface of the soil to protect the soil and plant roots from the effects of raindrops, soil crusting, freezing, evaporation, etc.

Mulch tillage: Tillage or preparation of the soil in such a way that plant residues or other materials are left to cover the surface; also called mulch farming, trash farming, stubble mulch tillage, ploughless farming.

Mutation: Sudden, heritable variation in an animal resulting from an abrupt change in the genotype nature.

Mycorrhiza: The association, usually symbiotic, of fungi with the roots of seed plants. Based on association they may classify as ectotrophic mycorrhiza (EcM); endotrophic mycorrhiza (EdM); vesiculararbuscular mycorrhiza (VaM).

National income accounts: Dollar-term measures of the naion's economic performance.

National income: Net national income minus indirect taxes, after accounting for depreciation; it is thus labor, management, and proprietary earnings plus net interest, net rent, and corporate profits.

Natural resource: A factor of production provided by nature. Examples are land, water, minerals, wind, tides, and so on.

Natural selection: An important feature of certain theories of evolution, according to which agents other than man determine which members of a population, will survive.

Nematodes: Very small worms abundant in many soils and important because some of them attack and destroy plant roots.

Net national product: The net value of all goods and services produced in the economy after deducting depreciation from GNP, to account for resources used up during the accounting period.

Net worth: The excess of assets over liabilities representing the owner's residual claim to assets.

Neutral soil: A soil in which the surface layer, at least to normal plow depth, is neither acid nor alkaline in reaction. In practice this means the soil is within the ph range of 6.6-7.3.

Nick: In hybrid seed production, the parents are said to nick when they produce higher yield as a result of synchronized flowering.

Nicotine: $C_{10}H_{14}N_{2.}$ Alkaloid commonly found in tobacco leaves and stem, commercially used as insecticide as nicotine sulphate.

Nicotinic Acid: $C_5H_5N_2$, obtained by oxidation of nicotine

Nitrification: The biochemical oxidation of ammonium to nitrate, predominantly by chemoautotrophic bacteria whose energy requirements come from these exergonic reactions.

Nitrogen assimilation: The incorporation of nitrogen into organic cell substances by living organisms.

Nitrogen cycle: The sequence of chemical and biological changes undergone by nitrogen as it moves from the atmosphere into water, soil, and living organisms, and upon death of these organisms (plants and animals) is recycled through a part or all of the entire process.

Nitrogen fixation: The biological conversion of elemental nitrogen (N_2) to organic combinations or to forms readily utilize in biological processes.

Nonrecourse loan: A loan in which the lender, such as the CCC, has no recourse beyond the physical commodity in payment for that loan.

Normal profit: A situation in which each of the firm's resources is earning a return just equal to its opportunity cost.

No-tillage system: A procedure whereby a crop is planted directly into a seedbed not tilled since harvest of the previous crop; also zero tillage.

Nucleic acid: A long molecule formed by lining many nucleotide molecules. The 2 kinds occurring in cells single or double stranded *i.e.* DNA and RNA.

Nucleus seed: The original seed produced for the first time by the plant breeder. It is 100 per cent pure in all genetically and physical quantities and used as a parent for the multiplication of breeder's stock seed.

Nursery bed sterilizations: Sterilization of nursery beds take care of weeds, seeds of other undesired crops, seed borne diseases and pests, thus it provide protection to young tiny and tender tobacco seedlings. This can be done by rabbing and chemical treatments.

Ochric epipedon: A surface horizon of mineral soil that is too light in color, too high in chroma, too low in organic carbon, or too thin to be a plaggen, mollie, umbric, anthropic, or histic epipedon, or that is both hard and massive when dry.

Off-type: Plants or seeds deviating significantly from the characteristics of a variety as described by the breeder in any observable respect.

Oligopolistic market: A market characterized by the small number of firms producing for that market.

Oligopoly: A firm in an oligopolistic market.

Open market operations: The purchase and sale in the open market of government securities by the Open Market Committee.

Opportunity cost: The value of other opportunities given up in order to produce or consume any good.

Options: Contracts that give the holder the right to buy or sell a property.

Osmotic potential: That portion of the total soil water potential due to the presence of solutes in soil water. It will generally be negative.

Osmotic pressure: Pressure exerted in living bodies as a result of unequal concentrations of salts on both sides of a cell wall or membrane. Water moves from the area having the lower salt concentration through the membrane into the area having the higher salt concentration and, therefore, exerts additional pressure on the side with higher salt concentration.

Out cross: The mating of a hybrid with a third parent, also an off type plant resulting from pollen of a different sort contaminating a seed field.

Outwash plain: A deposit of coarse-textured materials (e.g., sands, gravels) left by streams of melt water flowing from receding glaciers.

Oxisols: Soils with residual accumulations of low-activity clays, free oxides, kaolin, and quartz. They are mostly in tropical climates.

Papery tobacco leaves: Early planting results in production of papery leaves accompanied with on severe leaf spot infestation, while delay in planting results in production of small, leathery and brittle leaves.

Parity: That price which gives a unit of agricultural commodity the same purchasing power as it had in a specified base period.

Passive absorption: Movement of ions and water into the plant root as a result of diffusion along a gradient.

Peat soil: An organic soil containing more than 50 per cent organic matter. The word peat means an unconsolidated soil material consisting largely of undecomposed or only slightly decomposed organic matter accumulated under conditions of excess moisture. Highly decomposed organic material is referred to as 'muck'.

Peat soil: An organic soil containing more than 50 per cent organic matter. "Peat" referring to the slightly decomposed or undecomposed deposits and "muck" to the highly decomposed materials.

Peat: Unconsolidated soil material consisting largely of undecomposed or only slightly decomposed, organic matter accumulated under conditions of excessive moisture. *See also* organic soil materials; peat soil.

Ped: A unit of soil structure such as an aggregate, crumb, prism, block, or granule, formed by natural processes (in contrast to a clod, which is formed artificially).

Pedon: The smallest volume that can be called "a soil" It has three dimensions. It extends downward to the depth of plant roots or to the lower limit of the genetic soil horizons. Its lateral cross section is roughly hexagonal and ranges from 1 to 10 m^2 in size depending on the variability in the horizons.

Penetrability: The ease with which a probe can be pushed into the soil. May be expressed in units of distance, speed, force, or work depending on the type of penetrometer used.

Perched Water table: The surface of a local zone of saturation held above the main body of groundwater by an impermeable layer of stratum, usually clay, and separated from the main body of ground water by an unsaturated zone.

Percolation, soil water: The downward movement of water through soil. Especially, the downward flow of water in saturated or nearly saturated soil at hydraulic gradients of the order of 1.0 or less.

Perfect complements: Resources that must be used in a given ratio in order to produce a product.

Perfect substitutes: Resources that substitute for one another in production at a constant rate.

Permafrost: The permanently frozen soil in arctic and sub-arctic regions.

Permanent wilting point: The moisture content expressed in percentage of oven dried soil at which nearly all plants wilt and do not recover in a humid chamber unless water is added from an external source. This is the lower limit of the available soil moisture.

Permeability: Denotes the property of a porous medium to transmit the fluids. It is a broad term and can be further specified as hydraulic conductivity and intrinsic permeability. Hydraulic conductivity is the proportionality factor 'K' in Darcy's law and represents the rate of flow of water in a porous medium under a unit hydraulic gradient. It changes with the quality and quantity of water, but the values of the intrinsic: permeability, the property of the porous medium, remains constant for any fluid.

Personal income: The total income of people before the payment of income taxes.

pH: It is the negative logarithm of hydrogen ion concentrations.

Phenology: The study of the timing of recurring biological events, the causes of their timing with regard to biotic and abiotic factors and the interrelation among the phases of the same or different species.

Phenotype: Represents an organism bearing a physical or external resemblance as contrasted with its genetic composition. That is a group of organisms manifesting identical outward or physical traits.

Photoperiodism: Means the response of plants to a photoperiod or the length of exposure to light.

Physical function: The processing, storage, and transportation of commodities in the marketing system.

Physical supply: The total available quantity of a good or resource.

Place utility: The utility or satisfaction derived from the transportation function in marketing that delivers products in the place desired by consumers.

Plaggen epipedon: A man-made surface horizon more than 50 cm thick that is formed by longcontinued manuring and mixing.

Plant breeding: The applied branch of botany dealing with the crop improvement and production of new improved crop varieties far-better than original in all aspects.

Plant population: Plant population defines the number of plants per unit area or the size of the area Available to the individual plants.

Plastic soil: A soil capable of being molded or deformed continuously and permanently, by relatively moderate pressure, into various shapes. *See also* consistence.

Plough layer: The soil ordinarily moved when land is plowed; equivalent to surface soil.

Plough pan: Represents the sub-surface soil layer having a higher bulk density and lower total porosity than the soil layers above as well as below this layer, as a result of the pressure applied by normal ploughing or other tillage operations.

Plough pan: A subsurface soil layer having a higher bulk density and lower total porosity than layers above or below it, as a result of pressure applied by normal plowing and other tillage operations.

Plough –planting: The plowing and planting of land in a single trip over the field by drawing both plowing and planting tools with the same power source.

Plowing: A primary broad-base tillage operation that is performed to shatter soil uniformly with partial to complete inversion.

Positive economics: A scientifically testable conclusion of what *is or what can be.*

Potentially arable land: land that can be brought into production after undergoing sufficient physical development.

Poverty level: Official designation of an income below that which is needed to provide the kind of living our society considers to be a basic right.

Present value: The present value of money or an economic good to be received sometime in the future.

Price ceiling: A government decreed maximum price for a commodity or service.

Price elasticity of demand: The percentage change in quantity demanded due to a one percent change in price, ceteris omnibus.

Price elasticity of supply: The percentage change in quantity supplied due to a one percent change in price, ceteris omnibus.

Price floor: A government decreed minimum price for a commodity.

Price: The rupees value per unit of a resource, good or service as determined in the market or by other means.

Primary tillage: Tillage that contributes to the major soil manipulation, commonly with a plow.

Puddled soil: Dense, massive soil artificially compacted when wet and having no aggregated structure. The condition commonly results from the tillage of a clayey soil when it is wet.

Residual material: Unconsolidated and partly weathered mineral materials accumulated by disintegration of consolidated rock in place.

Rhizobia: Bacteria capable of living symbiotically with higher plants, usually in nodules on the roots of legumes, from which they receive their energy, and capable of converting atmospheric nitrogen to combined organic forms; hence, the term symbiotic nitrogen-fixing bacteria.

Rhizome: Underground stem, usually horizontal and capable of producing new shoots and roots at the nodes.

Rhizosphere: That portion of the soil in the immediate vicinity of plant roots in which the abundance and composition of the microbial population are influenced by the presence of roots.

Rill: A small, intermittent water course with steep sides; usually only a few centimeters deep and hence no obstacle to tillage operations.

Rock: The material that forms the essential part of the Earth's solid crust, including loose incoherent masses such as sand and gravel, as well as solid masses of granite and limestone.

Roguing: Removal of weeds, off-type and diseased plants from a standing field crops is known as roguing.

Root knot (*Meloidogyne javanica*): This is a common disease of tobacco characterized by knotting of roots, stunting and chlorosis of the plant and reduction in crop yield, as is common for similar root-knot diseases on other crops. It is caused by the root-knot nematode *Meloidogyne incognita* and by M. Chitwood in some places.

Rotary tillage: An operation using a power-driven rotary tillage tool to loosen and mix soil.

Rural development: All the developmental activities that result in making rural America a better place in which to live and work.

Saline soil: A nonsodic soil containing sufficient soluble salts to impair its productivity. The conductivity of a saturated extract is >4 ds/m, the exchangeable sodium adsorption ratio is less than about 15, and the pH is <8.

Salinization: The process of accumulation of salts in soil.

Saltation: Particle movement in water or wind where particles skip or bounce along the stream bed or soil surface.

Sand leaves: Inferior quality bottom leaves are termed as sand leaves, they are very trashy and thus it is uneconomical to use them hence 2 or 3 leaves from bottom must be removed once when the plants are 30 cm tall and again when plants are 60 cm tall to maintain quality of tobacco leaves.

Scarcity: A basic economic condition in which our wants exceed the resources available to satisfy those wants.

Secondary Tillage: Any tillage operations following primary tillage designed to prepare a satisfactory seedbed for planting.

Sedimentary rock: A rock formed from materials deposited from suspension or precipitated from solution and usually being more or less consolidated. The principal sedimentary rocks are sandstones, shales, limestones, and conglomerates.

Seed certification: A means to maintain and make available to the public, sources of high quality seeds and propagating materials of superior varieties so grown and distributed as to insure genetic identity. This is done by means of inspections of fields and seeds and by regulations for checking on the production, harvesting and cleaning of each lot of seed.

Seed control: Seed law enforcement which applies to all seeds sold commercially. Correct labeling with respect to the kind of seed, germination, purity and other quality factors are ensured.

Seed priming: Soaking of seeds in drying in water or any other chemical to improve germination is known as priming. In case of tobacco freshly harvested seed of tobacco may include a high proportion of seeds, which can only be germinated in the light.The light requirements can be removed by priming the seeds in dark (soaking and drying).

Seed rate of tobacco: A seed rate of 4 g should be sufficient for $10m^2$ nursery bed. Prior to sowing the seed should be mixed with fine sand and are sown uniformly by an experienced person and then leveled in order to cover the optimum depth of seeding at 0.25 cm. After seeding and covering compact the seed beds by planking. Higher seed rate should be avoided as it results in over crowding of seedlings which in turn lowers the quality of seedlings and induces diseases like damping off.

Seed: Sexually or vegetative propagated planting materials which are used for seeding and planting and as such should be free from (pests and diseases) any infection and should give a good crop stand by good seeding.

Seedbed: The soil prepared to promote the germination of seed and the growth of seedlings.

Seeds treatment: Tobacco seeds should treated before sowing with 2.5 per cent formalin solution or 0.25 per cent solution of Dithen Z-78 or Dithen M–45.

Selection: In a population the preservation of certain individuals that have desirable characteristics.

Selective controls: Federal Reserve System controls over interest rates, margins, and installment credit in order to achieve federal policy objectives.

Self-mulching soil: A soil in which the surface layer becomes so well aggregated that it does not crust and seal under the impact of rain but instead serves as surface mulch upon drying.

Self-pollination: Transfer of pollen from the stigma of the same flower or another flower on the same plant.

Semiarid: Term applied to regions or climates where moisture is more plentiful than in arid regions but still definitely limits the growth of most crop plants. Natural vegetation in uncultivated areas is short grasses.

Sequential cropping with tobacco: Growing at least two crops in sequence with tobacco on the same field in a year is commonly known as Sequential cropping.

Sewage sludge: Settled sewage solids combined with varying amounts of water and dissolved materials, removed from sewage by screening, sedimentation, chemical precipitation, or bacterial digestion.

Shelterbelt: A wind barrier of living trees and shrubs established and maintained for protection of farm fields.

Shifting cultivation: A farming system in which land is cleared, the debris burned, and crops grown for 2-3 years. When the farmer moves on to another plot, the land is then left idle for 5-15 years; then the burning and planting process is repeated.

Short position: Sales of futures contracts not offset by purchases.

Short run: **A** time period where one or more of the factors of production are fixed.

Short-day plants: The plants that normally develop flower with photoperiod (day-light period) of 8-10 hours, i.e. less than the critical day-length.

Side dressing: The application of fertilizer alongside row crop plants, usually on the soil surface. Nitrogen materials are most commonly side-dressed.

Silica/alumina ratio: The molecules of silicon dioxide (SiO_2) per molecule of aluminum oxide (Al_2O_3) in clay minerals or in soils. Used in the manufacture of glass and refractory materials.

Silica/sesquioxide ratio: The molecules of silicon dioxide (SiO_2) per molecule of aluminum oxide (Al_2O_3) plus ferric oxide (Fe_2O_3) in clay minerals or in soils.

Silk (corn): The stigma and style of the female corn flower through which the pollen-tube grows to reach the embryo sac.

Silting: The deposition of water-borne sediments in stream channels, lakes, reservoirs, or on floodplains, usually resulting from a decrease in the velocity of the water.

Sink capacity: The capacity of the spikelets or reproductive or economic plant part like earhead, cob, tuber etc. to accept photosyntheses.

Sink: The term usually applied to refer to the plant parts that are involved in utilization and storage of photosyntheses, for example grains, fruits etc.

Site index: A quantitative evaluation of the productivity of a soil for forest growth under the existing or specified environment.

Slag: A product of smelting, containing mostly silicates; the substances not sought to be produced as matte or metal and having a lower specific gravity.

Slip (propagation): A shoot or leaf cutting to be rooted for vegetative propagation.

Slope: The degree of deviation of a surface from horizontal, measured in a numerical ratio, percent, or degrees.

Social costs: The sum total of internal and external costs of an economic activity.

Socialism: A type of economic organization in which the government owns all resources and directs all economic activity.

Sodic soil: A soil that contains sufficient sodium to interfere with the growth of most crop plants, and in which the sodium adsorption ratio is 13 or greater.

Soil acidity: Active the activity of hydrogen ion in the aqueous phase of a soil. It is measured and expressed as a ph value.

Soil Aeration: The process' by which air in the soil is replaced by air from the atmosphere. In a well aerated soil, the soil air is similar in composition to the atmosphere above the soil. Poorly aerated soils usually contain more carbon dioxide and correspondingly less oxygen than the atmosphere above the soil.

Soil Aeration: The process by which air and other gases in the soil are renewed. The rate of soil aeration depends largely on the size and number of soil pores and on the amount of water clogging to permit rapid aeration, is said to be well aerated, while a poorly aerated soil either has few large pores or has most of its pores blocked by water.

Soil alkalinity: The degree or intensity of alkalinity of a soil, expressed by a value >7.0 on the pH scale.

Soil amendment: Any material, such as lime, gypsum, sawdust, or synthetic conditioner that is worked into the soil to make it more amenable to plant growth.

Soil amendment: Any substance other than fertilizers, such as lime, sulfur, gypsum, and sawdust, used to alter the chemical or physical properties of a soil, generally to make it more productive.

Soil association: A group of defined and named taxonomic soil units occurring together in an individual and characteristic pattern over a geographic region, comparable to plant associations in many ways.

Soil class: A group of soils having a definite range in a particular property such as acidity, degree of slope, texture, structure, land-use capability, degree of erosion, or drainage. *See also* soil structure; soil texture.

Soil classification: The systematic arrangement of soils into groups or categories on the basis of their characteristics.

Soil complex: A mapping unit used in detailed soil surveys where two or more defined taxonomic units are so intimately intermixed geographically that it is undesirable or impractical, because of the scale being used, to separate them.

Soil conditioner: Any material added to a soil for the purpose of improving its physical condition.

Soil conservation: A combination of all management and land-use methods that safeguard the soil against depletion or deterioriation caused by nature and/or humans.

Soil correlation: The process of defining, mapping, naming, and classifying the kinds of soils in a specific soil survey area, the purpose being to ensure that soils are adequately defined, accurately mapped, and uniformly named.

Soil genesis: The mode of origin of the soil, with special reference to the processes or soil-forming factors responsible for the development of the solum, or true soil, from the unconsolidated parent material.

Soil geography: A sub specialization of physical geography concerned with the areal distributions of soil types.

Soil management: The sum total of all tillage operations, cropping practices, fertilizer, lime, and other treatments conducted on or applied to a soil for the production of plants.

Soil map: A map showing the distribution of soil types or other soil mapping units in relation to the prominent physical and cultural features of the Earth's surface.

Soil mechanics and engineering: A subspecialization of soil science concerned with the effect of forces on the soil and the application of engineering principles to problems involving the soil.

Soil monolith: A vertical section of a soil profile removed from the soil and mounted for display or study.

Soil morphology: The physical constitution, particularly the structural properties, of a soil profile as exhibited by the kinds, thicknesses, and arrangement of the horizons in the profile, and by the texture, structure, consistence, and porosity of each horizon.

Soil organic matter: Represents the organic fraction of the soil that includes plant and animal residues at various stages of decomposition, cells and tissues of the soil organisms, and the substances synthesized by the soil population. Commonly, it is determined as the amount of the organic material contained in a soil sample passed through a 2 millimeter sieve.

Soil organic matter: The organic fraction of the soil that includes plant and animal residues at various stages of decomposition, cells and tissues of soil organisms, and substances synthesized by the soil population. Commonly determined as the amount of organic material contained in a soil sample passed through a 2 mm sieve.

Soil Productivity: The capacity of a soil for producing a specified plant or sequence of plants under a specified system of management. Productivity emphasizes the capacity of soil to produce crops and should be expressed in terms of yields.

Soil profile: A vertical section of the soil from the surface through all its horizons, including C horizons.

Soil salinity: The amount of soluble salts in a soil, expressed in terms of percentage, milligrams per kilogram, parts per million (ppm), or other convenient ratios.

Soil structure: The combination or arrangement of primary soil particles into secondary particles, units, or peds. These secondary units may be, but usually are not, arranged in the profile in such a manner as to give a distinctive characteristic pattern. The secondary units are characterized and classified on the basis of size, shape, and degree of distinctness into classes, types, and grades, respectively.

Soil survey: The systematic examination, description, classification, and mapping of soils in an area. Soil surveys are classified according to the kind and intensity of field examination.

Soil textural class: A grouping of soil textural units based on the relative proportions of the various soils separates (sand, silt, and clay). These textural classes, listed from the coarsest to the. Finest in texture, are sand, loamy sand, sandy loam, loam, silt loam, silt, sandy clay loam, clay loam, silty clay loam, sandy clay, silty clay, and clay. There are several subclasses of the sand, loamy sand, and sandy loam classes based on the dominant particle size of the sand fraction (e.g., loamy fine sand, coarse sandy loam).

Soil: A dynamic natural body composed of mineral and organic materials and living forms in which plants grow. (2) The collection of natural bodies occupying parts of the earth's surface that support plants and that have properties due to the integrated effect of climate and living matter acting upon parent material, as conditioned by relief, over periods of time.

Solarization: Extremely high light intensities which have inhibitory effect on photosynthesis.

Solum (plural sola): The upper and most weathered part of the soil profile; the A, E, and B horizons.

Speculation: The purchase or sale of title to goods or financial obligations in the expectation of favorable price movements.

Speculative middlemen: Own and hold commodities in the expectation of a price change in their favor.

Spillover effect: Benefits or costs that are external to the decision maker and "spill over" to others in the economy.

Spodosols: Soils with subsurface illuvial accumulations of organic matter and compounds of aluminum and usually iron. These soils are formed in acid, mainly coarse-textured materials in humid and mostly cool or temperate climates.

Sprinkler irrigation method: The water is sprayed over the soil surface through nozzles from a pressure system.

Stagflation: A situation in which a high rate of inflation and high unemployment occur simultaneously.

State farm: Large farms owned and operated by the government (sovkoz in Soviet Russia).

Stubble mulch: The stubble of crops or crop residues left essentially in place on the land as a surface cover before and during the preparation of the seedbed and at least partly during the growing of a succeeding crop.

Sub irrigation methods: The water is applied in open ditches or tile lines until the water table is raised sufficiently to wet the soil.

Subsoil: That part of the soil below the plow layer. Subsoiling Breaking of compact subsoils, without inverting them, with a special knife-like instrument (chisel), which is pulled through the soil at depths usually of 30-60 cm and at spacing usually of 1-2 m.

Substitutes Two different goods (or resources) between which a choice is made to satisfy human wants (or to produce a product).

Substitution effect: The substitution of one good for another as their relative prices change.

Subsurface tillage: Tillage with a special sweep-like plow or blade that is drawn beneath the surface, cutting plant roots and loosening the soil without inverting it or without incorporating residues of the surface cover.

Succulent plants: Succulents contribute a considerable proportion of the vegetation of the semi-arid regions and are frequently found in locally dry habitats. The most conspicuous succulents belong to the cactus family (*Cactaceae*). The other important families are *Euphorbiaceae, Liliaceae*, etc. which include a number of such plants. These succulents are of a distinct group of plants not only in structure but in metabolism and water economy. Such plants are able to survive dry periods because of the relatively more reserves of water accumulated in the inner tissues of the fleshy stems or in the fleshy leaves.

Summer deep ploughing: summer deep ploughing (twice) in the month of april-may, found highly beneficial in minimizing weeds and *Orobanche menace*, reducing insect pests and diseases problems and improving water and nutrient conserving capacity of the soil.

Summer fallow: A cropping system that involves management of uncropped land during the summer to control weeds and store moisture in the soil for the growth of a later crop.

Supply curve: The amount of a good or service a producer is willing to offer for sale at different prices, holding everything else constant.

Symbiosis: The living together in intimate association of two dissimilar organisms, the cohabitation being mutually beneficial.

Systems of cropping: System of cropping is the way in which different crops are grown. Sometimes a number of crops are grown together or they are grown separately at short intervals in the same field, *e.g.*, mixed cropping and Intensive cropping.

Tariff: A tax on imported goods.

Technical efficiency: The physical input-output relationships with which marketing functions are performed.

Tensiometer: A device for measuring the negative pressure (or tension) of water in soil in situ; a porous, permeable ceramic cup connected through a tube to a manometer or vacuum gauge.

Terrace: A level, usually narrow, plain bordering a river, lake, or the sea. Rivers sometimes are bordered by terraces at different levels. (2) A raised, more or less level or horizontal strip of earth usually constructed on or nearly on a contour and designed to make the land suitable for tillage and to prevent accelerated erosion by diverting water from undesirable channels of concentration; sometimes called diversion terrace.

Thermoperiodism: Refers to the response of plants to rhythmic fluctuation in temperature. The diurnal temperature regimes are called thermoperiodicity. *Three-way hybrid* is a hybrid evolved by crossing a single cross hybrid with an inbred line.

Thermophilic organisms: Organisms that grow readily at temperatures above 45°C.

Tillage: It means the mechanical manipulation of soil mechanically to provide and promote good tilth for crop production. It includes primary tillage which involves lifting, twisting and turning of soil to incorporate crop residues and animal waste, but the secondary tillage involves harrowing to kill weeds and to break the clods.

Tiller: An erect or semi-erect secondary stem which emerges from a basal auxiliary or adventitious bud or an erect shoot that grows from the crown of grass.

Tilth: The physical condition of the soil in relation to plant growth and thus includes all the physical conditions which influence the crop growth and development.

Time preference: Our desire for economic goods now rather than in the future.

Time utility: The utility or satisfaction derived from the storage function in marketing that delivers a product at the desired time.

Top cross: A cross between an inbread line and an open pollinated strain of any crop, it is very common in maize breeding programme.

Top dressing: An application of fertilizer to a soil after the crop stand has been established.

Top soil: The layer of soil moved in cultivation. See also surface soil. (2) Presumably fertile soil material used to top-dress roadbanks, gardens, and lawns.

Toposequence: A sequence of related soils that differ, one from the other, primarily because of topography as a soil formation factor.

Topping: Topping is the process of removal of flower heads either or alone with some upper leaves from plants. Topping increase the leaf size, thickness and dry weight of leaves the main objectives of topping are to improve the size, body texture and quality of leaves of tobacco.

Total cost (TC): The sum total of all the firm's fixed and variable costs at each level of output.

Total fixed cost (TFC): The implicit cost of fixed resources used to produce a good or service.

Total physical product (TPP): The total quantity of a product that is produced at each level of variable input use.

Total revenue (TR): A firm's total value of output, obtained by multiplying the units of product produced by the price of that product.

Total Soil water potential: A measure of the difference between the free energy state of soil water and that of pure water. Technically it is defined as "that amount of work that must be done per unit quantity of pure water in order to transport reversibly and isothermically an infinitesimal quantity of water from a pool of pure water, at a specified elevation and at atmospheric pressure to the soil water (at the point under consideration)." This *total* potential consists of the following potentials.

Total value product (TVP): The total value of output produced at each level of variable input use.

Total variable cost (TVC): Total of the costs of all the variable resources used in production.

Transfer costs: The costs arising from moving commodities from one market to another.

Trehalose: Trehalose, a non-reducing disaccharide, protects biological molecules in response to different stress conditions in many microorganisms.

Turgid: Describing a cell or tissue which is firm and plump because of the internal pressure resulting from the osmotic uptake of water.

Type (plant): A group of varieties of similar characters or an admixture of seed of indistinguishable varieties but having some characteristics in common like grain type sorghum, yellow maize etc.

Uitisols: Soils that are low in bases and have subsurface horizons of illuvial clay accumulations. They are usually moist, but during the warm season of the year some are dry part of the time.

Umbric epipedon: A surface layer of mineral soil that has the same requirements as' the mollic epipedon with respect to color, thickness, organic carbon content, consistence, structure, and P_20_5 content, but that has a base saturation of less than 50 per cent diatoms Algae having siliceous cell walls that persist as a skeleton after death; any of the microscopic unicellular or colonial algae constituting the class Bacillariaceae. They occur abundantly in fresh and salt waters and their remains are widely distributed in soils.

Unsaturated flow: The movement of water in a soil that is not filled to capacity with water.

Utility: The satisfaction an individual gets from consuming goods and services.

Value added: The contribution to final product value by each stage in the production process.

Value judgment: A conclusion that is based upon the values that one holds.

Value: The worth, in the market or to a person, of a good or service.

Values: Group or personally held principles regarding qualities which are worthwhile or desirable.

Variable costs: Those costs which increase or decrease as varying amounts of the resources are used.

Variance: Mean square deviations of variates from their mean. The square of the standard deviation.

Variety: A subdivision or a group of plants within a species which is characterized by growth, plant fruit, seed or other characters by which it can be differentiated from other seeds of the same kind.

Vegetative stage: Growth stage of a crop plant prior to flowering.

Vermiculite: A 2: I-type silicate clay usually formed from mica that has a high net negative charge stemming mostly from extensive isomorphous substitution of aluminum for silicon in the tetrahedral sheet.

Vernalization: The promotion of flowering by naturally or artificially applied period of extended low temp. Seeds, bulbs or entire plants may be so treated.

Vertical coordination: The linkage of successive stages in producing or marketing a product in a single decision entity.

Vertical integration: the linkage of firms in different stages of production or marketing under the ownership of a single firm.

Vertical merger: Combining two or more firms in different stages of production or marketing into one firm.

Vertisols: Clayey soils with high shrink-swell potential that have wide, deep cracks when dry. Most of these soils have distinct wet and dry periods throughout the year.

Vesicular arbuscular mycorrhiza: A common endomycorrhizal association produced by phycomycetous fungi of the genus *Endogone* and characterized by the development of two types of fungal structures: (a) within root cells small structures known as arbuscles and (b) between root cells storage organs known as vesicles. Host range includes many agricultural and horticultural crops. *See also* endomycorrhiza.

Viability of seeds: It is represented by germination percentage which expresses the number of seedlings which can be produced by a given number of seeds times 100.

Viability: The capability of a plant structure (seed, cutting etc.) to show living properties like germination and growth.

Virgin soil: A soil that has not been significantly disturbed from its natural environment.

Vitamins: Substances required in minute amount for the normal metabolism of plants and animals. They are synthesized by plants and micro-organisms. The role of vitamins is largely catalytic. They serve as coenzymes, and prosthetic groups of enzymes.

Water killing: Relatively high water transpiration rates in evergreens during a period when absorption of water can proceed only at a relatively slow rate may lead to a type of cold injury called winter killing.

Water logging: It denotes the state of soil saturation with free water and exclusion of the air from the soil. The water accumulation on the soil surface may occur due to inadequate surface as well as sub-surface drainage.

Water requirement of tobacco: The total water requirements (ETm) for maximum yield vary with climate and length of growing period from 400 to 600mm.

Water table: The upper surface of groundwater or that level below which the soil is saturated with water.

Water use efficiency: Dry matter or harvested portion of crop produced per unit of water consumed.

Weathering: All physical and chemical changes produced in rocks, at or near the Earth's surface, by atmospheric agents.

Weed: It represents an alien plant introducing itself and growing at a place (field *agronomically*) where it is unwanted. It encompasses all kinds of unwanted plants *viz.* trees, shrubs, broad and narrow leaved plants, sedges., aquatic as well as parasitic plants. In a wheat field, barley is a weed.

Wet year: It means the year in which the rainfall exceeds twice the normal.

Wettable powder: It is a solid formulation which when mixed with water forms a suspension of finely divided solid particles dispersed in the liquid.

Wheel track planting: A practice of planting in which the seed is planted in tracks formed by wheels rolling immediately ahead of the planter.

Wild-flooding irrigation methods: The water is released at high points in the field and distribution is uncontrolled.

Wilt: Loss of freshness and drooping of foliage of a plant due to inadequate supply of moisture, excessive transpiration or by a disease which interferes with the utilization of water by the plant.

Wilting point (permanent wilting point): The moisture content of soil, on an oven-dry basis, at which plants wilt and fail to recover their turgidity when placed in a dark humid atmosphere.

Wilting point: It is soil-moisture condition at which the release of water.to the plant roots is just barely too small to counterbalance the transpiration losses. On an oven-dry basis it is the moisture content of soil at which plants wilt and fail to recover their turgidity when placed in a dark humid atmosphere. It is also called wilting coefficient or permanent wilting point.

Windbreak: Planting of trees, shrubs, or other vegetation perpendicular, or nearly 50, to the principal wind direction to protect soils, crops, homesteads, etc., from wind and snow.

Xanthophylls: A yellow or orange carotenoid pigment associated with chlorophyll in chloroplasts, also present in certain chromoplasts.

Xerophytes: A plant with structural and physiological features which permit it to grow in a extremely dry soils or soil materials under water-stress conditions.

Zinc deficiency: Symptoms appear 2-4 weeks after sowing as blanching of the midrib of the emerging leaf, especially at base. Brown spots appear on the older leaves, which later on coaheses give brown color. Tillering and growth are depressed. In severe deficiency the plant dies. Zinc deficiency is associated with calcareous, alkali. Peat, and volcanic soils and soils that remains wet or waterlogged most of the year. Incidence is more severe where high rates of nitrogen and phosphorus are applied.

Zygote: The cell that results from the fusion of two gametes.

Index

Y

Z